INTERNATIONAL CENTRE FOR THEORETICAL PHYSICS, TRIESTE

CONTROL THEORY AND TOPICS IN FUNCTIONAL ANALYSIS

LECTURES PRESENTED AT
AN INTERNATIONAL SEMINAR COURSE
AT TRIESTE FROM 11 SEPTEMBER TO 29 NOVEMBER 1974
ORGANIZED BY THE
INTERNATIONAL CENTRE FOR THEORETICAL PHYSICS, TRIESTE

In three volumes

VOL. I

INTERNATIONAL ATOMIC ENERGY AGENCY
VIENNA, 1976

THE INTERNATIONAL CENTRE FOR THEORETICAL PHYSICS (ICTP) in Trieste was established by the International Atomic Energy Agency (IAEA) in 1964 under an agreement with the Italian Government, and with the assistance of the City and University of Trieste.

The IAEA and the United Nations Educational, Scientific and Cultural Organization (UNESCO) subsequently agreed to operate the Centre jointly from 1 January 1970.

Member States of both organizations participate in the work of the Centre, the main purpose of which is to foster, through training and research, the advancement of theoretical physics, with special regard to the needs of developing countries.

CONTROL THEORY AND TOPICS IN FUNCTIONAL ANALYSIS
IAEA, VIENNA, 1976
STI/PUB/415
ISBN 92-0-130076-X

Printed by the IAEA in Austria
March 1976

FOREWORD

The International Centre for Theoretical Physics has maintained an interdisciplinary character in its research and training programmes in different branches of theoretical physics and related applied mathematics. In pursuance of this objective, the Centre has – since 1964 – organized extended research courses in various disciplines; most of the Proceedings of these courses have been published by the International Atomic Energy Agency.

The present three volumes record the Proceedings of the 1974 Autumn Course on Control Theory and Topics in Functional Analysis held from 11 September to 29 November 1974. The first volume consists of fundamental courses on differential systems, functional analysis and optimization in theory and applications; the second contains lectures on control theory and optimal control of ordinary differential systems; the third volume deals with infinite-dimensional (hereditary, stochastic and partial differential) systems. The programme of lectures was organized by Professors R. Conti (Florence, Italy), L. Markus (Warwick, United Kingdom) and C. Olech (Warsaw, Poland).

Abdus Salam

CONTENTS OF VOL. I

BASIC CONCEPTS OF CONTROL THEORY

L. MARKUS
School of Mathematics,
University of Minnesota,
Minneapolis, Minnesota,
United States of America

Abstract

BASIC CONCEPTS OF CONTROL THEORY.

After a philosophical introduction on control theory and its position among various branches of science, mathematical control theory and its connection with functional analysis are discussed. A chapter on system theory concepts follows. After a summary of results and notations in the general theory of ordinary differential equations, a qualitative theory of control dynamical systems and chapters on the topological dynamics, and the controllability of linear systems are presented. As examples of autonomous linear systems, the switching locus for the synthesis of optimal controllers and linear dynamics with quadratic cost optimization are considered.

I. WHAT IS CONTROL THEORY - AND WHY?

Just what is control theory? Who or what is to be controlled and by whom or by what, and why is it to be controlled? In a nutshell, control theory, sometimes called automation, cybernetics or systems theory, is a branch of applied mathematics that deals with the design of machinery and other engineering systems so that these systems work, and work better than before.

As an example, consider the problem of controlling the temperature in a cold lecture hall. This is a standard engineering problem familiar to us all. The thermal system consists of the furnace as the heating source, and the room thermometer as the record of the temperature of the hall. The external environment we assume fixed and not belonging to the thermodynamic system under analysis. The basic heating source is the furnace, but the control of the furnace is through a thermostat. The thermostat device usually contains a thermometer to measure the current room temperature and a dial on which we set the desired room temperature. The control aspect of the thermostat is that it compares the actual and the desired temperatures at each moment and then it sends an electric signal or control command to the furnace to turn the fire intensity up or down. In this case, the job of the control engineer is to invent or design an effective thermostat.

Let us next look at a control problem from biology. Parts of the world are being overrun by an increasing population of rats. Here the system consists of the living population of rats and the environmental parameters that affect that population. The natural growth of the rat population is to be controlled towards some desired number, say, zero. Here the job of the control engineer is to build a better mouse-trap.

From this viewpoint control theory does not appear too sinister. On the other hand, it does not seem too profound. So let me elaborate on the structure of control theory to indicate the reasons why UNESCO and the ICTP believe this subject is important. To organize these ideas, I shall discuss control theory from four viewpoints:

(i) as an intellectual discipline within science and the philosophy of science;
(ii) as a part of engineering, with industrial applications;
(iii) as a part of the educational curriculum at university;
(iv) as a force in the world related to technological, economic and social problems of the present and the future.

First consider the philosophical position of the discipline of control theory. Within the framework of metaphysics, control theory is a teleological science. That is, the concepts of control theory involve ideas such as purpose, goal-seeking and ideal or desirable norms. These are terms of nineteenth century biology and psychology, terms of volition, will and motivation such as were introduced by Aristotle to explain the foundations of physics, but then carefully exorcized by Newton when he constructed an unhuman geometric mechanics. So control theory represents a synthesis of the philosophies of Aristotle and Newton showing that inanimate deterministic mechanisms can function as purposeful self-regulating organisms. Recall how the inanimate thermostat regulates the room temperature towards the agreed ideal.

Another philosophical aspect of control theory is that it avoids the concepts of energy but, instead, deals with the phenomenon of information in physical systems. If we compare the furnace with the thermostat we note a great disparity of size and weight. The powerful furnace supplies quantities of energy: a concept of classical physics. However, the tiny but ingenious thermostat deals with information – an aspect of modern statistical physics and mathematics. Thus control theory rests on a new category of physical reality, namely information, which is distinct from energy or matter. Possibly, this affords a new approach to the conundrum of mind versus matter, concerning which the philosophical journal Punch once remarked,

"What is matter? – Never mind.
What is mind? – No matter."

But what are the problems, methods and results of control theory as they are interpreted in modern mathematical physics or engineering? In this sense control theory deals with the inverse problem of dynamical systems. That is, suppose we have a dynamical system, for example many vibrating masses interconnected by elastic springs. Such a dynamical system is described mathematically by an array of ordinary differential equations that predict the evolution of the vibrations according to Newton's laws of motion. The customary or direct problem of dynamics is the analysis of the physical system (to obtain the array of mathematical differential equations) and then the analysis of the differential equations to compute the solutions describing the vibrations. For example, we might try to locate all equilibrium states of the dynamical system and to compute which of these are stable.

The inverse problem, that is, control theory applied to the vibrating system, concerns the question of synthesis as the inverse of analysis. Here we specify a goal and seek to modify the physical and mathematical systems to incorporate this goal. For instance, we might pick one of the known equilibrium states and insist that this must be stable so that, whatever the current state of the vibrating system, it must tend towards the desired equilibrium. In terms of the physics this means that we seek to synthesize new forces (for instance, frictional damping) into the system to achieve the goal. In terms of mathematics we seek to add or synthesize new coefficients (of some specified type) into the differential equations to produce the required type of stability.

In the simplest case the vibrating system consists of a single mass on a linear spring. If the displacement from equilibrium is x at time t, then Newton's law asserts that the acceleration d^2x/dt^2 is equal to the spring force -x (using units in which the mass is unity) and we obtain the differential equation of motion

$$\frac{d^2x}{dt^2} = -x$$

Can we introduce a control force u, depending on x and dx/dt, so that every solution of

$$\frac{d^2x}{dt^2} = -x + u$$

returns to the rest equilibrium state $x = 0$, $dx/dt = 0$, after a finite duration? This is the problem of controllability. Can we find a control law $u^*(x, dx/dt)$ such that the time of return to rest is of minimal duration, as compared with the corresponding duration for all control forces u with magnitudes limited by a prescribed bound? This is the problem of optimal control.

Thus the mathematical problem of control theory concerns the modification of differential equations, within prescribed limitations, so that the solutions behave in some desired specified manner. Moreover, control dynamics can refer to ordinary differential equations as in the case of the finite set of vibrating masses, or partial differential equations as in the case of temperature variations throughout a lecture hall. In fact, more complicated types of functional equations, linear or non-linear, deterministic or stochastic, finite- or infinite-dimensional can arise in control theory. The common feature of all these problems within control theory is that we prescribe the desired behaviour of the solutions and then we seek to modify the coefficients of the dynamical equations so as to induce this behaviour. Hence control theory takes its format from mathematics but also it gives to mathematics new types of stimulating questions and challenging problems.

In classical mathematical physics, we know all about the physical laws guiding the development of the phenomenon we may be studying; we know all the rules of the game and we want to predict the outcome. That is physics. In control theory we do not know the rules, in fact we can change them within certain limitations as we proceed, but we do know exactly how we want the game to end. Thus the mathematical problems of control theory are inverse to the usual problems of mathematical physics. We assume only a limited knowledge of our differential equations, which are partially subject to our control, but we specify definitively the development of the system from the present state to the future goal.

Now let us turn to the engineering side of control theory. Control theory is concerned with doing whereas classical physics is primarily interested in understanding. Naturally, the best procedure would involve comprehension first and action second, but we cannot always hope for the best.

Often we must act to guide some process when we do not understand it thoroughly. For instance, we do not have a complete knowledge of the love-life of rats, yet we can hope to control the rat population effectively by a cleverly chosen system of traps and poisons. If one method is not effective, we can always try another until we meet success. You know the old saying, "If at first you don't succeed, try, try again - and then give up. There's no use in making a fool of yourself". But mathematical control theoreticians do not mind being fools provided their successive trials gradually converge to an effective functional control synthesis. The process of measuring just how far short of perfection each method falls, and then correcting the next method accordingly, is called the principle of feedback, a repulsive word for an important concept.

Scientists did not invent the feedback control process; it is a phenomenon of nature. Almost all biological processes involve self-regulation through some biochemical feedback controls. In physiology this concept is called homeostasis.

Using feedback control we can solve problems that we do not understand. This circumstance is what makes control theory such a powerful tool for technological problems in engineering and for economic and social problems in urban planning.

Suppose that an engineer wishes to supervise an industrial plant making glass, or paper or steel. In practice, this is the main application of control theory, in controlling the quality of the product of a manufacturing process.

Consider the process of making paper. Scientists know very little about the basic chemistry of wood pulp or fibre sludge, yet we demand that the final paper sheeting should roll out with a very precisely controlled thickness. So we monitor the thickness as the paper rolls out and, if there is an error, we can make successive corrections to the temperature of the pulp tank and the speed of the rollers, until the correct thickness is achieved. Since very rapid adjustments of the operating conditions are required, automatic measurements and computer regulation of the plant must replace inaccurate and awkward human beings. Thus automatic factories arise involving no labourers - but, instead, hordes of computer repairmen.

How can universities teach this science of trial-and-error, of input-output comparison and of control action upon the pragmatic principle of feedback rather than upon the rational methods of analyses? In the first place, we usually have some general or qualitative ideas concerning the dynamics of the plant we plan to control. Thus the basic engineering and mathematical techniques that must be taught are found within the qualitative theory of dynamical systems.

Dynamical-system theory, or the study of changing physical quantities such as the positions of a system of vibrating masses, has been a central part of mathematical science since Newton. From the purely mathematical viewpoint, control theory falls under the domain of dynamical-system theory, with heavy overtones of statistics and probability and with some aspect of optimization, linear programming and the calculus of variations. Naturally, there is also a close relation to computer science since engineering control processes often involve computer machines as part of the feedback loop through which information and control commands flow.

Finally, I shall discuss some applications of control methods to socio-economic problems. Let us take first an example from economics where control methods are well accepted and are used as an effective practical procedure, namely in the Keynesian macro-economics of a nation state.

In the Keynesian fiscal theory all data, significant for the economic description of a nation, are aggregated into a few categories. These economic levels are such things as gross national product, unemployment, investment of heavy industry, production of consumer goods, and their numerical measures define the current economic state of the nation. The government seeks to control the economic state of the nation by regulating the level of taxation and the prime interest rate on money. If the economy sags, then this information is fed back to the government who then adjusts the controls of taxation and interest to revive and improve the economic state.

From the economic viewpoint the Keynesian control theory seems to work, but there are political and moral decisions that must be resolved. What is a desirable economic state? If there is an economic squeeze who should suffer during the re-adjustment? Should the burden be placed on poor unemployed young people, or on old-age pensioners crushed by inflation? This political or social decision overrides the economic programme. Nevertheless, the engineering control theory approach to macro-economics, as outlined by Keynes, gives us a method of attacking national economic problems.

In the micro-economic theory of a business company or firm we encounter other control methods and concepts involving personnel organization, inventory, distribution and production. In fact, the interaction between manufacturing firms and the national government will become more and more important as we require pollution control to become a part of the production process.

Keynesian fiscal theory and the organizational theory of the firm are relative success stories for control methods within economic systems. But now we come to a tragic failure.

Lake Erie is one of the Great Lakes of the United States. It is also one of the great sinks of industrial and human wastes. A major governmental plan to clean up Lake Erie was initiated several years ago, but the plan failed. The technological difficulties were overcome with new types of industrial and sanitary purification plants, the economic problems were met with adequate million dollar budgets from national and state agencies, but the political problems were unsolvable. Lake Erie is surrounded by a variety of countries, districts, towns and cities each with a different political structure. Co-operation and compromise were necessary for the salvation of Lake Erie, but co-operation proved politically difficult. What is needed to clean up Lake Erie is a new commission with appropriate political connections and authority. Just how this commission should be constituted is not clear. This is a problem for control theoreticians of political science.

New towns are now being planned in England. For each town the technological structure (road, transport, sewage etc.) is carefully planned. The economic viability (industrial plants, railroads etc.) is considered. But is the political organization of the town and regional government analysed as an integral segment of the over-all system design problem? Such political design questions sometimes are faced, but in a rather unsystematic way.

In summary, the influence of control theory on significant social problems proceeds through several levels indicated as follows:

POLITICS
  Feedback Synthesis with Logical Decisions
ECONOMICS
  Optimal control with cost criteria
ENGINEERING
  Control input-output design
BASIC SCIENCE
  Dynamical Systems

On the most fundamental or bottom level of the four layers encountered in the control problem, we require basic science. This level tells us what is possible. For instance, we know the physical or biological or economic laws describing the dynamical systems we seek to control. The more we understand about the world the better we can control our environment.

The next level involves engineering theory and practice as a method of improving technological processes by controlling production. This level tells us what is feasible.

In the third level of activity the economics of production must be analysed to find the best or optimal control, or at least a satisfactory control. This level tells us what is efficient. But we must still decide on the goal of our control system. The optimal must be defined in terms of a cost criterion that rests on human judgements and political and moral principles. But more than this, the implementation of the economic-engineering-scientific control design depends on the decision-making capability of the commission (its logical powers of analysis) and on the political strength of its authority.

Each person has his own limitations and his own interests. No one can range over the totality of the topics in the four levels of problem solving described above. Each has his own narrow expertise. Yet each one can appreciate how his efforts fit into a wider pattern, a continuum of knowledge. This interdisciplinary approach may offer a means of effectively attaching the behavioural sciences to the natural sciences with the glue of control theory. We can only try and wait and hope.

II. MATHEMATICAL CONTROL THEORY AND FUNCTIONAL ANALYSIS

1. CONTROL OF ORDINARY DIFFERENTIAL SYSTEMS

The mathematical formulation of control theory, within the framework of dynamical systems, involves an inverse problem. In the standard theory of differential equations, we are given the coefficients and we seek to compute the solutions; in control theory the solutions are prescribed (or some aspect of their behaviour is prescribed) and we seek the corresponding restrictions on the coefficients. From this general viewpoint, control theory resembles a boundary value problem rather than an initial value problem.

The physical foundation for this inverse nature of control theory is that the descriptions of science according to Newton, Maxwell, and others, are posed in terms of differential equations. The state of the physical system (mechanical displacements and velocities, or electrodynamical field strengths, etc.) are computed upon integrating differential equations whose coefficients contain physical quantities such as force and charge. In other words, we must invert the differential operators to find the state produced by the controlling forces or electric charges.

A different approach is often encountered in engineering, biological, economic, and social systems where the plant dynamics are unknown – no equations of Newton or Maxwell are available. In this case, we take a phenomenological approach and seek to investigate and control the system after preliminary experimentation or calibration. That is, we put in various control signals or forces, and then we measure or observe the state or some function of the state that is put out. By comparing the input with the output, we try to build up a model of the internal plant dynamics, and to determine the appropriate control. This is an approach of identification theory and adaptive control. When we further seek the best control for some purpose, we encounter the theory of optimal control. It will be a principle goal of this course to compare these two approaches to control theory:

dynamical systems and optimal control
input-output analysis and adaptive control

The physical or engineering problem of control design is delimited only by experience, tradition, practicability, and ingenuity. However, in mathematical control theory we must clarify the class of admissible controllers a priori. We can seek controllers only within certain selected classes such as linear, or piecewise linear, which are often suggested by engineering examples and motivations.

As remarked above, control theory, as interpreted within the framework of dynamical systems or differential equations, leads to problems that are the inverses of the classical mathematical investigations. The classical theory of differential equations deals with analysis whereas control theory deals with synthesis. In the classical approach to dynamical systems we are given the differential equations of motion, and then we try to analyse the behaviour of the resulting motions or solutions. In control theory we prescribe the desired behaviour of the solutions, and then we try to synthesize the differential equations to yield these motions. Of course, the procedure of synthesis means, in mathematical terms, that the basic form of the underlying differential equations can be modified by adjustment of certain control parameters or functional coefficients which are selected from certain admissible classes; whereas the synthesis means, in engineering terms, that the primary machine or plant can be modified by the adjustment of gains in feedback loops or the insertion of auxiliary devices of certain practical types.

Hence, to each part of classical theory of differential equations, say stability or oscillation theory, there corresponds a field of control theory with inverse problems.

For instance, consider the classical stability analysis of the damped linear oscillator

$$\ddot{x} + 2b\dot{x} + k^2x = 0$$

with constant coefficients. This oscillator is asymptotically stable (in the sense that all solutions approach $x = \dot{x} = 0$ as $t \to +\infty$) if and only if $b > 0$ and $k^2 > 0$. As an inverse problem, assume $k > 0$ fixed and try to choose $b > 0$ so that the solutions are damped at the maximal rate. That is, define the cost or efficiency of the control parameter b to be

$$C(b) = \max\{\operatorname{Re}\lambda_1, \operatorname{Re}\lambda_2\}$$

where λ is any eigenvalue satisfying $\lambda^2 + 2b\lambda + k^2 = 0$.
Then we seek to select b to minimize C(b).
An easy calculation shows that the optimal control b^* minimizing C(b) is $b^* = k$. It is interesting to note that this is the standard value for critical damping, and hence we see that this familiar physical adjustment is explained as an elementary result in control theory.

As another illustration consider the forced oscillator

$$\ddot{x} + 2b\dot{x} + k^2x = \sin \omega t$$

for constants $b > 0$, $k > 0$, $\omega > 0$. Classical analysis shows that there is a unique periodic solution

$$x = A \sin(\omega t + \phi)$$

with amplitude

$$A = [(k^2 - \omega^2)^2 + 4b^2\omega^2]^{-\frac{1}{2}}$$

For the inverse problem, fix $k > \sqrt{2}$ and $b > 0$, and try to choose the frequency $\omega > 0$ of the control input $\sin \omega t$ so as to maximize the amplitude $A(\omega)$ of the response output. An easy calculation shows that the optimal control ω^* maximizing $A(\omega) = [(k^2-\omega^2)^2 + 4b^2\omega^2]^{-\frac{1}{2}}$ is $\omega^* = (k^2 - 2b^2)^{\frac{1}{2}}$, which is assumed positive. Again we find this value familiar since ω^* is the resonating frequency, and hence this basic engineering tuning is explained as an elementary result in control theory.

These control-theoretic results are interesting in that they illuminate well-known physical and engineering practice. Yet, they are not typical of the modern theory of control. In the next section, we comment on a standard formulation of modern control theory, as interpreted within the framework of ordinary differential equations. Later we examine other approaches through partial differential equations and other functional equations.

In control theory we consider a process or plant or dynamical system described by a differential system,

$$\dot{x} = f(x, u)$$

where x is the real-state n-vector at time t, and the coefficient f is an n-vector function of the present state x and the control m-vector u. For simplicity we assume the process is autonomous (time-independent) and that f is continuous with continuous first derivatives for all $x \in R^n$ and $u \in R^m$, that is

$$f: R^n \times R^m \to R^n$$

is in class C^1, so that there exists a unique response x(t) from an arbitrary initial state x_0 for each controller u(t) on $0 \leq t \leq T$.

We might seek to control x(t) between given initial and final states in some fixed duration $0 \leq t \leq T$,

$$x(0) = x_0, \quad x(T) = x_1$$

by choosing a control function u(t) from some admissible function class (say $u \in L_\infty[0, T]$, that is, u(t) is a bounded measurable function on $0 \leq t \leq T$). Hence x(t) is a solution of the two-point boundary value problem, with separated end conditions,

$$\dot{x} = f(x, u(t)), \quad x(0) = x_0, \quad x(T) = x_1$$

This constitutes the basic problem of controllability in control theory.

Among all solutions x(t) to this boundary value problem, that is for all admissible control functions, we might try to select and describe the optimal solution $x^*(t)$ for the optimal controller $u^*(t)$ which minimizes some given cost or performance functional C(u). This leads to the central problem of optimal control theory, for which there is a vast literature.

Next let us examine these general concepts for more specific problems and particular examples.

Example 1. Consider a linear vibrator, say, an airplane wing whose vertical vibrations are to be controlled by aerodynamical wing-tabs (Fig. 1). We select suitable units (m = k = 1) so the free-dynamics system is $\ddot{x} + x = 0$ and the controlled system is $\ddot{x} + x = u$. This is equivalent to the differential system

$$\dot{x} = y$$

$$\dot{y} = x + u$$

where x is the displacement, y is the velocity, and u is the control force. Here the state $\begin{pmatrix} x \\ \dot{x} \end{pmatrix}$ or $\begin{pmatrix} x \\ y \end{pmatrix}$ is a vector in the real number plane R^2, and

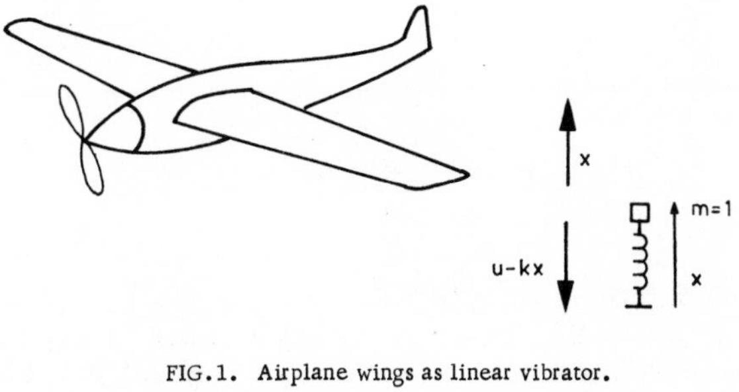

FIG. 1. Airplane wings as linear vibrator.

the control u is a real scalar in $R^1 = R$ at each time $t \in R$. In vector notation, we write $x = x^1$, $y = x^2$,

$$\dot{x}^1 = x^2 = f^1(x^1, x^2, u)$$

$$\dot{x}^2 = -x^1 + u = f^2(x^1, x^2, u)$$

and the matrix notation for this linear system is

$$\begin{pmatrix} \dot{x}^1 \\ \dot{x}^2 \end{pmatrix} = \begin{pmatrix} 0 & 1 \\ -1 & 0 \end{pmatrix} \begin{pmatrix} x^1 \\ x^2 \end{pmatrix} + \begin{pmatrix} 0 \\ 1 \end{pmatrix} u = A \begin{pmatrix} x^1 \\ x^2 \end{pmatrix} + bu$$

For each initial state $\begin{pmatrix} x_0 \\ y_0 \end{pmatrix}$ in R^2 at time $t = 0$, and each controller $u(t)$ on $t \geqq 0$, there is a unique solution or response $x(t, x_0, y_0)$, $y(t, x_0, y_0)$. The choice of u as a function of t, i.e. a pre-programmed control strategy, corresponds to open-loop control.

A more important physical approach, although more difficult mathematically, is to use closed-loop control where $u(x, y)$ depends only on the state at each moment. In this case, the control depends on the feedback loop carrying the current state $\begin{pmatrix} x \\ y \end{pmatrix}$ back to influence the future evolution of the differential system, and the physical system is self-correcting in case unforeseen disturbances arise from time to time.

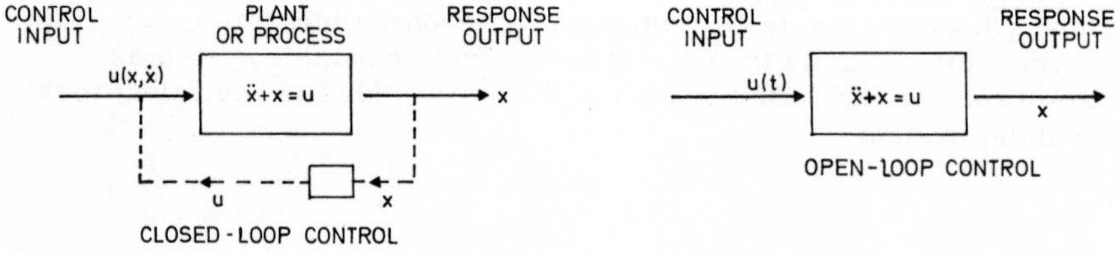

FIG.2. Closed-loop and open-loop control.

Let us examine the simple regulator problem of controlling an initial state $(x_0, y_0) \in R^2$ to the origin $(0, 0)$. Using a closed-loop feedback control, say $u = -y$, we can make the origin asymptotically stable for

$$\ddot{x} + x = -\dot{x}$$

or

$$x = y$$

$$\dot{y} = -x - y$$

That is, for each solution $x(t)$, $y(t) = \dot{x}(t)$ the curve $x(t)$, $y(t)$ approaches the origin in R^2 as $t \to +\infty$. Yet the origin is not attained in any finite time. Moreover, using pure feedback control $u(x, y)$, say $u(0, 0) = 0$ and $u(x, y) \in C^1$,

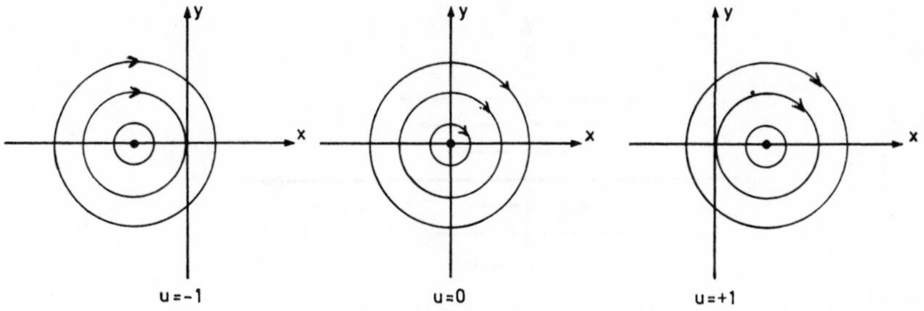

FIG.3. Phase-plane system with u = 0, u = +1 and u = -1.

it is impossible to steer (x_0, y_0) to the origin in a finite time (since the origin is then an equilibrium or critical point for the autonomous differential system $\ddot{x} + x = u(x, \dot{x})$). A more difficult mathematical analysis provides a discontinuous feedback controller that regulates all solutions to the origin in finite duration, but we defer this discussion until later in the course (Fig. 2).

Now, try to regulate the linear oscillator by an open-loop controller u(t), which can be discontinuous if necessary. If we sketch the phase-plane system with u = 0, we find a family of periodic orbits encircling the origin. For u = +1 or u = -1 we find similar portraits but merely displaced one unit toward the right or left, respectively (Fig. 3).

With a simple scheme of switching between the values u = +1 and u = -1 each initial point can be steered to the origin in finite time.

A more involved study shows that each (x_0, y_0) in R^2 can be steered to any required final state (x_1, y_1) in R^2. In this sense the system is controllable in the plane.

Example 2. Consider the double integrator describing a free accelerator according to the linear control system

$$\ddot{x} = u \text{ or } \begin{matrix} \dot{x} = y \\ \dot{y} = u \end{matrix} \text{ or } \begin{pmatrix} \dot{x} \\ \dot{y} \end{pmatrix} = \begin{pmatrix} 0 & 1 \\ 0 & 0 \end{pmatrix} \begin{pmatrix} x \\ y \end{pmatrix} + \begin{pmatrix} 0 \\ 1 \end{pmatrix} u$$

If $u \equiv 0$, then the velocity y is constant and the displacement x increases linearly with time t. It is easy to verify that this system is also controllable in the (x, y)-plane and that each initial state (x_0, y_0) can be regulated to the origin in finite time by some open-loop controller.

Now, consider the control u applied directly to the velocity rather than the acceleration according to the system (Fig. 4):

$$\dot{x} = y + u$$

$$\dot{y} = 0$$

This system is not too easily interpreted in physical terms but it sometimes is used to describe the motion of a particle along a moving platform, where y is the velocity relative to the platform and u is the velocity of the platform along the ground.

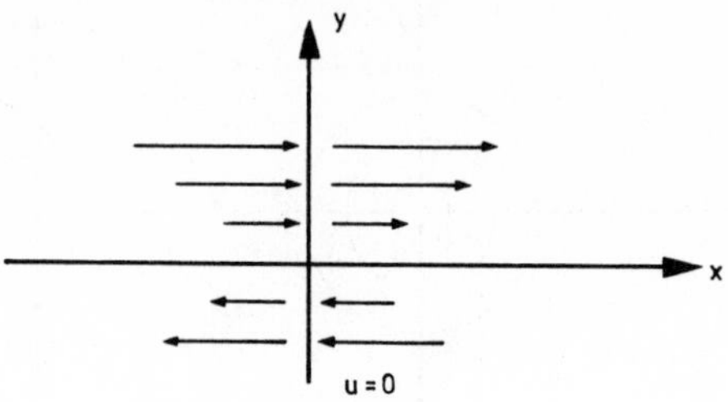

FIG.4. Control applied to velocity.

It is easy to see that this system is not controllable since $y = y_0$ regardless of the control u. However, suppose that only the displacement x is to be controlled, and the velocity y ignored. That is, there are observations z on the state $\begin{pmatrix} x \\ y \end{pmatrix}$ according to

$$z = (1, 0) \begin{pmatrix} x \\ y \end{pmatrix} = x$$

Then the observed output z can be controlled by u.

Once again let us look at these control problems from a more general viewpoint. Consider the control system

$$\frac{dx^i}{dt} = f^i(t, x^1, \ldots, x^n, u^1, \ldots, u^m) \quad i = 1, \ldots, n$$

or, in vector language, $\dot{x} = f(t, x, u)$. We often concentrate on the autonomous case where f does not depend explicitly on t so

$$\dot{x} = f(x, u)$$

where the smooth (highly differentiable) coefficients f(x, u) define the basic dynamics of a physical process or plant. The state of the process is the real n-vector x, and the control is the real m-vector u, at each time t.

Consider the regulator problem of choosing a control u(x, t) so that every solution of

$$\dot{x} = f(x, u(x, t))$$

tends to the origin, $x(t) \to 0$ as $t \to +\infty$. If u = u(t), the control is called open-loop and corresponds to a pre-programmed strategy, which usually depends on knowledge of the initial state. If u = u(x) this is closed-loop feedback control and is sometimes called a synthesis for a stabilizing control.

In case the control system is linear, we write

$$\dot{x} = A\,x + Bu$$

for $x \in R^n$, $u \in R^m$, and the real coefficient matrices A(t) and B(t) are known. Let us discuss here the autonomous linear control problem where A and B are real constant matrices. The feedback regulator problem consists in finding some real constant matrix K (often called the feedback gain) such that u = Kx stabilizes the linear system, i.e.

$$\dot{x} = Ax + BK\,x = (A + BK)\,x$$

has all solutions $x(t) \to 0$ in R^n as $t \to +\infty$. It is well-known that a necessary and sufficient condition for such stability is that all eigenvalues λ of $(A + BK)$ have negative real parts (lie in left-half complex number plane). Later we shall find conditions on the matrix pair (A, B) that guarantee the existence of some stabilizing gain matrix K.

Suppose the linear autonomous control system

$$\dot{x} = A\,x + Bu$$

is accompanied by a given observation relation

$$z = Cx$$

for the observed output z specified by the given real constant matrix C. Let us examine this total system from the input-output analysis of experimental investigations.

For each input u(t), in some appropriate set of control functions, we generate a state response x(t) and then an observed output $z(t) = C\,x(t)$. The input-output relation or transfer function is described by

$$u(t) \to z(t)$$

To be more explicit, let us restrict u(t) to periodic functions, say complex exponentials, with amplitude $U(\omega)$ for frequency ω

$$u = U\,e^{i\omega t}$$

and take the corresponding steady-state periodic response

$$x = X\,e^{i\omega t}, \quad z = Z\,e^{i\omega t}$$

Here the differential system becomes

$$(i\omega)\,X = A\,X + BU$$

or the complex response amplitude, for U = 1, is

$$X = (i\,\omega I - A)^{-1}\,B, \quad Z = C\,(i\omega I - A)^{-1}\,B$$

This is the amplitude-frequency response relation.
If we allow the frequency to be any complex number p, then

$$X = (pI - A)^{-1}\,B, \quad Z = C(pI - A)^{-1}\,B$$

which is the transfer function (or matrix). Note that

$$T(p) = C(pI - A)^{-1}\,B$$

is a rational function with poles at the eigenvalues of A. The basic problem of system identification concerns the computation of the matrices A, B, C from the transfer function T(p). In particular, if for all complex p

$$C_1(pI - A_1)^{-1}B_1 = C(pI - A)^{-1}B$$

when can we conclude that $A = A_1$, $B = B_1$, $C = C_1$?

2. FUNCTIONAL ANALYSIS APPLIED TO CONTROL THEORY

Functional analysis is a method of treating analytical problems in differential and integral equations by means of geometric methods. While the same calculations might be performed without using the language of functional analysis, the geometric concepts prove very useful in suggesting various possible approaches to the problems.

For example, consider the collection of all real-valued continuous functions of a real variable on the unit interval, say f(t) on $0 \leq t \leq 1$. That is, we consider the space C [0, 1] of all continuous maps of [0, 1] into R. By classical analysis, C [0, 1] is a real vector space. If we define the distance between two members f and g of the space C [0, 1] by

$$\|f - g\| = \max_{0 \leq t \leq 1} |f(t) - g(t)|$$

then C [0, 1] becomes a complete metric space, in fact, a Banach space with the norm $\|f\| = \|f - 0\|$. The Weierstrass approximation theorem asserts that the collection P of real polynomials on $0 \leq t \leq 1$ is dense in C [0, 1]. Thus the language of functional analysis permits us to interpret this classical uniform approximation result as the assertion that P is a dense linear subspace of C [0, 1]. Note that this geometric approach is convenient but must be followed with care – since in R^n a dense linear subspace must coincide with all R^n.

Functional analysis is particularly useful in control theory since we must often deal with a space $\mathscr{U}$ of control functions u(t) and a space $\mathscr{X}$ of response functions x(t). The transfer function is then some map

$$\mathscr{U} \to \mathscr{X}: u(.) \to x(.)$$

which can only be studied precisely when the detailed properties of the spaces $\mathscr{U}$ and $\mathscr{X}$ are specified. For instance, since discontinuous controllers are often employed the space $\mathscr{U}$ might be L_1 [0, 1] (integrable functions on $0 \leq t \leq 1$) or perhaps some other class.

Let us briefly consider the famous linear-quadratic problem of optimal control theory in order to appreciate the connections with functional analysis.

Consider the linear autonomous control system

$$\dot{x} = Ax + Bu$$

for the state $x \in R^n$ and control $u \in R^m$ at each time t. We seek an open loop controller u(t) on $0 \leq t \leq 1$ to steer an initial state $x(0) = x_0$ to the origin $x(1) = 0$. Moreover, we judge the performance of each such controller, steering x_0 to 0, by a cost functional

$$C(u) = \int_0^1 |u(t)|^2 \, dt$$

which depends on a quadratic integrand $|u|^2 = \sum_{j=1}^{m} |u^j|^2$.

The optimal controller $u^*(t)$ is required to minimize this cost among all competing controllers.

Here, $u(t)$ ranges over some subset S of the linear space $L_2[0,1]$ of all m-vector functions for which $\int_0^1 |u(t)|^2 \, dt < \infty$. For each $u(t)$ in S we have the real-valued cost $C(u)$, and it seems reasonable to expect the optimal controller $u^*(t)$ to satisfy the necessary condition

$$\frac{dC}{du} = 0 \quad \text{at} \quad u = u^*(t)$$

But how is this functional derivative to be defined or computed? We shall answer these questions later in the paper.

The response $x(t)$ to any controller in S is

$$x(t) = e^{At} x_0 + e^{At} \int_0^t e^{-As} Bu(s) \, ds$$

so

$$0 = e^{A} x_0 + e^{A} \int_0^1 e^{-As} Bu(s) \, ds$$

If we define the linear operator L carrying $L_2[0,1]$ to R^n,

$$Lu = \int_0^1 e^{-As} Bu(s) \, ds$$

then our problem is to solve the linear system

$$Lu = x_0$$

while minimizing the norm $\|u\| = \left[\int_0^1 |u(t)|^2 \, dt \right]^{1/2}$.

Since the set of all possible solutions $u \in L_2[0,1]$ of the linear system $Lu = x_0$ is some linear sub-manifold S (if not empty), we must find the point $u^* \in S$ nearest to the origin of $L_2[0,1]$. That is, u^* is at the point of S which is tangent to the sphere of radius $\|u^*\|$ about the origin. This geometric approach will prove useful in solving this important problem later in the course.

III. SYSTEM THEORY CONCEPTS

The basic stages in the construction and application of any theory of control systems are:

(i) Modelling
(ii) Identification
(iii) Controllability and Stability Analysis
(iv) Optimization for Optimal Control
(v) Implementation.

In the first stage (i) we model the physical process by a mathematical dynamical or control system and in the last stage (v) we reinterpret our mathematical results in terms of engineering hardware. Thus these opening and closing stages are not strictly within the domain of mathematics and we concentrate on the middle three stages, identification, control and stability analysis and optimization, in this course.

(i) Naturally the physically significant variables and relations affect the choice of the mathematical model and all the further control analysis. For instance, should a certain process be described by a linear autonomous control system in R^n

$$\dot{x} = A\,x + Bu \qquad 0 \leq t < \infty$$

or, perhaps, a discrete-time difference scheme

$$x_{t+1} = Ax_t + Bu_t \quad t = 0, 1, 2, \ldots$$

would be more appropriate?

In the modelling process the structure of the particular type of dynamical system is selected, say, ordinary or partial differential system, linear or non-linear, deterministic or stochastic. For example, the vibrations of a string can be studied by the wave partial differential equation

$$\frac{\partial^2 w}{\partial t^2} = \frac{\partial^2 w}{\partial x^2}$$

where $w(x, t)$ is the displacement of the string at position $0 \leq x \leq 1$ at time $t \geq 0$. Suppose one end is fixed $w(0, t) = 0$ and the initial state is given $w(x, 0) = w_0(x)$, $(\partial w/\partial t)(x, 0) = v_0(x)$ on $0 \leq x \leq 1$.

Then we can apply the control $u(t)$ at the endpoint $x = 1$ by, say,

$$w(1, t) = u(t)$$

and seek to bring the vibrating string to the rest state $w(x, T) = 0$, $(\partial w/\partial t)(x, T) = 0$ at a later time $T > 0$. This is a form of boundary control, as distinct from distributed control where we modify the partial differential equation according to

$$\frac{\partial^2 w}{\partial t^2} = \frac{\partial^2 w}{\partial x^2} + u(x, t)$$

Another approach would be to consider the string as a chain with a finite number of rigid links, and to treat the dynamics as a high-order system of ordinary differential equations. The choice of the mathematical model depends on the physical system, and on some decision as to what are the significant features and the negligible aspects.

In another type of problem, say, the population dynamics of a community we might decide that the population P(t) grows according to the law

$$\dot{P}(t) = a\, P(t), \quad P(0) = P_0$$

where $a > 0$ is the net birth-rate. A control might be introduced by an emigration rate u(t) so

$$\dot{P} = a\, P - u(t)$$

However, let us suppose that the birth-rate is not constant, but depends on the size of the previous generation $a = \alpha - \beta\, P(t-1)$, for constants $\alpha > 0$, $\beta > 0$. Then we have a differential-delay equation

$$P(t) = [\alpha - \beta\, P(t-1)]\, P(t) - u(t)$$

A still more complicated assumption on the birth-rate leads to other differential-functional equations such as

$$\dot{P}(t) = \int_{-1}^{0} f(P(t),\ P(t+\theta), \theta)\, d\theta + \int_{-1}^{0} g(u(t+\theta), \theta)\, d\theta$$

Let us note that partial differential equations, such as the wave equation or heat equation, and differential-delay or differential-functional equations can often be analysed as dynamical systems in appropriate infinite-dimensional spaces. For example, the initial state $W_0 = (w_0(x),\ v_0(x))$ of

$$\frac{\partial^2 w}{\partial t^2} = \frac{\partial^2 w}{\partial x^2}, \quad w(0,t) = w(1,t) \equiv 0$$

becomes after time $t > 0$ the state $W_t = (w(x,t),\ (\partial w/\partial t)(x,t))$, a pair of real-valued functions on $0 \leq x \leq 1$.

Hence, the map T_t of the state space is defined by

$$T_t : W_0 \to W_t$$

and the basic flow condition is satisfied

$$T_{t_2} \circ T_{t_1} = T_{t_2+t_1} \quad \text{for } t_1 \geqq 0,\ t_2 \geqq 0$$

For the differential-delay equation in R

$$\dot{x}(t) = f(x(t),\ x(t-1))$$

the initial state is a given continuous function $x_0(\theta)$ on $-1 \leqq \theta \leqq 0$. Then the future solution x(t) is defined for $t \geqq 0$ and the state x at time $t > 0$ is the past unit-time segment of x(t) that is, $x_t(\theta) = x(t+\theta)$ on $-1 \leq \theta \leq 0$. Hence the solution x(t) in R can be defined by a flow T_t from $x_0(\theta)$ to $x_t(\theta)$, say elements of the function space C [-1, 0]. That is, consider

$$T_t : x_0(.) \to x_t(.)$$

again satisfying the basic flow condition for future times $t \geq 0$.

Of course, the precise specification of the function space of states and the appropriate metric or topology is important in any serious study of these flows.

(ii) From input-output experiments (perhaps analysed statistically) we identify the control system, within some accepted class of systems. For instance, from a rational transfer function matrix T(p) can we compute the matrices (A, B, C) such that

$$T(p) = C(pI - A)^{-1} B$$

Note that two systems for $T(p) = \frac{1}{p} = \frac{p}{p^2}$

are

$$\dot{x} = u \qquad \ddot{x} = u$$

and

$$z = x \qquad z = \dot{x}$$

Yet, the first is a minimal model, for a state space $x \in R$, and seems preferable. It turns out that the minimal model, for a given transfer function, is characterized by the properties of being controllable and observable. Roughly speaking, the system is controllable in case the origin can be controlled to every point of the state space. Also it is observable in case no two different state trajectories $x_1(t) = x_2(t)$, for fixed control, can yield the same observation z(t).

(iii) and (iv). We omit any further discussion of these topics here since they have been illustrated by some earlier examples, and they will form the central part of the development of the course.

IV. GENERAL THEORY OF ORDINARY DIFFERENTIAL EQUATIONS, SUMMARY OF RESULTS AND NOTATIONS

We shall primarily be interested in first-order ordinary differential systems

$$\frac{dx^i}{dt} = f^i(t, x^1, x^2, \ldots, x^n) \qquad i = 1, \ldots, n$$

or, in vector notation,

$$\dot{x} = f(t, x)$$

Here, x is a real n-vector at each time t and we seek a solution $x(t) = \varphi(t, t_0, x_0)$, with initial data $\varphi(t_0, t_0, x_0) = x_0$ in R^n, satisfying the system

$$\frac{\partial \varphi}{\partial t} = f(t, \varphi(t, t_0, x_0))$$

This is the Cauchy or initial-value problem, which has a unique solution x(t) under very general conditions on the coefficient n-vector f(t, x), as stated later.

If the system is autonomous, so the coefficient vector f = f(x) is independent of the time t, then we usually denote the solution initiating at x_0 when $t_0 = 0$, by $x = \varphi(t, x_0)$. Then $\varphi(t - t_0, x_0)$ is the solution of the autonomous system

$$\dot{x} = f(x)$$

satisfying the initial data x_0 when $t = t_0$.

In the special case when f(t, x) is linear in x we can use matrix notation

$$\dot{x}^i = a^i_j(t)\, x^j + b^i(t) \qquad \text{(sum on repeated index j)}$$

or

$$\dot{x} = A(t)\, x + b(t)$$

When $b(t) \equiv 0$ we have a homogeneous linear system. If A and b are constant matrices, the linear system is autonomous, and the solution is

$$\varphi(t, x_0) = e^{At}\, x_0 + \int_0^t e^{A(t-s)}\ b\ ds$$

Let us note that a scalar higher-order ordinary differential equation can be reduced to the study of a first-order vector system by a suitable change of notation. That is, the scaler equation, (with the real-valued function $\mathcal{f}$)

$$\frac{d^n x}{dt^n} = \mathcal{f}(t, x, \dot{x}, \ddot{x}, \ldots, x^{(n-1)})$$

can be written in terms of the variables $x = x^1$ and

$$\begin{aligned} \dot{x}^1 &= x^2 \\ \dot{x}^2 &= x^3 \\ &\vdots \qquad\qquad (\text{or } \dot{x} = x^2,\ \ddot{x} = \dot{x}^2 = x^3,\ \text{etc.}) \\ \dot{x}^{n-1} &= x^n \\ \dot{x}^n &= \mathcal{f}(t, x^1, x^2, \ldots, x^n) \end{aligned}$$

Then the n-vector

$$\underline{x} = \begin{pmatrix} x^1 \\ x^2 \\ \vdots \\ x^n \end{pmatrix}$$

has a time-derivative denoted by the n-vector

$$\begin{pmatrix} x^2 \\ x^3 \\ \\ x^n \\ \mathscr{f}(t, x^1, \ldots, x^n) \end{pmatrix}$$

which we can write as

$$\underline{f}(t, x) = \begin{pmatrix} f^1(t, x) \\ f^2(t, x) \\ \\ f^n(t, x) \end{pmatrix}$$

(and we usually omit the underlining of the vectors).

Note that a real-valued function x(t) whose n-th derivative exists and satisfies the scalar equation always determines a n-vector $\underline{x}(t)$ whose first-derivative exists satisfying the vector system, and vice versa. Also note that the degree of differentiability of the coefficients and the properties of being autonomous or linear are transferred straightforwardly from the n-th-order scalar ordinary differential equation to the vector system. Because of the greater symmetry and generality of the vector system, we make this format the principal mathematical object for study.

In any careful analysis of ordinary differential equations we must declare the hypotheses of differentiability on the coefficient vector f(t, x), and the domain in which the solutions are defined. For instance, consider the non-linear differential equation in R^1

$$\dot{x} = (x)^2$$

with initial condition $x = x_0 > 0$ at $t = 0$. Note that the coefficient is a quadratic polynomial, hence differentiable to all orders everywhere in R^1. Yet the solution is given by

$$-\frac{1}{x} + \frac{1}{x_0} = t \quad \text{so} \quad x = \frac{1}{\dfrac{1}{x_0} - t} \quad \text{for} \quad t < \frac{1}{x_0}$$

(upon separating the variables, as in $\frac{dx}{x^2} = dt$). Hence the solution x(t) exists only on a time duration $\tau_- < t < \tau_+$ where $\tau_- = -\infty$ and $\tau_+ = 1/x_0$.

This elementary example shows that we can only expect local solutions, even when the coefficients are very smooth. In control theory, where the coefficients may be discontinuous in t, other complications arise. We first state a "smooth" existence theorem, and then indicate the modifications that must be made when incorporating control dynamics. The proofs of these basic theorems are found in advanced texts.

We first consider a vector autonomous differential system

$$\dot{x} = f(x)$$

with f(x) in class C^k in an open set $\mathcal{O} \subset R^n$, for some differentiability order $k = 1, 2, \ldots, \infty$. Here a vector (or a matrix) is in a class C^k in case each of its components is continuous in $\mathcal{O}$, and there has continuous partial derivatives of all orders $\leq k$ (or all orders if we take $k = \infty$). Of course, a real analytic function (written $f \in C^\omega$) specified in each locality in $\mathcal{O}$ by an absolutely convergent power series, is necessarily in class C^∞.

Theorem. Consider the vector ordinary differential equation in an open set $\mathcal{O} \subset R^n$,

$$\mathcal{S})\ \dot{x} = f(x)$$

with f(x) in $C^k(\mathcal{O})$ for some fixed $k = 1, 2, 3, \ldots, \infty, \omega$. Let x_0 be a given initial point in $\mathcal{O}$.

Then there exists a solution $x = \varphi(t, x_0)$ of $\mathcal{S}$ in $\mathcal{O}$, initiating at $\varphi(0, x_0) = x_0$, and defined on some duration $\tau_- < t < \tau_+$, which we can take maximal. Furthermore, this solution $\varphi(t, x_0)$ is then unique, and is itself in class C^k in an open neighbourhood of $t = 0$ within the space $R^1 \times \mathcal{O} \subset R^{1+n}$.

Also the $n \times n$ matrix $Z(t) = (\partial\varphi/\partial x_0)(t, x_0)$, treated as a function of t, satisfies the linear variational system

$$\dot{Z}(t) = \frac{\partial f}{\partial x}(\varphi(t, x_0))\ Z(t), \quad Z(0) = I$$

Remarks on global solutions. Let $x = \varphi(t, x_0)$ on $\tau_- < t < \tau_+$ be a solution of $\mathcal{S}$ in $\mathcal{O} \subset R^n$. If $\tau_+ < +\infty$, then x(t) must eventually leave and stay out of any prescribed compact set $K \subset \mathcal{O}$ as $t \to \tau_+$.

In particular, when $\mathcal{O} = R^n$ and x(t) remains bounded for $t > 0$, then we can conclude that $\tau_+ = +\infty$, and a similar condition enforces $\tau_- = -\infty$. Thus if $|f(x)|$ is bounded (or even grows linearly with $|x|$) then every solution x(t) is defined for all times $t \in R^1$ in $\mathcal{O} = R^n$.

Remarks on dependence on parameters

Next consider the differential system in R^n

$$\dot{x} = f(t, x, \lambda)$$

for a constant parameter λ in R^m. We reduce this non-autonomous system to an autonomous differential system in a space of $1 + n + m$ dimensions. That is, let (t, x, λ) be the co-ordinates of a vector in R^{1+n+m} and consider

$$\frac{dx}{d\tau} = f(t - t_0, x, \lambda)$$

$$\frac{dt}{d\tau} = 1$$

$$\frac{d\lambda}{d\tau} = 0$$

for $f(s, x, \lambda) \in C^k$ in an open set, say R^{1+n+m}. Here we prescribe the initial data $t = t_0$, $x = x_0$, $\lambda = \lambda_0$ at $\tau = 0$, and denote the unique solution by

$$x = \psi(\tau, t_0, x_0, \lambda_0), \quad t = \tau + t_0, \quad \lambda = \lambda_0$$

Then $x = \psi(t - t_0, t_0, x_0, \lambda_0)$ is the required solution of

$$\dot{x} = f(t, x, \lambda_0)$$

passing through x_0 when $t = t_0$. Note that this vector function of (t, t_0, x_0, λ_0) is in class C^k in an open subset of $R^{1+1+n+m}$. It is this theorem which justifies differentiation of the solution with respect to initial data and parameter components.

However, for even the simplest control systems we encounter more technical problems. Consider in R^n

$$\dot{x} = f(x, u)$$

where the n-vector $f(x, u)$ is in class C^k for the state n-vector x and the control m-vector u. Now if $u(t)$ is a controller in $L_\infty[0, T]$ (bounded measurable on the interval $0 \leq t \leq T$) then $f(x, u(t))$ is not necessarily continuous in t. However $f(x, u(t))$ and $(\partial f/\partial x)(x, u(t))$ lie in $L_\infty[0, T]$ for each fixed $x \in R^n$ (since $|f(x, u)|$ and $|(\partial f/\partial x)(x, u)|$ are bounded when $|x|$ and $|u|$ are bounded) and $f(x, u(t))$ is in C^k for each fixed t. These conditions are enough to ensure the existence of an absolutely continuous solution $x(t)$, whose derivative exists everywhere on $0 \leq t \leq \tau_+ \leq T$ excepting a set of times of measure zero, and which then satisfies the differential system. We emphasize that $x(t)$ is a continuous curve in R^n and the possibility of corners, or discontinuities of $x(t)$, causes no special difficulties.

We state the fundamental existence and uniqueness theorem needed in control theory, and refinements can be found in the text of Lee-Markus.

Theorem. Consider the n-vector differential system

$$\mathscr{S}) \quad \dot{x}^i = f^i(t, x^1, x^2, \ldots, x^n) \qquad i = 1, \ldots, n$$

where $f(t, x)$ is defined for t in an open interval $\mathscr{J} \subset R^1$ and x in an open set $\mathscr{O} \subset R^n$.

Assume

(a) For each fixed $t \in \mathscr{J}$ the functions $f^i(t, x)$ are in $C^1(\mathscr{O})$,

(b) For each fixed $x \in \mathscr{O}$ the functions $f^i(t, x)$ are measurable in t in $\mathscr{J}$,

(c) Given any compact sets $\mathscr{J}_c \subset \mathscr{J}$ and K $\mathscr{O}$ there exists an integrable function $m(t)$ on $\mathscr{J}_c$ such that

$$|f(t, x)| \leq m(t) \quad \text{and} \quad \left|\frac{\partial f}{\partial x}(t, x)\right| \leq m(t)$$

for all $(t, x) \in \mathscr{J}_c \times K$ (note $\int_{\mathscr{J}_c} m(t)dt < \infty$).

Then for each initial point (t_0, x_0) in $\mathscr{J} \times \mathscr{O}$ there exists a unique solution of $\mathscr{S}$:

$$x = \varphi(t, t_0, x_0) \quad \text{with} \quad \varphi(t_0, t_0, x_0) = x_0$$

defined for a maximal time duration in $\mathcal{J}$

$$\tau_-(t_0, x_0) < t < \tau_+(t_0, x_0)$$

Moreover, the n-vector function $\varphi(t, t_0, x_0)$ is then defined and continuous in an open set $D \subset R^{1+1+n}$. For each fixed (t_0, x_0) the function $\varphi(t, t_0, x_0)$ is absolutely continuous in t and satisfies the vector differential equation

$$\frac{d\varphi}{dt} = f(t, \varphi(t, t_0, x_0))$$

almost everywhere on $\tau_- < t < \tau_+$, and hence the integral equation

$$\varphi(t) = x_0 + \int_{t_0}^{t} f(s, \varphi(s))\, ds$$

for all $\tau_- < t < \tau_+$. For each fixed (t, t_0) the function $\varphi(t, t_0, x_0)$ is in class C^1 in x_0, and the vector $\partial\varphi / \partial x_0^j\ (t, t_0, x_0)$, for each $j = 1, \ldots, n$, is absolutely continuous in t and satisfies the linear differential system

$$\frac{d}{dt}\left(\frac{\partial\varphi^i}{\partial x_0^j}\right) = \sum_{\ell=1}^{n} \frac{\partial f^i}{\partial x^\ell}(t, \varphi(t, t_0, x_0)) \left(\frac{\partial\varphi^\ell}{\partial x_0^j}\right)$$

<u>Example.</u> <u>Control of non-linear vibrator with one degree of freedom</u>

Consider a particle with displacement x along a straight track at time t. Newton's equations of motion for the particle of mass m assert

$$m\ddot{x} = F$$

We assume that the total force F is the superposition of an elastic spring force $F_s = -g(x)$, a frictional force $F_f = -f(x, \dot{x})$, and an external control force $F_c = u(t)$. Then, taking units where $m = 1$ for convenience, the control dynamical system becomes

$$\ddot{x} + f(x, \dot{x}) + g(x) = u(t)$$

or introducing the velocity y,

$$\dot{x} = y$$

$$\dot{y} = -f(x, y) - g(x) + u(t)$$

We impose the following physically motivated assumptions:

(i) $f(x, y)$ and $g(x)$ are in C^1 in the (x, y)-phase plane R^2 and $u(t)$ is bounded and measurable on $0 \leq t < \infty$.

(ii) $xg(x) > 0$ for all $x \neq 0$ (restoring-force condition) and $\lim_{|x|\to\infty} G(x) = \infty$ where

$$G(x) = \int_0^x g(s)ds.$$

(iii) $yf(x, y) \geqq 0$ for all (x, y) (frictional-damping condition.)

By assumption i) there exists a unique solution $x(t, x_0, y_0)$, $y(t, x_0, y_0)$ initiating at (x_0, y_0) in R^2, and defined for some maximal future time $\tau_+ > 0$. We shall next show that conditions ii) and iii) guarantee the global existence of all such future trajectories, that is, $\tau_+ = +\infty$. For this proof we pick a finite time $T > 0$ and show that the trajectory must remain within a certain compact subset $K \subset R^2$ on the duration $0 \leqq t \leqq \min(\tau_+, T)$. From this it follows that $\tau_+ > T$, according to the general theory summarized above.

In the (x, y) phase-plane define the (energy or Liapunov) function

$$V(x, y) = \frac{y^2}{2} + G(x)$$

Then each level curve $V(x, y) = c$ (where c is a positive constant) is a simple closed curve encircling the origin in R^2. We compute the time derivative of $V(x, y)$ along a solution curve $x(t)$, $y(t)$ by

$$\frac{dV}{dt} = \frac{\partial V}{\partial x}\dot{x} + \frac{\partial V}{\partial y}\dot{y} = g(x)y + y[-f - g + u(t)]$$

Hence

$$\dot{V} = -yf(x, y) + yu(t)$$

Note that the inequality $\dot{V} \leqq 0$, which holds for the free motion where $u(t) \equiv 0$, already shows that each free trajectory lies within the level curve $V(x, y) = V(x_0, y_0)$ and so is defined for all future times. But we must allow controlled motion with $|u(t)| \leqq u_0$ (the essential least upper bound on $0 \leqq t < \infty$).

For any controlled trajectory

$$\dot{V} \leqq |y|u_0 \leqq \left(\frac{y^2}{2} + 1\right)u_0 \leqq (V + 1)\,|u_0|$$

The along the controlled trajectory

$$\frac{d}{dt}(V + 1) \leqq (V + 1)\,u_0$$

so by the usual inequalities we find, for all times in $0 \leqq t \leqq T$,

$$V + 1 \leqq [V(x_0, y_0) + 1]\,e^{u_0 T} = c_1$$

Hence

$$V(x(t), y(t)) \leqq [V(x_0, y_0) + 1]\,e^{u_0 T} = c_1$$

and so the future trajectory lies within the simple closed level curve $V(x, y) = c_1$ and must be defined on the full interval $0 \leq t \leq T$. Thus $\tau_+ > T$ for each $T > 0$ and so $\tau_+ = +\infty$.

V. QUALITATIVE THEORY OF CONTROL DYNAMICAL SYSTEMS

Let us return to the study of autonomous differential systems in R^n,

$$\mathscr{S}) \; \dot{x} = f(x)$$

For simplicity of exposition we assume throughout this section that $f(x) \in C^1$ in all R^n, and that each solution $x(t) = \varphi(t, x_0)$, initiating at $x_0 \in R^n$ when $t = 0$, is defined for all times $t \in R$.

In the state space R^n, we picture the dynamical system $\mathscr{S}$ as a vector field. That is, at each point x we attach the vector f(x). Then a solution $x(t) = \varphi(t, x_0)$ initiating at x_0 can be pictured as a time-parametrized curve in R^n whose tangent vector $\dot{x}(t)$ coincides with the given vector field f(x) at each point on the curve x(t). We call the solution curve x(t) the trajectory or the orbit through x_0 (sometimes these terms are used more stringently to distinguish between the map $t \to x(t)$ and the point set $\bigcup_{t \in R} x(t)$, but we shall assume the meaning is clear from the context).

There are basically three types of orbits, a single point x_0 (a critical point where $f(x_0) = 0$), a simple closed curve (a periodic orbit), and a one-to-one differentiable image of a line R in R^n. The proof of this assertion, and other statements of this section, will be given later in our discussion of topological dynamics. In this section, we merely introduce concepts and present plausible arguments to motivate the later abstract reasoning.

Let us consider the map of R^n into R^n defined by following each trajectory of $\mathscr{S}$ for a specified time duration t,

$$\Phi_t: \; R^n \to R^n: \; x_0 \to x_t = \varphi(t, x_0)$$

By the uniqueness and regularity theorems of the previous section we know that Φ_t is a bijective map of R^n onto R^n, and both Φ_t and its inverse map Φ_{-t} are differentiable of class C^1. Thus Φ_t is a C^1-diffeomorphism of R^n onto R^n for each given real number t (assuming that each trajectory is defined for all times).

Moreover, because the system $\mathscr{S}$ is autonomous we can conclude that

$$\varphi(t_2, \varphi(t_1, x_0)) = \varphi(t_2 + t_1, x_0)$$

or the composition of maps satisfies $\Phi_{t_2} \circ \Phi_{t_1} = \Phi_{t_2 + t_1}$ for all $t_1, t_2 \in R$. This is known as the "flow condition" or "group homomorphism condition" since it asserts that $t \to \Phi_t$ is a homomorphism of the additive group R into the group Diff(R^n) of all C^1-diffeomorphisms of R^n onto itself.

Incidentally, let us note that for periodic non-autonomous systems

$$\dot{x} = f(t, x) \quad \text{with} \quad f(t + P, x) \equiv f(t, x)$$

for some period $P > 0$, the map

$$x_0 \to \varphi(P, x_0)$$

is a C^1-diffeomorphism of R^n. Also the correspondence

$$k \to \Phi_{kP} = (\Phi_P)^k \quad \text{for} \quad k = 0, \pm 1, \pm 2, \ldots$$

is a homomorphism of the integer group $\mathbb{Z}$ into Diff (R^n).

By means of the geometry of the flow Φ_t (or the discrete flow $(\Phi_P)^k$ for periodic systems) we can analyse the qualitative properties of the solutions of the autonomous system $\mathcal{S}$ in R^n. For example, let us consider two kinds of stability for $\mathcal{S}$, Liapunov and structural stability, as discussed next.

Let us study the behaviour of the solutions of

$$\mathcal{S}) \quad \dot{x} = f(x) \quad \cdot (f(x) \text{ in } C^1 \text{ in } R^n)$$

near a critical (or equilibrium) point $\hat{x}_0$ where $f(\hat{x}_0) = 0$. For simplicity take $\hat{x}_0$ to be the origin so $\varphi(t, 0) \equiv 0$. We shall call the origin stable (future Liapunov stable) in case $|\varphi(t, x_0)|$ stays small forever on $t \geq 0$ provided $|x_0|$ is sufficiently small. If also $\varphi(t, x_0) \to 0$ as $t \to +\infty$, then the origin is asymptotically stable. Thus Liapunov stability refers to the tendency of the solution to remain near to the equilibrium state $\hat{x}_0 = 0$ provided the initial state x_0 suffers a suitably small perturbation away from $\hat{x}_0 = 0$.

To analyse the possible Liapunov stability of the origin, we construct the linear variational equation for small deviations $v = \varphi(t, x_0)$ with x_0 near 0. That is,

$$\frac{d}{dt} v = f(\varphi(t, x_0)) = f(0) + \frac{\partial f}{\partial x}(0) v + \ldots,$$

so the linearized system is defined by

$$\dot{v}^i = \frac{\partial f^i}{\partial \dot{x}^j}(0) v^j \qquad i = 1, \ldots, n$$

or in matrix notation

$$\dot{v} = A v \quad \text{where} \quad A = \frac{\partial f}{\partial x}(0)$$

It seems plausible that if all linearized solutions $v(t) \to 0$, then the non-linear system will be asymptotically stable about the origin. This conjecture is true (we return to the details later) and hence we conclude that $\mathcal{S}$ is asymptotically stable at the origin provided all eigenvalues λ_j of the matrix A lie in the left-half complex plane, that is Re $\lambda_j < 0$.

Next let us contrast Liapunov stability with structural stability of a system $\mathcal{S}$ in R^n: The first concept deals with stability of some equilibrium point (or an invariant set) under perturbations of the initial data, and the second refers to stability of the family of all solutions of $\mathcal{S}$ under perturbations of the coefficients of the differential system.

Let us compare the two differential systems in the (x, y) plane

$$\mathcal{S}_1) \begin{array}{l} \dot{x} = y \\ \dot{y} = -x \end{array} \qquad \text{and} \qquad \mathcal{S}_2) \begin{array}{l} \dot{x} = y \\ \dot{y} = -x - y \end{array}$$

The first is the conservative harmonic oscillator $\ddot{x} + x = 0$, and the second is a damped oscillator $\ddot{x} + \dot{x} + x = 0$. Both are stable about the critical point at the origin, in fact the damped oscillator is globally asymptotically stable. Yet the stability of $\mathscr{S}_1$ can be destroyed by an arbitrarily small perturbation, say

$$\begin{aligned} \dot{x} &= y + \epsilon x^3 \\ & \qquad\qquad \epsilon > 0 \\ y &= -x + \epsilon y^3 \end{aligned}$$

where the polar radius $r = (x^2 + y^2)^{1/2}$ increases,

$$\frac{d}{dt}\left(r^2/2\right) = x\dot{x} + y\dot{y} = xy + \epsilon x^4 - xy + \epsilon y^4 = \epsilon(x^4 + y^4) > 0$$

But consider a small C^1-perturbation of $\mathscr{S}_2$, say,

$$\dot{x} = y + \epsilon_1(x, y)$$

$$\dot{y} = -x - y + \epsilon_2(x, y)$$

where $\sum_{j=1}^{2} |\epsilon_j(x, y)| + |\partial\epsilon_j/\partial x| + |\partial\epsilon_j/\partial y| < \epsilon(x, y)$ for a suitable continuous bound $\epsilon(x, y) > 0$ in R^2. It can be shown that when $\epsilon(x, y)$ is suitably small, every such perturbation of $\mathscr{S}_2$ still has the same qualitative form for its solution curve family in R^2; namely, a single critical point towards which all solution curves tend as t increases.

To summarize the above comparisons, we define $\mathscr{S}$ to be structurally stable in R^n if its solution family is topologically unchanged whenever $\mathscr{S}$ is modified by a suitably small C^1-perturbation. Then $\mathscr{S}_2$ is structurally stable in R^2, but $\mathscr{S}_1$ is not structurally stable.

Finally, let us consider a control system in R^n

$$\dot{x} = f(x, u)$$

for state $x \in R^n$ and control $u \in R^m$ at each time t. We assume $f(x, u)$ in C^1 in all R^{n+m} and use controllers $u(t)$ in $L_\infty[0, T]$ for some finite duration $0 \leq t \leq T$. We shall define the control system to be (completely) controllable on $[0, T]$ in case: for each pair of points x_0 and x_1 in R^n there exists a controller $u(t)$ for which the response initiating at x_0 terminates at x_1. That is, the solution $x(t) = \varphi(t, 0, x_0)$ of $\dot{x} = f(x, u(t))$ satisfies the two endpoint conditions

$$x(0) = x_0, \quad x(T) = x_1$$

so $u(t)$ steers or controls $x(t)$ from x_0 to x_1.

We shall show later that a linear autonomous control system in R^n

$$\dot{x} = Ax + Bu$$

is controllable if and only if

$$\text{rank}\,[B, AB, A^2B, A^3B, \ldots, A^{n-1}B] = n$$

For general non-linear systems in R^n no such useful controllability criterion is known.

<u>Example.</u> Consider the system in the (x, y) plane R^2

$$\dot{x} = y$$

$$\dot{y} = x \quad \text{(corresponding to } \ddot{x} - x = 0\text{)}$$

The origin is the unique critical point in R^2. An elementary sketch of the solutions shows that the origin is a saddle point and hence it is not Liapunov stable.

It seems plausible that this linear system has its qualitative appearance unchanged when the coefficients are slightly perturbed. This assertion is correct (although a difficult theorem) and so the linear saddle system is structurally stable in R^2.

Next consider the control system

$$\ddot{x} - x = u$$

or

$$\begin{pmatrix}\dot{x}\\ \dot{y}\end{pmatrix} = \begin{pmatrix}0 & 1\\ 1 & 0\end{pmatrix}\begin{pmatrix}x\\ y\end{pmatrix} + \begin{pmatrix}0\\ 1\end{pmatrix}u$$

Here $A = \begin{pmatrix}0 & 1\\ 1 & 0\end{pmatrix}$, $B = \begin{pmatrix}0\\ 1\end{pmatrix}$, and

$$\text{rank}\,[B, AB] = \text{rank}\begin{pmatrix}0 & 1\\ 1 & 0\end{pmatrix} = 2$$

Thus the linear system is controllable in R^2.

VI. TOPOLOGICAL DYNAMICS. FLOWS IN METRIC SPACES

To construct a general theory of qualitative dynamics applying to ordinary, delay, and partial differential equations we allow the state space M to be an arbitrary metric space on which the dynamics evolves continuously with the time $t \in R$. Thus, there is a distance metric in M satisfying the usual axioms.

<u>Definition.</u> A topological flow Φ in a metric space M is a map (continuous in two arguments):

$$\Phi : R \times M \to M : (t, x_0) \to x_t = \varphi(t, x_0)$$

such that: for each $t \in R$ we note the restriction to M

$$\Phi_t : x_0 \to x_t$$

is a topological map of M onto M and further

$$\Phi_0 = \text{identity}, \quad \Phi_{-t} = (\Phi_t)^{-1}$$

and more generally

$$\Phi_{t_1+t_2} = \Phi_{t_1} \circ \Phi_{t_2} \quad \text{(group property)}$$

Remark. $t \to \Phi_t$ is a homomorphism of R into the group Top(M) of all topological maps of M onto itself.

(a) If this homomorphism is defined only on the semi-group $R_+ (t \geq 0)$ into the group Cont(M), then this defines a (future) semi-flow in M.

(b) If Φ is defined only on a neighbourhood of the slice $t = 0$ in the product space $R \times M$, then Φ is a local flow. Note that a local flow Φ is defined for $|t| < \epsilon$ uniformly for all x_0 in a prescribed compact subset $K \subset M$. Of course, the concept of a future local flow is immediate.

Examples.

1) Consider the autonomous ordinary differential system in R^n

$$\dot{x} = f(x)$$

with f(x) in C^1 in R^n. Then the solutions

$$\Phi_t : x_0 \to x_t = \varphi(t, x_0)$$

define a local flow. If $|f(x)| < k|x|$ in all R^n, then the flow is globally defined for all $t \in R$ (and this also happens in important cases which do not satisfy a linear growth hypothesis).

2) The non-autonomous ordinary differential system in R^n

$$\dot{x} = f(t, x)$$

with f(t, x) in C^1 in R^{1+n} defines an autonomous system and flow in R^{1+n}. Namely, write

$$\frac{dx}{d\tau} = f(t, x)$$

$$\frac{dt}{d\tau} = 1$$

and then the solutions yield a local flow in R^{1+n}:

$$\Phi_\tau : (t_0, x_0) \to (\tau + t_0, \varphi(\tau + t_0, t_0, x_0))$$

If $f(t + P, x) \equiv f(t, x)$ is periodic in t, then the flow can be defined in the product of the circle S^1 with R^n.

3) A differentiable n-manifold M is a metric space which is covered by a family of local coordinate charts, each of which maps an open set of M topologically onto an open set in R^n, with the additional requirement that the coordinate transformations defined between overlapping charts should be in class C^∞. Thus differential systems can be defined globally on M by defining them in each coordinate chart $(x^1, \ldots, x^n)$ as

$$\dot{x}^i = f^i(x), \quad i = 1, \ldots, n$$

with the usual (contravariant or tangent) vector transformation rule in overlapping $(\bar{x}^1, \ldots, \bar{x}^n)$ to yield

$$\dot{\bar{x}}^i = \bar{f}^i(\bar{x}) \qquad i = 1, \ldots, n$$

with

$$\bar{f}^i = \frac{\partial \bar{x}^i}{\partial x^j} f^j$$

We can always picture an n-manifold as an n-surface differentiably embedded in some Euclidean space (actually R^{2n}), and then an ordinary differential system on M is merely a C^1-vector field f everywhere tangent to M. Then the solution trajectories of f define a local flow on M. If M is a compact manifold (say a sphere S^n or a torus T^n), then each vector field f defines a global flow on M.

4) Consider the ordinary differential-delay system in R^n

$$\dot{x}(t) = f(x(t), x(t-1))$$

with the n-vector $f(x, y)$ in C^1 in R^{2n}. For each initial state $x_0(\theta) \in C_n[-1, 0]$ (the space of continuous n-vectors on $-1 \leq \theta \leq 0$) let $x(t)$ be the corresponding solution in R^n for some maximal duration $0 \leq t < \tau_+$. Define the future states by

$$x_t(\theta) = x(t+\theta) \quad \text{in} \quad C_n[-1, 0]$$

Take the Banach space $C_n[-1, 0]$ as the state space M and then the solutions define a future local flow in $C_n[-1, 0]$,

$$\Phi : R \times C_n[-1, 0] \to C_n[-1, 0] : (t, x_0) \to x_t$$

If $|f(x, y)| \leq k(|x| + |y|)$ is a linear growth condition in R^{2n}, then Φ is a global semi-flow – for instance in the case of a linear differential-delay system

$$\dot{x}(t) = Ax(t) + Bx(t-1)$$

A more general hereditary process in R^n is defined by a differential-functional system

$$\dot{x}(t) = f(x_t) \quad \text{for} \quad t \geq 0$$

where

$$f : C_n[-1, 0] \to R^n$$

satisfies a Lipschitz condition. Again there is a future local flow (global if f satisfies a linear growth condition) in the state space $C_n[-1, 0]$.

5) Consider the linear parabolic partial differential equation

$$\frac{\partial w}{\partial t} = a^{ij}(x) \frac{\partial^2 w}{\partial x^i \partial x^j} + b^i(x) \frac{\partial w}{\partial x^i} + c(x)w + f(x)$$

with real-valued solutions $w(x, t)$, for x in a bounded open domain $D \subset R^n$ (with smooth boundary ∂D) and $t \geq 0$. Assume the coefficients $a^{ij}(x)$, $b^i(x)$, $c(x)$, $f(x)$ are C^∞ in $\overline{D}$ and that $a^{ij}(x) = a^{ji}(x)$ is strictly positive definite in $\overline{D}$. Then a continuous solution $w(x, t)$ in $\overline{D} \times R_+$ will be determined by suitable Cauchy data:

$w(x, 0) = w_0(x)$ initial function in $\overline{D}$ vanishing on ∂D

$w(x, t) = 0$ for $x \in \partial D$ and all $t \geq 0$

In more detail, if $w_0(x) \in C_0(\overline{D})$ (continuous on $\overline{D}$ but vanishing on ∂D), then $w(x, t) \in C^\infty(D \times (t > 0))$ is the unique classical solution. In particular, there is a future semi-flow in the state space $C_0(\overline{D})$ (Banach space with max norm)

$$\Phi_t : w_0 \to w_t \quad (\text{where } w_t = w(x, t)) \quad \text{for} \quad t \geq 0$$

Take the same parabolic partial differential equation but use the state space $\mathring{H}^1(D)$, the Hilbert space arising as the completion of $\mathring{C}^\infty_0(D)$, C^∞ functions on $\overline{D}$ vanishing in a neighbourhood of ∂D, under the Sobolev norm $|v|_1 = \left[\int_D \Sigma \left(|\partial v|^2 + |v|^2\right) dx\right]^{1/2}$. Then, if $w_0(x) \in \mathring{H}^1(D)$, there is a unique weak solution $w_t \in \mathring{H}^1(D)$ and a corresponding semi-flow

$$\Phi_t : w_0 \to w_t \quad \text{in} \quad \mathring{H}^1(D) \quad \text{for} \quad t \geq 0$$

Let us return to the general theory of a flow Φ in a metric space M

$$\Phi : R \times M \to M, \quad \Phi_t : M \to M$$

Define the future trajectory or orbit of a point $P_0 \in M$ to be the map $R_+ \to M : P_0 \to P_t$ or the point set $\bigcup_{t \geq 0} P_t$. The past trajectory is similar, and the full trajectory of P_0 consists of both past and future trajectories of P_0.

<u>Theorem.</u> The orbit of $P_0 \in M$ is one of the three types

i) a single point P_0, that is a critical or stationary or equilibrium point where $\varphi(t, P_0) \equiv P_0$.

ii) a simple closed curve (topological circle), that is $\varphi(t+\tau, P_0) \equiv \varphi(t, P_0)$ with smallest positive period $\tau > 0$

iii) an injective continuous image of a line R.

Proof

Consider the set $T \subset R$ of all times for which $\varphi(t, P_0) = P_0$. Then T is a closed subgroup of R and so either

i) $T = R$

ii) $T = \{k\tau\}$ for $k \in \mathbb{Z}$ and some positive τ

iii) $T = 0$

In case i) $\varphi(t, P_0) \equiv P_0$ is a critical point. In case ii) $\varphi(\tau, P_0) = P_0$ but $\varphi(s, P_0) \neq P_0$ for $0 < s < \tau$. Then the map of the compact interval $0 \leq t \leq \tau$ is one-to-one, except for the endpoints and is a topological map of S^1 into M. Clearly, each point on the orbit of P_0 has the same period τ.

In case iii) $P_t \neq P_0$ for $t \neq 0$ so the map $t \rightarrow P_t$ is a 1-to-1 continuous map of R into M. (Note this map might not have a continuous inverse, as is the case for the Kronecker irrational flow on a torus). Q.E.D.

Definition. A set $S \subset M$ is invariant under the flow Φ_t in case: $P_0 \in S$ implies the orbit $P_t \in S$ for all t. That is, an invariant set S is the union of complete orbits. Usually we study closed invariant sets, since $\overline{S}$ is invariant.

We next describe the various types of (Liapunov, future) stability that might hold for an invariant set.

Definition. Let $S \subset M$ be an invariant set for a flow Φ_t. Then S is stable in case: for each neighbourhood U of S there exists a smaller neighbourhood V of S so $S \subset V \subset U$, such that $P_t \in U$ for all $t \geq 0$ provided $P_0 \in V$. If, further, $\lim_{t \to +\infty} \text{dist}(P_t, S) = 0$, then S is asymptotically stable. If, still further, $Q_t \rightarrow S$ regardless of the initial point $Q_0 \in M$, then S is globally asymptotically stable.

Example. In R^2 the systems

$$\dot{x} = y \qquad\qquad \dot{x} = y$$
$$\dot{y} = -x \quad \text{(centre), and} \quad \dot{y} = -x - y \quad \begin{matrix}\text{spiral}\\ \text{(focus)}\end{matrix}$$

are stable about the critical point $S = (0, 0)$, but

$$\begin{aligned} \dot{x} &= y \\ \dot{y} &= x \end{aligned} \qquad \text{(saddle)}$$

is not stable about S.

In polar co-ordinates in R^2, consider the system

$$\dot{r} = (1 - r), \qquad \dot{\theta} = -1$$

Then the periodic orbit $r(t) = 1$, $\theta(t) = -t$ is a stable invariant set.

Definition. Let Φ_t define a flow in M. The future (or positive or ω) limit set of a point P_0 (or the orbit P_t) is the set

$$\omega(P_0) = \bigcap_{t_1>0} \overline{\bigcup_{t\geq t_1} P_t}$$

and the past (or negative or α) limit set is

$$\alpha(P_0) = \bigcap_{t_1<0} \overline{\bigcup_{t\leq t_1} P_t}$$

It is easy to show that $\omega(P_0)$ can be characterized as the set of all points $Q \in M$ for which some sequence of points on the future orbit P_t approach Q. That is, $Q \in \omega(P_0)$ just in case:

$$\exists\ t_k \to +\infty \quad \text{for which} \quad \lim_{t_k\to\infty} P_{t_k} = Q$$

From the definitions it is easy to verify that $\omega(P_0)$ and $\alpha(P_0)$ are connected, closed, invariant sets. In case the future orbit of P_0 lies in some compact subset of M (say M itself is compact), then $\omega(P_0)$ is a nonempty compact set.

Definition. $P_0 \in M$ is future-recurrent (or Poisson stable) in case $P_0 \in \omega(P_0)$. That is, P_0 is future-recurrent in case the future orbit P_t recurs back to an arbitrarily small neighbourhood of P_0 for some arbitrarily late times. Past recurrence is similar, and recurrence means both past and future recurrence.

The set of all periodic points of Φ_t is an invariant set, and is included in the invariant set of all recurrent points. But these invariant sets may not be closed in M. We next define a larger invariant set Ω that is closed in M.

Definition. Let Φ_t define a flow in M. The non-wandering (or regionally recurrent) set $\Omega \subset M$ consists of all points $P_0 \in M$ such that each neighbourhood N of P_0 in M recurs to meet itself at arbitrarily large times. That is, $P_0 \in \Omega$ just in case: for each neighbourhood N of P_0 there exist arbitrarily large times $t_k \to +\infty$ when Φ_{t_k} N meets N.

It is easy to prove that Ω is a closed invariant subset of M which contains all recurrent points. While we require only future regional recurrence for $P_0 \in \Omega$, this automatically implies past regional recurrence.

VII. TOPOLOGICAL DYNAMICS. MINIMAL SETS AND STABILITY THEORY

In the study of a flow Φ_t on a metric space M we often search for an invariant set $S \subset M$. Then Φ_t defines a flow on S which might be rather simpler and which helps to understand the total flow on M.

Example. In the space R^{2n} with coordinates $(x^1, x^2, \ldots, x^n, y_1, y_2, \ldots, y_n)$ we take any real C^2 function H(x, y), and then determine a dynamical system of Hamiltonian form

$$\dot{x}^i = \frac{\partial H}{\partial y_i}(x, y)$$

$$\dot{y}_i = -\frac{\partial H}{\partial x^i}(x, y)$$

Then, for each constant C, the level hypersurface

$$H(x, y) = C$$

is an invariant subset of R^{2n}. This is just the assertion of classical mechanics that a Hamiltonian dynamical system preserves the energy H. Namely,

$$\frac{dH}{dt} = \frac{\partial H}{\partial x}\dot{x} + \frac{\partial H}{\partial y}\dot{y} = H_x H_y - H_y H_x \equiv 0$$

Definition. Let Φ_t be a flow on a metric space M. A non-empty compact invariant set Σ of Φ_t is called minimal in case Σ contains no proper compact invariant subset.

Theorem. Let S be a non-empty compact invariant set for a flow Φ_t in a metric space M. Then S contains some minimal set Σ for Φ_t.

Proof

Partially order the compact invariant subsets of Φ_t by the usual relation of inclusion. Use Zorn's lemma to select a maximal linearly ordered chain of such sets, and then Σ is the intersection of the members of this chain. Thus this result is proved by Zorn's lemma or by use of the axiom of choice.

Q. E. D.

Theorem. Let Σ be a minimal set for a flow Φ_t in a metric space M. Then the flow Φ_t in Σ is such that:

1) each point $P_0 \in \Sigma$ is topological-transitive in Σ, that is, $\alpha(P_0) = \omega(P_0) = \Sigma$.
2) each point $P_0 \in \Sigma$ is recurrent in Σ, and with bounded time gaps between recurrences to any prescribed neighbourhood.

Proof

Take $P_0 \in \Sigma$ and then $\omega(P_0)$ is a nonempty compact invariant set in Σ. Since Σ is assumed minimal, $\omega(P_0) = \Sigma$. A similar argument proves $\alpha(P_0) = \Sigma$.

Next take any neighbourhood N of P_0 in Σ and suppose P_t lies outside N for a sequence of future time intervals L_k with lengths tending toward infinity. Let t_k be the mid-time of L_k and consider the sequence P_{t_k} in the compact set Σ. Select a subsequence t_{k_i} such that $P_{t_{k_i}} \to Q$ in Σ. But $\omega(Q) = \Sigma$ so the orbit through Q meets N after some finite duration L. By continuity when $P_{t_{k_i}}$ is sufficiently near Q, the orbit of $P_{t_{k_i}}$ must also meet N after duration L. But this contradicts the supposition that $L_{k_i} \to \infty$. Q. E. D.

Now we return to the study of ordinary differential systems in R^n in order to make some applications of our abstract theory of topological dynamics.

<u>Theorem.</u> Consider a differential system in R^n

$$\mathscr{S}) \quad \dot{x} = f(x), \quad f(x) \in C^1 \text{ in } R^n$$

with a critical point, say at the origin, $f(0) = 0$. Assume there exists a real function $V(x) \in C^1(R^n)$, called Liapunov function, satisfying

i) $V(x) > 0$ in R^n except $V(0) = 0$

ii) $\frac{\partial V}{\partial x^i} f^i(x) \leq 0$ in R^n.

Then the origin is (future, Liapunov) stable.

Furthermore assume that the set

$$Z = \left\{ x : \frac{\partial V}{\partial x^i} f^i(x) = 0 \right\}$$

in R^n is such that; in some neighbourhood N_0 of the origin, $Z \cap N_0$ contains no minimal set other than the origin. Then the origin is asymptotically stable.

Moreover, if we can take $N_0 = R^n$ and also assume $V(x) \to \infty$ as $|x| \to \infty$, then the origin is globally asymptotically stable.

<u>Proof</u>

Take a neighbourhood N of the origin, say with compact closure $\overline{N}$, and define

$$\epsilon = \min_{x \in \partial N} V(x) > 0$$

Take a subneighbourhood U wherein $V(x) < \epsilon/2$. Then for $x_0 \in U$ the solution curve of $\mathscr{S}$ never meets ∂N, since along $x(t, x_0)$ we compute

$$\frac{dV}{dt} = \frac{\partial V}{\partial x} f(x) \leq 0$$

so

$$V(x, (t, x_0)) < \epsilon/2 \quad \text{for} \quad t \geq 0$$

Thus the origin is stable.

Now assume $Z \cap N_0$ contains no minimal set other than the origin. Take a subneighbourhood $U_0 \subset N_0$ so $x_0 \in U_0$ has a future trajectory in N_0, and suppose $\omega(x_0)$ contains a point $Q \neq 0$. Note $V(x) = V(Q) > 0$ must be constant on $\omega(x_0)$, since the montonic function $V(x(t, x_0)) \to V(Q)$. So $\omega(x_0)$ does not contain the origin.

Hence $\omega(x_0)$ contains some minimal set K, not lying in Z. Then we choose a point $y \in K - Z$ so $V(y(t)) < V(y(0)) = V(Q)$, for small $t > 0$. But this contradicts the constancy of $V(x)$ on $\omega(x_0)$. Thus we must conclude that $\omega(x_0) = \{0\}$, and so $\mathscr{S}$ is asymptotically stable.

Finally, assume Z contains no minimal set other than the origin and also $V(x) \to \infty$ as $|x| \to \infty$. Then each simple closed curve $V(x)$ = constant encloses an invariant region which can play the role of N_0 above. Hence $x_0 \in N_0$ has a future trajectory that remains forever in the compact set N_0, and $\omega(x_0) = \{0\}$. Q.E.D.

Example. Consider an autonomous system of non-linear mechanics with 1-degree of freedom

$$\ddot{x} + f(x)\dot{x} + g(x) = 0$$

with $f(x)$ and $g(x)$ in C^1 for x in R, and we assume

$$f(x) > 0, \quad \text{and} \quad xg(x) > 0 \quad \text{for} \quad x \neq 0$$

Then the system in the (x, y) phase plane R^2 is

$$\dot{x} = y, \quad \dot{y} = -g(x) - yf(x)$$

which we show to be asymptotically stable about the origin. If also $G(x) = \int_0^x g(s)ds \to \infty$ as $|x| \to \infty$, then the system can be seen to be globally asymptotically stable.

To verify these assertions, consider the Liapunov function

$$V(x, y) = \frac{y^2}{2} + G(x)$$

Note $V > 0$ except at the origin. Also

$$\dot{V} = g(x)y + y[-g(x) - yf(x)] = -y^2 f(x) \leq 0$$

Thus the origin is stable.

Next consider the set in R^2

$$Z = \{(x, y) : -y^2 f(x) = 0\}$$

so Z is just the x-axis where $y = 0$. On the set Z we observe the dynamics $\dot{y} = -g(x)$, so the only invariant subset of Z is the origin. By the above theorem the origin is then asymptotically stable.

If, further, $G(x) \to \infty$ as $|x| \to \infty$ then the level curves $V(x, y)$ = const. are closed curves encircling the origin. The theorem then asserts that the system is globally asymptotically stable about the origin in R^2.

Note that the asymptotic stability is a consequence of the positive damping $f(x) > 0$. If instead we have a conservative system with $f(x) \equiv 0$, then every solution lies on some level curve of $V(x, y)$ and so the origin is stable. If $V \to \infty$ as $|x| + |y| \to \infty$, then each level curve of V is a closed curve encircling the origin, so each solution is periodic.

VIII. CONTROLLABILITY OF LINEAR SYSTEMS

For an autonomous linear differential equation in R^n,

$$\dot{x} = Ax$$

the origin is an asymptotically stable critical point if and only if all the eigenvalues λ_j of A lie in the left-hand complex plane, that is, $\operatorname{Re} \lambda_j < 0$.

To prove this make a linear co-ordinate transformation in R^n, $y = Px$, and compute the linear system

$$\dot{y} = P\dot{x} = PAx = (PAP^{-1})y = \Lambda y$$

Choose P so that $PAP^{-1} = \Lambda$ is in suitable Jordan canonical form, and for simplicity of exposition we assume

$$\Lambda = \operatorname{diag}\{\lambda_1, \lambda_2, \ldots, \lambda_n\} \quad \text{with real } \lambda_j < 0$$

Then the solution initiating at y_0 is

$$y(t) = e^{\lambda_1 t} y_0^1 + \ldots e^{\lambda_n t} y_0^n$$

Thus $x(t) = P^{-1}y(t) \to 0$ exponentially as $t \to +\infty$, so the origin is globally asymptotically stable.

Even for the non-linear differential equation in R^n,

$$\dot{x} = f(x) = Ax + h(x)$$

where $h(x) \in C^1$ is of higher order near $x = 0$, the condition $\operatorname{Re} \lambda_j < 0$ implies (local) asymptotic stability. The proof again uses the linear co-ordinates $y = Px$ and the Liapunov function

$$V(y) = -y'\Lambda y > 0, \text{ for } y \neq 0$$

Then

$$\frac{dV}{dt} = -y'\Lambda\,[\Lambda y + h(P^{-1}y)]$$

and the negative-definite quadratic form $-y'\Lambda^2 y$ dominates the higher-order terms near $y = 0$.

However, note that the solution $x(t) = P^{-1}y(t)$ merely tends asymptotically towards the origin as $t \to +\infty$, and we might wish to steer an initial state x_0 to $x = 0$ in a finite time duration by applying a suitable controller $u(t)$.

<u>Definition.</u> The autonomous linear control system

$$\mathscr{L})\quad \dot{x} = Ax + Bu, \quad x \in R^n, \quad u \in R^m$$

is (completely) controllable in R^n on the duration $0 \leq t \leq T$ in case: for each pair of states x_0 and x_1 in R^n there exists a piecewise continuous controller $u(t)$ on $0 \leq t \leq T$ steering the response $x(t)$ from $x(0) = x_0$ to $x(T) = x_1$.

If the duration $T < \infty$ is allowed to vary with the choice of points x_0, x_1, then $\mathscr{L}$ is controllable on finite durations.

Theorem

$$\mathscr{L})\quad \dot{x} = Ax + Bu, \quad x \in R^n, \quad u \in R^m$$

is controllable on a given finite duration $0 \leq t \leq T$ if and only if

$$\operatorname{rank}\,[B, AB, A^2 B, \ldots, A^{n-1} B] = n$$

Proof

Fix x_0 and choose a controller $u(t)$ to get a response

$$x(t) = e^{At} x_0 + e^{At} \int_0^t e^{-As} Bu(s)\, ds$$

Then $x(T) = e^{AT} x_0 + \{L\}_T$, for a linear subspace $\{L\}_T \subset R^n$.

Assume that rank $[B, AB, A^2B, \ldots, A^{n-1}B] = n$ so this $n \times nm$ matrix has n independent rows. Then a row n-vector η annihilates $[B, AB, A^2B, \ldots, A^{n-1}B]$ if and only if $\eta = 0$. That is, $\eta B = 0$, $\eta AB = 0$, $\eta A^2B = 0, \ldots, \eta A^{n-1}B = 0$ hold if and only if $\eta = 0$. Hence

$$\eta e^{-As} B = \eta\,[I - As - \frac{s^2}{2!} A^2 - \ldots]\, B \equiv 0 \text{ on } 0 \leq s \leq T$$

implies $\eta = 0$.

Suppose $\dim \{L\}_T < n$ so there exists $\eta \neq 0$ for which $\eta\{L\}_T = 0$ or $\eta e^{AT} \int_0^T e^{-As} Bu(s)\, ds \equiv 0$, for all $u(s)$. Then $(\eta e^{AT})\, e^{-As} B \equiv 0$ so $\eta e^{AT} = 0$. This contradicts the supposition $\eta \neq 0$ so we conclude $\{L\}_T = R^n$ and $\mathscr{L}$ is controllable on $0 \leq t \leq T$.

Conversely, assume $\mathscr{L}$ is controllable – or only the weaker assertion that the origin can be steered to all points of R^n in various finite durations. Let $\{L\}_\infty = \bigcup_{k>0} \{L\}_k = R^n$. Since the linear spaces $\{L\}_k$ are nested increasing for $k > 0$ (insert a zero control for duration σ before a controller on $0 \leq t \leq k$ to construct a controller on $0 \leq t \leq k + \sigma$), the reachable set $\{L\}_\infty$ is a linear space with the same dimension as $\{L\}_N$, for some large N. That is, $\dim \{L\}_N = n$ and so only the zero vector is orthogonal to $\{L\}_N$.

Thus, for any row n-vector η,

$$\eta e^{AN} \int_0^N e^{-As} Bu(s)\, ds \equiv 0 \qquad \text{implies } \eta = 0$$

Hence, writing $\eta_1 = \eta e^{AN}$,

$$\int_0^N \eta_1 e^{-As} Bu(s)\, ds \equiv 0 \qquad \text{implies } \eta_1 = 0$$

Now suppose rank $[B, AB, A^2B, \ldots, A^{n-1}B] < n$ so there exists $\eta_1 \neq 0$

$$\eta_1 B = 0,\ \eta_1 AB = 0,\ \eta_1 A^2B = 0,\ \ldots,\ \eta_1 A^{n-1}B = 0$$

By the Cayley Hamilton Theorem A^n is a real linear combination of $(I, A, A^2, \ldots, A^{n-1})$ so $\eta_1 A^n B = 0$. Continuing we have $\eta_1 e^{-As}B \equiv 0$ with $\eta_1 \neq 0$. This contradicts the assertion that $\{L\}_N = R^n$, so we conclude rank $[B, AB, A^2B, \ldots, A^{n-1}B] = n$, as required. Q.E.D.

Remarks. $\mathscr{L}$ is controllable on $[0, T]$ if and only if $\mathscr{L}$ is controllable the unit duration $[0,1]$. Also the proof of the theorem shows that $\{L\}_\infty = R^n$ implies that $\mathscr{L}$ is controllable on $[0, T]$. Hence controllability on variable finite durations is the same requirement as controllability on a fixed duration for autonomous linear systems $\mathscr{L}$.

This last remark is not valid for time-dependent linear systems. However, the same methods as above (Hermes-La Salle) prove that a linear system with C^∞ coefficients

$$\dot{x} = A(t)x + B(t)u$$

is controllable in R^n on any fixed duration $0 \leq t \leq T$ provided

$$\text{rank}\ [B, \Gamma B, \Gamma^2 B, \ldots, \Gamma^{k-1}B]_{t=0} = n$$

Here $k \geq 1$ is any integer and the differential operator Γ is given by

$$\Gamma B(t) = -A(t)\,B(t) + \dot{B}(t)$$

Remark. Note that the class of controllers $u(t)$ on $0 \leq t \leq T$ is not particularly significant in the controllability of $\mathscr{L}$. In fact, we could take $u(t)$ belonging to $L_1[0, T]$, or any dense linear subspace thereof.

Corollary. For scalar controllers, so $m = 1$ and $B = b$ is a column vector, $\mathscr{L}$ is controllable in R^n if and only if

$$\det |b, Ab, A^2b, \ldots, A^{n-1}b| \neq 0$$

Theorem. An autonomous linear scalar process

$$x^{(n)} + a_1 x^{(n-1)} + \ldots + a_n x = u$$

or the corresponding control system in R^n

$$\mathscr{D})\quad \begin{aligned} \dot{x}^1 &= x^2 \\ \dot{x}^2 &= x^3 \\ &\vdots \\ \dot{x}^n &= -a_n x^1 - a_{n-1}x^2 - \ldots - a_1 x^n + u \end{aligned}$$

is controllable in R^n with $u \in R^1$.

Moreover, every autonomous linear process in R^n

$$\mathcal{L})\quad \dot{x} = Ax + bu, \quad u \in R^1$$

which is controllable, is linearly equivalent to some scalar control process $\mathcal{D}$.

<u>Proof</u>

For the matrices

$$A_1 = \begin{pmatrix} 0 & 1 & 0 & 0 & \dots & 0 \\ 0 & 0 & 1 & 0 & \dots & 0 \\ \vdots & & & & & \\ 0 & 0 & 0 & & \dots & 1 \\ -a_n & \dots & & & & -a_1 \end{pmatrix} \quad \text{and} \quad b_1 = \begin{pmatrix} 0 \\ 0 \\ \vdots \\ 0 \\ 1 \end{pmatrix}$$

the controllability criterion is easily verified by computation.

Next consider the controllable process $\mathcal{L}$ in R^n and consider the non-singular $n \times n$ matrix

$$P = [A^{n-1}b, A^{n-2}b, \dots, A^2b, Ab, b]$$

Introduce new linear co-ordinates in R^n by $\overline{x} = P^{-1}x$ so

$$\dot{\overline{x}} = P^{-1}AP\overline{x} + P^{-1}bu$$

By direct matrix multiplication verify

$$b = [A^{n-1}b, \dots, Ab, b]\begin{pmatrix} 0 \\ 0 \\ \vdots \\ 0 \\ 1 \end{pmatrix} \quad \text{or} \quad P^{-1}b = b_1$$

Also $AP = PN$ or $P^{-1}AP = N$ where we define

$$N = \begin{pmatrix} \alpha_1 & 1 & 0 & \dots & 0 \\ \alpha_2 & 0 & 1 & \dots & 0 \\ \vdots & \vdots & \vdots & & \vdots \\ \alpha_{n-1} & 0 & 0 & \dots & 1 \\ \alpha_n & 0 & 0 & \dots & 0 \end{pmatrix}$$

Here the real constants $\alpha, \alpha_2, \dots, \alpha_n$ are uniquely specified by the characteristic equation for A,

$$A^n = \alpha_1 A^{n-1} + \alpha_2 A^{n-2} + \dots + \alpha_n$$

Thus $\mathscr{L}$ is linearly equivalent to a system in R^n

$$\dot{\bar{x}} = N\bar{x} + b_1 u$$

But if we begin with a scalar process $\mathscr{D}$ with matrix A_1, and corresponding characteristic equation

$$A_1^n = -a_1 A_1^{n-1} - a_2 A_1^{n-2} - \ldots - a_n$$

then $\mathscr{D}$ is linearly equivalent to $\mathscr{L}$ provided we take

$$-a_1 = \alpha_1, \; -a_2 = \alpha_2, \; \ldots, \; -a_n = \alpha_n$$

Since linear equivalence is a transitive relation, $\mathscr{L}$ is linearly equivalent to the scalar system $\mathscr{D}$, as required. Q.E.D.

Theorem. Consider a controllable linear autonomous process

$$\mathscr{L}) \; \dot{x} = Ax + bu, \quad x \in R^n, \quad u \in R^1$$

Then there exists a linear feedback $u = kx$, for constant row vector k, such that

$$\dot{x} = (A + bk)x$$

is asymptotically stable towards the origin.

Proof

It is easy to see that the property of possessing a linear feedback which yields asymptotic stability, is invariant under the relation of linear equivalence or change of linear co-ordinates $\bar{x} = P^{-1}x$ in R^n.

But $\mathscr{L}$ is linearly equivalent to a scalar process $\mathscr{D}$

$$\bar{x}^{(n)} + a_1 \bar{x}^{(n-1)} + \ldots + a_n \bar{x} = u$$

and we can stabilize $\mathscr{D}$ by taking

$$u = (a_n - 1)\bar{x} + (a_{n-1} - n)\dot{\bar{x}} + \ldots + (a_1 - n)\bar{x}^{(n-1)}$$

to obtain

$$\bar{x}^{(n)} + n\bar{x}^{(n-1)} + \ldots + n\dot{\bar{x}} + \bar{x} = 0$$

But this has the characteristic equation

$$\lambda^n + n\lambda^{n-1} + \ldots + n\lambda + 1 = (\lambda + 1)^n$$

with all eigenvalues $\lambda = -1 < 0$. Q.E.D.

IX. LINEAR SYSTEMS WITH CONTROL RESTRAINTS

Consider an autonomous linear control process in R^n

$\mathscr{L}$) $\dot{x} = Ax + Bu$

where the controllers are in $L_2[0, T]$ with value $u(t)$ restrained to some set $\Omega \subset R^m$. We shall always assume that Ω is a convex compact set which contains the origin of R^m in its interior. For instance Ω could be the cube $|u^j| \leq 1$ for $j = 1, 2, \ldots, m$.

Definition. Consider an autonomous linear process in R^n

$\mathscr{L}$) $\dot{x} = Ax + Bu$

with $u(t) \subset \Omega$, a convex compact neighbourhood of the origin in R^m. The set of attainability from x_0 at time $T > 0$ is

$$K_{x_0}(T) = \left\{ x(T) = e^{AT} x_0 + e^{AT} \int_0^T e^{-As} Bu(s)\, ds \;\middle|\; \text{all admissible } u(s) \subset \Omega \right\}$$

Theorem. $K_{x_0}(T)$ is a compact convex subset of R^n which varies continuously with $T > 0$. Also $K_{x_0}(T)$ has a nonempty interior if and only if $\mathscr{L}$ is controllabl[e]

Proof

The subset of $L_2[0, T]$ of admissible controllers is a convex set which is weakly compact (note: a closed ball in L_2 is weakly compact – also the weak limit of a non-negative function is non-negative and this can be used to prove that the weak limit of admissible controllers is itself admissible). Since integration is a linear compact operator from L_2 to R^n, we find that $K_{x_0}(T)$ is convex and compact. The continuity (Hausdorff metric) is evident from the boundedness of $\dot{x}(t)$.

The controllability of $\mathscr{L}$ yields an interior of $K_{x_0}(T)$ just as is proved in earlier theorems where there is no restraint on u. For details see Lee-Markus.

Definition. The set $\mathscr{C}$ of null-controllability for

$\mathscr{L}$) $\dot{x} = Ax + Bu, \quad u(t) \subset \Omega \subset R^m$

is the subset of R^n consisting of all states that can be steered to the origin in finite times.

Theorem. Consider as above,

$\mathscr{L}$) $\dot{x} = Ax + Bu, \quad x \in R^n, \quad u \subset \Omega \subset R^m$

Then the set of null controllability is all R^n provided both

i) All eigenvalues of A have negative real parts
ii) rank $[B, AB, A^2B, \ldots, A^{n-1}B] = n$.

Example. Consider the damped linear oscillator with control

$$\ddot{x} + 2\beta\dot{x} + k^2 x = u(t), \quad \beta > 0, \ k > 0, \ |u(t)| \leq 1$$

Then every initial state $(x_0, y_0 = \dot{x}_0)$ lies in $\mathscr{C} = R^2$.

X. TIME-OPTIMAL CONTROL OF AUTONOMOUS LINEAR SYSTEMS

Again we consider an autonomous linear control process

$$\mathscr{L}) \quad \dot{x} = Ax + Bu, \quad x \in R^n, \quad u \in \Omega$$

where Ω is a convex compact neighbourhood of the origin in R^m. Let an initial state x_0 lie in the null-controllability set $\mathscr{C}$, so x_0 can be steered to the origin in finite time by some admissible controller $u(t) \subset \Omega$ on $0 \leq t \leq t_1$.

Definition. Among all admissible controllers $u(t) \subset \Omega$ on various $[0, t_1]$ steering x_0 to $x_1 = 0$, a time-optimal controller $u^*(t)$ on $[0, t^*]$ is defined by the requirement $t^* = \inf t_1$.

Theorem. Consider the autonomous linear process

$$\mathscr{L}) \quad \dot{x} = Ax + Bu, \quad x \in R^n, \quad u \in \Omega$$

with initial state $x_0 \in \mathscr{C}$. Then there exists a time-optimal controller $u^*(t)$ on $0 \leq t \leq t^*$ steering x_0 to the origin in minimal time t^*.

Proof

Take $x_0 \neq 0$ and consider the moving set of attainability $K_{x_0}(t)$ for $t > 0$. Since $K_{x_0}(t)$ is always a compact set in R^n, and it moves continuously with t, we find that $K_{x_0}(t)$ meets $x_1 = 0$ for a first time $t^* > 0$. The corresponding controller $u^*(t)$ which yields this point $x_1 \in K_{x_0}(t^*)$ is the required optimal controller. Q.E.D.

If Ω is a segment $|u| \leq 1$, and $\mathscr{L}$ is controllable, then $u^*(t)$ is the unique (almost everywhere) controller steering x_0 to $x_1 = 0$ in minimal time t^*. This will follow from the characterization of $u^*(t)$ by means of the maximal principle of Pontryagin, as proved below.

Theorem. (Max. Principle). Consider the autonomous linear control system in R^n

$$\mathscr{L}) \quad \dot{x} = Ax + Bu$$

with controllers $u(t) \subset \Omega$, a convex compact neighbourhood of the origin in R^m. Let $x_0 \in \mathscr{C}$ so that x_0 can be steered to the origin $x_1 = 0$ in finite time, and so there exists an optimal controller $u^*(t)$ with response $x^*(t)$ for minimal time $0 \leq t \leq t^*$.

Then there exists a nonzero solution $\eta^*(t)$ of the adjoint linear system

$$\dot{\eta} = -\eta A$$

for which

$$\eta^*(t)\,B\,u^*(t) = \max_{u\in\Omega} \eta^*(t)\,B\,u \qquad \text{almost everywhere on } 0 \leq t \leq t^*$$

Thus if we define the Hamiltonian function

$$H(\eta, x, u) = \eta[Ax + Bu]$$

then $\eta^*(t)$, $x^*(t)$ satisfy

$$\dot{x} = \frac{\partial H^*}{\partial \eta}, \quad \dot{\eta} = -\frac{\partial H^*}{\partial x}(\eta, x, t)$$

where

$$H^*(\eta, x, t) = H(\eta, x, u^*(t))$$

and

$$H(\eta^*(t), x, u^*(t)) = \max_{u\in\Omega} H(\eta^*(t), x, u)$$

Proof

We need only show that

$$\eta^*(t)\,B\,u^*(t) = \max_{u\in\Omega} \eta^*(t)\,B\,u$$

almost everywhere, for a non-trivial adjoint row vector $\eta^*(t)$, since the other assertions concerning the Hamiltonian are merely re-statements. This maximal principle will be an analytical statement of the geometric condition that $x_1 = 0$ necessarily lies on the boundary ∂K of the convex compact set of attainability $K_{x_0}(t^*)$.

Since $K_{x_0}(t^*)$ is convex, choose an outward unit normal vector η_1 (normal to a supporting hyperplane) at $x_1 \in \partial K$. Then x_1 is the point of $K_{x_0}(t^*)$ farthest in the direction of η_1, that is,

$$\eta_1 x_1 \geqq \eta_1 x \quad \text{for all } x \in K_{x_0}(t^*)$$

or

$$\eta_1 e^{At^*} + \eta_1 e^{At^*}\int_0^{t^*} e^{-As}\,Bu^*(s)\,ds \geqq \eta_1 e^{At^*} + \eta_1 e^{At^*}\int_0^{t^*} e^{-As}\,Bu(s)\,ds$$

for all admissible controllers $u(s)$ on $0 \leq s \leq t^*$.

Let

$$\eta^*(t) = (\eta_1 e^{At^*})\,e^{-At}, \text{ so } \eta^*(t^*) = \eta_1$$

be a non-trivial solution of the adjoint differential system $\dot{\eta} = -\eta A$, for a row n-vector η. Then we obtain the maximal principle:

$$\int_0^{t^*} \eta^*(s)\, Bu^*(s)\, ds \geqq \int_0^{t^*} \eta^*(s)\, Bu(s)\, ds$$

But this yields the pointwise result:

$$\eta^*(s)\, Bu^*(s) = \max_{u \in \Omega} \eta^*(s)\, Bu$$

Since if such an equality failed on some positive duration S we could define a new controller

$$\hat{u}(s) = \begin{cases} \max_{u \in \Omega} \eta^*(s)\, Bu & \text{for } s \in S \\ u^*(s) & \text{otherwise} \end{cases}$$

Then $\hat{u}(s)$ (which is an admissible controller) would contradict the above integral inequality for the maximal principle. Q.E.D.

Corollary. For the case of a scalar controller where $m = 1$ so Ω is a segment, say $|u| \leq 1$, the maximal principle asserts

$$u^*(t) = \operatorname{sgn} \eta^*(t)\, b$$

Hence $u^*(t)$ is a bang-bang controller with a finite number of switches, and $u^*(t)$ is uniquely specified by the maximal principle, provided

$$\det[b, Ab, A^2b, \ldots, A^{n-1}b] \neq 0$$

Proof

For a time-optimal controller $u^*(t)$ on $0 \leq t \leq t^*$ we must have

$$\eta^*(t)\, b\, u^*(t) = \max_{u \in \Omega} \eta^*(t)\, b\, u = |\eta^*(t)\, b|$$

Thus

$$u^*(t) = \operatorname{sgn} \eta^*(t)\, b$$

Now assume that $\mathscr{L}$ is controllable so

$$\det[b, Ab, A^2b, \ldots, A^{n-1}b] \neq 0$$

Let $u_1(t)$ be any admissible controller steering x_0 to $x_1 = 0$ on $0 \leq t \leq t^*$. Then $u_1(t)$ also satisfies the maximal principle so

$$\eta^*(t)\, b\, [u^*(t) - u_1(t)] \equiv 0 \qquad \text{almost everywhere}$$

If $u^*(t) \not\equiv u_1(t)$ on some positive duration Σ, then

$$\eta^*(t)\, b \equiv 0 \quad \text{for} \quad t \in \Sigma$$

Since $\eta^*(t) = (\eta_1 e^{At^*})\, e^{-At}$ is a real analytic vector, and since Σ contains an infinite set of points,

$$\eta^* e^{-At} b \equiv 0 \text{ for all } 0 \leq t \leq t^*$$

where we set $\eta^* = \eta_1 e^{At^*}$ as a nonzero row vector. Then at $t = 0$ we have $\eta^* b = 0$. Differentiate and set $t = 0$ to get $\eta^* Ab = 0$ and continue to find

$$\eta^* [b, Ab, A^2 b, \ldots, A^{n-1} b] = 0$$

But this contradicts the controllability assumption for $\eta^* \neq 0$. Thus

$$u_1(t) \equiv u^*(t) \quad \text{almost everywhere}$$

In other words, there is a unique optimal controller for the minimal time t^*. Moreover any controller $u_1(t)$ on $0 \leq t \leq t^*$ which satisfies the maximal principle must be this optimal $u^*(t)$. Q.E.D.

More general bang-bang theorems are available for the restraint set Ω as a polyhedron in R^m – see the text of Lee-Markus.

XI. SWITCHING LOCUS FOR THE SYNTHESIS OF OPTIMAL CONTROLLERS

Consider an autonomous linear system in R^n

$$\mathscr{L})\quad \dot{x} = Ax + Bu$$

for $u \subset \Omega$, a convex compact neighbourhood of the origin in R^m. For any initial state x_0 in $\mathscr{C}$, the set of null controllability in R^n, there exists an optimal controller $u^*(t)$ steering $x^*(t)$ from x_0 to $x_1 = 0$ in minimal time t^*. Also $u^*(t)$ satisfies the maximal principle

$$\eta^*(t)\, Bu^*(t) = \max_{u \in \Omega} \eta^*(t)\, Bu \qquad \text{almost everywhere on } 0 \leq t \leq t^*,$$

for some nontrivial adjoint vector $\eta^*(t)$ solving

$$\dot{\eta} = -\eta A$$

In the case when $B = b$ is a vector, so $m = 1$ for scalar controllers $|u(t)| \leq 1$, and when $\mathscr{L}$ is controllable, the optimal controller is characterized by the maximal principle as a bang-bang controller with switches only at the finite set of zeros of $\eta^*(t)\, b$. That is,

$$u^*(t) = \operatorname{sgn} \eta^*(t)\, b$$

We illustrate this theory by constructing the switching locus, and synthesizing the optimal controller as a feedback control, for the case of the linear oscillator

$$\ddot{x} + x = u$$

or the system in R^2

$$\mathscr{L}) \quad \begin{aligned} \dot{x} &= y \\ \dot{y} &= -x + u \end{aligned}$$

with Ω : $|u| \leq 1$
Here

$$\begin{pmatrix} \dot{x} \\ \dot{y} \end{pmatrix} = A \begin{pmatrix} x \\ y \end{pmatrix} + bu, \quad A = \begin{pmatrix} 0 & 1 \\ -1 & 0 \end{pmatrix}, \quad b = \begin{pmatrix} 0 \\ 1 \end{pmatrix}$$

Thus the adjoint equation for $\eta = (\eta_1, \eta_2)$ is

$$\dot{\eta} = -\eta A \quad \text{or} \quad (\dot{\eta}_1, \dot{\eta}_2) = -(\eta_1, \eta_2) \begin{pmatrix} 0 & 1 \\ -1 & 0 \end{pmatrix}$$

Thus

$$\dot{\eta}_1 = \eta_2, \quad \dot{\eta}_2 = -\eta_1 \quad \text{so} \quad \ddot{\eta}_2 + \eta_2 = 0$$

Hence $\eta_2^*(t)$ is a sinusoid with a constant duration of π between successive zeros.

The maximal principle now asserts that

$$u^*(t) = \text{sgn}\,(\eta_1^*(t), \eta_2^*(t)) \begin{pmatrix} 0 \\ 1 \end{pmatrix} = \text{sgn}\,\eta_2^*(t)$$

Hence the optimal controller $u^*(t)$ switches between the extreme values +1 and -1 every π units of time. The main difficulty in finding $u^*(t)$, steering a given point $\begin{pmatrix} x_0 \\ y_0 \end{pmatrix}$ to $\begin{pmatrix} 0 \\ 0 \end{pmatrix}$ in minimal time t^*, is that we do not have the initial (or terminal) data for $\eta_2^*(t)$. But we do know that each extremal controller (satisfying the maximal principle) acts as the unique optimal controller for some initial point in R^n. Thus, we proceed to construct all possible extremal controllers, and observe the corresponding extremal responses, to find the correct optimal controller for $\begin{pmatrix} x_0 \\ y_0 \end{pmatrix}$.

The optimal response from $\begin{pmatrix} x_0 \\ y_0 \end{pmatrix}$ to the origin must follow arcs of the solutions of the extremal differential systems

$$\mathscr{S}_-) \quad \begin{aligned} \dot{x} &= y \\ \dot{y} &= -x - 1 \end{aligned} \qquad \text{and} \qquad \mathscr{S}_+) \quad \begin{aligned} \dot{x} &= y \\ \dot{y} &= -x + 1 \end{aligned}$$

Since the extremal differential systems $\mathscr{S}_-$ and $\mathscr{S}_+$ are autonomous, we can construct extremal responses that terminate at the origin by a process of backing out of the origin as -t increases. That is, we start the extremal response at t = 0 at the origin and follow the appropriate solution curves of $\mathscr{S}_-$ and $\mathscr{S}_+$ backwards in time (switching every π units of time) to reach the point $\begin{pmatrix} x_0 \\ y_0 \end{pmatrix}$ at some negative value of $t = -t^*$. Then reverse the time sense and start from $\begin{pmatrix} x_0 \\ y_0 \end{pmatrix}$ at t = 0 to arrive at the origin at $t = t^*$, thus obtaining the optimal response $\begin{pmatrix} x^*(t) \\ y^*(t) \end{pmatrix}$, the optimal time t^*, and the optimal controller $u^*(t)$.

If we start at the origin at t = 0 with $\eta_1 = 1$, $\eta_2 = 0$, then $\eta_2(t) = -\sin t$ and, on the interval $-\pi < t < 0$ we have sgn $\eta_2(t) = +1$. The corresponding extremal response traces out the solution curve of $\mathscr{S}_+$ through the origin, that is,

$$\Gamma_+ : \begin{array}{l} x = -\cos t + 1 \quad \text{on} \quad -\pi < t < 0 \\ y = \sin t \end{array}$$

or

$$(x-1)^2 + y^2 = 1, \quad y < 0$$

On the other hand if we start with $\eta_1 = -1$, $\eta_2 = 0$, then $\eta_2(t) = \sin t$ and, on $-\pi < t < 0$, we have sgn $\eta_2(t) = -1$. The corresponding extremal response curve of $\mathscr{S}_-$ is

$$\Gamma_- : \begin{array}{l} x = \cos t - 1 \quad \text{on} \quad -\pi < t < 0 \\ y = -\sin t \end{array}$$

or

$$(x+1)^2 + y^2 = 1, \quad y > 0$$

However, for any other choice of (η_1, η_2) at t = 0 with $\eta_2 > 0$, the extremal response traces back from the origin along Γ_+ until $\eta_2(t) = 0$. At this point the extremal response switches to a solution of $\mathscr{S}_-$ which it follows for a duration of length π before switching again to a solution of $\mathscr{S}_+$. A similar process occurs for initial data with $\eta_2(0) < 0$ but here the extremal response starts back from the origin along Γ_-.

The switching locus $\mathscr{W}$, which consists of all the points where all the above extremal responses switch between the families $\mathscr{S}_-$ and $\mathscr{S}_+$ is not difficult to describe in this example. In fact, $\mathscr{W}$ is made up of the arcs Γ_+ and Γ_- and their successive transportations along the appropriate solution families of $\mathscr{S}_-$ and $\mathscr{S}_+$ for durations of length π. Note that such a transportation is just a rigid rotation of the phase plane through π radians about the corresponding centre (1, 0) or (-1, 0). This process of successive switching leads to a final construction of $\mathscr{W}$ as a continuous curve over the entire x-axis, composed of semi-circles of radius one, as pictured in Fig. 5.

Define the synthesis for real $(x, y) \neq (0, 0)$ by

$$\Psi(x,y) = \begin{cases} -1 & \text{if } (x,y) \text{ lies above } \mathscr{W} \text{ or on } \Gamma_- \\ 0 & \text{if } (x,y) \text{ lies on } \mathscr{W} \text{ otherwise} \\ +1 & \text{if } (x,y) \text{ lies below } \mathscr{W} \text{ or on } \Gamma_+ \end{cases}$$

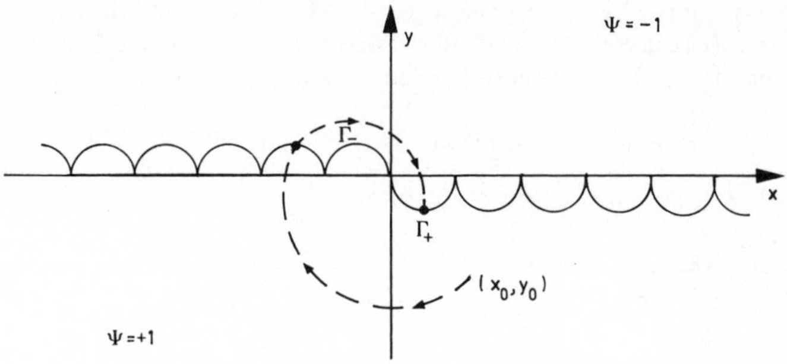

FIG.5. Switching locus $\mathscr{W}$ for $\ddot{x} + x = u$.

Then the optimal control is given by the feedback

$$u = \Psi(x, y)$$

and each optimal response is a solution of

$$\ddot{x} + x = \Psi(x, y)$$

XII. LINEAR DYNAMICS WITH QUADRATIC COST OPTIMIZATION

Consider an autonomous linear control system in R^n

$$\mathscr{L}) \quad \dot{x} = Ax + Bu$$

with admissible controllers u(t) as m-vectors which are square-integrable on a given finite duration $0 \leq t \leq T$. That is, the only restraint on u(t) is that

$$\int_0^T \|u(s)\|^2 \, ds < \infty, \quad \text{where } \|u\|^2 = u'u \text{ in } R^m$$

Then for a given initial state $x_0 \in R^n$ there exists a response x(t) on $0 \leq t \leq T$ and we define the cost of the controller u(t) to be

$$C(u) = \int_0^T (\|x(s)\|_W^2 + \|u(s)\|_U^2) \, ds$$

where $W = W' > 0$ and $U = U' > 0$ are positive-definite constant matrices used to define the norms $\|x\|_W^2 = x'Wx$, $\|u\|_U^2 = u'Uu$. We seek an optimal controller $u^*(t)$ minimizing the cost or performance functional C(u) so

$$C(u^*) = \inf C(u)$$

By techniques similar to those of the minimal time-optimal problem it can be shown (see Lee-Markus) that there exists a unique optimal controller $u^*(t)$ and that it is characterized by the maximal principle:

$$u^*(t) = U^{-1} B' \eta^*(t)' \qquad \text{almost everywhere on } 0 \leq t \leq T$$

where $\eta^*(t)$ is a row n-vector solution of

$$\dot{\eta} = x'W - \eta A, \quad \eta(T) = 0$$

with

$$\dot{x} = Ax + BU^{-1}B'\eta', \quad x(0) = x_0$$

We can express this optimal control through a feedback gain matrix $E^*(t)$, independent of x_0, so

$$u^*(t) = E^*(t)\, x^*(t)$$

Here (see Lee-Markus for details) it can be established that

$$E^*(t) = U^{-1}B'E(t)$$

where $E(t)$ is the solution of the matrix Riccati differential equation

$$\dot{E} = W - A'E - E'A - EBU^{-1}BE$$

with the terminal data $E(T) = 0$. Using this gain matrix $E^*(t) = U^{-1}B'E(t)$, which can be computed in advance of the control programme, we obtain the optimal response $x^*(t)$ as a solution of

$$\dot{x} = Ax + B[U^{-1}B'E(t)x], \quad x(0) = x_0$$

In the limiting case where $T = +\infty$ the Riccati differential equation is replaced by the quadratic equation

$$A'E + E'A + EBU^{-1}B'E = W$$

which has a unique symmetric negative definite solution E. Then a unique optimal controller $u^*(t)$ exists (here we must assume that $\mathscr{L}$ is controllable in R^n) and can be determined by a feedback synthesis

$$u^*(t) = U^{-1}B'E\, x^*(t)$$

so $x^*(t)$ satisfies

$$\dot{x} = (A + BU^{-1}B'E)x, \quad x(0) = x_0$$

The minimal cost for

$$C(u) = \int_0^\infty \|x(s)\|_W^2 + \|u(s)\|_U^2\, ds$$

is just

$$C(u^*) = -x_0'E\,x_0$$

which is a positive definite quadratic form in x_0.

It is interesting to derive this result by the intuitive methods of dynamic programming. Let us call the minimal cost from x_0 on $[t_0, T]$ to be

$$V(x_0, t_0) = \min_{u \in L_2[t_0, T]} \int_{t_0}^{T} \|x(t)\|_W^2 + \|u(t)\|_U^2 \, dt$$

Then each controller u(t) on $[t_0, T]$ with corresponding response $x_u(t)$ initiating at x_0 at $t = t_0$ has cost

$$\int_{t_0}^{t_0+\delta} (\|x_u\|_W^2 + \|u\|_U^2)\, dt + \int_{t_0+\delta}^{T} (\|x_u\|_W^2 + \|u\|_U^2)\, dt$$

where $\delta > 0$ is an arbitrarily small number. By modifying u(t) to become optimal on $[t_0 + \delta, T]$ we obtain the cost

$$\int_{t_0}^{t_0+\delta} (\|x_u\|_W^2 + \|u\|_U^2)\, dt + V(x_u(t_0+\delta),\ t_0 + \delta)$$

Thus

$$V(x_0, t_0) = \min_{u(t) \in L_2[t_0, T]} \left\{ \int_{t_0}^{t_0+\delta} (\|x_u\|_W^2 + \|u\|_U^2)\, dt + V(x_u(t_0+\delta),\ t_0+\delta) \right\}$$

This equation illustrates the idea of dynamic programming in that we think of the optimal-control program decomposed into the sum of two programs, on $[t_0, t_0 + \delta]$ and on $[t_0 + \delta, T]$ with the dynamics indicated by the possibility of varying δ. We ignore problems of differentiability and convergence and proceed formally.

Using a Taylor series expansion in terms of the small number $\delta > 0$ we obtain

$$V(x_0, t_0) = \min_{u(t)} \Big\{ \delta\, [\, \|x_0\|_W^2 + \|u(t_0)\|_U^2] + V(x_0, t_0)$$

$$+ \left[\frac{\partial V}{\partial x}(x_0, t_0)\, \frac{dx_u}{dt}(t_0) + \frac{\partial V}{\partial t}(x_0, t_0) \right] \delta + o(\delta) \Big\}$$

Write

$$\frac{dx_u}{dt}(t_0) = Ax_0 + Bu(t_0)$$

and let $\delta \to 0$ to obtain

$$-\frac{\partial V}{\partial t}(x, t) = \min_{u} \left\{ \|x\|_W^2 + \|u\|_U^2 + \frac{\partial V}{\partial x}[Ax + Bu] \right\}$$

where we have written the generic point (x_0, t_0) as (x, t).

Thus we must compute the minimum of the real function

$$h(u) = \|u\|_U^2 + \frac{\partial V}{\partial x}[Ax + Bu]$$

for each fixed (x, t). But

$$\frac{\partial h}{\partial u} = 2u'U + \frac{\partial V}{\partial x} B = 0$$

so take

$$u' = -\frac{1}{2}\frac{\partial V}{\partial x} BU^{-1} \quad \text{or} \quad u = -\frac{1}{2} U^{-1}B'\left(\frac{\partial V}{\partial x}\right)'$$

Hence $V(x, t)$ must be the solution of the non-linear partial differential equation

$$-\frac{\partial V}{\partial t} = \|x\|_W^2 - \frac{1}{4}\frac{\partial V}{\partial x} BU^{-1}B'\left(\frac{\partial V}{\partial x}\right)' + \frac{\partial V}{\partial x} Ax$$

with terminal data

$$V(x, T) = 0$$

In the case $T = +\infty$ we can verify that

$$V(x, t) = -x'Ex$$

is a solution of this Bellman dynamic programming partial differential equation just in case E satisfies

$$W = EBU^{-1}B'E + A'E + E'A$$

as specified earlier.

CONCEPTS OF STABILITY AND CONTROL

P.C. PARKS
Control Theory Centre,
University of Warwick,
Coventry, United Kingdom

Abstract

CONCEPTS OF STABILITY AND CONTROL.

The first part of the paper deals with the role transfer functions play in control problems (closed loop, Nyquist stability criterion, sampled data systems and z-transforms; the "hog cycle"; spring oscillations; Lyapunov functions; the Zubov method; positive-real functions and the Popov criterion; the circle criterion; linear time-delay systems; equations with periodic coefficients; stability of repeated processes).
In the second part the author considers the control of systems which are described by partial differential equations (heat-conduction equation; wave equation; control of the heat and the wave equations; parasitic oscillations; noise in linear systems; discrete noise processes). Many examples are given and briefly discussed.

1. CONCEPTS OF STABILITY AND CONTROL; TRANSFER FUNCTIONS

The mathematical theory of control has always been prompted by practical considerations and much of the development of theory has been carried out by electrical and mechanical engineers, who have introduced a number of concepts and words which are not always familiar to the mathematician brought up on traditional courses in pure and applied mathematics. The mathematician has made important contributions at critical stages in the development of control theory — for example Maxwell (stability, 1868), Hurwitz (stability, 1895), Wiener (random processes, 1941), Pontryagin (optimal control, 1950) — but these developments have depended on a good liaison between the engineer and the mathematician — for example between the turbine engineer Stodola and the mathematician Hurwitz. It is important therefore for mathematicians interested in control theory to familiarize themselves with these control engineering concepts.

One such concept which appears in any book or paper on control engineering is that of the transfer function.

Consider the tower crane shown in Fig.1.1. The load B is suspended freely by a cable of length ℓ from a horizontally moving trolley A. Considering movements in the plane of the tower and jib we wish to set up a mathematical relationship between the "input" to the system which is the distance CA = x of the trolley from the datum point C, and the horizontal movement of the load B measured as DB = y as shown. Considering small motions of the load (so that $y - x \ll \ell$) we obtain the equation of motion

$$m \frac{d^2y}{dt^2} + mc \frac{dy}{dt} = - T \frac{(y-x)}{\ell}$$

where m is the mass of the load, T = mg is the tension in the cable and mc is a damping constant representing aerodynamic damping of the motion (probably rather small).

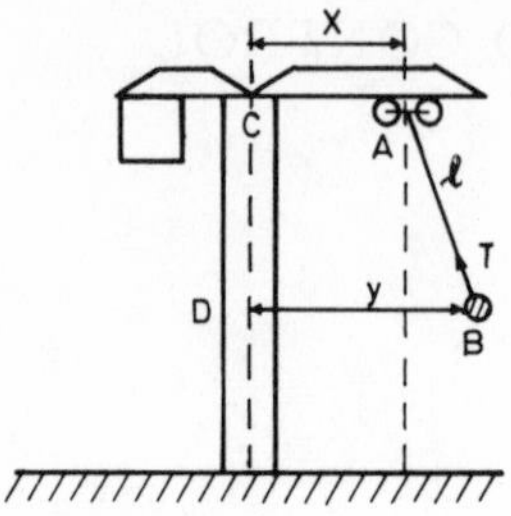

FIG. 1.1. Tower crane.

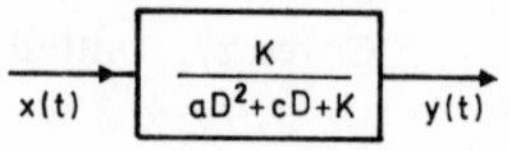

FIG. 1.2. Transfer function (1.2).

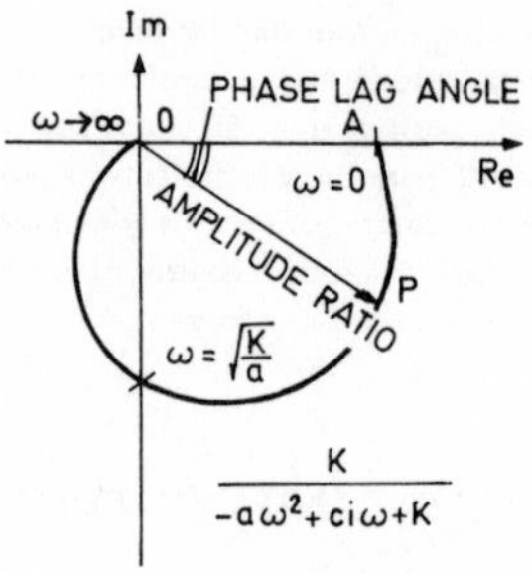

FIG. 1.3. Frequency response of (1.1).

We obtain

$$\frac{d^2y}{dt^2} + c\frac{dy}{dt} + \frac{g}{\ell}y = \frac{g}{\ell}x \tag{1.1}$$

as the second order differential equation relating the "output" y(t) to the "input" x(t). Given x(t) and initial conditions on y(t) and dy(t)/dt, Eq. (1.1) may be solved by classical methods.

Now the control engineer has developed the useful concept of the transfer function to represent this input/output relationship. Writing D for d/dt we can say that

$$\frac{y}{x} = \frac{K}{aD^2+cD+K} \tag{1.2}$$

where $K = g/\ell$, $a = 1$, and the expression $K/(aD^2+cD+K)$ is the transfer function between y and x. It is also usual to draw a "block diagram" of this relationship as shown in Fig. 1.2. It is only necessary to cross-multiply Eq. (1.2) to recover the differential equation (1.1).

The transfer function plays two other useful roles — if D is replaced by $i\omega$ we obtain the system frequency response which relates the output to input when the input is a sine-wave at frequency ω (rad/s). This is shown plotted on the Argand diagram in Fig. 1.3. For a given ω the radius 0P gives the ratio of output amplitude to input amplitude and the angle A0P represents the phase lag angle of the output sinusoid relative to the input sinusoid.

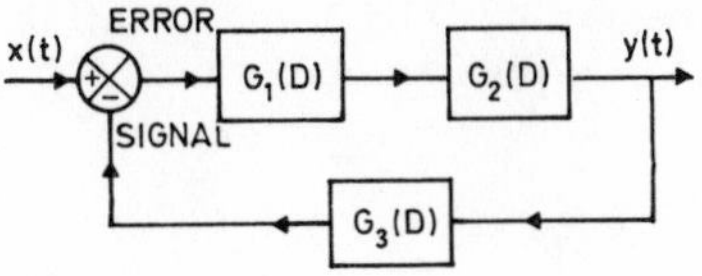

FIG. 1.4. Feedback system.

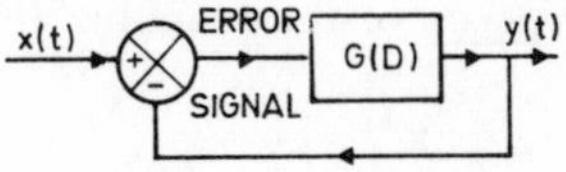

FIG. 1.5. Unit feedback.

If D is replaced by s as used in Laplace transform notation, the transfer function then represents the Laplace transform of the unit impulse response of the system, that is the Laplace transform of y(t) when x(t) is a unit impulse at t = 0+, the system starting with zero initial conditions on y(t) and dy/dt. The unit impulse response h(t) (t > 0) is a useful concept since it may be used to write down the solution y(t) for a very general input x(t) in the form of a "convolution integral"

$$y(t) = \int_{\tau=0}^{\infty} h(\tau)\, x(t-\tau)\, d\tau$$

It often happens that we build up a feedback system involving a number of transfer functions $G_1(D)$, $G_2(D)$, etc., where each $G_i(D)$ is a ratio of polynomials in D as in Fig. 1.4. Here we may operate by the ordinary rules of algebra to obtain the "closed-loop transfer function" relating y(t) to x(t) which is, in this case,

$$\frac{y}{x} = \frac{G_2(D)\ G_1(D)}{(1 + G_2(D)\ G_1(D)\ G_3(D))}$$

(The circle ⊗ represents a subtraction to form the "error signal" which is $x(t) - G_3(D)y(t)$ hence $y(t) = G_2(D)\, G_1(D)\, [x(t) - G_3(D)y(t)]$.)

A special case is shown in Fig. 1.5 where we call G(D) the open-loop transfer function and $G(D)/(1 + G(D))$ is the closed loop transfer function. Returning to the differential equation (1.1) and thinking about the classical method of solution where y(t) = particular integral + complementary function, we observe the complementary functions, or transient solutions in the language of control engineers, determined by

$$\frac{d^2y}{dt^2} + c\frac{dy}{dt} + \frac{g}{\ell}y = 0$$

This is satisfied by $y = Ae^{\lambda t}$ where

$$\lambda^2 + c\lambda + \frac{g}{\ell} = 0$$

or, looking at the transfer function,

$$\frac{g/\ell}{D^2 + cD + g/\ell}$$

by the denominator of the transfer function replacing D by λ and equating to zero. Replacing D by s and regarding s as a complex number the roots in λ are poles of the transfer function

$$\frac{g/\ell}{s^2 + cs + g/\ell}$$

If the transient solutions are to die away as time increases, then the roots in λ must be real and negative, or complex with negative real parts. Alternatively "the poles must lie in the left-hand half of the complex plane". This is a general requirement for stability of the transfer function, and especially when it is the transfer function of a closed-loop system.

The behaviour of the system also depends on the transient solutions, for example, these may be very oscillatory even though eventually damped out. A useful test, theoretically and practically, is the unit-step response of the system. The unit-step responses of the second-order system

$$(D^2 + 2\zeta\omega_n D + \omega_n^2)y = \omega_n^2 x$$

(a standard form for second-order systems) for various values of ζ is shown in Fig.1.6.

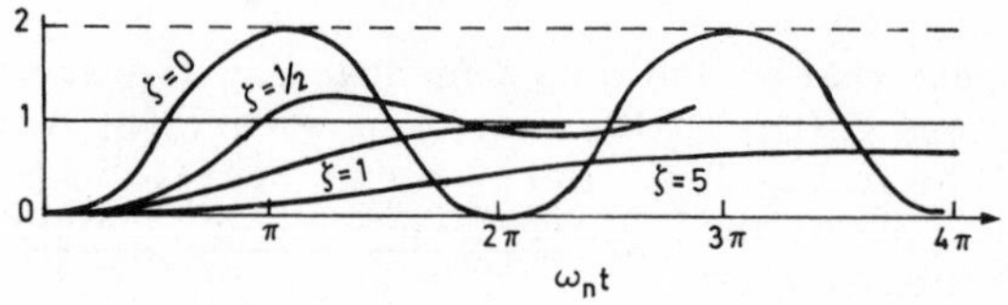

FIG. 1.6. Step responses of the second-order system $(D^2 + 2\zeta\omega_n D + \omega_n^2)y = x$.

A classical stability problem solved by Hermite (1854), Routh (1877) and Hurwitz (1895) was to find necessary and sufficient conditions on the coefficients of the linear differential equation

$$\frac{d^n y}{dt^n} + a_1 \frac{d^{n-1}y}{dt^{n-1}} + a_2 \frac{d^{n-2}y}{dt^{n-2}} + \ldots + a_{n-1}\frac{dy}{dt} + a_n y = 0$$

for damped or stable transients.

Hermite's solution, which deserves to be better known, is that the $n \times n$ matrix

$$\underset{\sim}{H} = \begin{Bmatrix} a_n a_{n-1} & 0 & a_{n-3}a_n & \cdots & \cdots & \cdots \\ 0 & a_{n-2}a_{n-1} - a_{n-3}a_n & 0 & \cdots & \cdots & \cdots \\ \cdots & \cdots & \cdots & \cdots & \cdots & \cdots \\ \cdots & \cdots & \cdots & a_5 - a_1 a_4 + a_2 a_3 & 0 & a_3 \\ \cdots & \cdots & & 0 & a_1 a_2 - a_3 & 0 \\ \cdots & \cdots & & a_3 & 0 & a_1 \end{Bmatrix}$$

be positive definite (that is the matrix of a positive-definite quadratic form).

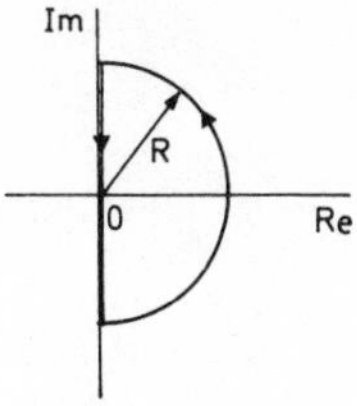

FIG. 1.7. Contour C.

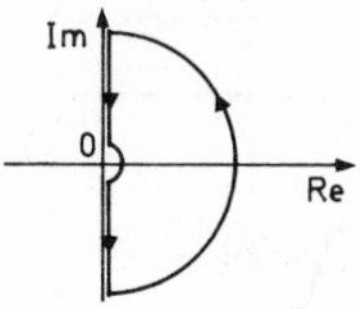

FIG. 1.8. Indented contour.

In the 1930s another technique was developed to investigate stability of closed-loop systems from a knowledge of the open-loop frequency response. This is the celebrated Nyquist stability criterion. If G(D) is the open-loop transfer function then the closed-loop transfer function is $G(D)/(1+G(D))$ and we are thus investigating roots in s of the equation $1+G(s) = 0$. Now $G(s) \equiv q(s)/p(s)$, where p(s) and q(s) are polynomials, and so we are interested in the roots of $p(s) + q(s) = 0$. We now consider the contour C on the Argand diagram shown in Fig. 1.7. Applying "the principle of the argument" to the function $1+G(s)$ of the complex variable s

$$\Big[\arg[1+G(s)]\Big]_C = 2\pi(N-P)$$

where N = number of zeros of $1+G(s)$ and P = number of poles of $1+G(s)$ within C. Here we have assumed no zeros or poles actually lie on C itself. If the degree of p(s) is greater than that of q(s), then the contribution to $[\arg(1+G(s))]$ on the arc of radius R is zero as $R \to \infty$ and we have to consider the change in argument of $1+G(i\omega)$ as ω varies from $+\infty$ through 0 to $-\infty$, that is $1+G(s)$ for $s = i\omega$ on the imaginary axis. If there are no poles of $1+G(s)$, that is zeros of p(s) in the right-hand half-plane so that $P = 0$, and furthermore we require $N = 0$ for stability (so that no roots of $1+G(s) = 0$ have positive real parts), then $1+G(i\omega)$ must not encircle the origin as ω varies form $+\infty$ through 0 to $-\infty$. Alternatively, $G(i\omega)$ must not encircle the point (-1, 0). This is the Nyquist stability criterion in its simplest form.

A modification that is quite often necessary occurs when p(s) has a zero at the origin so that G(s) has a pole there. It is then necessary to "indent" the contour C by a small semicircle as shown in Fig. 1.8. The behaviour of G(s) at infinity is determined by what happens as s describes the small semi-circle surrounding the origin. An example is shown in Fig. 1.9.

2. TRANSFER FUNCTIONS AND STABILITY

2.1. Sampled data systems and z-transforms

Discussion hitherto has concerned ordinary differential equations and continuous functions of time t. An important class of control systems uses discrete time inputs and outputs: these are known as sampled data control systems and a theory analogous to the Laplace transform for continuous systems has been built up based on the so-called z-transform. This theory may be applied also to difference equations which occur directly in some applications, for example, economic models.

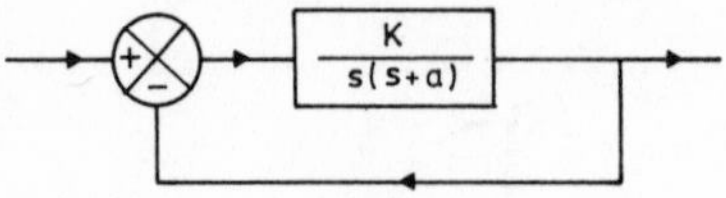

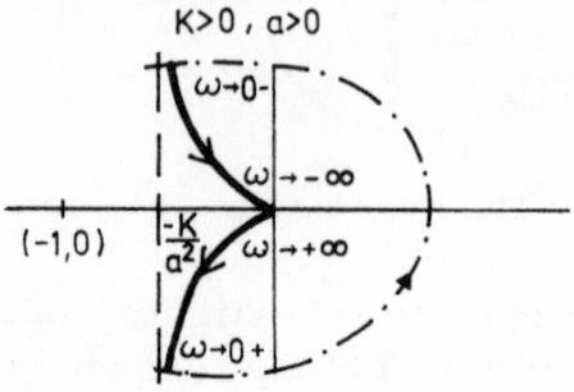

Closed-loop system stable, P = 0, N = 0 within indented contour.

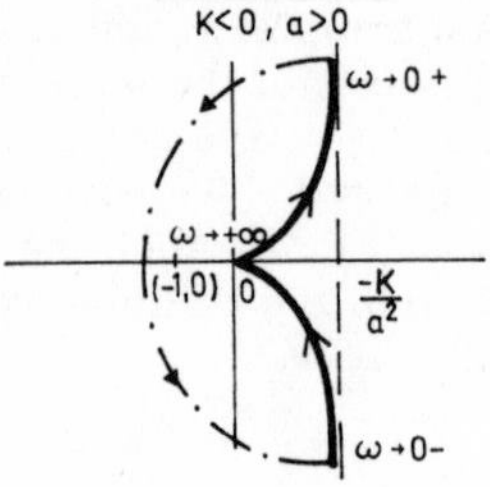

Closed-loop system unstable, P = 0, N = 1 within indented contour.

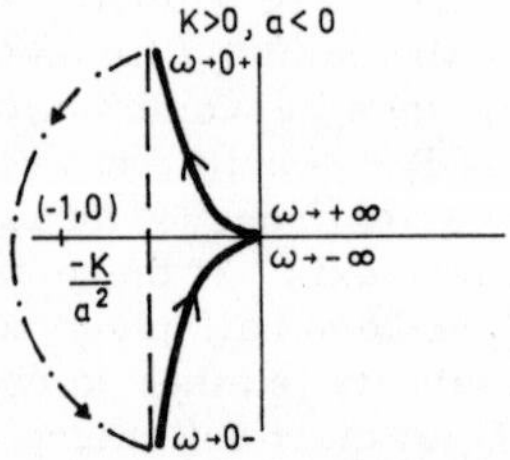

Closed-loop system unstable, P = 1, N = 2 within indented contour.

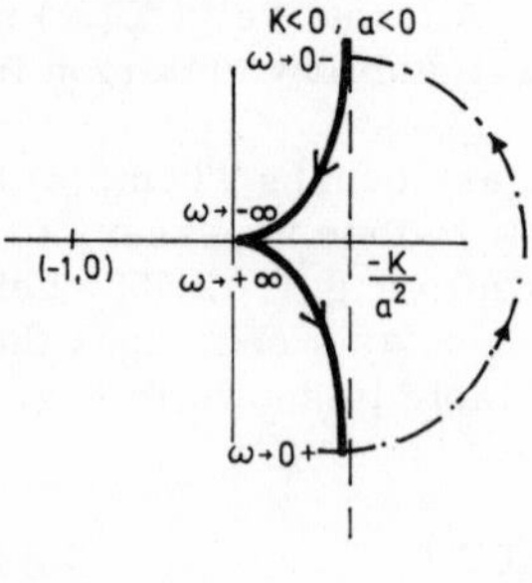

Closed-loop system unstable, P = 1, N = 1 within indented contour.

FIG. 1.9. Nyquist criterion with an open-loop pole at s = 0.

Difference equations describe the behaviour between input and output sequences, for example

$$y_{n+1} + \tfrac{1}{2} y_n = x_n; \quad n = 0, 1, 2, 3, \ldots$$

where the initial condition is $y_0 = 1$ and x_n is the sequence (1, 1, 1, 1, ...). By a step-by-step process we obtain the solution as

$$y_1 = \tfrac{1}{2}, \; y_2 = \tfrac{3}{4}, \; y_3 = \tfrac{5}{8}, \; y_4 = \tfrac{11}{16}, \; \ldots$$

The z-transform of a sequence $\{a_r\} = \{a_0, a_1, a_2, \ldots\}$ is by definition the series

$$\sum_{r=0}^{\infty} a_r z^{-r} = \overline{A}(z)$$

The z-transform of a shifted sequence $\{a_{r+1}\} = \{a_1, a_2, a_3, \ldots\}$ is

$$\sum_{r=0}^{\infty} a_{r+1} z^{-r} = (\overline{A}(z) - a_0) \times z$$

The difference equation may be solved by "taking sequences" and then "taking z-transforms" to obtain

$$(\overline{Y}(z) - y_0)\, z + \tfrac{1}{2}\, \overline{Y}(z) = 1 + z^{-1} + z^{-2} + \ldots = \frac{1}{1-z^{-1}} = \frac{z}{z-1}$$

Hence

$$\overline{Y}(z) = \frac{y_0 z}{z+\frac{1}{2}} + \frac{z}{(z-1)(z+\frac{1}{2})}$$

$$= \frac{y_0 z}{z+\frac{1}{2}} + \frac{2}{3}\left[\frac{z}{z-1} - \frac{z}{z+\frac{1}{2}}\right]$$

These two terms correspond to the "complementary function" and "particular integral" terms that we encounter in both differential and difference equations.

Now the z-transform of the geometric sequence

$$\{1, a^2, a^3, \ldots\} \quad \text{is} \quad 1 + az^{-1} + a^2 z^{-2} + \ldots = \frac{1}{1 - az^{-1}} = \frac{z}{z-a}$$

so we may write the inverse transform of $\overline{Y}(z)$ to give the solution

$$\{y_n\} = (y_0 - \tfrac{2}{3})\,\{1, -\tfrac{1}{2}, \tfrac{1}{4}, -\tfrac{1}{8}, \ldots\} + \tfrac{2}{3}\,\{1, 1, 1, 1, \ldots\}$$

or when

$$y_0 = 1 \;\; \{y_n\} = \{1, \tfrac{1}{2}, \tfrac{3}{4}, \tfrac{5}{8}, \ldots\}$$

(Note that $\lim_{n\to\infty} y_n = \frac{2}{3}$).

2.2. Sampled data systems

Consider the system shown in Fig. 2.1. A continuous input $x(t)$ is sampled by the switch S_1 and the samples modelled as impulses are fed to the first-order system with the transfer function $K/(s+a)$. The switch S_1 closes for a short time Δ at equally spaced times h apart. An impulse of magnitude $x(nh)\Delta$ is thus fed to the first-order system. The sampling may be organized so that $\Delta = 1$. The response of the first-order system to this impulse is $x(nh)\Delta\, k(t-nh)$ for $t > nh$, where $k(t)$ for $t > 0$ is the unit impulse

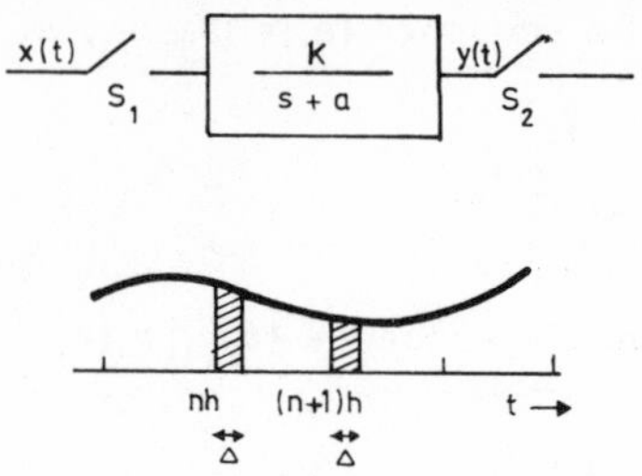

FIG. 2. 1. System with sampled input.

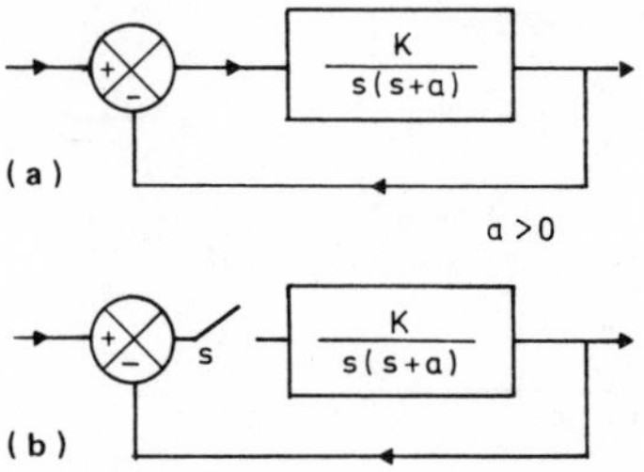

FIG. 2. 2. Second-order continuous linear closed-loop system (a) stable for all $K > 0$, (b) stable for $0 < K < 2a \coth(ah/2)$.

response of the first-order system. If we consider the output sampled by a second synchronized switch S_2 then we find that

$$\begin{aligned} y(nh) = x(nh)\Delta\ k(0) &+ x((n-1)h)\Delta k(h) \\ &+ x((n-2)h)\Delta k(2h) + \ldots \\ &+ x(0)\Delta k(nh) \end{aligned}$$

Considering now the z-transform of $\{y(nh)\}$ we find that

z-transform of sampled output = (z-transform of sampled input)

× (z-transform of the sampled unit impulse response) × Δ

Another arrangement is the "sample-and-hold" device which has $\Delta = 1$ followed by the transfer function $(1 - e^{-sh})/s$ which multiplies any subsequent transfer function such as $K/(s+a)$ above.

We note that sampling can change the character of the system, for example the second-order continuous linear closed-loop system shown in Fig. 2. 2a is stable for all $K > 0$, whereas 2. 2b, with $\Delta = 1$ and sampling internal h, is stable for $0 < K < 2a \coth(ah/2)$ only.

This requires some analysis as follows:
For Fig. 2. 2b:

(z-transform of output) = z transform of unit impulse response of $K/s(s+a)$

× (z-transform of input − z-transform of output)

and we are interested in the "closed-loop z-transfer function" $H(z)/[1+H(z)]$ whose $H(z)$ is the z-transform of the unit impulse response. Now the (continuous) unit impulse response of $K/s(s+a)$ is the inverse Laplace transform of this function of s which is $(K/a)(1-e^{-at})$. The function $H(z)$ is given by

$$\frac{K}{a}\left(\frac{z}{z-1} - \frac{z}{z-e^{-ah}}\right)$$

Stability of the closed loop depends on the zeros in z of $1+H(z) = 0$ which must lie inside the unit circle.

We have

$$z^2 - z(1+e^{-ah}) + e^{-ah} + \frac{K}{a}(1-e^{-ah})z = 0$$

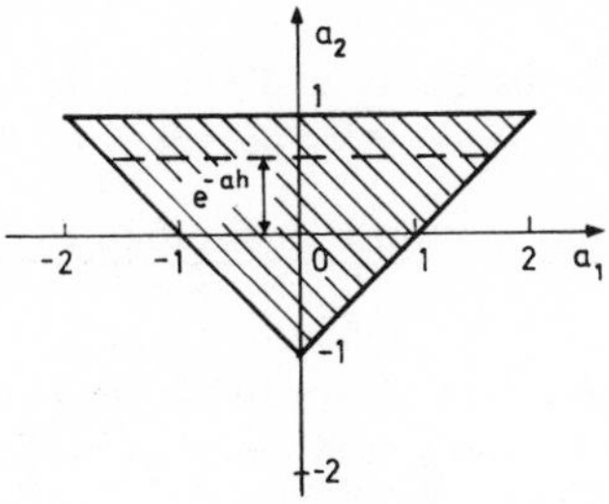

FIG. 2.3. Unit triangle.

Now the roots of $z^2 + a_1z + a_2 = 0$ lie inside the unit circle if (a_1, a_2) lies inside the triangle shown in Fig. 2.3. We have $a_2 = e^{-ah}$ and

$$a_1 = \frac{K}{a}(1-e^{-ah}) - (1+e^{-ah})$$

and so

$$-1-e^{-ah} < \frac{K}{a}(1-e^{-ah}) - (1+e^{-ah}) < 1+e^{-ah}$$

or

$$0 < \frac{K}{a} < \frac{2(1+e^{-ah})}{(1-e^{-ah})} = 2\coth(ah/2)$$

The general stability criterion for roots of polynomial equations in z to lie within the unit circle on the complex plane is associated with the names of Schur, Cohn and Jury. One version is given below (Jury):

Let the polynomial be

$$F(z) \equiv a_0 + a_1z + a_2z^2 + \ldots + a_nz^n, \quad a_n > 0$$

We construct the following table of coefficients and 2×2 determinants formed by cross-multiplication:

$$
\begin{array}{lllllll}
 & a_n & a_{n-1} & a_{n-2} & \cdots\cdots & a_2 & a_1 & a_0 \\
 & a_0 & a_1 & a_2 & \cdots\cdots & a_{n-2} & a_{n-1} & a_n \\
\Delta_1 = & b_0 & b_1 & b_2 & \cdots\cdots & b_{n-2} & b_{n-1} \\
 & b_{n-1} & b_{n-2} & b_{n-3} & \cdots\cdots & b_1 & b_0 \\
\Delta_2 = & c_0 & c_1 & c_2 & \cdots\cdots & c_{n-2} \\
 & c_{n-2} & c_{n-3} & c_{n-4} & \cdots\cdots & c_0 \\
\cdots & \cdots & \cdots & \cdots & \cdots & \cdots & \cdots \\
\Delta_{n-1} = & t_0 & t_1
\end{array}
$$

where

$$b_k = \begin{vmatrix} a_n & a_k \\ a_0 & a_{n-k} \end{vmatrix}, \quad c_k = \begin{vmatrix} b_0 & b_{n-k-1} \\ b_{n-1} & b_k \end{vmatrix}, \quad d_k = \begin{vmatrix} c_0 & c_{n-k-2} \\ c_{n-2} & c_k \end{vmatrix}, \quad \text{etc.}$$

then the conditions for all the roots of $F(z) = 0$ to lie within the unit circle are $F(1) > 0$, $(-1)^n F(-1) > 0$, $\Delta_1 > 0$, $\Delta_2 > 0$, ... $\Delta_{n-1} > 0$.

2.3. Example of dynamical systems described by difference equations — the "hog cycle"

Difference equation models occur in various fields of applications. A famous model appearing in many textbooks of mathematical economics is the "hog cycle" ("hog" (USA) = "pig" (UK)).

We suppose there are two curves describing "demand" D and "production" P versus price p as in Fig.2.4. We suppose that farmers base their production on the previous year's price and that the price adjusts itself so that supply equals demand. We have:

$$g(p_n) = D(n) = P(n) = f(p_{n-1})$$

If we approximate the curves of f and g by straight lines so that

$$P = ap + b$$

$$D = -cp + d; \quad a, c > 0$$

then

$$-cp_n + d = ap_{n-1} + b$$

or

$$p_n = -\frac{a}{c} p_{n-1} + \frac{d-b}{c}$$

the hog price cycle will be oscillatory as $-a/c < 0$, and stable or unstable depending on whether $a/c < 1$ or > 1 respectively.

If the cycle is stable it tends to the equilibrium point where the production and demand curves intersect. The diagram in Fig.2.4 which traces the movement of prices is known as a "cob-web diagram".

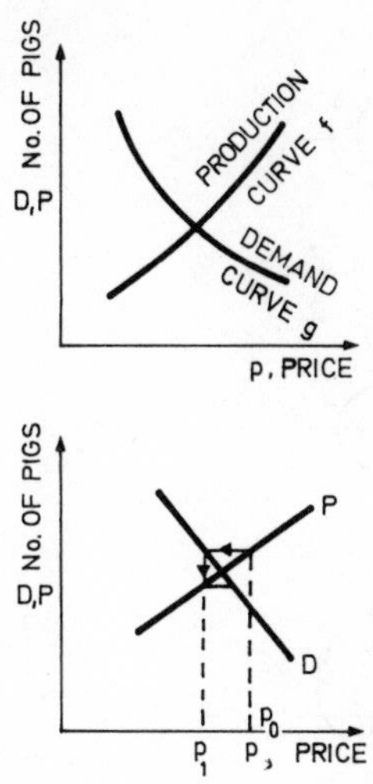

FIG. 2.4. The hog cycle.

2.4. Stability

In the material considered so far stability has involved the decay of the complementary functions as time tends to infinity. This has involved roots in the left-hand half of the complex plane or roots inside the unit circle for the continuous time and discrete time systems, respectively.

More precise definitions of stability can be given when the systems are expressed in phase space or in state space form.

Consider the unforced second-order differential equation

$$\frac{d^2y}{dt^2} + a_1 \frac{dy}{dt} + a_2 y = 0 \tag{a}$$

By introducing the phase-plane co-ordinates $x_1 = y$ and $x_2 = dy/dt$ we may replace the single second-order equation by two first-order equations:

$$\left.\begin{aligned} \frac{dx_1}{dt} &= x_2 \\ \frac{dx_2}{dt} &= -a_2 x_1 - a_1 x_2 \end{aligned}\right\}$$

or in matrix form

$$\begin{bmatrix} \dot{x}_1 \\ \dot{x}_2 \end{bmatrix} = \begin{bmatrix} 0 & 1 \\ -a_2 & -a_1 \end{bmatrix} \begin{bmatrix} x_1 \\ x_2 \end{bmatrix} \tag{b}$$

Instead of plotting the solutions of (a), i.e. $y(t)$ or dy/dt with respect to time, we consider plotting the phase-plane diagram of (b) in which we plot x_2 (= dy/dt) against x_1 (= y) where, for varying t, the point (x_1, x_2) describes a curve or trajectory on the phase-plane. Figure 2.5 shows some of the possibilities for various values of a_1 and a_2, where the arrows indicate

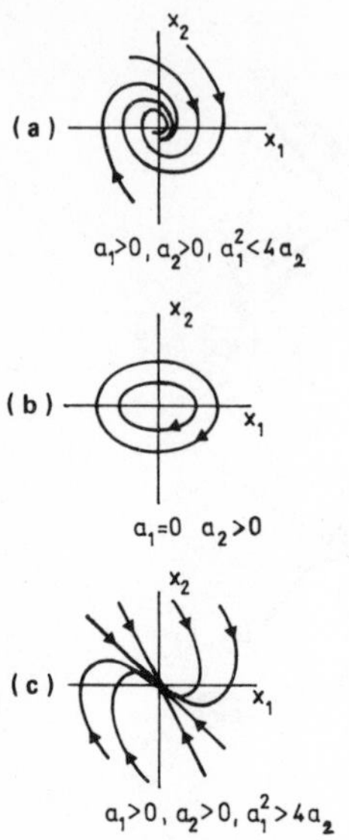

FIG. 2.5. Phase-plane diagrams: (a) $a_{1,2} > 0,\ a_1^2 < 4a_2$; (b) $a_1 = 0,\ a_2 > 0$; (c) $a_{1,2} > 0,\ a_1^2 > 4a_2$.

increasing time. The origin of the phase-plane is itself a solution and trajectory and is a singular point or equilibrium point. The stability of an equilibrium point may be defined by considering the behaviour of trajectories in its neighbourhood. If from all starting points in a small spherical neighbourhood of the equilibrium point the trajectories subsequently stay close to the equilibrium point, then the equilibrium point is stable. If, in addition, the trajectories tend to the origin as time tends to infinity then we say the equilibrium point is asymptotically stable. Thus the origin is asymptotically stable in Figs 2.5a and 2.5c but stable only in Fig. 2.5b.

A more formal mathematical definition of the local stability properties of an equilibrium point represented by the vector $\underline{x}_0$ is as follows:

If given any $\epsilon > 0$ there exists a $\delta = \delta(\epsilon) > 0$ such that for all $||\underline{x}(t_0) - \underline{x}_0|| < \delta$, $||\underline{x}(t) - \underline{x}_0|| < \epsilon$ for all $t \geq t_0$, then the equilibrium point $\underline{x}_0$ is stable.

If, in addition, $\lim_{t\to\infty} ||\underline{x}(t) - \underline{x}_0|| = 0$ then $\underline{x}_0$ is asymptotically stable.

In this definition $||\cdot||$ represents the Euclidean norm.

A number of more sophisticated definitions such as "uniform asymptotic stability" may be developed from examination of the limiting process $||\underline{x}(t) - \underline{x}_0|| \to 0$ as $t \to \infty$.

So far we have considered as an example a linear system only (Eqs (a), (b) and Fig. 2.5). A non-linear system may have a number of equilibrium points as shown in Fig. 2.6 for soft-spring oscillations (Fig. 2.6a):

$$m\ddot{x} + c\dot{x} + k_1x - k_3x^3 = 0$$

$$\left.\begin{aligned} \dot{x}_1 &= x_2 \\ \dot{x}_2 &= -\frac{k_1}{m}x_1 + \frac{k_3}{m}x_1^3 - \frac{c}{m}x_2 \end{aligned}\right\}$$

$$m,\ c,\ k_1,\ k_3 > 0$$

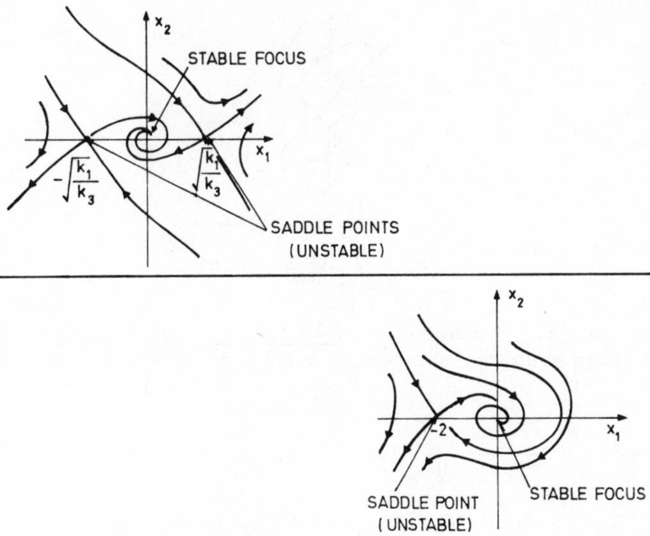

FIG. 2.6. Equilibrium points of simple non-linear systems; (a) soft spring, (b) asymmetric spring.

and an asymmetric spring (La Salle) (Fig. 2.6b):

$$\ddot{x} + \tfrac{1}{4}\dot{x} + 2x + x^2 = 0$$

$$\left.\begin{aligned}\dot{x}_1 &= x_2\\ \dot{x}_2 &= -2x_1 - x_1^2 - \tfrac{1}{4}x_2\end{aligned}\right\}$$

3. STABILITY. I.

3.1. Liapunov functions

An important idea for investigating the stability of equilibrium points of linear and non-linear systems is the concept of the Liapunov function (Liapunov, 1892). This is illustrated in Fig. 3.1 where the origin of the system shown is surrounded by a nest of closed curves given by

$$V(x_1, x_2) = b_{11}x_1^2 + 2b_{12}x_1x_2 + b_{22}x_2^2 = \text{constant} > 0$$

where

$$\left.\begin{aligned}\dot{x}_1 &= x_2\\ \dot{x}_2 &= -a_2x_1 - a_1x_2\end{aligned}\right\}$$

are the system equations discussed in Section 2.4, with $a_1 > 0$, $a_2 > 0$.

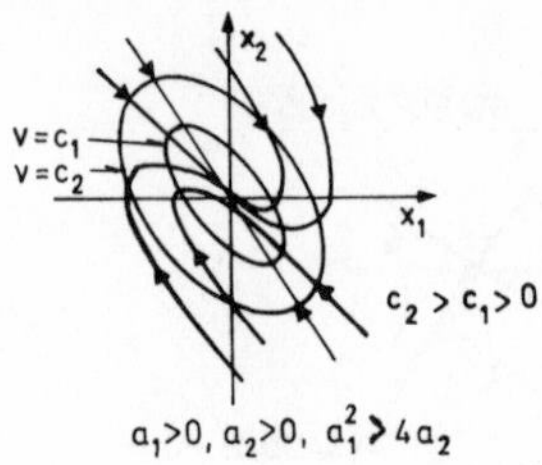

FIG. 3.1. Liapunov function.

We examine the rate of change of V following the trajectories of the system equations. This quantity, sometimes called the Eulerian derivative of V, is given by

$$\frac{dV}{dt} = \frac{\partial V}{\partial x_1}\frac{dx_1}{dt} + \frac{\partial V}{\partial x_2}\frac{dx_2}{dt}$$

$$= (2b_{11}x_1 + 2b_{12}x_2)\, x_2 + (2b_{12}x_1 + 2b_{22}x_2)\,(-a_2x_1 - a_1x_2)$$

on differentiation by parts and substitution for $\dot{x}_1$ and $\dot{x}_2$ from the system equations. We now try to choose b_{11}, b_{12} and b_{22} so that (i) the contours of V form closed curves surrounding the origin, (ii) $dV/dt \leqq 0$, but $\neq 0$ except at $\underline{x} = \underline{0}$. If this is possible we can deduce that the origin is an asymptotically stable equilibrium point, for V will decrease continually along any trajectory, which must be moving inwards through the contours of V and thus tending to the origin as $t \to \infty$.

In the case of a linear system such as that given in Fig. 3.1 we can make $dV/dt \leqq 0$ by equating it identically to a negative-definite quadratic form (e.g. $-2x_1^2 - 2x_2^2$) which will in general give rise to 3 equations for the 3 unknowns b_{11}, b_{12}, b_{22}. We may then examine V for positive definiteness for which necessary and sufficient conditions are $b_{11} > 0$, $b_{11}\, b_{22} > b_{12}^2$. These conditions will also be necessary and sufficient for asymptotic stability of the origin.

We note that $V \to 0$ as $t \to \infty$ and since V majorizes the norm $\|\underline{x}\| = \sqrt{x_1^2 + x_2^2}$, this proves that $\|\underline{x}\| \to 0$ which is what we mean by asymptotic stability of the point $\underline{x} = \underline{0}$. This idea may be generalized into general situations, for example to discuss the stability of systems governed by partial differential equations.

With non-linear systems we may still be able to find Liapunov functions and deduce stability properties, even though we cannot find any explicit solutions of the non-linear differential equations themselves. This makes the technique particularly attractive. Certain systems, e.g. model-reference adaptive control systems, may be synthesized as stable systems by choosing V positive definite and making dV/dt negative by suitable feedback arrangements.

3.2. Construction of Liapunov functions

Several methods for constructing Liapunov functions have been devised, some of which are briefly listed below:

The Liapunov matrix equation for linear systems

Given the linear constant systems $\dot{\underline{x}} = \underline{A}\underline{x}$ and the quadratic form $V = \underline{x}^T \underline{P}\, \underline{x}$ then $dV/dt = \underline{x}^T (\underline{PA} + \underline{A}^T\underline{P})\underline{x}$ and, by equating dV/dt to $-\underline{x}^T\underline{Q}\underline{x}$, we obtain the Liapunov matrix equation

$$\underline{P}\,\underline{A} + \underline{A}^T\underline{P} = -\underline{Q}$$

The variable-gradient method

Given the system of equations $\dot{\underline{x}} = \underline{f}(\underline{x})$, we consider the gradient vector

$$\text{grad } V \equiv \left(\frac{\partial V}{\partial x_1}, \frac{\partial V}{\partial x_2}, \ldots \frac{\partial V}{\partial x_n} \right)^T$$

and try to construct its elements so that

(i) $(\text{grad } V)^T \cdot \underline{f} \equiv \frac{dV}{dt} \leq 0$

(ii) grad V is in fact the gradient of a scalar.

If the vector to be found is $\underline{g} = (g_1, \ldots, g_n)^T$ then

(i) $\underline{g}^T \cdot \underline{f} \leq 0$

(ii) $\text{curl}\, \underline{g} = 0$

where the $n \times n$ matrix curl $\underline{g}$ is defined as a matrix with its (i, j)th element g_{ij} given by

$$g_{ij} = \frac{\partial g_j}{\partial x_i} - \frac{\partial g_i}{\partial x_j}$$

The curl matrix is antisymmetric by definition, and so there are $\frac{n(n-1)}{2}$ conditions on the g_i to be satisfied, namely

$$\frac{\partial g_j}{\partial x_i} = \frac{\partial g_i}{\partial x_j} \quad i, j = 1, 2, \ldots, n, \quad i \neq j$$

A certain arbitrariness is present in this technique.

The Zubov method

This is of particular interest when the f_i (elements of $\underline{f}$) are algebraic forms with linear, quadratic, cubic, ..., parts. The Liapunov function V is taken as a sum of V_2, quadratic, V_3, cubic, V_4 quartic, ... forms. First the linear part of $\underline{f}$ is used to construct V_2 with a negative quadratic form as derivative, using the technique for linear systems. Then V_3 is found using V_2, the linear and quadratic parts of $\underline{f}$ to keep dV/dt unchanged. V_4, ... are found similarly in succession in a logical but increasingly complex process. There may be problems of convergence and sign-definiteness of the resulting function V.

Integration by parts

This is an important idea which has been exploited by Brockett to provide an important connection between Liapunov functions and frequency response methods. The following is a basic theorem:

Theorem (Brockett, 1964)

Suppose

$$p(D) \equiv D^n + a_1 D^{n-1} + a_2 D^{n-2} + \ldots + a_n$$

where $D \equiv d/dt$ and the a_i are real.
Let $q(D)$ be another polynomial in D of degree less than or equal to that of $p(D)$ and such that $q(z)/p(z)$ is a positive real function. If $q(D)$, $p(D)$ and $[\,\text{Ev. } q(D)p(-D)]^-$ do not have a common factor then

$$V(\underline{x}) = \int q(D)y \; p(D)y - \{[\,\text{Ev. } q(D)p(-D)]^- y\}^2 \, dt$$

is a positive definite Liapunov function for $p(D)y = 0$ with a non-positive derivative, where $\underline{x}$ is the phase-space vector

$$\underline{x} = [y, Dy, \ldots., D^{n-1}y]^T$$

The theorem requires some explanatory notes:

(i) A positive real function $\phi(z)$ is such that $\phi(z)$ is real if z is real and $\text{Re}\,\phi(z) \geqq 0$ for $\text{Re}\, z \geqq 0$. In particular $\phi(i\omega) \geqq 0$ for real ω which enables the important connection with frequency response to be made.

(ii) The notation (Ev. $(\cdot)$ means the "even part of", e.g. $\text{Ev.}\,(z^3 + z^2 + z - 1) = z^2 - 1$.

(iii) The notation $[\,\cdot\,]^-$ means that part of an even function which has zeros in the left-hand half of the z-plane, e.g. $[z^2 - 1]^- = [(z+1)(z-1)]^- = z + 1$.

(iv) $V(\underline{x})$ is in fact a quadratic form in the phase-space variables $x_1 = y$, $x_2 = Dy$, ..., $x_n = D^{n-1}y$.

An example of this technique is given below:

$$p(D) = (D+2)(D+3) = D^2 + 5D + 6, \quad q(D) = (D+1)$$

We note that $\dfrac{q(z)}{p(z)}$ is positive real (see Fig. 3.2).

$$\begin{aligned} \text{Ev. } q(D)\,p(-D) &= \text{Ev. } (D+1)(D^2 - 5D + 6) \\ &= \text{Ev. } (D^3 - 4D^2 + D + 6) \\ &= -4D^2 + 6. \end{aligned}$$

$$[\,\text{Ev. } q(D)p(-D)]^- = [-4D^2 + 6]^- = [(2D + \sqrt{6})(-2D + \sqrt{6})]^- = (2D + \sqrt{6})$$

Figure 3.2 shows a plot of $\dfrac{q(i\omega)}{p(i\omega)} = \dfrac{6 + 4\omega^2 + i(\omega - \omega^3)}{(6 - \omega^2)^2 + 25\omega^2}$

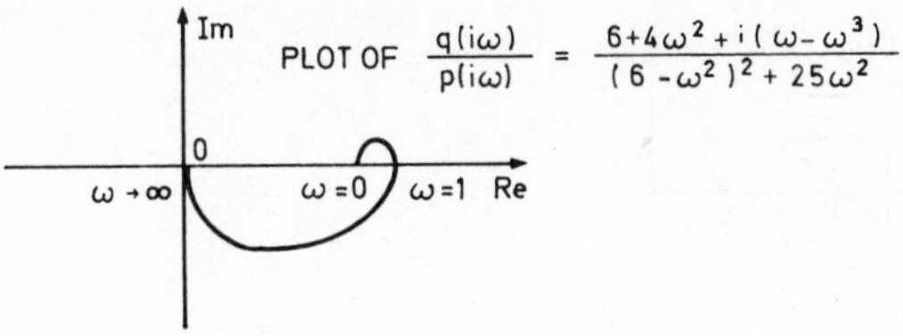

FIG. 3.2. Plot of q(iω)/p(iω).

$$V(\underline{x}) = \int (D+1)\,y\,(D^2+5D+6)\,y - \{(2D+\sqrt{6})y\}^2\,dt$$

$$= \int (\dot{y}+y)\,(\ddot{y}+5\dot{y}+6y) - (2\dot{y}+\sqrt{6}y)^2\,dt$$

$$= \int (\dot{y}\ddot{y}+\ddot{y}y+5\dot{y}^2+5y\dot{y}+6y\dot{y}-4\dot{y}^2-4\sqrt{6}\,y\dot{y}\,dt$$

$$= \tfrac{1}{2}\,\dot{y}^2+\dot{y}y+(11-4\sqrt{6})\,\frac{y^2}{2}$$

$$= [\,y\ \dot{y}\,]\begin{pmatrix} \frac{11-4\sqrt{6}}{2} & \frac{1}{2} \\ \frac{1}{2} & \frac{1}{2} \end{pmatrix}\begin{pmatrix} y \\ \dot{y} \end{pmatrix} \quad \text{which is positive-definite.}$$

$$\frac{dV}{dt} = -(2\dot{y}+\sqrt{6}y)^2 \quad \text{which is negative-definite.}$$

4. STABILITY. II

4.1. Positive-real functions and the Popov criterion

The positive real-function idea may be extended to the non-linear feedback arrangement shown in Fig. 4.1 a to give the following result:
Theorem (Brockett): "The zero solution of Fig. 4.1 a is asymptotically stable for all admissible f(y) if there exists an $\alpha > 0$ such that $(1+\alpha s)\,G(s)$ is a positive real function."

By an admissible non-linearity we mean (i) f(y) defined, continuous and single-valued for all y, (ii) $f(0) = 0$, $yf(y) > 0$, $y \neq 0$,

(iii) $\int_0^{\pm\infty} f(y)dy$ divergent.

The proof of this result uses the Liapunov function formed by the integration-by-parts technique explained in Section 3. We consider the variable x in Fig. 4.1 b which satisfies the equation

$$p(D)x = -f\,\{q(D)x\}$$

We multiply by $(1+\alpha D)\,q(D)x$ and form the new equation

$$(1+\alpha D)\,q(D)x\;p(D)x + \alpha Dq(D)x\;f\{q(D)x\} = -q(D)x\;f\{q(D)x\}$$

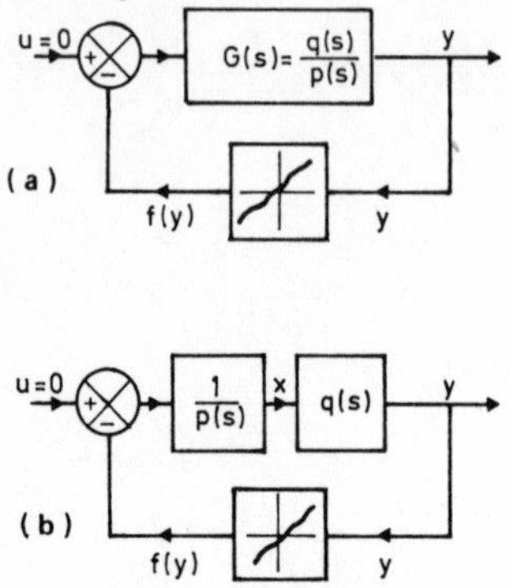

FIG. 4. 1. Non-linear feedback system.

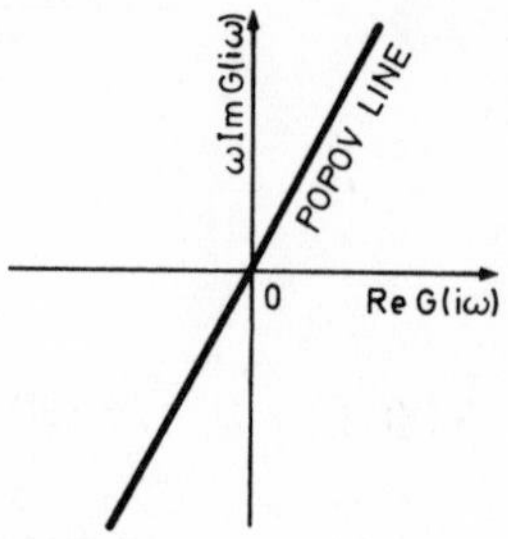

FIG. 4. 2. Modified polar plot.

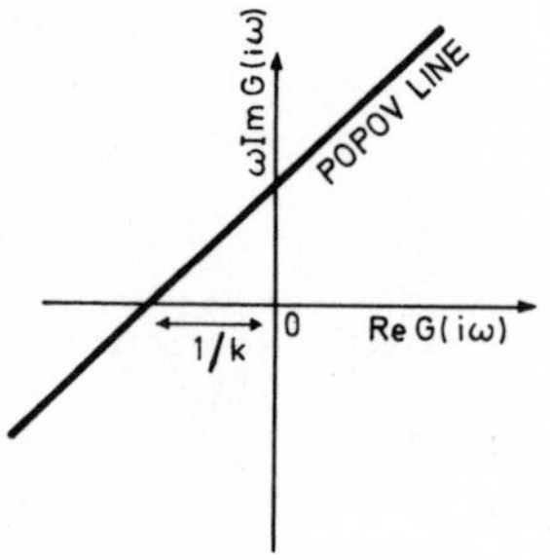

FIG. 4. 3. Shifted Popov line in modified polar plot.

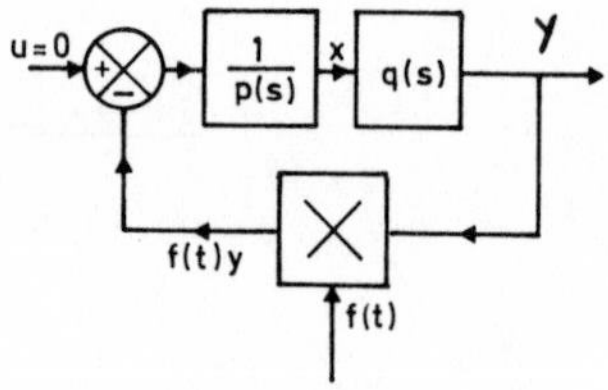

FIG. 4. 4 Time-varying feedback.

We take V($\underline{x}$) as the time integral of the left-hand side having first subtracted $\{[\text{Ev.}\ (1+\alpha D)q(D)p(-D)]^{-}x\}^2$ from both sides.

The left-hand side integrates into a positive-definite quadratic form in the phase-space variables by the result for the linear system p(D)y = 0, given in Section 3, plus

$$\alpha \int_0^{q(D)x} f(\sigma)d\sigma$$

which is also non-negative for $\alpha > 0$. The derivative

$$\frac{dV}{dt} = -q(D)x\, f\{q(D)x\} \ -\{[\text{Ev.}\ \ldots]^{-} x\}^2 \leq 0$$

The condition of the theorem that $(1+\alpha s)\,G(s)$ is positive-real has an important graphical interpretation if we consider s = iω, for then the plot of ω Im G(iω) against Re G(iω) must lie to the right of the straight line through the origin with slope $1/\alpha$ (see Fig.4.2). The diagram is called a "modified polar plot" and the straight line the "Popov line". Of course the multiplication of Im G(iω) by ω considerably changes the diagram from the more familiar frequency response diagram where Im G(iω) is plotted against Re G(iω).

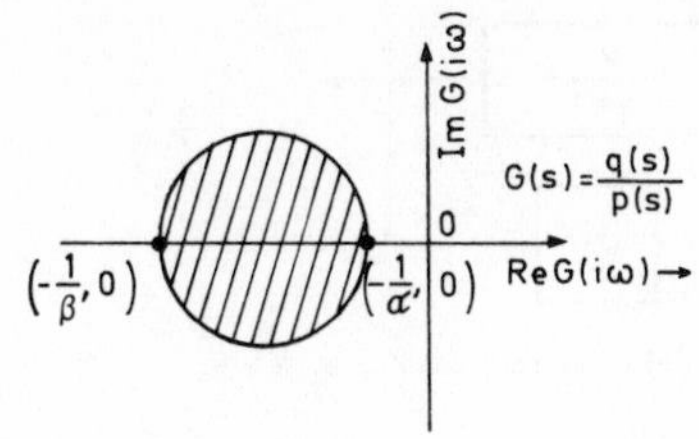

FIG. 4.5. Circle theorem disc.

A useful extension to this result in which further restrictions on the non-linearity are traded in for relaxations on the transfer function is embodied in the following theorem:

"The system of Fig. 4.1 a has an asymptotically stable null solution for all admissable f(y) such that $0 \leq y\, f(y) \leq ky^2$ if there exists an $\alpha > 0$ such that $(1 + \alpha s)\, G(s) + 1/k$ is positive real."

This yields a new Popov line as shown in Fig. 4.3, the modified polar plot of G(iω) having to lie to the right of this line.

4.2. The circle criterion

This technique can be extended to the time varying system shown in Fig. 4.4 when the "circle theorem" is obtained: that is the loop of Fig. 4.4 is asymptotically stable provided the Nyquist locus $G(i\omega) = q(i\omega)/p(i\omega)$ does not encircle or enter the circular disc with diameter $(-1/\beta, 0)$, $(-1/\alpha, 0)$ as shown in Fig. 4.5, where

$$0 \leq \beta < f(t) < \alpha < \infty$$

(Notice that here we are using the usual Nyquist diagram, not the modified polar plot of the previous discussion.)

The proof is indirect, using first the following theorem:

"If $\dfrac{kq(s) + p(s)}{p(s)}$ is positive real, the loop of Fig. 4.4 is asymptotically stable for $0 < f(t) < k$."

Adding and subtracting $\beta q(D)x$ to the basic equation

$$p(D)x + f(t)\, q(D)x = 0$$

to obtain

$$p(D)x + \beta q(D)x + (f(t) - \beta)\, q(D)x = 0$$

we can deduce that the loop is asymptotically stable for $0 < f(t) - \beta < \alpha - \beta$ if $\alpha q(s) + p(s)/\beta q(s) + p(s)$ is positive-real. The bilinear mapping $W = (\alpha z + 1)/(\beta z + 1)$, which maps the disc of Fig. 4.5 into the left-hand half plane and $q(s)/p(s)$ into $[\alpha q(s) + p(s)]/[\beta q(s) + p(s)]$, completes the proof.

4.3. Linear time-delay systems

A number of control processes involve pure time delays in the forward loop or in the feedback loop. An example is given in Fig. 4.6. (The transfer-function of a time-delay T is e^{-sT}.)

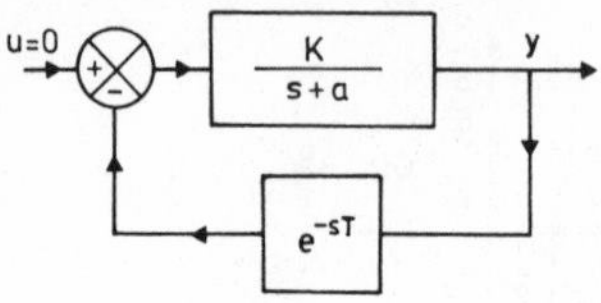

FIG.4.6. Time delay system for K > 0, a > 0.

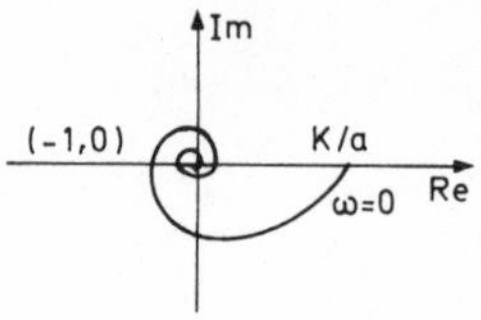

FIG.4.7. Open-loop frequency response.

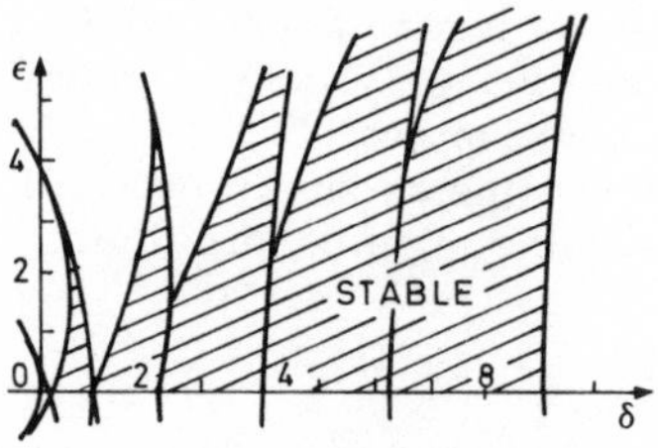

FIG.4.8. Stability boundaries for Mathieu equation.

The stability of the loop is conveniently handled by the Nyquist criterion, bearing in mind the frequency response function of the time-delay T is $e^{-i\omega T}$, a unit circle in the complex plane described in a clockwise direction as ω increases from 0 to $+\infty$. The open-loop frequency response is sketched in Fig.4.7. The closed loop will be unstable for sufficiently large K/a, since the (-1,0) point will be encircled.

Conditions on the characteristic equation (which, for the example of Fig.4.6, is $s + a + Ke^{-sT}$) for stability and instability have been given by Pontryagin (1942).

4.4. Equations with periodic coefficients

These occur naturally in a number of problems, for example the whirling of unsymmetrical shafts in unsymmetrical bearings or in the analysis of helicopter blade dynamics, but more especially when examining the stability of non-linear oscillations since the linearized equation for small disturbances occurring on top of the "steady-state" oscillation will be a linear differential equation with periodic coefficients. The calculation of stability boundaries of these equations is usually difficult and the boundaries are complicated. The Mathieu equation

$$\ddot{x} + (\delta + \epsilon \cos t)x = 0$$

is a famous example with a stability diagram in terms of δ and ϵ shown in Fig.4.8.

Use of Liapunov functions or the circle theorem can sometimes give useful sufficient (but not necessary) stability criteria.

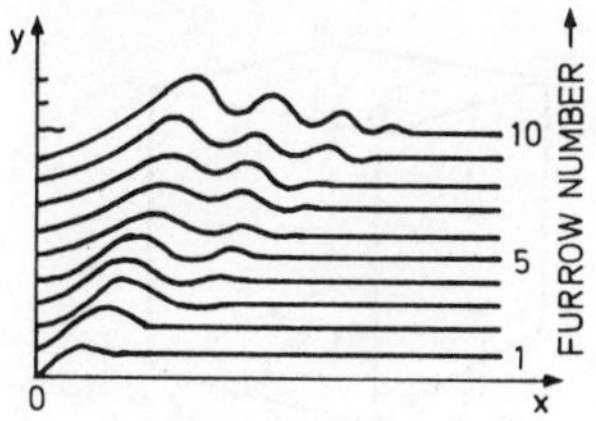

FIG. 4. 9. Growing disturbance in repeated process (ploughing).

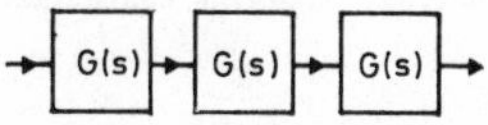

FIG. 4. 10. Mathematical representation of process of Fig. 4. 9.

4.5. Stability of repeated processes

A topic of recent interest is the stability of repeated processes, for example coal-cutting, agricultural ploughing, machining of metals, and control of strings of urban transport vehicles by automatic means. If a slight irregularity occurs on one run of the tractor ploughing a field will the disturbance be magnified subsequently (see Fig. 4. 9), which depicts a series of plough furrows following a step disturbance in the first furrow.

Mathematically the process is that of Fig. 4. 10 where an initial wave form is fed through the transfer function G(s) many times.

The stability criterion is

(i) that G(s) itself is stable (that is with all its poles in the left-hand half plane),

(ii) that $|G(i\omega)| \leq 1$ for all real ω.

The Fourier transform of the output after n stages will be $(G(i\omega))^n \overline{U}(i\omega)$ where $\overline{U}(i\omega)$ is the Fourier transform of the input disturbance, and so the Fourier transform tends to zero as $n \to \infty$ for all ω if $|G(i\omega)| \leq 1$.

5. CONTROL OF SYSTEMS DESCRIBED BY PARTIAL DIFFERENTIAL EQUATIONS

In recent years there has been a growth in the theory of control of "distributed parameter systems", that is systems governed by partial differential equations, as opposed to "lumped parameter systems" governed by ordinary differential equations. Before discussing control of distributed parameter systems it is necessary to discuss first two particularly important examples of the governing equations — the heat conduction equation and the wave equation.

5.1. The heat conduction equation

Consider the three-dimensional Cartesian frame and a small cubic element of a conducting solid as shown in Fig. 5. 1. The heat flow across the face ADEF is $-k\,(\partial T/\partial x) \times \delta y\, \delta z$ into the element where k is the thermal

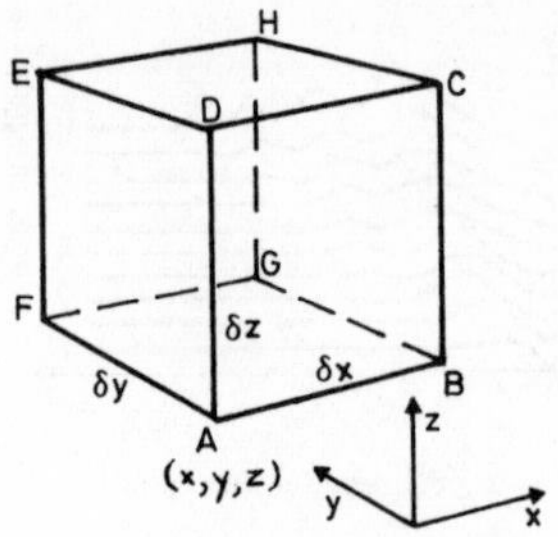

FIG. 5.1. Cube for which heat conduction is considered.

conductivity and $T = T(x, y, z, t)$ is the temperature. The flow out across BCHG is

$$-\left(k\frac{\partial T}{\partial x} + \frac{\partial}{\partial x}\left[k\frac{\partial T}{\partial x}\right]\delta x\right)\delta y\,\delta z$$

Considering the other two pairs of faces the total heat input into the element is

$$\delta x\,\delta y\,\delta z\left(\frac{\partial}{\partial x}\left[k\frac{\partial T}{\partial x}\right] + \frac{\partial}{\partial y}\left[k\frac{\partial T}{\partial y}\right] + \frac{\partial}{\partial z}\left[k\frac{\partial T}{\partial z}\right]\right)$$

or, if k is constant,

$$k\,\delta x\,\delta y\,\delta z\left(\frac{\partial^2 T}{\partial x^2} + \frac{\partial^2 T}{\partial y^2} + \frac{\partial^2 T}{\partial z^2}\right)$$

This heat flow has increased the temperature so that, if ρ is the density and c the specific heat of the material,

$$k\left(\frac{\partial^2 T}{\partial x^2} + \frac{\partial^2 T}{\partial y^2} + \frac{\partial^2 T}{\partial z^2}\right) = \rho c\,\frac{\partial T}{\partial t}$$

on cancelling out $\delta x\,\delta y\,\delta z$ on both sides. This is the highly important heat conduction equation. In the steady state, when $\partial T/\partial t = 0$, we obtain Laplace's equation

$$\frac{\partial^2 T}{\partial x^2} + \frac{\partial^2 T}{\partial y^2} + \frac{\partial^2 T}{\partial z^2} = 0$$

which has also other important applications, for example in potential theory and fluid-flow problems.

It is usual to introduce a new constant $K = k/\rho c$ called the "thermal diffusivity" so that the heat conduction equation can be written neatly as

$$K\,\nabla^2 T = \frac{\partial T}{\partial t}$$

∇^2 being known as the Laplace operator

$$\frac{\partial^2}{\partial x^2} + \frac{\partial^2}{\partial y^2} + \frac{\partial^2}{\partial z^2}$$

Some classical solutions of the heat equation may be found by separation of the variables: for example, a thin rod lying on the x-axis gives rise to the one-dimensional heat equation

$$K \frac{\partial^2 T}{\partial x^2} = \frac{\partial T}{\partial t}$$

If we assume $T(x,t) = X^*(x)\, T^*(t)$, where X^* is a function of x only and T^* a function of t only, then

$$\frac{1}{X^*} \frac{d^2X^*}{dx^2} = \frac{1}{KT^*} \frac{dT^*}{dt} = p^2, \text{ constant}$$

giving solutions in X^* and T^* so that

$$T(x,t) = (A \cos px + B \sin px)\, e^{-Kp^2t} \qquad p \neq 0$$

$$= A + Bx \qquad p = 0$$

The boundary and initial conditions determine admissible values for p and the constants A and B for each p - the original equation being linear we may add different solutions together.

For example: "The rod is of length ℓ and at time $t = 0$ is at a uniform temperature T_0. One end $x = 0$ is reduced to zero temperature, the other end $x = \ell$ is held at T_0. Find the general temperature $T(x,t)$".

We look at solutions $T(x,t)$ above. When $x = 0$ then $T(0,t) = 0$ for all t so $A = 0$. When $x = \ell$, $T(\ell,t) = T_0$ and so we may have as a possible solution

$$T(x,t) = T_0 \frac{x}{\ell} + \sum_{r=1}^{\infty} B_r \sin \frac{r\pi x}{\ell}\, e^{-Kr^2\pi^2 t/\ell^2}$$

where the possible values of p are $r\pi/\ell$ ($r = 1, 2, 3, \ldots$). When $t = 0$ we have

$$T(x,t) = T_0 = \frac{T_0 x}{\ell} + \sum_{r=1}^{\infty} B_r \sin \frac{r\pi x}{\ell}$$

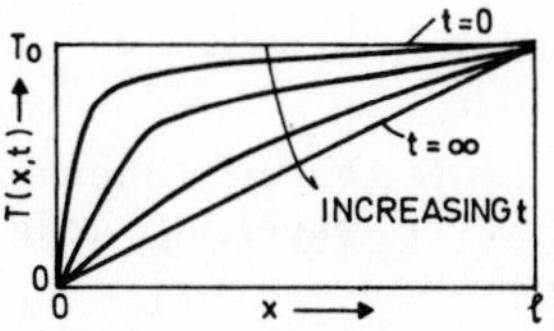

FIG. 5.2. Diagram of solution to heat-conduction equation.

and so a Fourier analysis yields

$$B_r = \frac{2}{\ell} \int_0^{\ell} T_0 \left(1 - \frac{x}{\ell}\right) \sin \frac{r\pi x}{\ell} \, dx = \frac{2T_0}{r\pi}$$

and

$$T(x,t) = \frac{T_0 x}{\ell} + \frac{2T_0}{\pi} \sum_{r=1}^{\infty} \frac{1}{r} \sin \frac{r\pi x}{\ell} \, e^{-Kr^2\pi^2 t/\ell^2}$$

The solution is sketched in Fig.5.2.

An alternative boundary condition to a given temperature on the boundary is a given heat input. This corresponds to specifying $\partial T/\partial x$ rather than T itself in the one-dimensional case. In particular, if one end of the rod is insulated $\partial T/\partial x = 0$ there.

The heat equation may be expressed in other co-ordinates depending on the problem in hand, for example cylindrical polar co-ordinates and spherical polar co-ordinates.

Besides describing heat flow the equation

$$K \nabla^2 \phi = \frac{\partial \phi}{\partial t}$$

represents the "consolidation equation" in soil mechanics, the diffusion of chemicals ("Fick's law"), the "skin-effect equation" in electrical field theory, the behaviour of an electrical cable with capacitance and resistance but with negligible inductance, and incompressible viscous flow.

5.2. The wave equation

Waves occur all around us — in the air as sound waves, on the surface of water, in space as radio and light waves and in elastic bodies as mechanical vibrations.

The simplest mathematical model is the one-dimensional wave equation which may be used to describe planar vibrations of a string under tension (Fig.5.3). If the tension is T and the mass per unit length is m, we obtain by resolving perpendicularly to the x-axis

$$-T \frac{\partial y}{\partial x} + T\left(\frac{\partial y}{\partial x} + \delta x \frac{\partial^2 y}{\partial x^2}\right) = m \, \delta x \frac{\partial^2 y}{\partial t^2}$$

where y(x, t) is the displacement of the string and $\partial y/\partial x$ is assumed to be small. We obtain the <u>one-dimensional wave equation</u>

$$\frac{\partial^2 y}{\partial x^2} = \frac{1}{c^2} \frac{\partial^2 y}{\partial t^2}$$

where in this case $c^2 = T/m$; (c has the dimension of velocity and will be found to be the velocity of waves travelling along the string). The most general solution of the equation is

$$y(x,t) = f(x - ct) + g(x + ct)$$

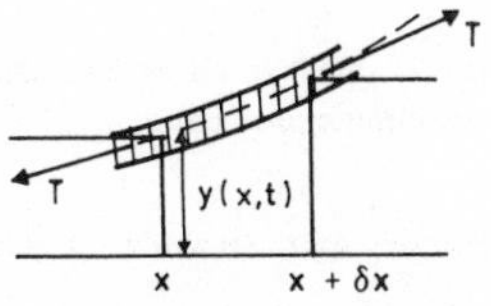

FIG. 5.3. Equation of motion of a string.

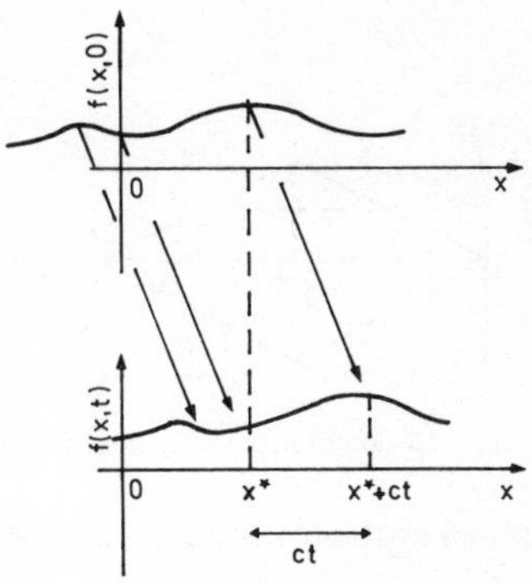

FIG. 5.4. Waves travelling through a string.

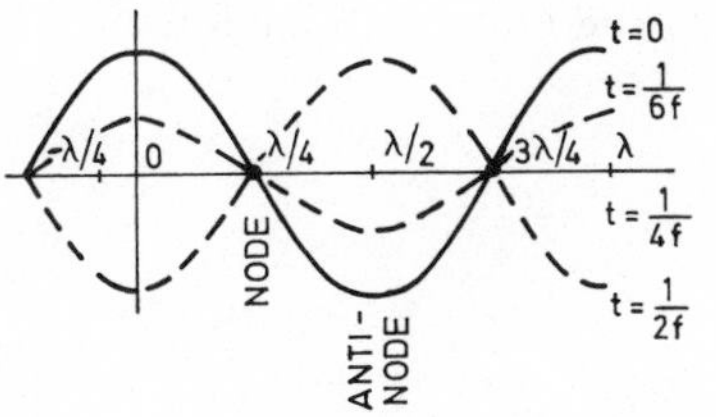

FIG. 5.5. Motion in standing wave.

These are arbitrary continuous functions. As t increases, the wave form f moves to the right at a velocity c (Fig. 5.4) (and similarly the wave form g moves to the left with velocity c).

Of particular importance are progressive harmonic waves

$$y(x,t) = A \cos \frac{2\pi(x-ct)}{\lambda} = A \cos 2\pi(\frac{x}{\lambda} - ft)$$

where $c = f\lambda$. Here A is the amplitude, λ the wavelength, f the frequency in Hz, t is the time (s) and c the wave velocity. The period of the wave is $1/f$ and the wave number $2\pi/\lambda$.

The harmonic progressive waves travelling in opposite directions may be combined to form a standing or stationary wave:

$$y(x,t) = A \cos \frac{2\pi}{\lambda}(x-ct) + A \cos \frac{2\pi}{\lambda}(x+ct)$$

$$= 2A \cos \frac{2\pi x}{\lambda} \cos \frac{2\pi}{\lambda} ct$$

$$= 2A \cos \frac{2\pi x}{\lambda} \cos 2\pi ft$$

Figure 5.5 illustrates the motion. The points $x = \lambda/4 \pm r\lambda/2$ ($r = 0, 1, 2, \ldots$) are all called nodes and the points $x = r\lambda/2$ ($r = 0, 1, 2, \ldots$) antinodes.

Example: A string of length ℓ is fixed at each end. It is plucked at its mid-point which is displaced through a distant h and released. Determine the subsequent motion.

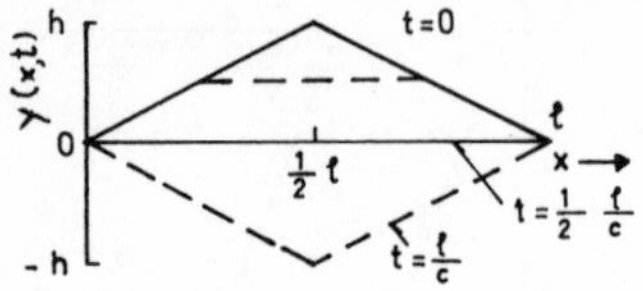

FIG. 5.6. Motion of string with amplitude h.

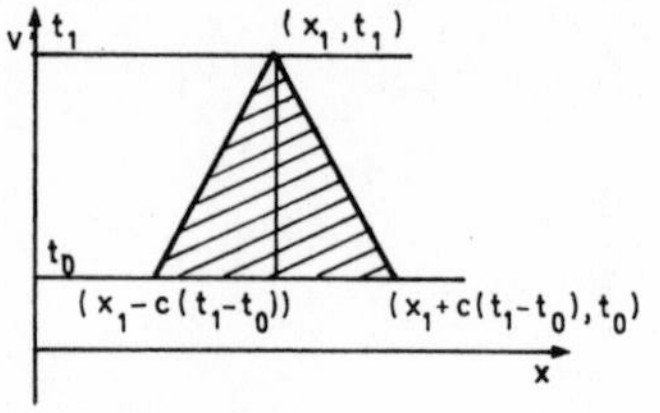

FIG. 5.7. Domains resulting from characteristics in the one-dimensional case.

The solution may be expressed as a sum of harmonic standing waves

$$y(x, t) = \sum_{r=1}^{\infty} B_r \sin \frac{r\pi x}{\ell} \cos \frac{r\pi ct}{\ell}$$

for which $\partial y/\partial t\ (x, 0) = 0$. The B_r must be found by Fourier analysis of the initial wave form to give

$$B_r = \frac{8h}{r^2\pi^2} (-1)^{(r-1)/2}, \quad r \text{ odd}$$

$$= 0 \qquad , \quad r \text{ even}$$

The motion is illustrated in Fig. 5.6.

A useful property of the wave equation is the existence of characteristics. In the case of the one-dimensional wave equation with solution $y(x, t)$ we can set up a plane with co-ordinates x and t. In this plane $c(\partial y/\partial x) + \partial y/\partial t$ is constant along straight lines with equation $ct + x =$ constant, and $c(\partial y/\partial x) - \partial y/\partial t$ is constant along lines $ct - x =$ constant. These lines are called "characteristics" and the constancy of the two quantities $c(\partial y/\partial x) + \partial y/\partial t$ and $c(\partial y/\partial x) - \partial y/\partial t$ on these lines is useful in constructing general solutions, especially when boundary controls are acting.

Consideration of the characteristics gives rise to the concepts of domains of determinancy and dependence illustrated in Fig. 5.7. The motion at the point (x_1, t_1) can be found from a knowledge of $y(x, t_0)$ and $\partial y/\partial t\ (x, t_0)$ for $x_1 - c(t_1 - t_0) < x < x_1 + c(t_1 - t_0)$ on the line $t = t_0$ which is a "domain of dependence". On the other hand, knowledge of $y(x, t_0)$ and $\partial y/\partial t\ (x, t_0)$ on this interval determines $y(x, t)$ for all points x, t inside the shaded triangle which is the "domain of determinancy" of the interval $x_1 - c(t_1 - t_0) < x < x_1 + c(t_1 - t_0)$ on $t = t_0$.

6. CONTROL CONCEPTS FOR PARTIAL DIFFERENTIAL EQUATIONS. I

6.1. Control of the heat equation

Figure 6.1 shows a feedback control arrangement for heating a metal bar. The behaviour of temperature $T(x, t)$ in the bar may be described by the heat equation

$$K \frac{\partial^2 T}{\partial x^2} = \frac{\partial T}{\partial t}$$

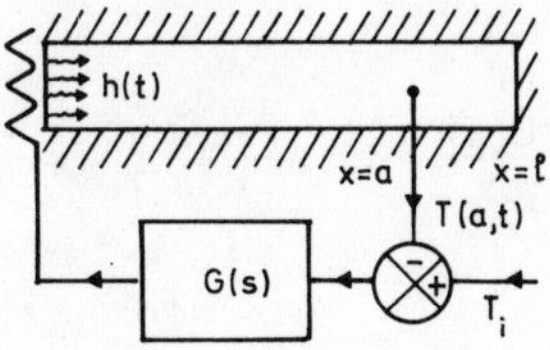

FIG. 6.1. Feedback control for heating a metallic bar.

and the heat input h(t) is given by the boundary condition

$$-k \left.\frac{\partial T}{\partial x}\right|_{x=0} = h$$

If the bar is insulated as shown, there is a second boundary condition at $x = \ell$:

$$\left.\frac{\partial T}{\partial x}\right|_{x=\ell} = 0$$

The temperature is measured by a sensor at $x = a$.

We observe that the control is a "boundary control" and that we have one pointwise sensor. (In general, we might have a distributed heat input, and in theory we might be able to measure the temperature at many points simultaneously.)

Taking the Laplace transform of the heat equation we obtain

$$K \frac{d^2\overline{T}}{dx^2} (x, s) = s\overline{T} (x, s) - T(x, 0) \tag{6.1}$$

where $\overline{T}(x, s)$ is the Laplace transform of $T(x, t)$, given by

$$\overline{T}(x, s) = \int_{t=0}^{\infty} T(x, t)e^{-st}\, dt$$

We have also the transformed boundary conditions

$$\frac{d\overline{T}}{dx} (\ell, s) = 0$$

$$-k \frac{d\overline{T}}{dx} (0, s) = \overline{H}(s) = G(s) (\overline{T}_i (s) - \overline{T}(a, s))$$

where $\overline{H}(s)$ is the Laplace transform of h(t), and $\overline{T}_i(s)$ that of $T_i(t)$.

The second-order ordinary differential equation (6.1) for $\overline{T}(x, s)$ may be solved by the "method of variation of parameters" in which we assume

$$\overline{T}(x, s) = A(x) \cosh\sqrt{s/K}\, x + B(x) \sinh \sqrt{s/K}\, x$$

so that

$$\frac{d\overline{T}(x, s)}{dx} = \sqrt{s/K}\, (A(x) \sinh \sqrt{s/K}\, x + B(x) \cosh \sqrt{s/K}\, x)$$

provided

$$\frac{dA(x)}{dx} \cosh \sqrt{s/K}\, x + \frac{dB(x)}{dx} \sinh \sqrt{s/K}\, x = 0 \tag{6.2}$$

Now

$$\frac{d^2T(x,s)}{dx^2} = \frac{s}{K} (A(x) \cosh \sqrt{s/K}\, x + B(x) \sinh \sqrt{s/K}\, x)$$

$$+ \sqrt{s/K} \left(\frac{dA(x)}{dx} \sinh \sqrt{s/K}\, x + \frac{dB(x)}{dx} \cosh \sqrt{s/K}\, x \right)$$

or, substituting into (6.1),

$$\sqrt{sK} \left(\frac{dA(x)}{dx} \sinh \sqrt{s/K}\, x + \frac{dB(x)}{dx} \cosh \sqrt{s/K}\, x \right) = - T(x,0) \tag{6.3}$$

Equations (6.2) and (6.3) enable us to get solutions for $dA(x)/dx$ and $dB(x)/dx$ giving

$$\frac{dA(x)}{dx} = \frac{T(x,0)}{\sqrt{sK}} \sinh \sqrt{s/K}\, x$$

$$\frac{dB(x)}{dx} = - \frac{T(x,0)}{\sqrt{sK}} \cosh \sqrt{s/K}\, x$$

At $x = 0$

$$- k \sqrt{s/K}\, B(0) = \overline{H}(s)$$

and at $x = \ell$

$$A(\ell) \sinh \sqrt{s/K}\, \ell + B(\ell) \cosh \sqrt{s/K}\, \ell = 0$$

so that

$$B(x) = \frac{-\overline{H}(s)}{k\sqrt{s/K}} - \int_0^x \frac{T(\xi,0)}{\sqrt{sK}} \cosh \sqrt{s/K}\, \xi \, d\xi$$

$$A(x) = \int_0^x \frac{T(\xi,0)}{\sqrt{sK}} \sinh \sqrt{s/K}\, \xi \, d\xi + C$$

where

$$\left[\int_0^\ell \frac{T(\xi,0)}{\sqrt{sK}} \sinh \sqrt{s/K}\, \xi \, d\xi + C \right] \sinh \sqrt{s/K}\, \ell$$

$$+ \left[\frac{-\overline{H}(s)}{k\sqrt{s/K}} - \int_0^\ell \frac{T(\xi,0)}{\sqrt{sK}} \cosh \sqrt{s/K}\, \xi \, d\xi \right] \cosh \sqrt{s/K}\, \ell = 0$$

or

$$C \sinh \sqrt{s/K}\,\ell = \frac{\overline{H}(s)}{k\sqrt{s/K}} \cosh \sqrt{s/K}\,\ell + \int_0^{\ell} \frac{T(\xi,0)}{\sqrt{sK}} \cosh \sqrt{s/K}\,(\ell-\xi)\,d\xi$$

giving, finally,

$$\overline{T}(x,s) = \cosh \sqrt{s/K}\,x \left\{ \int_0^x \frac{T(\xi,0)}{\sqrt{sK}} \sinh \sqrt{s/K}\,\xi\,d\xi \right.$$

$$\left. + \frac{1}{\sin \sqrt{s/K}\,\ell} \int_0^{\ell} \frac{T(\xi,0)}{\sqrt{sK}} \cosh \sqrt{s/K}\,(\ell-\xi)\,d\xi + \frac{\overline{H}(s) \cosh \sqrt{s/K}\,\ell}{k\sqrt{s/K} \sinh \sqrt{s/K}\,\ell} \right\}$$

$$- \sinh \sqrt{s/K}\,x \left\{ \int_0^x \frac{T(\xi,0)}{\sqrt{sK}} \cosh \sqrt{s/K}\,\xi\,d\xi + \frac{\overline{H}(s)}{k\sqrt{s/K}} \right\}$$

We notice that this is made up of two parts — one depending on the initial temperature distribution $T(x,0)$ and one depending on $\overline{H}(s)$, the heat input. The transfer function relating $\overline{T}(a,t)$ to $h(t)$ or $\overline{T}(a,s)$ to $\overline{H}(s)$ is therefore

$$\frac{\cosh \sqrt{s/K}\,\ell}{k\sqrt{s/K}} \frac{\cosh \sqrt{s/K}\,a}{\sinh \sqrt{s/K}\,\ell} - \frac{\sinh \sqrt{s/K}\,a}{k\sqrt{s/K}} = \frac{\cosh \sqrt{s/K}\,(\ell - x)}{k\sqrt{s/K} \sinh \sqrt{s/K}\,\ell} \tag{6.4}$$

If $a = 0$ this reduces to

$$\frac{\cosh \sqrt{s/K}\,\ell}{h\sqrt{s/K}} \tag{6.5}$$

and if $a = \ell$ we have the transfer function

$$\frac{1}{k\sqrt{s/K} \sinh \sqrt{s/K}\,\ell} \tag{6.6}$$

We may analyse the stability of the closed loop using the Nyquist criterion in which case we are interested in the frequency response of Eqs (6.4) - (6.6). The frequency responses of (6.5) and (6.6) putting $s = i\omega$ are shown in Fig. 6.2a and b.

We notice that if $\overline{H}(s)$ is a constant gain K then the open loop frequency response will encircle the (-1, 0) point for sufficiently large K unless a = 0.

(From a practical point of view this problem is somewhat artificial as it is unlikely that h(t) can go negative, unless heating and cooling arrangements are available at x = 0.)

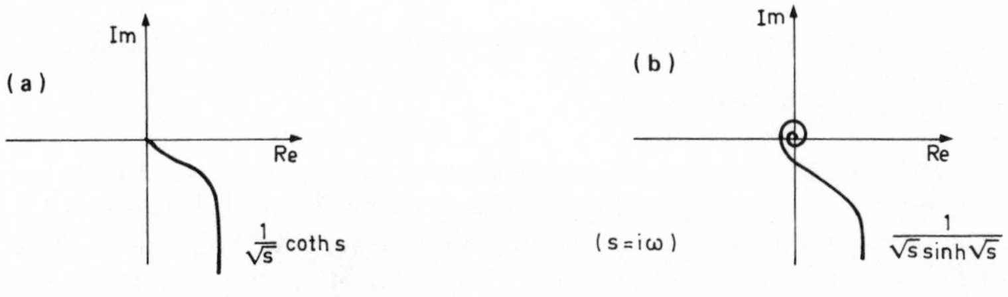

FIG. 6.2. Frequency responses of Eq. (6.5) (a) and (6.6) (b).

Another way of relating T(x, t) to h(t) is by assuming a Fourier series for T(x, t) in the form

$$T(x,t) = \sum_{r=0}^{\infty} A_r \cos\frac{r\pi x}{\ell}$$

which satisfies the boundary condition at $x = \ell$. However, to satisfy the boundary condition at x = 0 we actually arrange for it to be satisfied at x = 0+ by considering the modified differential equation

$$K\frac{\partial^2 T}{\partial x^2} + \frac{K}{k}\,\delta(x)h(t) = \frac{\partial T}{\partial t}$$

which contains a Dirac delta function at x = 0.

On substitution of the Fourier series for T(x, t), multiplication by $\cos\frac{r\pi x}{\ell}$ and integration from x = 0 to $x = \ell$ we obtain

$$\dot{A}_r + \frac{Kr^2\pi^2}{\ell^2}\,A_r = \frac{2K}{k\ell}\,h(t); \quad r = 1, 2, 3, \ldots$$

$$\dot{A}_0 = \frac{K}{k\ell}\,h(t); \quad r = 0$$

If h(t) is a unit impulse we obtain

$$T(x,t) = \frac{2K}{k\ell}\sum_{r=1}^{\infty} e^{-r^2\pi^2Kt/\ell^2}\cos(r\pi x/\ell) + \frac{K}{k\ell}$$

(notice that this is not convergent for t = 0).
Taking the Laplace transform

$$\overline{T}(x,s) = \frac{2K}{k\ell}\sum_{r=1}^{\infty}\frac{\cos(r\pi x/\ell)}{(s + r^2\pi^2K/\ell^2)} + \frac{K}{k\ell s}$$

It is not immediately obvious that this is identical to Eq.(6.4) obtained earlier which was (putting a = x)

$$\overline{T}(x,s) = \frac{\cosh\sqrt{s/K}\,(\ell - x)}{k\sqrt{s/K}\,\sinh\sqrt{s/K}\,\ell}$$

The first expression may be regarded as a partial fraction expansion of the second, using the relationship

$$\sinh z = z\prod_{r=1}^{\infty}\left(1 + \frac{z^2}{r^2\pi^2}\right)$$

We notice that extension of feedback control ideas to partial differential equations involve a number of difficulties: more complex solutions, more elaborate Laplace transforms, convergence of series, and use of Dirac delta functions. A comprehensive account of control of distributed parameter systems still has to be written.

7. CONTROL CONCEPTS FOR PARTIAL DIFFERENTIAL EQUATIONS. II

7.1. Control of the wave equation

Figure 7.1 illustrates an angular position control of a uniform flexible shaft. The torsion of the shaft gives rise to the one-dimensional wave equation

$$I\frac{\partial^2\theta}{\partial t^2} = GJ\frac{\partial^2\theta}{\partial x^2} \quad \text{or} \quad \frac{\partial^2\theta}{\partial t^2} = c^2\frac{\partial^2\theta}{\partial x^2} \qquad \text{where } c^2 = \frac{GJ}{I}$$

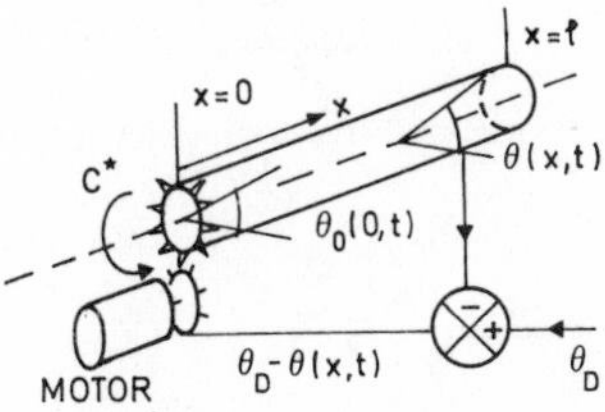

FIG. 7.1. Angular position control of flexible shaft.

I is the moment of inertia per unit length in the x-direction and GJ is the torsional stiffness of the shaft (in conventional notation). The boundary conditions are

$$C^* = -GJ\frac{\partial\theta}{\partial x}\Bigg|_{x=0}$$

$$0 = \frac{\partial\theta}{\partial x}\Bigg|_{\ell}$$

where C^* is the torque produced by the motor.

While we could develop a transfer function approach which would be quite similar to the procedure in Section 6, we shall consider here a different problem, that of "controllability". In particular, we shall consider what torque C^* should be applied to bring the shaft to rest from any given initial condition which involves specification of $\theta(x,0)$ and $\partial\theta/\partial t\,(x,0)$ for each x, $0 \leq x \leq \ell$. Let us suppose we require $\partial\theta/\partial x\,(x,T) = 0$ and $\partial\theta/\partial t\,(x,T) = 0$ for all $0 \leq x \leq \ell$ and T to be as small as possible.

In this problem the characteristics in the (x, t)-plane are very useful. We know from Section 5 that

$$\frac{\partial\theta}{\partial t} + c\frac{\partial\theta}{\partial x} \text{ is constant on lines } x + ct = \text{constant}$$

and

$$\frac{\partial\theta}{\partial t} - c\frac{\partial\theta}{\partial x} \text{ is constant on lines } x - ct = \text{constant.}$$

Figure 7.2 shows the characteristics diagram when $T = 2\ell/c$. Let us suppose that the objectives have been achieved at $t = T = 2\ell/c$. Then at P, both $\partial\theta/\partial t$ and $\partial\theta/\partial x$ are zero, and so $\partial\theta/\partial t + c\,\partial\theta/\partial x$ is zero along PA and $\partial\theta/\partial t - c\,\partial\theta/\partial x$ is zero along PC.

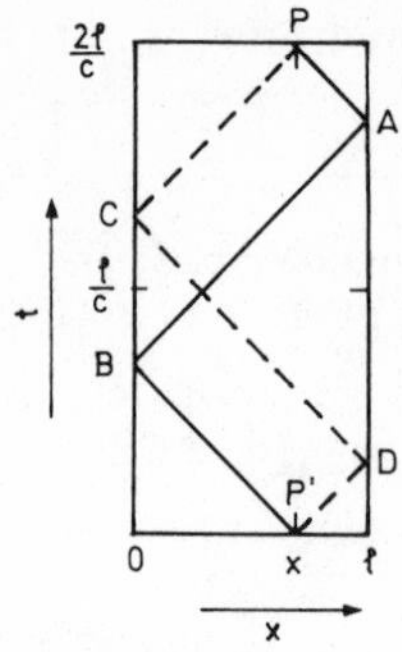

FIG. 7.2. Characteristics diagram for $T = 2\ell/c$.

At C the torque $C^*(t)$ is given by

$$- GJ \left.\frac{\partial\theta}{\partial x}\right|_{x=0} = C^*\left(\frac{2\ell - x}{c}\right)$$

and so

$$\frac{\partial\theta}{\partial t} = c\frac{\partial\theta}{\partial x} = -c\frac{C^*}{GJ}\left(\frac{2\ell - x}{c}\right)$$

Along CD

$$\frac{\partial\theta}{\partial t} + c\frac{\partial\theta}{\partial x} = \text{constant} = \frac{-2cC^*}{GJ}\left(\frac{2\ell - x}{c}\right)$$

At D

$$\frac{\partial\theta}{\partial x} = 0 \text{ so } \frac{\partial\theta}{\partial t} = -\frac{2cC^*}{GJ}\left(\frac{2\ell - x}{c}\right)$$

and along DP' and at P'

$$\frac{\partial\theta}{\partial t} - c\frac{\partial\theta}{\partial x} = \text{constant} = -\frac{2cC^*}{GJ}\left(\frac{2\ell - x}{c}\right) \tag{7.1}$$

Similarly, following the path PABP', we obtain that at P'

$$\frac{\partial\theta}{\partial t} + c\frac{\partial\theta}{\partial x} = -\frac{2cC^*}{GJ}\left(\frac{x}{c}\right) \tag{7.2}$$

Thus the torque $C^*(t)$ is uniquely determined from (7.1) for $\ell/c \leq t \leq 2\ell/c$ and from (7.2) for $0 \leq t \leq \ell/c$ given the values of $\partial\theta/\partial t$ and $\partial\theta/\partial x$ at $t = 0$ and for $0 \leq x \leq \ell$.

The angle turned through at $x = 0$ may be found by taking the integral of $\partial\theta/\partial t\,(0, t)$ which is equal to

$$\frac{-cC^*(x/c)}{GJ} \quad \text{and} \quad \frac{-cC^*((2\ell - x)/c)}{GJ}$$

at B and C, respectively. Hence, from (7.1) and (7.2),

$$\theta(0, T) = \theta(0, 0) + \int_0^{\ell} \frac{\partial\theta}{\partial t}(x, 0)\frac{dx}{c}$$

A more general problem is how to move the shaft from an undeformed position with $\theta(x, 0) = 0$, $\partial\theta/\partial t(x, 0) = 0$, $0 \leq x \leq \ell$, to a new position with $\theta(x, T) = \theta$, $\partial\theta/\partial t(x, T) = 0$, $0 \leq x \leq \ell$.

We may apply a pulse torque of magnitude C^* constant over the time interval $(-h, 0)$ giving initial conditions

$$\frac{\partial\theta}{\partial t} = \frac{cC^*}{GJ}, \quad \frac{\partial\theta}{\partial x} = -\frac{C^*}{GJ} \quad \text{for } 0 \leq x \leq ch \text{ at } t = 0$$

To bring this to rest we require a torque of magnitude $-C^*$ for $(2\ell/c) - h < t < 2\ell/c$. The total angle turned by the shaft is $2C^* h/Ic$, the total time $(2\ell/c) + h$. By increasing C^* and decreasing h we can achieve a given angle in time $2\ell/c$, by use of two impulses at $t = 0$ and $t = 2\ell/c$.

We have concentrated on the $T = 2\ell/c$ case as an examination of the characteristics diagram corresponding to Fig.7.2, which reveals that in general the problem of bringing the shaft to rest in a time less than $2\ell/c$ is impossible, no matter what size the torque $C^*(t)$ is permitted. This is in marked contrast to controllability of lumped systems which may be brought to rest instantaneously if large enough controls are applied (using, of course, Dirac delta functions and their derivatives if need be).

7.2. Parasitic oscillations

Many flexible devices having stiffness and inertia and obeying the wave equation or similar equations (the beam equation, for example) exhibit unstable parasitic oscillations when they are included in a feedback loop, that is when they form a link between an actuator and a feedback instrument. These parasitic instabilities may be analysed by use of the Nyquist diagram and may sometimes be avoided by careful siting of the feedback instruments. In particular, if it is possible to site the feedback instrument close to the actuator, the time delay involved in waves travelling from one to the other is avoided. It is really this time delay which causes the instability.

8. NOISE IN LINEAR SYSTEMS

An important part of linear system theory is concerned with the treatment of random inputs and outputs. Given the "power spectral density" of the input signal, then the power spectral density of the output signal is easily calculated, and other quantities such as the mean square output may then be found. Examples of random inputs to systems include the motion of a motor car on its suspension when running over a rough road, and glint noise entering an automatic tracking radar scanner or a homing missile guidance system.

Given a collection or ensemble of records of a random signal plotted as functions of time (Fig.8.1), we distinguish two kinds of average of the properties of the records:

(i) a time average taken along a particular record,
(ii) an ensemble average taken at a particular time across the collection of records.

If the ensemble averages of various properties are constant with time the records are said to be "stationary".

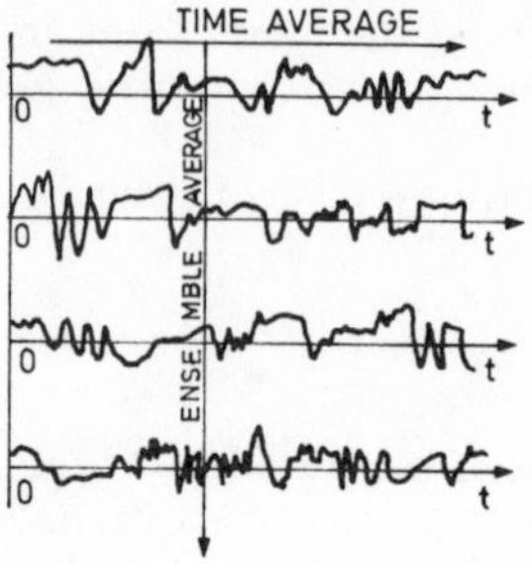

FIG. 8.1. Ensemble and time averages.

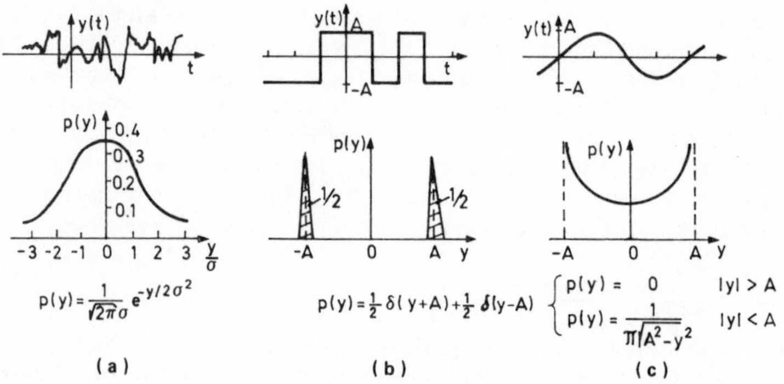

FIG. 8.2. Amplitude probability distributions; (a) Gaussian noise, (b) random square wave, (c) sine wave.

If,in addition,the time average equals the ensemble average, then the records satisfy the "ergodic hypothesis". We thus have the set-theoretical formulation:

non-stationary signals $\supset$ stationary signals $\supset$ ergodic signals

The "ergodic hypothesis" is often made in engineering problems in the absence of evidence to the contrary.

The most important properties of a random signal are

(i) its amplitude probability distribution
(ii) its frequency content or power spectral density.

The amplitude probability distribution p(y) may be measured as an ensemble average, or on a particular record by considering the proportion of time $p(y)\delta y$ spent by the signal in a small interval $(y, y + \delta y)$ as $\delta y \to 0$. Some examples are given in Fig. 8.2.

The frequency content or power spectral density $S^*(\omega)$ may be regarded as the power or mean square output $S^*(\omega)\,\delta\omega$ from an ideal filter with centre frequency ω and small bandwidth $\delta\omega$ as $\delta\omega \to 0$. Mathematically it is usual to distribute the power spectral density equally over positive and negative ω, taking $S(\omega) = \frac{1}{2} S^*(\omega)$. While there are spectrum analysers which work in this way using analogue or recorded signals, another approach to spectral analysis is via the autocorrelation function of the signal y(t).

The autocorrelation function $R(\tau)$ is the time average $\widetilde{y(t)\, y(t-\tau)}$. Note that R(0) is the mean square of the signal and that $R(\tau) = R(-\tau) \leq R(0)$. The autocorrelation function $R(\tau)$ and the power spectral density function $S(\omega)$ are Fourier transforms one of the other:

$$\left.\begin{aligned} S(\omega) &= \frac{1}{2\pi}\int_{\tau=-\infty}^{\infty} R(\tau)\, e^{i\omega\tau}\, d\tau \\ R(\tau) &= \int_{\omega=-\infty}^{+\infty} S(\omega) e^{-i\omega\tau}\, d\omega \end{aligned}\right\}$$

Since $S(\omega)$ and $R(\tau)$ are both even functions we may rewrite these relationships as

$$S(\omega) = \frac{1}{\pi}\int_{\tau=0}^{\infty} R(\tau)\cos\omega\tau\, d\tau = \tfrac{1}{2} S^*(\omega)$$

$$R(\tau) = 2\int_{0}^{\infty} S(\omega)\cos\omega\tau\, d\omega = \int_{0}^{\infty} S^*(\omega)\cos\omega\tau\, d\omega$$

If we put $\omega = 2\pi f$, where f is in cycles per second (Hz) (ω being in radians per second), the integrals become as shown, where $S^*(f) = 2\pi S^*(\omega)$ is measured in power units per cycle per second

$$\left.\begin{aligned} S^*(f) &= 4\int_{\tau=0}^{\infty} R(\tau)\cos 2\pi f\tau\, d\tau \\ R(\tau) &= \int_{f=0}^{\infty} S^*(f)\cos 2\pi f\tau\, df \end{aligned}\right\}$$

known as the Wiener-Khintchine relations.

This relationship may be demonstrated by taking a finite length 2T of the random signal, carrying out a Fourier analysis (regarding it as periodic with period 2T), calculating the power from the Fourier coefficients and an estimate of the power spectral density by distributing this power over the frequency interval $2\pi/2T$. The resulting expression may be written as a double integral. This integral may be evaluated in a different way giving an expression involving $\widetilde{y(t)\, y(t+\tau)}$ with $-T < \tau < T$. By letting $T \to \infty$ the first of the Fourier transform pair is obtained.

Some examples of the Fourier-transform relationship are given in Fig.8.3.

The reciprocal relationship between $R(\tau)$ and $S(\omega)$ should be noted: a "narrow" $R(\tau)$ corresponds to a "broad" $S(\omega)$, and vice versa. Extreme cases of this are shown in Fig.8.4 where we have y(t) as "direct current" and "white noise" signals. While "direct current" is a realistic signal, "white noise" is in fact a useful mathematical concept which cannot, however, exist in reality, since its mean square or power is infinite.

We now have to consider the response of a stable linear system with transfer function H(s) and impulse response h(t) to a random input of known power spectral density $G_{11}(\omega)$, say. There are two ways of thinking about

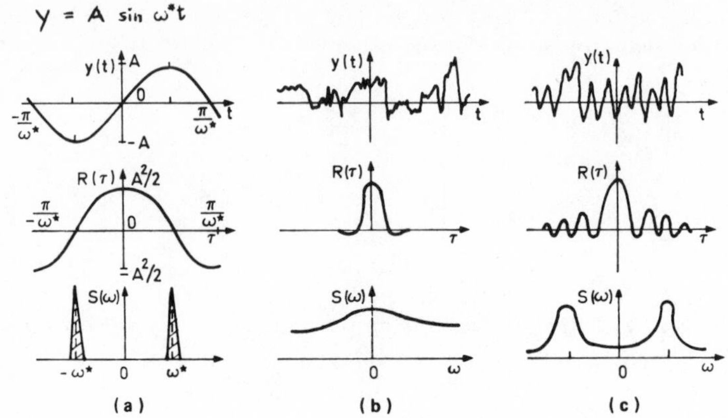

FIG. 8.3. Autocorrelation functions and power spectral density functions; (a) sine wave, (b) wide-band noise, (c) narrow-band noise.

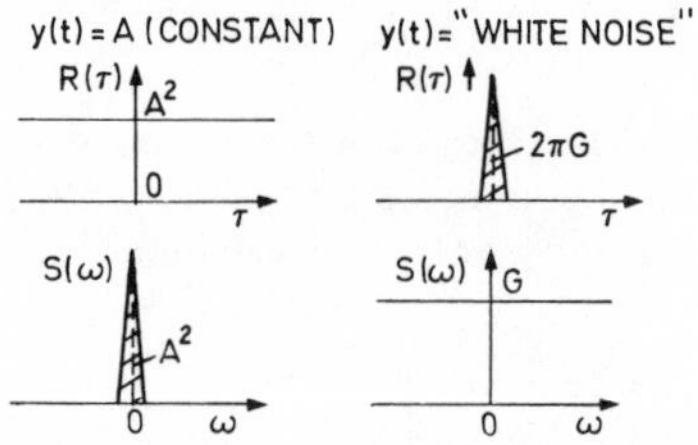

FIG. 8.4. "Direct current" and "white noise".

this — first considering the effect of the transfer function on sine waves, we can deduce heuristically that the power spectral density of the output $G_{00}(\omega)$ is given by

$$G_{00}(\omega) = |H(i\omega)|^2 \; G_{11}(\omega)$$

A more sophisticated approach is via the cross-correlation function $R_{01}(\tau) = \overline{y_0(t)\; y_1(t-\tau)}$ and its Fourier transform, the cross-power spectrum $G_{01}(\omega)$. Now

$$y_0(t) = \int_{\tau=0}^{\infty} h(\tau)\; y_1(t-\tau)\; d\tau$$

from which it follows, on multiplying through by $y_1(t-\tau_1)$ and taking time averages, that

$$R_{01}(\tau_1) = \int_{\tau=0}^{\infty} h(\tau)\; R_{11}(\tau_1 - \tau)\; d\tau$$

where $R_{11}(\tau)$ is the autocorrelation function of $y_1(t)$. Taking Fourier transforms yields

$$G_{01}(\omega) = H(i\omega)\; G_{11}(\omega)$$

Multiplying by $y_0(t - \tau_1)$ and time-averaging yields

$$R_{00}(\tau_1) = \int_{\tau=0}^{\infty} h(\tau)\, R_{10}(\tau_1 - \tau)\, d\tau$$

Hence

$$G_{00}(\omega) = H(i\omega)\, G_{10}(\omega) = H(i\omega)\, G_{01}(-\omega) = H(i\omega)\, H(-i\omega)\, G_{11}(\omega)$$
$$= |H(i\omega)|^2\, G_{11}(\omega)$$

which is the result obtained above.

The mean square output, σ^2, may then be calculated as

$$\sigma^2 = \int_{\omega=-\infty}^{\infty} G_{00}(\omega)\, d\omega = \int_{\omega=-\infty}^{\infty} |H(i\omega)|^2\, G_{11}(\omega)\, d\omega$$

$$= \int_{\omega=-\infty}^{\infty} H(i\omega)\, H(-i\omega)\, G_{11}(\omega)\, d\omega$$

This integral may be calculated by the residue theorems of the complex integral calculus, and standard forms for polynomial fractions in ω are available up to degree 9.

For example: if $G_{11} = G$ constant ("white noise" input) and

$$H(i\omega) = \frac{K}{(i\omega + a)\,(i\omega + b)} \qquad K,\ a,\ b > 0$$

then we can consider the integral

$$\int_C \frac{K^2 G\, dz}{(z + a)\,(z + b)\,(-z + a)\,(-z + b)}$$

where C is a large semicircle in the left-hand half-plane and its diameter (the imaginary axis). The integrand has simple poles at $z = -a$ and $z = -b$ inside C and so by the residue theorem

$$\int_C \frac{K^2 G\, dz}{(z + a)\,(z + b)\,(-z + a)\,(-z + b)} = \int_{\omega=-\infty}^{\infty} \frac{K^2 G\, i\, d\omega}{(i\omega + a)\,(i\omega + b)\,(-i\omega + a)\,(-i\omega + b)} = i\sigma^2$$

$$= 2\pi i \text{ (sum of residues at poles within C).}$$

$$= 2\pi i \left(\lim_{z \to -a} \frac{K^2 G}{(z + b)\,(-z + a)\,(-z + b)} + \lim_{z \to -b} \frac{K^2 G}{(z + a)\,(-z + a)\,(-z + b)} \right)$$

$$= 2\pi i\, K^2 G \left(\frac{1}{(b - a)\, 2a\, (a + b)} + \frac{1}{(a - b)\,(b + a)\, 2b} \right)$$

$$= 2\pi i\, K^2 G\, \frac{1}{(a + b)\, 2ab}$$

Hence it follows that

$$\sigma^2 = \frac{\pi K^2 G}{(a+b)\, ab}$$

An analogous theory exists for discrete time systems described by difference equations. The output sequence $\{y_0(t)\}$ is related to the input sequence $\{y_1(t)\}$ (where t takes discrete values ...-3, -2, -1, 0, 1, 2, 3, ...) by the relationship

$$y_0(t) = \sum_{n=0}^{\infty} h(n) y_1(t-n)$$

where the sequence $\{h(n)\}$ corresponds to the unit impulse response h(t) in the continuous case described above. With a stationary random input with the autocorrelation function $R_{11}(\tau) = \overline{y_1(t) y_1(t-\tau)}$, defined for integer values of τ, we obtain

$$R_{00}(\tau) = \sum_{k=0}^{\infty} \sum_{\ell=0}^{\infty} h(k) h(\ell) R_{11}(\tau + \ell - k)$$

The spectral density function is defined as

$$S(\omega) = \frac{1}{2\pi} \sum_{n=-\infty}^{\infty} R(n) e^{in\omega}$$

where

$$R(n) = \int_{-\pi}^{\pi} S(\omega) e^{-in\omega} d\omega$$

(The R(n) defined for integer n are the Fourier coefficients of the periodic function $S(\omega)$ defined on the basic interval $(-\pi, \pi)$.)

Now the pulse transfer function using z-transform notation is

$$H(z) = \sum_{n=0}^{\infty} h(n) z^{-n}$$

and so taking the transform of $R_{00}(\tau)$ we obtain

$$S_{00}(\omega) = \frac{1}{2\pi} \sum_{n=-\infty}^{\infty} \sum_{k=0}^{\infty} \sum_{\ell=0}^{\infty} e^{ik\omega} h(k) e^{-i\ell\omega} h(\ell)\, e^{+i(n+\ell-k)\omega} R_{11}(n+\ell-k)$$

$$= H(e^{-i\omega}) H(e^{i\omega}) S_{11}(\omega)$$

To find the mean square output $R_{00}(0)$ we need to evaluate

$$\int_{-\pi}^{\pi} H(e^{-i\omega}) H(e^{i\omega}) S_{11}(\omega)\, d\omega$$

Such integrals can often be evaluated by considering a contour integral around the unit circle (where $z = e^{i\omega}$, $dz = iz\, d\omega$) and evaluating this by use of the residue theorem. If the discrete system is stable, which is a necessary requirement, then H(z) will have all its poles inside the unit circle.

9. DISCRETE NOISE PROCESSES

To simulate noisy systems on a digital or analogue computer it is useful to be able to make up random signals with prescribed amplitude probability distributions and with prescribed spectral density properties. Such signals are also useful for system identification using cross-correlation techniques. We shall discuss a number of examples.

9.1. The random square wave

Consider time divided up into a sequence of intervals each of length h. The random square wave y(t) takes the constant value +A or - A for $nh \leq t < (n+1)h$ for $n = \{\ldots, -3, -2, -1, 0, 1, 2, 3, \cdots\}$ so that (i) there is an equal probability of +A or of -A in each interval and (ii) there is no correlation between the values +A and - A in different intervals (Fig. 9.1).

The amplitude probability distribution function, p(a), is clearly two Dirac delta functions at ±A (Fig. 9.2) each of magnitude $\frac{1}{2}$.

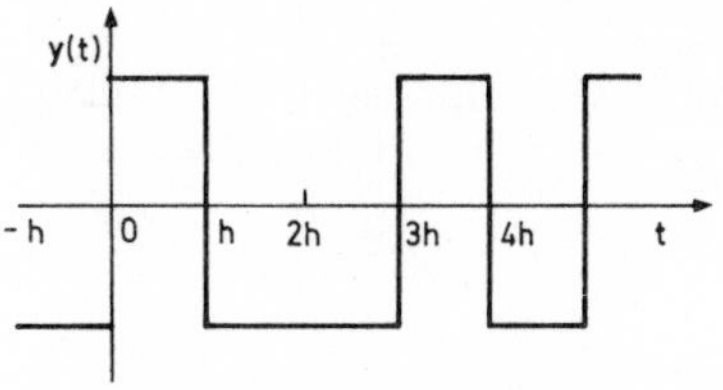

FIG. 9.1. Random square wave.

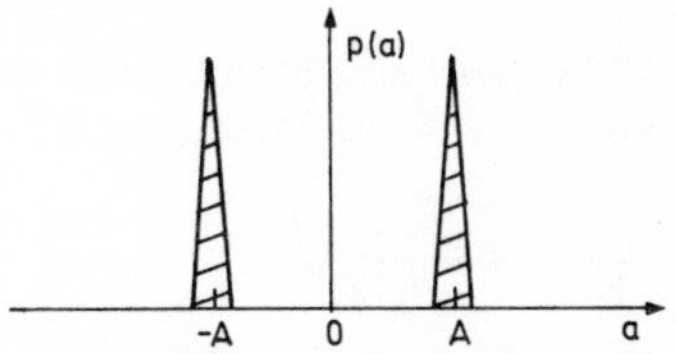

FIG. 9.2. Amplitude probability distribution of Fig. 9.1.

The spectral density of the random square wave is obtained from its autocorrelation function. This may be calculated by considering pairs of points, a distance τ apart, with the left-hand point placed at random on the t-axis. When $\tau = 0$ we shall obtain the mean square which is clearly A^2. When $\tau > h$, the mean value of $y(t)y(t+\tau)$ calculated from lots of pairs of points is zero, because of the zero correlation in different intervals. When $0 < \tau < h$ the two points of one pair sometimes lie in the same interval and sometimes in adjacent intervals in proportion $1 - \tau/h$ to τ/h. From this we deduce the autocorrelation function $R(\tau)$ shown in Fig. 9.3.

With this the power spectral density function $S(\omega)$ is obtained from the Fourier transform of $R(\tau)$ as

$$S(\omega) = \frac{1}{2\pi} \int_{-\infty}^{\infty} R(\tau) e^{-i\omega\tau} d\tau$$

$$= \frac{A^2 h}{2\pi} \frac{\sin^2(\omega h/2)}{(\omega h/2)^2}$$

sketched in Fig. 9.4.

A modification of this random square wave is to replace the values of ±A by a sequence of random numbers with a zero mean and given variance

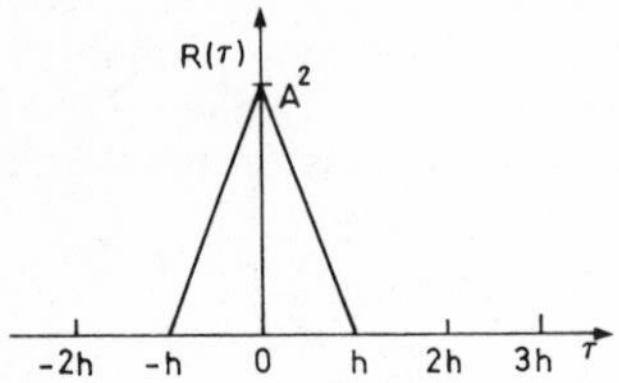

FIG. 9.3. Autocorrelation function of the random square wave.

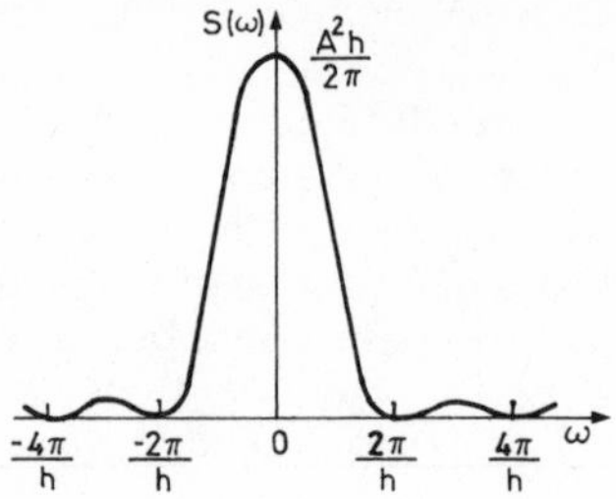

FIG. 9.4. Power spectral density of the random square wave.

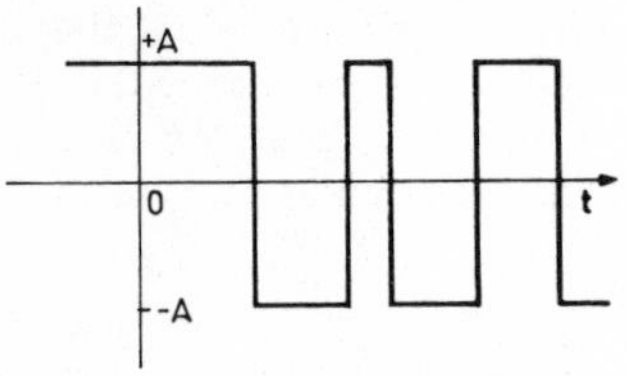

FIG. 9.5. Random telegraph signal.

σ^2. This will produce a random stepping signal with a prescribed amplitude probability distribution and (by similar arguments) a power spectral density

$$\frac{\sigma^2 h}{2\pi} \; \frac{\sin^2(\omega h/2)}{(\omega h/2)^2}$$

9.2. The random telegraph signal

Random telegraph signal is the name given to a signal taking the two values $+A$ and $-A$ with equal probability where the switching points form a Poisson process in time, with an average frequency of ν switches per unit time (Fig. 9.5). Its amplitude probability distribution is clearly the same as that shown in Fig. 9.2, but its autocorrelation function and power spectral density are quite different from the random square wave. If we consider many pairs of points, the two points of each pair being a distance τ apart, and then suppose τ to be increased to $\tau + \delta\tau$ then, by considering the number of switching points which will be included in the many $\delta\tau$ intervals, we deduce that

$$R(\tau + \delta\tau) = -R(\tau)\,\nu\,\delta\tau + R(\tau)\,(1 - \nu\,\delta\tau)$$

giving the differential equation for $R(\tau)$

$$\frac{dR(\tau)}{d\tau} = -2\nu R(\tau)$$

Now $R(0) = A^2$ so $R(\tau) = A^2\,e^{-2\nu|\tau|}$. By Fourier transformation

$$S(\omega) = \frac{2\nu A^2}{\pi(4\nu^2 + \omega^2)}$$

See Figs 9.6 and 9.7.

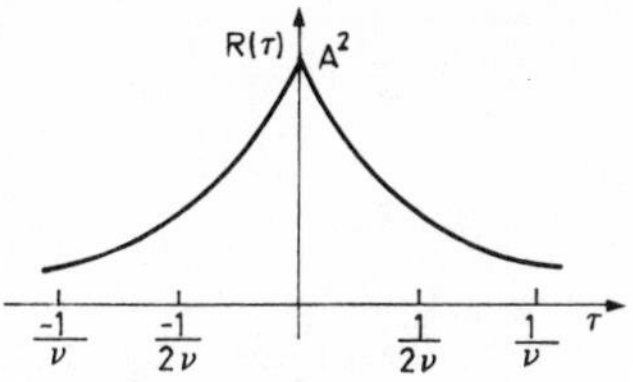

FIG. 9.6. Autocorrelation function for the random telegraph signal.

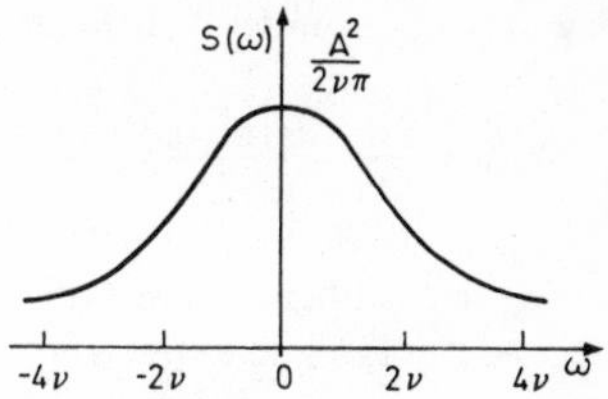

FIG. 9.7. Power spectral density function for the random telegraph signal.

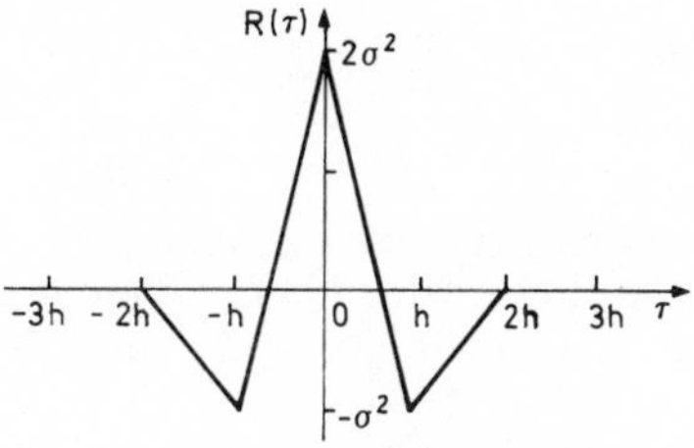

FIG. 9.8. Autocorrelation function for the differenced random square wave.

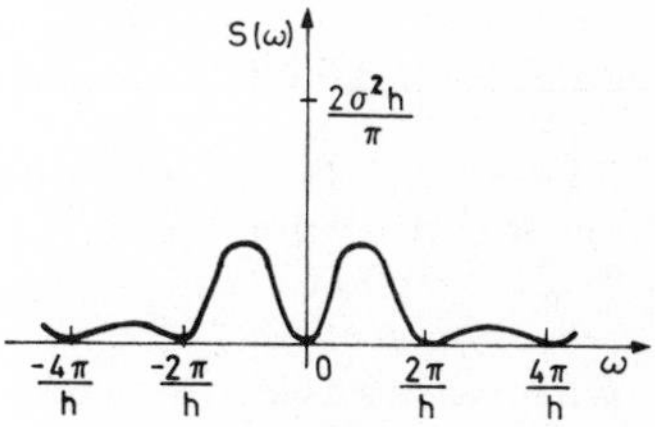

FIG. 9.9. Power spectral density of Fig. 9.8.

9.3. "Differentiated" random sequences

One may "tailor" power spectral densities by differencing given random sequences. For example, if we consider the random square wave or the random stepping wave formed from the random sequence $\{y_n\}$ say, with mean 0 and variance σ^2, then we can form a new sequence $\{z_n\}$ given by $z_n = y_n - y_{n-1}$ and a random stepping wave z(t) by putting $y(t) = z_n$ for $nh \leq t < (n+1)h$.
The new wave form has the autocorrelation function shown in Fig. 9.8.
From the Fourier transform, after some calculation, we find

$$S(\omega) = \frac{2\sigma^2 h}{\pi} \frac{(\sin \omega h/2)^4}{(\omega h/2)^2}$$

See Fig. 9.9.
The fact that S(0) = 0 could be deduced directly from the form of $R(\tau)$ since

$$S(0) = \frac{1}{2\pi} \int_{-\infty}^{\infty} R(\tau)\, d\tau = 0$$

from Fig. 9.8.

9.4. An approximation to "white noise"

If we examine the random stepping wave with an autocorrelation function

$R(\tau) = \sigma^2 (1 - |\tau|/h)$ for $0 \leq |\tau| \leq h$

$R(\tau) = 0, |\tau| > h$

and a power spectral density

$$S(\omega) = \frac{\sigma^2 h}{2\pi} \frac{\sin^2(\omega h/2)}{(\omega h/2)^2}$$

We may consider letting $\sigma^2 \to \infty$ and $h \to 0$ in such a way that $\sigma^2 h$ = constant = $2\pi G$, say. This means that the autocorrelation function becomes a delta function at $\tau = 0$ of magnitude $\sigma^2 h = 2\pi G$, and the power spectral density becomes a constant (G) at all frequencies.

In practice we would make h small compared with system time constants (or $2\pi/h$ large compared with the system pass-band), and σ^2 large so that $\sigma^2 h/2\pi = G$ (given).

9.5. Pseudo-random binary sequences

These signals, which are in fact periodic, are particularly useful for system identification using cross-correlation of output with input. There are two important classes — quadratic residue sequences and shift register or M-sequences.

Quadratic residue sequences are based on work in the theory of numbers due to Gauss. We take a prime number $N = 4k - 1$ where k is an integer, and calculate the numbers $1^2, 2^2, 3^2, \ldots, \{4k-2/2\}^2$ modulo N. We form a signal y(t) which is +A in intervals corresponding to these N - 1 numbers and is -A in the other intervals making up N intervals in all. The signal is then repeated to form a periodic signal of period Nh where h is the length of the basic interval.

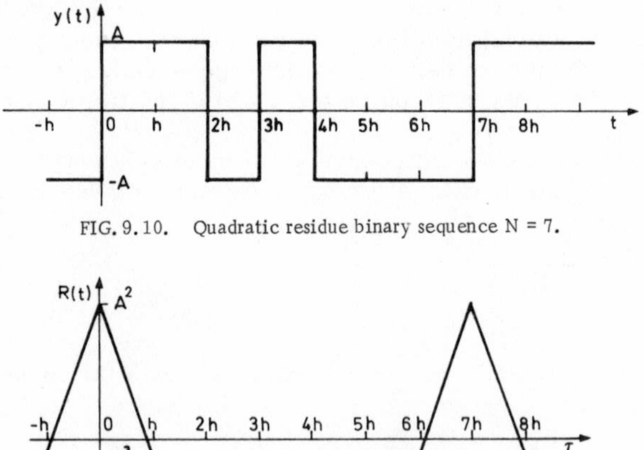

FIG. 9.10. Quadratic residue binary sequence N = 7.

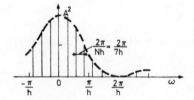

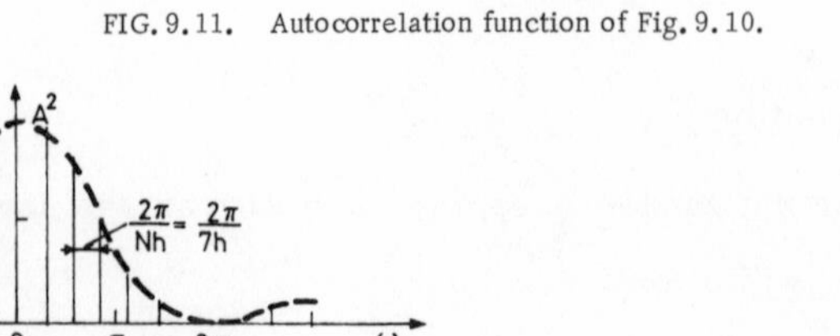

FIG. 9.11. Autocorrelation function of Fig. 9.10.

FIG. 9.12. Line spectrum of Fig. 9.10.

1	-1	1	1	1	-1	-1	-1	1	-1	1	1	-1	1	1
1	1	-1	1	1	-1	-1	1	-1	1	1	1	-1	1	-1
1	-1	1	-1	-1	1	-1	-1	-1	-1	-1	1	1	1	1
-1	-1	1	1	-1	1	-1	-1	-1	1	1	1	-1	-1	1
-1	-1	-1	1	-1	-1	-1	1	1	1	-1	-1	1	1	1
1	-1	1	1	1	1	1	-1	-1	1	-1	1	-1	1	-1
1	1	1	1	1	-1	1	-1	1	1	1	-1	-1	1	-1
1	-1	1	1	-1	-1	-1	1	-1	1	1	-1	1	-1	-1
1	1	-1	1	1	1	1	-1	1	1	-1	1	-1	1	1
-1	1	-1	-1	-1	-1	1	-1	1	1	-1	1	-1	1	-1

1	-1	1	1	1	-1	-1	-1	-1	-1	-1	-1	-1	-1	1
1	1	1	1	-1	1	1	-1	-1	-1	1	-1	-1	-1	1
1	-1	-1	-1	-1	-1	1	-1	-1	-1	-1	1	1	-1	-1
1	1	-1	1	-1	-1	-1	-1	1	-1	1	1	-1	1	1
1	-1	-1	1	1	-1	-1	-1	1	1	-1	1	-1	1	1
-1	1	1	1	-1	-1	-1	-1	1	-1	-1	1	1	1	1
-1	-1	1	-1	1	1	-1	-1	-1	-1	-1	-1	1	-1	1
1	1	-1	1	-1	1	1	1	-1	-1	1	1	-1	1	-1
-1	1	1	-1	1	1	1	1	1	1	-1	1	1	-1	1
1	1	-1	1	-1	1	1	1	-1	-1	1	1	-1	-1	1

-1	-1	1	1	-1	-1	1	-1	-1	-1	1	1	-1	1	-1
1	-1	1	1	1	-1	1	1	1	-1	1	1	1	1	-1
-1	-1	-1	-1	-1	-1	1	-1	1	1	1	-1	-1	-1	1
1	-1	1	-1	1	-1	-1	-1	1	-1	1	-1	1	-1	-1
1	-1	1	1	1	1	1	1	-1	1	1	1	1	-1	1
1	-1	1	1	1	1	1	1	-1	-1	-1	1	1	-1	1
1	-1	-1	-1	1	1	1	-1	-1	1	-1	1	-1	1	1
1	-1	-1	-1	1	1	-1	1	-1	-1	1	-1	-1	1	-1
-1	-1	-1	1	1	1	1	-1	-1	-1	-1	1	1	1	1
1	1	-1	-1	-1	1	1	-1	-1	1	1	1	-1	1	-1

-1	1	1	-1	1	-1	-1	-1	1	1	1	1	1	1	1
1	1	-1	1	1	-1	1	1	-1	1	-1	-1	-1	-1	-1
1	1	-1	1	1	-1	1	1	-1	-1	-1	-1	1	1	1
1	1	1	-1	1	1	1	-1	1	1	-1	1	1	1	-1
-1	-1	1	-1	-1	1	-1	-1	-1	1	-1	-1	-1	-1	-1
-1	1	1	1	1	-1	-1	1	-1	-1	1	-1	-1	1	1
1	1	1	-1	1	-1	-1	1	-1	-1	1	-1	-1	-1	-1
-1	-1	-1	-1	-1	1	1	1	-1	1	-1	-1	1	1	-1
1	-1	-1	-1	1	1	-1	-1	1	1	1	-1	-1	-1	-1
-1	-1	1	1	1	1	-1	-1	-1	-1	1	1	1	1	-1

1	1	-1	1	1	-1	1	-1	-1	1	1	1	-1	-1	-1
1	-1	1	-1	1	1	-1	-1	-1	1	1	1	-1	-1	1
-1	-1	1	1	1	-1	-1	-1	-1	-1	-1	1	-1	-1	1
-1	-1	-1	-1	1	-1	-1	-1	-1	-1	-1	1	-1	1	1
-1	1	-1	1	-1	1	1	1	-1	1	-1	1	-1	-1	1
1	1	-1	-1	-1	1	-1	1	1	1	1	1	1	1	-1
-1	-1	-1	1	-1	-1	-1	1	-1	-1	-1	1	-1	1	-1
1	-1	-1	1	1	1	-1	1	1	-1	-1	1	1	-1	1
1	-1	-1	1	1	-1	-1	-1	1	-1	1	-1	-1	-1	1
-1	-1	1	-1	-1	-1	-1	-1	-1	1	-1	-1	1	1	-1

1	-1	-1	1	1	-1	-1	-1	1	-1	1	-1	-1	-1	1
-1	1	1	1	1	1	1	-1	-1	1	-1	1	1	-1	-1
-1	-1	1	1	-1	1	1	1	1	-1	-1	-1	1	-1	-1
1	-1	1	-1	-1	1	1	1	-1	-1	1	1	-1	-1	-1
1	-1	-1	1	-1	1	1	1	1	-1	1	-1	-1	1	1
-1	-1	1	1	1	1	-1	1	1	1	1	1	-1	-1	1
1	1	-1	1	1	1	-1	-1	1	-1	-1	-1	-1	-1	1
1	1	1	1	1	1	1	1	-1	-1	-1	1	-1	1	-1
1	-1	1	-1	-1	1	-1	1	1	1	1	-1	1	-1	-1
1	-1	1	-1	-1	1	-1	-1	-1	-1	1	-1	-1	1	1

-1	1	-1	-1	1	-1	1	1	1	-1	-1	1	-1	1	-1
1	1	-1	-1	-1	1	-1	1	-1	-1	-1	-1	-1	1	-1
1	-1	1	-1	1	1	-1	-1	-1	-1	-1	1	-1	-1	-1
-1	1	1	-1	-1	-1	1	1	1	-1	1	1	1	-1	1
1	-1	-1	-1	1	1	1	-1	1	-1	-1	1	1	-1	-1
-1	-1	1	1	1	1	1	-1	1	1	-1	1	-1	1	-1
1	-1	-1	-1	1	-1	1	1	-1	-1	1	-1	-1	-1	-1
1	-1	-1	1	-1	1	1	1	-1	-1	-1	1	-1	-1	

FIG. 9.13. Quadratic residue binary sequence N = 1019

For example: When N = 7 (k = 2) we calculate 1^2, 2^2, 3^2 (modulo 7) = 1, 4, 2 and thus form the signal shown in Fig. 9.10. This signal has the autocorrelation function shown in Fig. 9.11. This gives a line spectrum (Fig. 9.12) with a frequency separation of lines equal to $2\pi/Nh$ rad/s. The envelope is of the form $\left(\frac{\sin(\omega h/2)}{\omega h/2}\right)^2$.

A longer quadratic residue sequence with N = 1019 and A = 1 is given in Fig. 9.13.

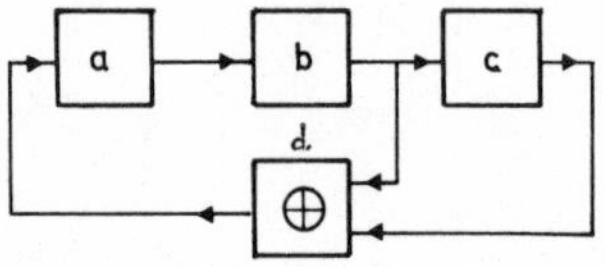

FIG. 9.14. Two-level shift register with binary adder.

Another form of pseudo-random binary sequence is the "M-sequence" which can be generated with a 2-level shift register having appropriate feedback arrangements and a clock with period h. Consider the system shown in Fig. 9.14, where a, b and c are registers and d is a binary adder. Let us initially set the binary numbers 1 in a, 0 in b and 0 in c. At each clock pulse the digits are shifted one to the right, whilst the vacant place in a is filled with the modulo-2 sum of the digits in b and c before the shift, this sum being calculated by d. The sequence of operations is as follows:

	Register	a	b	c
Time ↓	0	1	0	0
	h	0	1	0
	2h	1	0	1
	3h	1	1	0
	4h	1	1	1
	5h	0	1	1
	6h	0	0	1
	7h	1	0	0

This pattern is then repeated. If we tap the output of the register a and convert this into a square wave where 0 becomes +A and 1 becomes -A we obtain the signal shown in Fig. 9.15. (In this case this becomes equivalent to the waveform of Fig. 9.10 shifted by 2h to the left and the autocorrelation function of Fig. 9.15 is identical to Fig. 9.11.)

Any sequence of length $N = 2^k - 1$ generated by a k-stage shift register is called a maximal length sequence or "M-sequence". In such a sequence the registers go through all possible states except the all-zero state and it follows from this that there is one more 1 than 0 in such a sequence.

An important property of such sequences is that if a sequence is added modulo 2 to a delayed version of itself, then the original sequence with a new delay is formed; for example

Original sequence	(register a)	1	0	1	1	1	0	0	1	0	1	1	1
Single delay of h	(register b)	0	1	0	1	1	1	0	0	1	0	1	1
Addition modulo 2		1	1	1	0	0	1	0	1	1	1	0	0

(The resulting delay is 5h.)

Modulo 2 addition of the sequence is equivalent to autocorrelation of the signal y(t) in Fig. 9.15

e.g.

t	0+	h+	2h+	3h+	0+	h+	2h+	3h+
y(t)	- A	+ A	- A	- A	1	0	1	1
y(t + 2h)	- A	- A	- A	+ A	1	1	1	0
y(t) y (t + 2h)	$+A^2$	$-A^2$	$+A^2$	$-A^2$	0	1	0	1

It follows that the autocorrelation function has one more $-A^2$ than $+A^2$ adding over one period Nh. Hence $R(\tau) = -A^2/N$ for $h \leq \tau \leq (N-1)h$. This yields the now familiar form shown in Fig. 9.16.

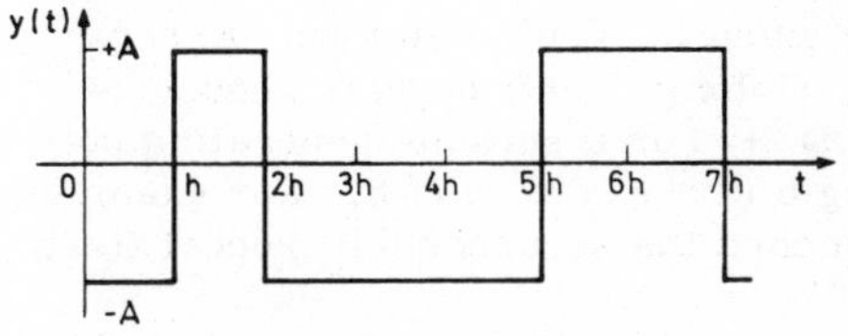

FIG. 9.15. Square-wave signal for register pattern.

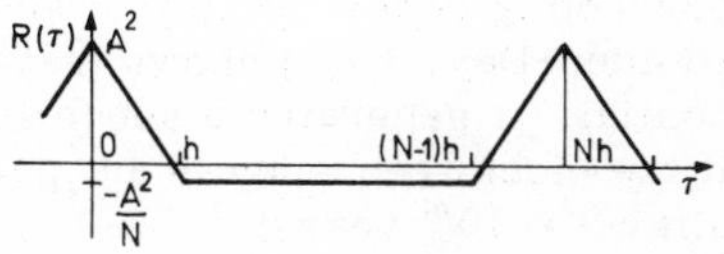

FIG. 9.16. Autocorrelation function.

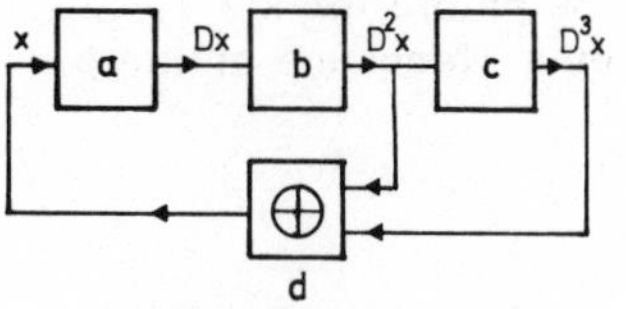

FIG. 9.17. Corresponding shift register as in Fig. 9.10.

The shift register operation may be described by a polynomial in the delay operator D. Figure 9.17 shows the corresponding shift register diagram.

$$x = D^3x \oplus D^2x$$

or

$$D^3x \oplus D^2x \oplus x = 0$$

For example

$$x = 0 \quad Dx = 1 \quad D^2x = 0 \quad D^3x = 0$$

$$x = 1 \quad Dx = 0 \quad D^2x = 1 \quad D^3x = 0$$

$$x = 1 \quad Dx = 1 \quad D^2x = 0 \quad D^3x = 1$$

looking at the table of contents of the registers a, b, c regarded as Dx, D^2x and D^3x.

The polynomial involved, $D^3 \oplus D^2 \oplus 1$, is known as the "generating polynomial". To generate an M-sequence, this polynomial must be (a) irreducible, that is it must have no factors; for example

$$D^4 \oplus D^3 \oplus D^2 \oplus 1 = (D \oplus 1)(D^3 \oplus D \oplus 1) \text{ is } \underline{\text{not}} \text{ irreducible;}$$

(b) primitive - that is it must not divide exactly into any polynomial of the form $D^n \oplus 1$ for any n less than $2^k - 1$ where k is the degree of the original polynomial. For example

$$D^4 \oplus D^3 \oplus D^2 \oplus D \oplus 1 = \frac{D^5 \oplus 1}{D \oplus 1} \text{ and } 5 < 2^4 \ominus 1$$

$$\text{so } D^4 \oplus D^3 \oplus D^2 \oplus D \oplus 1 \text{ is } \underline{\text{not}} \text{ primitive.}$$

Primitive polynomials up to k = 34 are given in W.W. Peterson "Error correcting codes" (Wiley, 1965); S.W. Golomb "Shift regular sequences" (Holden-Day, 1967) shows that $D^{127} \oplus D \oplus 1$ is a suitable generating polynomial: it generates a sequence of length $(2^{127} - 1)\, h \simeq 10^{37}\, h$! For example, if the clock frequency is 10^6 pulses a second the sequence will repeat itself after 3×10^{24} years!

We note that all irreducible generating polynomials of degree k divide $D^{2^k-1} + 1$ which means the period must be a factor of $2^k - 1$. If $2^k - 1$ is prime the period of the irreducible polynomial must be $2^k - 1$ and so it is also primitive. Such prime numbers are called Mersenne primes (Mersenne, 1644). $2^k - 1$ is known to be prime for k = 1, 2, 3, 5, 7, 13, 17, 19, 31, 61, 89, 107, 127 and 11213. Given a Mersenne prime $p = 2^k - 1$, then $D^k \oplus D \oplus 1$ is a primitive polynomial.

BIBLIOGRAPHY

KUO, B.C., Automatic Control Systems, Prentice-Hall (1962).

LEFSCHETZ, S., La SALLE, J.P., Stability by Liapunov's Direct Method with Applications, Academic Press (1961).

BROCKETT, R.W., Finite-dimensional Linear Systems, Wiley (1970).

OGATA, K., State Space Analysis of Control Systems, Prentice-Hall (1967).

BARNETT, S., STOREY, C., Matrix Methods in Stability Theory, Nelson (1970).

WILLEMS, J.L., Stability Theory of Dynamical Systems, Nelson (1970).

La SALLE, J.P., LEFSCHETZ, S., Stability by Liapunov's Direct Method with Applications, Academic Press (1961).

CESARI, L., Asymptotic Behaviour and Stability Problems in Ordinary Differential Equations, Springer-Verlag (1959).

PORTER, B., Stability Criteria for Linear Dynamical Systems, Oliver and Boyd (1967).

MINORSKI, N., Non-linear oscillations, Van Nostrand (1962).

WANG, P.K.C., "Control of distributed parameter systems", Advances in Control Systems (LEONDES, C.T., Ed.), Academic Press (1964).

BUTKOVSKII, A.G., Distributed Control Systems, American Elsevier (1969).

BUTKOVSKII, A.G., Control of Systems with Distributed Parameters, Nauka, Moscow (1975).

LIONS, J.L., Optimal Control of Systems Governed by Partial Differential Equations, Springer-Verlag (1971).

WANG, P.K.C., Theory of stability and control for distributed parameter systems, Int. J. Control 7 (1968) 101.

HAMZA, M.H. (Ed.), Proc. IFAC Conference on Control of Distributed Parameter Systems, Banff, Canada, June 1971, Vols 1 and 2.

PARKS, P.C., "On how to shake a piece of string to a standstill", Recent Mathematical Developments in Control (BELL, D.J. Ed.), Academic Press (1973) 267.

JAMES, H.M., NICHOLS, N.B., PHILLIPS, R.S., "Theory of Servomechanisms", McGraw Hill (1947).

ÅSTRÖM, K., "Introduction to Stochastic Control Theory", Academic Press (1970).

HOFFMANN de VISME, G., Binary Sequences, English Universities Press (1971).

PETERSON, W.W., Error Correcting Codes, Wiley (1961).

GOLOMB, S.W., Shift Register Sequences, Holden-Day (1967).

FOUNDATIONS OF FUNCTIONAL ANALYSIS THEORY

Ruth F. CURTAIN
Control Theory Centre,
University of Warwick,
Coventry, United Kingdom

Abstract

FOUNDATIONS OF FUNCTIONAL ANALYSIS THEORY.

This paper provides the basic analytical background for applications in optimization, stability and control theory. Proofs are generally omitted. The topics of the sections are the following: 1. Normed linear spaces; 2. Metric spaces; 3. Measure theory and Lebesgue integration; 4. Hilbert spaces; 5. Linear functionals, weak convergence and weak compactness; 6. Linear operators; 7. Spectral theory; 8. Probability measures; 9. Calculus in Banach spaces.

Some mathematical symbols

$\exists$	there exists	$A \subset B$	A is contained in B
$\in$	in, belongs to	$A \supset B$	A contains B
$\forall$	for all	E'	complement of E'
$\ni$	such that	$f \vee g$	minimum of f and g
$\mathbb{R}$	real numbers	$f \wedge g$	maximum of f and g
$\mathbb{C}$	complex numbers	$\perp$	orthogonal to
$\Rightarrow$	implies	$\overline{A}$	closure of A
iff	if and only if		

1. NORMED LINEAR SPACES

Definition 1. Linear vector space

A linear vector space is a set $\mathscr{V} = \{x, y, z, \ldots\}$ of elements with an operation $\oplus$ between any two elements such that

1. $x \oplus y = y \oplus x$ commutative property
2. $\exists\, e \ni x \oplus e = x, \ \forall x \in \mathscr{V}$ existence of the identity
3. $\exists\, -x \ni x \oplus -x = e$ existence of an inverse
4. $x \oplus (y \oplus z) = (x \oplus y) \oplus z$ associative property

(i.e. $\mathscr{V}$ is a commutative group under the operation $\oplus$) and there is an associated scalar multiplication by the real numbers $\mathbb{R}$ or the complex numbers $\mathbb{C}$ such that αx is an element of $\mathscr{V}$ for $x \in \mathscr{V}$ and

$$\left.\begin{array}{ll} 1. & \alpha(x \oplus y) = \alpha x \oplus \alpha y \\ 2. & (\alpha + \beta) x = \alpha x \oplus \beta x \\ 3. & (\alpha\beta) x = \alpha(\beta x) \\ 4. & 1 \cdot x = x \end{array}\right\} \quad \text{where } \alpha, \beta \text{ are scalars}$$

This concept is best understood from some examples:

Example 1. Take $\mathscr{V} = \mathbb{R}$ under addition and $\oplus$ as ordinary multiplication.

Example 2. Take $\mathscr{V}$ being a set of all polynomials of degree n with real coefficients and scalar multiplication by $\mathbb{R}$ — this is a real vector space, but if we consider complex coefficients and multiplication by $\mathbb{C}$, it is a complex vector space.

Example 3. $\mathbb{R}^n$ = set of all n-tuples, $\underset{\sim}{x} = \{x_1, x_2, \ldots, x_n\}$

with $\underset{\sim}{x} + \underset{\sim}{y} = \{x_1 + y_1, x_2 + y_2, \ldots, x_n + y_n\}$

and $\alpha \underset{\sim}{x} = \{\alpha x_1, \alpha x_2, \ldots, \alpha x_n\}$

Check that $\mathbb{R}^n$ is a vector space under scalar multiplication by $\mathbb{R}$. What if we take scalar multiplication by $\mathbb{C}$?

Example 4. $\mathscr{V}$ = set of all $m \times n$ matrices with real entries and scalar multiplication by $\mathbb{R}$.

Example 5. $\mathscr{V}$ = set of all scalar-valued functions $f: S \to \mathbb{R}$, where S is any non-empty set. For any $s \in S$, $f(s) \in \mathbb{R}$ we define

$$(f+g)(s) = f(s) + g(s), \text{ for all } s \in S$$

$$(\alpha f)(s) = \alpha f(s), \text{ for all } s \in S, \alpha \in \mathbb{R}.$$

Example 6. $\mathscr{V}$ = set of real-valued functions $f: [0,1] \to \mathbb{R}$ such that

$$\int_0^1 |f(s)|^2 \, ds < \infty$$

Define addition and scalar multiplication as in 5. We must verify that if $f, g \in \mathscr{V}$, then $f + g$ and $\alpha f \in \mathscr{V}$, i.e. $\int_0^1 |f(s) + g(s)|^2 \, ds < \infty$ and $\int_0^1 |\alpha f(s)|^2 ds < \infty$.
The last inequality is trivial and for the first we have

$$\int_0^1 |f(s)+g(s)|^2 ds \leq \int_0^1 [|f(s)|^2 + 2|f(s)||g(s)| + |g(s)|^2]ds$$

$$\leq \int_0^1 |f(s)|^2 ds + 2\left(\int_0^1 |f(s)|^2 ds \int_0^1 |g(s)|^2 ds\right)^{1/2} + \int_0^1 |g(s)|^2 ds$$

$$< \infty$$

where we have used Schwarz's inequality

$$\int_0^1 |f(s)\, g(s)|\, ds \leq \left(\int_0^1 |f(s)|^2 ds\right)^{1/2} \left(\int_0^1 |g(s)|^2 ds\right)^{1/2}$$

provided both sides exist.

This last example is a linear subspace of example 5 with $S = [0,1]$.

Definition 2. Linear subspace

If $\mathscr{V}$ is a linear vector space, then a subset S of $\mathscr{V}$ is a linear subspace if $x, y \in S \Rightarrow \alpha x + \beta y \in S$, i.e. S is closed under addition and scalar multiplication.

Other examples of linear subspaces are

Example 7. In example 3, let S be the set of n-tuples of the form $\underset{\sim}{x} = \{x_1, x_2, 0, \ldots, 0\}$

Example 8. In example 4, let S be the set of matrices with certain blocks zero.

Example 9. In example 2, let S be the set of all rth-order polynomials, where $r < n$.

Linear subspaces have the special property that they contain the zero element. A 'translated' subspace is given by

Definition 3. Affine subset

If $\mathscr{V}$ is a linear vector space, then an affine subset has the form

$M = \{x: x = c + x_0,$ where $x_0 \in S$ and c is fixed$\}$ for some $c \in \mathscr{V}$ and some linear subspace S of $\mathscr{V}$.

Example 7a. In example 3, let M be the set of n-tuples of the form $\underset{\sim}{x} = \{x_1, x_2, 1, \ldots, 1\}$.

Example 8a. In example 4, let M be the set of matrices with certain blocks of 1's.

Another very important type of subset of a vector space is a convex set.

Definition 4.

A subset A of $\mathscr{V}$ is convex, if $x, y \in A$ implies that $\lambda x + (1-\lambda) y \in A$ for all $\lambda \geqslant 0 \leq \lambda \leq 1$.

We now introduce the concept of the dimension of a vector space.

Definition 5. If $x_1, .., x_n \in \mathscr{V}$ and there are scalars $\alpha_1, .., \alpha_n$ not all zero such that $\alpha_1 x_1 + \alpha_2 x_2 + ... + \alpha_n x_n = 0$ then we say this is a linearly dependent set. If no such scalars $\alpha_1, .., \alpha_n$ exist, then $x_1, x_2, ..x_n$ is a linearly independent set.

For example, $1, x, x^2, .., x^n$ is a linearly independent set of nth-order polynomials, as is $1+x$, $\frac{1}{2}+3x$; however, $1+x$, $\frac{1}{2}+3x$, $2x$ are linearly dependent.

Definition 6. If $x_1, .., x_n$ is a linearly independent set in $\mathscr{V}$, then we say that $S = \mathrm{Sp}\{x_1, .., x_n\}$, the set of all linear combinations of $x_1, .., x_n$, has dimension n. If $\mathscr{V} = \mathrm{Sp}\{x_1, ..., x_k\}$ for some finite set of linearly independent elements, then $\mathscr{V}$ is of dimension k. If there exists no such set, then $\mathscr{V}$ is infinite-dimensional.

For example, the dimension of example 1 is 1, example 2 is $(n+1)$, example 3 is n, example 4 is mn and examples 5 and 6 are infinite-dimensional.

Definition 7. If $\mathscr{V} = \mathrm{Sp}\{x_1, .., x_n\}$, the set $\{x_1, ..., x_n\}$ is called a (Hamel) basis for $\mathscr{V}$. This basis is not unique, although the dimension of $\mathscr{V}$ is unique.

Example 10. A basis for example 2 is $\{1, x, ..., x^n\}$ or, equivalently,

$$\left\{1, x, \frac{3x^2-1}{2}, \;...,\; \frac{d^n(x^2-1)^n}{dx^n}\right\}$$

the Legendre polynomials.

A useful fact to remember is that all real vector spaces of dimension n are algebraically identical (or isomorphic to $\mathbb{R}^n$).

Definition 8. Vector spaces $\mathscr{V}$ and $\mathscr{W}$ are isomorphic if there is a bijective linear map $T: \mathscr{V} \to \mathscr{W}$, such that $T(\alpha x + \beta y) = \alpha Tx + \beta Ty$ for all $x, y \in \mathscr{V}$ and α, β scalars.

Another example of isomorphic spaces are all n-dimensional vector spaces over the complex numbers, which are all isomorphic to $\mathbb{C}^n$, the space of complex n-tuples.

Definition 9. A hyperplane, H, of a vector space $\mathscr{V}$ is a maximal proper affine subset of $\mathscr{V}$, i.e. the complement of H has dimension 1. For example, hyperplanes in $\mathbb{R}^n$ have dimension $(n-1)$ (i.e. $\mathscr{V} = H + S$, where H and S are disjoint and S has dimension 1).

Hyperplanes in $\mathbb{R}^n$ have the dimension $(n-1)$.

So far we have only considered algebraic properties of sets. In order to develop mathematical concepts for "nearness" or distance, we need some topology, namely metric spaces.

Definition 10. A metric space $\underline{X}$ is a set of elements $\{x, y, ...\}$ and a distance function $d(x, y)$ with the following properties:

1. $d(x, y) \geq 0$ for all $x, y \in \underline{X}$
2. $d(x, y) = 0$ if and only if $x = y$
3. $d(x, y) = d(y, x)$
4. $d(x, y) \leq d(x, z) + d(z, y)$ for all $x, y, z \in \underline{X}$

We call $d(., .)$ a metric on $\underline{X}$.
This is essentially a generalization of distance in the Euclidean plane.

Example 11. Let $\underline{X}$ be the set of 2-tuples $\underset{\sim}{x} = \{x_1, x_2\}$ or Cartesian co-ordinates and $d(\underset{\sim}{x}, \underset{\sim}{y}) = [(x_1 - y_1)^2 + (x_2 - y_2)^2]^{1/2}$. Then d satisfies all the properties 1 - 4 and property 4 is the familiar triangular inequality.

Example 12. Let $\underline{X}$ be as in ex. 11, but $d(\underset{\sim}{x}, \underset{\sim}{y}) = |x_1 - y_1| + |x_2 - y_2|$ or more generally, $d_p(\underset{\sim}{x}, \underset{\sim}{y}) = (|x_1 - y_1|^P + |x_2 - y_2|^P)^{1/P}$, $1 \leq p < \infty$, and

$$d'(\underset{\sim}{x}, \underset{\sim}{y}) = \max\{|x_1 - y_1|, |x_2 - y_2|\}.$$

So the same set $\underline{X}$ can generate different metric spaces.

Example 13. $\underline{X} = \mathscr{C}[a, b]$ the set of continuous functions on $[a, b]$ and

$$d(x, y) = \max\{|x(t) - y(t)|; \ a \leq t \leq b\}$$

or

$$d_p(x, y) = \left(\int_a^b |x(t) - y(t)|^P dt\right)^{1/P}; \qquad p \geq 1.$$

We sometimes use a pseudometric which satisfies conditions 1, 3, 4 of Definition 6, but instead of 3, we have only $d(x, x) = 0$. $d(x, y) = 0$ does not necessarily imply that $x = y$.

So far we have introduced the algebraic structure of a linear vector space which enables us to consider linear combinations of elements and then the topological concept of a metric space which enables us to measure nearness or distance and hence to consider the tools of analysis, such as open sets, closed sets, convergence and continuity. We now combine these two notions in a normed linear space.

Definition 11. A normed linear space $\underline{X}$ is a linear vector space with a norm on each element, i.e. to each $x \in \underline{X}$ corresponds a positive number $\|x\|$, such that

1. $\|x\| = 0$ iff $x = 0$
2. $\|\alpha x\| = |\alpha| \, \|x\|$, for all scalars α
3. $\|x + y\| \leq \|x\| + \|y\|$

If 1 is not necessarily true, we call it a seminorm. We note that if we define $d(x, y) = \|x - y\|$, then d is a metric on $\underline{X}$.

Example 14. Consider $\mathbb{R}^n$ again. This is already a vector space (see Ex.3). We can define several norms on $\mathbb{R}^n$ by

$$\|\underset{\sim}{x}\|_p = \left(\sum_{i=1}^{n} |x_i|^p\right)^{1/p}$$

where $1 \leq p < \infty$, p fixed, the so-called 'p-norm'. We denote the normed linear space with the p-norm by ℓ_p^n and

$$\ell_\infty^n = \left\{\underset{\sim}{x} \in \mathbb{R}^n \triangleright \|\underset{\sim}{x}\|_\infty = \max_{1\leq i\leq n} \{|x_i|\}\right\}$$

Example 15. Consider the space of infinite sequences, $\underset{\sim}{x} = \{x_1, x_2, \ldots\}$. Then this forms a vector space in a similar manner to $\mathbb{R}^n$ and we can again define a p-norm:

$$\|\underset{\sim}{x}\|_p = \left(\sum_{i=0}^{\infty} |x_i|^p\right)^{1/p} \quad \text{for } 1 \leq p < \infty.$$

Note that $\|\underset{\sim}{x}\|_p$ is not finite for all infinite sequences, so to define a normed linear subspace we take the subset with finite p-norm, i.e.

$$\ell_p = \{\infty\text{-tuples } \underset{\sim}{x} = \{x_1, x_2, \ldots\}, \text{ with } \|\underset{\sim}{x}\|_p < \infty\}.$$

This contrasts with the ℓ_p^n normed linear spaces where the sets of elements for each p are identical, although the norm is different. Now for $\|\cdot\|p$ the set of elements for each p are different. The actual proofs that $\|\cdot\|p$ is a norm rely on

Minkowski's inequality

$$\left(\sum_{i=1}^{n} |x_i + y_i|^p\right)^{1/p} \leq \left(\sum_{i=1}^{n} |x_i|^p\right)^{1/p} + \left(\sum_{i=1}^{n} |y_i|^p\right)^{1/p}$$

which holds for n finite or infinite.

Finally

$$\ell_\infty = \left\{\infty\text{-tuples } \underset{\sim}{x}, \text{ with } \|\underset{\sim}{x}\|_\infty = \sup_{1\leq i\leq\infty} \{|x_i|\} < \infty\right\}$$

We have the following inequalities:

$$\|\underset{\sim}{x}\|_1 \geq \|\underset{\sim}{x}\|_2 \geq \ldots \geq \|\underset{\sim}{x}\|_\infty$$

and so

$$\ell_1 \subset \ell_2 \subset \ldots \subset \ell_\infty.$$

Example 16. $\underline{X} = \mathcal{C}[a, b]$, the space of real continuous functions on $[a, b]$ with norm

$$\|x(\cdot)\| = \max_{a\leq t\leq b} |x(t)|$$

This is called the uniform or sup norm

Example 17. $\underline{X} = \mathcal{C}[a,b]$ under the p-norm

$$\|x(\cdot)\|_p = \left(\int_a^b |x(t)|^p \, dt\right)^{1/p}$$

That this satisfies the properties of a norm depends on the integral form of the Minkowski inequality:

$$\left(\int_a^b |x(t)+y(t)|^p \, dt\right)^{1/p} \leq \left(\int_a^b |x(t)|^p \, dt\right)^{1/p} + \left(\int_a^b |y(t)|^p \, dt\right)^{1/p}$$

Equality holds only if $x(t) = ky(t)$ almost everywhere on $[a, b]$.

We note that although $\underline{X} = \mathcal{C}[a,b]$ is the same linear vector space in Exs 16 and 17 by defining two different norms, we obtain two distinct normed linear spaces.

2. METRIC SPACES

We return to study the properties of metric spaces in more detail.

There are two common ways of creating new metric spaces from known ones — subspaces and product spaces.

Definition 12. Let $(\underline{X}, d)$ be a metric space and A a subset of $\underline{X}$, then we can consider (A, d) as a metric space in its own right. Then (A, d) is a subspace of $(\underline{X}, d)$.

There are, in general, several metrics we can put on A, but (A, d_1) is a subspace of $(\underline{X}, d)$ only when the metrics d_1 and d coincide on A.

Definition 13. Let $(\underline{X}, d_x)$, $(\underline{Y}, d_y)$ by two metric spaces, then the product set of ordered pairs $\underline{X} \times \underline{Y} = \{(x,y) : x \in \underline{X},\ y \in \underline{Y}\}$ may be defined to be a metric space, the product space, in several ways.

1. $d(u_1, u_2) = d_x(x_1, x_2) + d_y(y_1, y_2)$, where $u_i = (x_i, y_i)$, $i = 1, 2$.
2. $d_2(u_1, u_2) = (d_x^2(x_1, x_2) + d_y^2(y_1, y_2))^{1/2}$
3. $d_p(u_1, u_2) = (d_x^p(x_1, x_2) + d_y^p(y_1, y_2))^{1/p}$; $1 \leq p < \infty$
4. $d_\infty(u_1, u_2) = \max\{d_x(x_1, x_2), d_y(y_1, y_2)\}$

Under any of these metrics $(\underline{X} \times \underline{Y}, d)$ is called the product space of $(\underline{X}, d_x)$ and $(\underline{Y}, d_y)$. (There are infinitely many choices of metrics for $\underline{X} \times \underline{Y}$).

You can verify that if $\underline{X}$ and $\underline{Y}$ are normed linear spaces, then $(\underline{X}, \|\cdot\|_x) \times (\underline{Y}, \|\cdot\|_y)$ becomes a normed linear space using any of the product metrics derived from $\|\cdot\|_x$ and $\|\cdot\|_y$.

Continuity in metric spaces

The introduction of the distance function $d(.,.)$ allows us to generalize the definition of continuous functions on metric spaces.

Definition 14. Let $f: \underline{X} \to \underline{Y}$ be a map from the metric space $(\underline{X}, d_x)$ to the metric space $(\underline{Y}, d_y)$. f is continuous at x_0 in $\underline{X}$ if given $\epsilon > 0$, $\exists$ a real number $\delta > 0$, such that $d_y(f(x), f(y)) < \epsilon$, whenever $d_x(x, x_0) < \delta$. f is continuous if it is continuous at each point in its domain.

Definition 15. A map $f: \underline{X} \to \underline{Y}$ is uniformly continuous if for each $\epsilon > 0$, $\exists\, \delta = \delta(\epsilon) > 0$, such that for any x_0, $d_y(f(x_0), f(x)) < \epsilon$, whenever $d_x(x, x_0) < \delta$.

Example 18. $\underline{X} = \mathbb{R}^n$ under $d_2(\underset{\sim}{u}, \underset{\sim}{v}) = \left(\sum_{i=1}^{n} |u_i - v_i|^2\right)^{1/2}$

$$\underline{Y} = \mathbb{R}^m \text{ under } d_2(\underset{\sim}{w}, \underset{\sim}{z}) = \left(\sum_{i=1}^{m} |w_i - z_i|^2\right)^{1/2}$$

and $f: \underline{X} \to \underline{Y}$ is represented by the matrix $F = (f_{ij})$, i.e. $y = Fx$.

Let x_0 be fixed in $\underline{X}$ and $y_0 = Fx_0$.

Let x be an arbitrary point in $\underline{X}$ and $y = Fx$, then

$$d(y, y)^2 = \sum_{i=1}^{m} \left| \sum_{j=1}^{n} f_{ij}(x_j - x_{0j}) \right|^2$$

$$\leq \sum_{i=1}^{m} \left(\sum_{j=1}^{n} |f_{ij}|^2 \right) \left(\sum_{j=1}^{n} |x_j - x_{0j}|^2 \right)$$

by the Schwarz inequality

$$\leq c^2 d(x, x_0)^2, \text{ where } c^2 = \left(\sum_{i,j} |f_{ij}|^2 \right)$$

So if we are given $\epsilon > 0$ we may choose $\delta = \epsilon / c$, provided $c \neq 0$. So the map f is uniformly continuous.

Example 19. $\underline{X} = \underline{Y} =$ space of integrable functions on $[0, T]$ with the metric

$$d(x, y) = \left(\int_0^T [x(t) - y(t)]^2 dt \right)^{1/2}$$

Define

$$f: \underline{X} \to \underline{X} \quad \text{by} \quad fx = y, \quad \text{where } y(t) = \int_0^t x(s)\, ds.$$

Then

$$|y(t) - y(t_0)| = \left| \int_0^t x(s) - x_0(s)\, ds \right|$$

$$\leq \left(\int_0^t 1^2\,ds\right)^{1/2}\left(\int_0^T |x(s) - x_0(s)|^2\,ds\right)^{1/2}$$

$$\leq \sqrt{T}\, d(x, x_0)$$

$$\therefore\ d(y, y_0) = \left(\int_0^T |y(t) - y_0(t)|^2\,dt\right)^{1/2}$$

$$\leq T d(x, x_0)$$

so f is uniformly continuous.

If, however, we consider the interval $(-\infty, \infty)$, the map $f:\underline{X}\to\underline{X}$ is not continuous, where $fx = y$ is given by $y(t) = \int_{-\infty}^{t} x(s)\,ds$.

For let $x_0 \in \underline{X}$ be fixed and see k an $\epsilon > 0$, such that there is an $x \in \underline{X}$, such that $d(x, x_0) < \delta$ and $d(f(x), f(y)) \geq \epsilon$, for any choice of δ.

Let $y_0 = f(x_0)$, $y = f(x)$.

Then $y(t) - y_0(t) = \int_{-\infty}^{t} (x(s) - x_0(s))ds$.

Choose x such that $x(t) - x_0(t) = \begin{cases} c, & 0 \leq t \leq 3T^2 \\ -c, & 3T^2 < t \leq 6T^2 \\ 0 & \text{otherwise} \end{cases}$

$$\therefore\ d(x, x_0) = \sqrt{6}\, Tc$$

and

$$y(t) - y_0(t) = \begin{cases} ct, & 0 \leq t \leq 3T^2 \\ c(6T^2 - t), & 3T^2 < t \leq 6T^2 \\ 0 & \text{otherwise} \end{cases}$$

and

$$d(y, y_0) = \sqrt{18}\, T^3 c.$$

Let $\delta > 0$ be given and choose c, T, such that $\sqrt{18}\,T^3 c = 1$, $\sqrt{6}\, Tc < \delta$, i.e. $d(x, x_0) < \delta$ and yet $d(y, y_0) \geq 1$, for this particular x. So f is not continuous at x_0.

A fundamental concept in analysis is of course convergence and we now define convergence in metric spaces.

Definition 16. A sequence $\{x_n\} \subset$ metric space $(\underline{X}, d)$ converges to x_0 in $(\underline{X}, d)$ if $d(x_n, x_0) \to 0$ as $n \to \infty$.

Continuity and convergence are related concepts as is clearly seen from the following result:

Let $f:(\underline{X}, d_x) \to (\underline{Y}, d_y)$ be a map between two metric spaces and x_0 a given point in $\underline{X}$. Then the following two statements are equivalent:

(a) f is continuous at x_0.

(b) $\lim_{n\to\infty} f(x_n) = f(\lim_{n\to\infty} x_n)$, for every convergent sequence $x_n \to x_0$, i.e. a map is continuous iff it preserves convergent sequences.

We now define Cauchy sequences.

Definition 17. A sequence $\{x_n\}$ of elements in a metric space $(\underline{X}, d)$ is a Cauchy sequence if $d(x_n, x_m) \to 0$ as $m, n \to \infty$.

Metric spaces have the property that Cauchy sequences can have at most one limit, but they need not have a limit point in the metric space, as is seen from the following example.

Example 20. Consider $\mathscr{C}[0,1]$ under the 2-norm and take the sequence $\{x_n\}$ where we define

$$x_n(t) = \begin{cases} 0 & \text{for} \quad 0 \leq t \leq 1/2 - 1/n \\ (n/2)t - n/4 + 1/2 & 1/2 - 1/n \leq t \leq 1/2 + 1/n \\ 1 & 1/2 + 1/n \leq t \leq 1 \end{cases}$$

Graphically this looks like

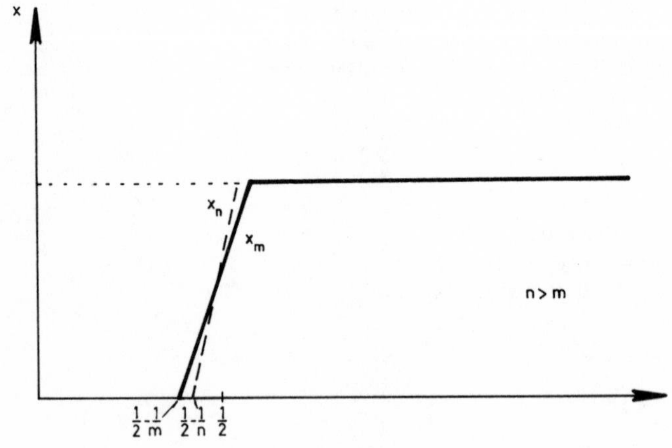

Now

$$\|x_m - x_n\|_2 = \left(\int_0^1 |x_m(t) - x_n(t)|^2 dt\right)^{1/2}$$

$$= \left(\int_{1/2-1/m}^{1/2} |x_m - x_n|^2\, dt + \int_{1/2}^{1/2+1/m} |x_m - x_n|^2\, dt\right)^{1/2}$$

$= \sqrt{2}$ (difference in the area of triangles).

$= \sqrt{2}\, |1/4n - 1/4m|$

$\to 0$ as $m, n \to \infty$.

so $\{x_n\}$ is a Cauchy sequence under the 2-norm.

We easily see that the pointwise limit of x_n is

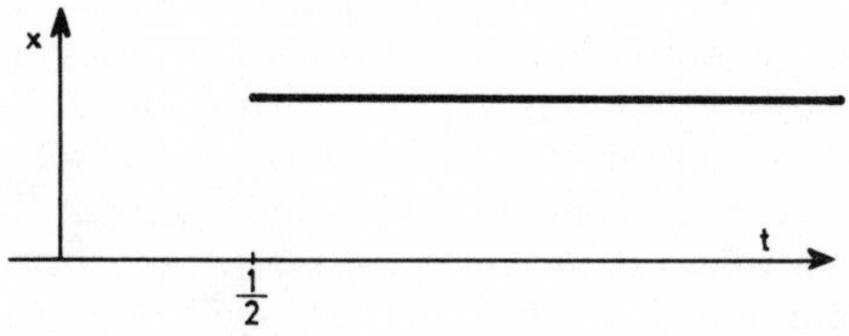

However, this function does not belong to $\mathcal{C}[0,1]$ because of its discontinuity at 1/2. It is rather awkward to use spaces which have Cauchy sequences whose limits do not belong to the space and so we define a class of metric spaces which always contain limit points of Cauchy sequences.

Definition 18. A metric space $(\underline{X}, d)$ is said to be complete if each Cauchy sequence in $(\underline{X}, d)$ is a convergent sequence in $(\underline{X}, d)$.

This concept of completeness is so important that it gives use to the definition of new types of spaces, as we shall see from the examples to follow.

First let us state two major results about complete metric spaces:

1. If $(\underline{X}, d)$ is a complete metric space and $(\underline{Y}, d)$ is a subspace of $(\underline{X}, d)$, $(\underline{Y}, d)$ is complete iff $\underline{Y}$ is a closed set in $(\underline{X}, d)$.

Definition 19. A set A in $\underline{X}$ is closed if all convergent sequences in A have their limits also in A.

2. Every metric space has a completion which is unique up to an isometry.

This last property allows us to define an important concept in normed linear spaces.

Definition 20. A Banach space is a complete normed linear space (B-space).

Some examples of Banach spaces are $\mathcal{C}[a,b]$ under the sup norm (since uniformly convergent sequences of continuous functions have continuous limit functions), ℓ_p^n, ℓ_∞^n, ℓ_p and ℓ_∞.

However, from Ex.20 we see that $\mathcal{C}[a,b]$ is not complete under the 2-norm. This is an important norm used in analysis for defining 'mean square convergence of functions':

$$\int_a^b (f_n(t) - f(t))^2\, dt \to 0, \text{ as } n \to \infty$$

and so we would like to identify what space we get when we complete $\mathcal{C}[a,b]$ under the p-norm.

At first guess, one would look at all Riemann integrable functions with finite p-norm, i.e. $x(\cdot)$ such that

$$\left(\int_a^b x(t)^p \, dt \right)^{1/p} < \infty$$

Unfortunately, a convergent sequence of Riemann integrable functions does not always converge to a Riemann integrable function. However, it happens that if you define integration in a more sophisticated way — called Lebesgue integration, you do get this property. We shall pursue this idea later on, but for the present we may consider Lebesgue integration as a generalization of Riemann integration and just think of Lebesgue integrable functions as Riemann integrable functions and some "mavericks". $L_p[a,b]$ denotes the space of Lebesgue integrable functions with finite p-norm, i.e.

$$\left(\int_a^b |x(t)|^p \, dt \right)^{1/p} < \infty$$

and it is a B-space. It may also be considered as the completion of $\mathcal{C}[a,b]$ under the p-norm.

The following examples of normed linear spaces are important in the theory of partial differential equations. Examples 21 and 23 are not complete, however, their completion under the $\|\cdot\|_{n,p}$ norm gives rise to the important Sobolev spaces, which are used in distribution theory for studying partial differential equations.

Example 21. Let $\mathcal{C}^\infty[a,b]$ be the space of infinitely differentiable functions on $[a,b]$. Then the following are well-defined norms

$$\|x\|_{n,p} = \left[\int_a^b \sum_{i=0}^{n} |D^i x(t)|^p \, dt \right]^{1/p}, \qquad 1 \leq p < \infty$$

where D^i denotes the ith derivative. So we can define infinitely many normed spaces on $\mathcal{C}^\infty[a,b]$. Since $\|x\|_{n+1,p} \geq \|x\|_{n,p}$, we have that $\mathcal{C}^\infty[a,b]$ under the $(n+1,p)$ norm $\subset \mathcal{C}^\infty[a,b]$ with the (n,p) norm.

Example 22. Consider $\mathcal{C}^k(\Omega)$ the space of real functions of n variables on Ω which are continuously differentiable up to order k. Ω is an open subset of $\mathbb{R}^n$. Let $\alpha = (\alpha_1, \ldots, \alpha_n)$ be a vector with positive integer entries and define $|\alpha| = \sum_{i=1}^{n} \alpha_i$. Then for $u \in \mathcal{C}^k(\Omega)$, the following derivative exists and is continuous:

$$D^\alpha u = \frac{\partial^{|\alpha|} u}{\partial x_1^{\alpha_1} \ldots \partial x_n^{\alpha_n}}$$

$\mathcal{C}^k(\Omega)$ is of course a linear vector space and we may take the sup norm or, alternatively, the norm

$$|||u||| = \max_{0 \leq \alpha \leq k} \left\{ \| D^\alpha u \| \right\}$$

where $\|D^\alpha u\|$ is the usual sup norm. $\mathcal{C}^k(\Omega)$ is a B-space under the $|||\cdot|||$ -norm.

Example 23. Let $\mathcal{C}^\infty(\Omega)$ be the space of infinitely differentiable functions on Ω, an open subset of $\mathbb{R}^n$. Let $\alpha = (\alpha_1 \ldots, \alpha_n)$ be as in Ex.22, and as before, define the space differential operator $D^\alpha u$. Then we may define the following norms

$$\|u\|_{k,p} = \left(\int_\Omega \sum_{|\alpha| \leq k} |D^\alpha u(x)|^p \, dx \right)^{1/p}$$

$\|\cdot\|_{k,p}$ is also a norm on $\mathcal{C}^k(\Omega)$.

Although the spaces in Exs 21 and 23 are not complete they do have the important property that they are dense in some underlying space, the relevant Sobolev space.

Definition 21. A linear subspace S of a metric space $\underline{X}$ is dense in $\underline{X}$ if the closure of S with respect to the metric $\supseteq \underline{X}$. This means that any element $x \in \underline{X}$ can be approximated by some element $s \in S$ as closely as we like, i.e. $d(s,x) < \epsilon$.

Example 24. Consider ℓ_2 and let

$$S = \{ \underset{\sim}{x} \in \ell_2 > \underset{\sim}{x} = \{x_1, x_2, \ldots, x_k, 0, \ldots 0 \ldots\}, \ k < \infty \}$$

Then S is a dense subspace of ℓ_2, but it is not a B-space itself, since the sequence $\{1, 0, \ldots\}$, $\{1, 1/2, 0, \ldots\}$, $\{1, 1/2, 1/2^2, 0, \ldots\}$, $\ldots \{1, 1/2, 1/2^2, \ldots, 1/2^n, 0, \ldots\}$ converges to $\{1, 1/2, 1/2^2, \ldots\}$ which is not in S.

Example 25. $L_2[0,1] = \overline{\mathcal{C}[0,1]}$ (closure under $\|\cdot\|_2$ norm) and

$$S = \{x(\cdot) \in L_2[0,1] \text{ with } x(0) = 0 \text{ and } \dot{x}(\cdot) \in L_2[0,1]\}.$$

This is not a closed subspace, since it is easy to construct sequences of functions in S whose derivatives converge to a function which is integrable, but its derivative is not. However, it can be shown that it is a dense subspace of $L_2[0,1]$.

If we take a new norm

$$|||x(\cdot)||| = \left[\int_0^1 \left(|x(t)|^2 + |\dot{x}(t)|^2 \right) dt \right]^{1/2}$$

for S, then S is a B-space with respect to this norm.

We also define the concept of connectedness.

Definition 22. A metric space $(\underline{X}, d)$ is disconnected if it is the union of two open, non-empty disjoint subsets. Otherwise $(\underline{X}, d)$ is said to be connected.

Example 26. $\underline{X} = [0,1] \cup [5,6]$ with the Euclidean metric is disconnected.

Contractions in metric spaces

This is an important concept which gives us the basic tool for proving existence and uniqueness of solutions of differential equations.

Definition 23. Let $(\underline{X}, d)$ be a metric space and $f: \underline{X} \to \underline{X}$. Then f is a contraction (mapping) if there is a real number K, $0 \leq K < 1$, such that

$$d(f(x), f(y)) \leq K d(x, y) \quad \forall x, y \in \underline{X}.$$

This implies that f is uniformly continuous.

Contraction mapping theorem

Let $(\underline{X}, d)$ be a complete metric space and $f: \underline{X} \to \underline{X}$ a contraction mapping, then there is a unique $x_0 \in \underline{X}$, such that $f(x_0) = x_0$. Here x_0 is called the fixed point of f.

Moreover, if x is any point in $\underline{X}$ and we define the sequence $\{x_n\}$ by

$$x_1 = f(x), \; x_2 = f(x_1), \ldots, x_n = f(x_{n-1}),$$

then $x_n \to x_0$ as $n \to \infty$.

Corollary

Let $(\underline{X}, d)$ be a complete metric space and $f: \underline{X} \to \underline{X}$ such that f^p is a contraction for some $p > 0$. Then f has a fixed point.

Example 27. Existence and uniqueness theorem for solutions of ordinary differential equations

Consider

$$\begin{cases} \dfrac{dy}{dt} = f(y, t) \\ \\ y(0) = 0 \end{cases} \tag{1}$$

where f is real-valued, continuous on $\mathbb{R} \times \mathbb{R}$.

Then (1) is equivalent to the solution of the integral equation

$$y(t) = \int_0^t f(y(s), s) ds$$

and this may be thought of as $z = F(y)$, where,

$$z(t) = \int_0^t f(y(s), s)\,ds$$

and $F : \mathscr{C} \to \mathscr{C}$, the space of real-valued continuous functions defined on $[0 - A,\ 0 + A]$.

Now $y(t)$ is a solution of (1) iff $y = Fy$, i.e. iff y is a fixed point of the map F.

We now show that under appropriate conditions on f, F is a contraction.

<u>Assumptions on f:</u> (a) $|f(y,t)| \leqq M$ for $-1 \leqq y \leqq 1, \quad -1 \leqq t \leqq 1$.

(b) Lipschitz condition $|f(y,t) - f(x,t)| \leqq K|y - x|$

for (y,t), (x,t) in $[-1,1] \times [-1,1]$.

Let $\underline{\overline{X}}$ = space of continuous real-valued functions $\phi(t)$, such that

$$|\phi(t)| \leqq M|t| \quad \text{on} \quad [-T,T], \quad 0 \leqq T \leqq 1.$$

$$MT \leqq 1, \quad KT < 1.$$

$\underline{\overline{X}}$ is a subspace of $\mathscr{C}[-T,T]$ and is closed under the sup metric d_∞, i.e. $(\underline{\overline{X}}, d_\infty)$ is a complete metric space.

(i) $F : \underline{\overline{X}} \to \underline{\overline{X}}$, since

$$|F(\phi)(t)| \leqq \int_0^{|t|} |f(\phi(s), s)|\,ds \leqq \int_0^{|t|} M\,ds = M|t|.$$

(ii) F is a contraction, since

for $t \geqq 0$ $$|F(x)(t) - F(y)(t)| = \left|\int_0^t [f(x(s), s) - f(y(s), s)]\,ds\right|$$

$$\leqq \int_0^t K|x(s) - y(s)|\,ds$$

$$\leqq K \int_0^t d_\infty(x,y)\,ds$$

$$\leqq Kt\,d_\infty(x,y).$$

for $t \leqq 0$, $|F(x)(t) - F(y)(t)| \leqq K|t|\,d_\infty(x,y)$

and so for

$$|t| \leq T, \quad d_\infty(F(x), F(y)) \leq KT\, d_\infty(x, y)$$

∴ F has a unique fixed point which is the solution of (1).

Compactness

Compactness is a very important concept in analysis, although it does not arise in finite-dimensional spaces as such. For finite-dimensional spaces all closed and bounded sets are compact. This is never true in infinite dimensions and we here define compactness for metric spaces.

Definition 24. A set A in a metric space $(\underline{X}, d)$ is compact if every sequence in A contains a convergent subsequence with a limit point in A.

We recall that in $\mathbb{R}^n$ every closed and bounded set contains a convergent subsequence.

For more general topological spaces there are several different kinds of compactness, the definition above corresponding to sequential compactness. However, for metric spaces all types of compactness are equivalent. Still we can define a weaker form of compactness, relative compactness.

Definition 25. A set $A \subset \underline{X}$ is relatively compact (or conditionally compact) if its closure $\bar{A}$ is compact.

This means that every sequence in A contains a convergent subsequence which converges to a point not necessarily in A.

We also note the following properties of compact sets:

1. If $(\underline{X}, d)$ is compact, it is a complete metric space.
2. A compact set is closed and bounded.
3. If A is compact, every infinite subset in A has at least one point of accumulation – the Bolzano-Weierstrass property. (This is also an equivalent definition of compactness.)
4. If $f: \underline{X} \to \underline{Y}$ is continuous and $A \subset \underline{X}$ is compact, f(A) is compact.
5. A continuous function $f: A \to \mathbb{R}$ achieves its minimum if A is compact.

Finally we shall state the Arzela-Ascoli theorem which will be used in the application sections.

Let $(\underline{X}, d_1)$ be a compact metric space and $(\underline{Y}, d_2)$ a complete metric space and $\mathscr{C}(\underline{X}, \underline{Y})$ the space of continuous functions on $\underline{X}$ with range in $\underline{Y}$.

$(\mathscr{C}(\underline{X}, \underline{Y}), \rho)$ is a metric space, where

$$\rho(f, g) = \sup\{d_2(f(x), g(x)),\ x \in \underline{X}\}$$

and in fact $(\mathscr{C}(\underline{X}, \underline{Y}), \rho)$ is a complete metric space.

Definition 26. $A \subset \mathscr{C}(\underline{X}, \underline{Y})$ is equicontinuous at $x_0 \in \underline{X}$. If given $\epsilon > 0$, $\exists \delta < 0$, such that $d_2(f(x) - f(x_0)) < \epsilon \ \forall f \in A$ whenever $d_1(x - x_0) < \delta$ and $x \in \underline{X}$.

A is equicontinuous on $\underline{X}$ if is is equicontinuous at all points in $\underline{X}$.

Arzela-Ascoli theorem

$A \subset \mathcal{C}(\underline{X}, \underline{Y})$ is relatively compact in $(\mathcal{C}(\underline{X}, \underline{Y}), \rho)$ iff

A is equicontinuous on $\underline{X}$ and for each $x \in \underline{X}$,

$A(x) = \{ f(x)$ where $f \in A\}$ is relatively compact.

Deeper topological notions in metric spaces

We have already mentioned that metric spaces are a special class of topological spaces and here we shall explore some of the fundamental topological concepts for metric spaces.

First we show how the metric space may be characterized in terms of neighbourhoods.

Definition 27. Let $(\underline{X}, d)$ be a metric space and x_0 an arbitrary point in $(\underline{X}, d)$. $B_r(x_0) = \{x \in \underline{X} : d(x, x_0) < r\}$,

$0 < r < \infty$ is the open ball of radius r centred at x_0.

$B_r[x_0] = \{x \in \underline{X};\ \ d(x, x_0) \leq r\}$ is the closed ball. ...

$Sr[x_0] = \{x \in \underline{X};\ \ d(x, x_0) = r\}$ is the sphere ...

Definition 28. Let x_0 be an arbitrary point in $(\underline{X}, d)$, then a subset N of $(\underline{X}, d)$ is a local neighbourhood of x_0 if $N = B_r(x_0)$ or $B_r[x_0]$ for some $r \neq 0$. $B_r(x_0)$ are called open local neighbourhoods and $B_r[x_0]$ closed local neighbourhoods.

The local neighbourhood system of $x_0 = \{B_r(x_0),\ B_r[x_0]\}$, $r > 0$

$= N(x_0)$.

It is now possible to redefine continuity in terms of local neighbourhood systems using the following theorem:

Theorem

A function $f : (\underline{X}, d_x) \to (\underline{Y}, d_y)$ is continuous at x_0 in $(\underline{X}, d_x)$ iff its inverse image of any local neighbourhood of $f(x_0)$ contains a local neighbourhood of x_0.

Similarly we can redefine convergence in terms of local neighbourhood systems.

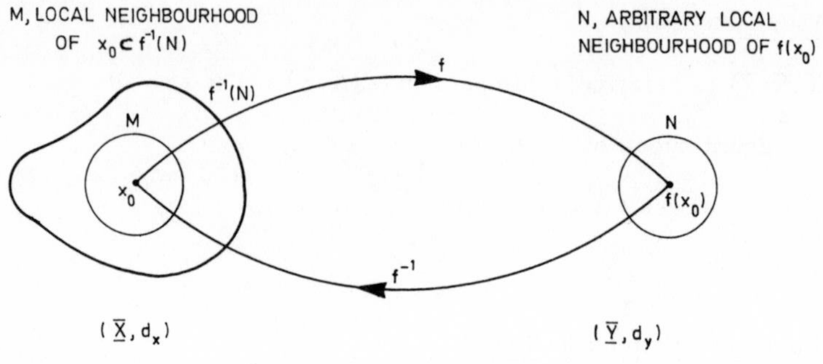

Theorem

A sequence $\{x_n\}$ in a metric space $(\overline{\underline{X}}, d)$ converges to x_0 iff $x_n \in N$, a local neighbourhood of x_0, for all $n \geq m(N)$, a number depending on N. To some extent our continuity and convergence may be defined independently of the metric, but we can even characterize continuity and convergence in terms of open sets. Topological spaces are defined in terms of open sets and this family is called "the topology" on the space. The topology or open sets of metric space are always definable in terms of the metric, but often several different metrics generate the same topology. In fact, equivalent metric spaces need not have the same local neighbourhood systems.

Definition 29. Let $(\overline{\underline{X}}, d_1)$, $(\overline{\underline{X}}, d_2)$ be two metric spaces with the same underlying set. Then the metrics d_1, d_2 are equivalent if

(a) $f: (\overline{\underline{X}}, d_1) \to (\overline{\underline{Y}}, d_y)$, an arbitrary metric space, is continuous iff $f: (\overline{\underline{X}}, d_2) \to (\overline{\underline{Y}}, d_y)$ is continuous, and
(b) a sequence $\{x_n\}$ converges to x_0 in $(\overline{\underline{X}}, d_1)$ iff $\{x_n\}$ converges to x_0 in $(\overline{\underline{X}}, d_2)$.

In fact, either of (a) or (b) ensures the equivalence of d_1 and d_2 or even the condition:

(c) $I: (\overline{\underline{X}}, d_1) \to (\overline{\underline{X}}, d_2)$ and $I^{-1}: (\overline{\underline{X}}, d_2) \to (\overline{\underline{X}}, d_1)$ are continuous (I is the identity map).

A more fundamental way of expressing equivalence of metric spaces is in terms of open sets.

Definition 30. A set A in a metric space $(\overline{\underline{X}}, d)$ is open if A contains a local neighbourhood of each one of its points. Note that $\emptyset$ and $\overline{\underline{X}}$ are always open sets.

Definition 31. The class of all open sets in $(\overline{\underline{X}}, d)$ is referred to as the topology (generated by the metric d) and is denoted by $\mathcal{J}$. We now state the basic result on equivalence of metric spaces.

Theorem

Let $(\underline{X}, d_1)$ and $(\underline{X}, d_2)$ be two metric spaces with the same underlying set $\underline{X}$. Then d_1 and d_2 are equivalent iff $\mathcal{I}_1 = \mathcal{I}_2$, i.e. iff they generate the same class of open sets.

We now restate the definitions of continuity and convergence in terms of open sets.

Definition 32. A map $f: (\underline{X}, d_1), (\underline{Y}, d_2)$ is continuous if the inverse image of each open set in $(\underline{Y}, d_2)$ is an open set in $(\underline{X}, d_1)$.

Definition 33. A sequence $\{x_n\}$ in a metric space $(\underline{X}, d)$ converges to x_0 in $\underline{X}$ iff x_n is in every open set containing x_0 for sufficiently large n.

These are the usual definitions one uses for more general topological spaces.

Previously we defined closed sets as those containing all their limit points, but this property may also be defined in terms of open sets.

Definition 34. Let $(\underline{X}, d)$ be a metric space. A subset $A \subset \underline{X}$ is closed if its complement $A' = \underline{X} - A$ is an open set.

Open sets and closed sets have the properties:

1. $\emptyset$ and $\underline{X}$ are closed and open.
2. If A_i are closed, then $\bigcap_i A_i$ is closed, but only a finite union is closed.
3. If A_i are open, then $\bigcup_i A_i$ is open, but only a finite intersection $\bigcap_{i=1}^{n} A_1$ is open.

Finally, we define a separable metric space.

Definition 35. A metric space $(\underline{X}, d)$ is separable if it contains a countable subset A which is dense in $\underline{X}$.

In the applications considered, most spaces will be separable Banach spaces, but in partial differential equation theory, you often need Fréchet spaces, which may be defined as complete metric spaces. Although they are not normed linear spaces, they do have similar properties and can be defined in terms of seminorms on locally converse topological spaces. Probably the best way of thinking of Fréchet spaces is as the inductive limit of normed linear spaces, because a more detailed explanation of these spaces is beyond the scope of this paper.

3. MEASURE AND INTEGRATION THEORY

We recall that $\mathcal{C}[0,1]$, the space of continuous functions, was not complete under the $\|\cdot\|_2$ norm

$$\|f\|_2 = \left(\int_0^1 f(t)^2 dt\right)^{1/2}$$

If we extend our class of functions to the Riemann integrable functions, then it is still not complete under this norm. Really, what we need is an integral which has the property that if f_n is square integrable and $f_n \to f$ in mean square, then f is also square integrable. The integral which does have this property is the Lebesgue integral. With our background of metric spaces and normed linear spaces we could define Lebesgue square integrable functions to be elements of the completion of $\mathscr{C}[0, 1]$ under the $\|\cdot\|_2$ norm, and similarly for general $p \geq 1$. However, we shall sketch the construction of the Lebesgue integral starting from basic measure theory.

Measure spaces

Definition 36. A measure space is a couple $(\overline{X}, \mathscr{B})$ consisting of a set $\overline{X}$ and a 6-algebra $\mathscr{B}$ of subsets of $\overline{X}$. A subset of $\overline{X}$ is called measurable if it is in $\mathscr{B}$.
(σ-algebra means closed under all countable set operations).

Definition 37. A measure μ on a measurable space $(\overline{X}, \mathscr{B})$ is a non-negative set function defined for all sets in $\mathscr{B}$ with the properties

$$\mu(\emptyset) = 0$$

$$\mu\left(\bigcup_{i=1}^{\infty} E_i\right) = \sum_{i=1}^{\infty} \mu E_i, \text{ where } E_i \text{ are disjoint sets in } \mathscr{B}.$$

(countably additive)

The triple $(\overline{X}, \mathscr{B}, \mu)$ is then a measure space.

The Lebesgue measure space

Example 28. $(\mathbb{R}, \mathscr{M}, m)$ where m is the outer measure defined by

$$mA = \inf_{\bigcup_n I_n \supset A} \Sigma \text{ length } (I_n)$$

where I_n is a countable collection of intervals covering the set A.

The outer measure is not countably additive over all possible subsets of $\mathbb{R}$, so we define a subcollection $\mathscr{M}$, the set of Lebesgue measurable sets, by

$$E \in \mathscr{M} \text{ if } \quad mA = m(A \cap E) + m(A \cap E') \qquad \forall A \subset \mathbb{R}$$

Then you can show that $\mathscr{M}$ is a σ-algebra of sets and it also contains all subsets of $\mathbb{R}$ with measure zero, i.e. $(\mathbb{R}, \mathscr{M}, m)$ is a well-defined measure space.

We remark that all intervals are in $\mathscr{M}$, including (a, ∞).

Further, all Borel sets are in $\mathscr{M}$ (this includes all open and closed sets). ($\mathscr{B}$, the set of Borel sets, is the smallest σ-algebra containing all the open sets of $\mathbb{R}$.)

Example 29. Another example is $([0,1], \mathscr{M}, m)$, where $\mathscr{M}$ is the Lebesgue measurable sets in $[0,1]$. Also $(\mathbb{R}, \mathscr{B}, m)$.

We say that μ is a finite measure if $\mu(\underline{X}) < \infty$.

Example: $([0,1], \mathscr{M}, m)$ with $m([0,1]) = 1$; all probability measures.

Definition 38. A measure space $(\underline{X}, \mathscr{B}, \mu)$ is complete if $\mathscr{B}$ contains all the subsets of sets of measure 0. (The Lebesgue measure is complete.) All measure spaces can be completed.

Definition 39. Measurable functions

Let $f : \underline{X} \to \mathbb{R} \cup \{\infty\}$, then f is measurable if $\{x : f(x) < \alpha\} \in \mathscr{B}$ for each α.

If f, g are measurable, so are $f+g$, cf, $f+c$, $f \cdot g$, $f \vee g$ where c is a constant. If $\{f_n\}$ is a sequence of measurable functions then so are $\sup f_n$, $\inf f_n$, $\lim f_n$, $\underline{\lim}\, f_n$, i.e. the set of measurable functions is closed under limiting operations.

A special type of measurable function is a simple function which we use to construct the Lebesgue integral.

Definition 40. A simple function is $g(x) = \sum_{i=1}^{n} c_i \chi_{E_i}(x)$, where c_i are constants and χ_{E_i} is the characteristic function of $E_i \in \mathscr{B}$.

$$\chi_{E_i}(x) = \begin{cases} 1 & \text{if } x \in E_i \\ 0 & \text{if } x \notin E_i \end{cases}$$

All non-negative measurable functions $f = \lim g_n$, where g_n is a monotonic increasing sequence of simple functions.

Example 30. $f : A \to \mathbb{R} \cup \{\infty\}$ is Lebesgue measurable if $A \in \mathscr{B}$ and for each α, $\{x : f(x) < \alpha\} \in \mathscr{B}$.

All continuous functions and piecewise continuous functions are Lebesgue measurable. All Riemann integrable functions are measurable.

We now build up our definition integration on $(\underline{X}, \mathscr{B}, \mu)$ by defining it just for non-negative simple functions.

$$\int_E g \, d\mu = \sum_{i=1}^{n} c_i \, \mu(E_i \cap E)$$

where $E \in \mathscr{B}$.

This integral is independent of the representation of g and has the usual properties of integrals:

$$\int_E (a g + b\psi) \, d\mu = a \int_E g d\mu + b \int_E \psi d\mu$$

for $a, b > 0$.

For non-negative measurable f on $(\underline{\overline{X}}, \mathscr{B}, \mu)$, define

$$\int_E f\,d\mu = \sup \int_E g\,d\mu$$

where g ranges over all simple functions with $0 \leq g \leq f$.

This is a well-defined integral for non-negative measurable functions with the properties

1. $\displaystyle\int_E (af + bg)\,d\mu = a\int_E f\,d\mu + b\int_E g\,d\mu$ for $a, b \geq 0$.

2. $\displaystyle\int_E f\,d\mu \geq 0$ with equality iff $f = 0$ 'almost everywhere', i.e. except on a set with measure zero.

3. If $f \geq g$ a.e, then $\displaystyle\int_E f\,d\mu \geq \int_E g\,d\mu$.

4. If $\{f_n\}$ is a sequence of non-negative measurable functions

$$\int_E \sum_{n=1}^{\infty} f_n\,d\mu = \sum_{n=1}^{\infty} \int_E f_n\,d\mu$$

This last property is an important advantage over the Riemann integral.

To extend the integral to arbitrary measurable functions we define:

Definition 41. A non-negative measurable function is integrable on E if

$$\int_E f\,d\mu < \infty.$$

An arbitrary measurable function is integrable on E if f^+ and f^- are both integrable and we define

$$\int_E f\,d\mu = \int_E f^+ d\mu - \int_E f^- d\mu.$$

f^+ and f^- are both non-negative functions defined by

$$f^+ = \tfrac{1}{2}(f + |f|)$$

$$f^- = \tfrac{1}{2}(|f| - f)$$

This integral has the properties 1 - 4 and, in addition, the very important Lebesgue dominated convergence property.

5. Let $\{f_n(x)\}$ be a sequence of measurable functions on E such that $f_n(x) \to f(x)$ almost everywhere on E and $|f_n(x)| \leq g(x)$, an integrable function on E, then

$$\int_E f\,d\mu = \lim_{n\to\infty} \int_E f_n d\mu$$

and

6. $$\left|\int_E f\,d\mu\right| \leq \int_E |f|\,d\mu$$

We have now defined integration over an arbitrary measure space $(\underline{X}, \mathscr{B}, \mu)$ and two important examples are

Example 31. Lebesgue integral on $(\underline{X}, \mathscr{B}, \mu)$ or $((a,b), \mathscr{M}, m)$

We note that if f is Riemann integrable on (a, b), then it is also Lebesgue integrable and the integrals agree.

Example 32. Probability measures $(\underline{X}, \mathscr{B}, \mu)$, where $\mu(\underline{X}) = 1$. In this case the integral on $(\underline{X}, \mathscr{B}, \mu)$ is the expectation of a random variable. (These spaces will be discussed in more detail later on.)

Example 33. Multiple integrals on $(\mathbb{R}^n, \mathscr{M}^n, m)$ or $(I_1 x \ldots x I_n, \mathscr{M}^n, m)$. If $\underline{X} = I_1 x \ldots x I_n$ measurable sets have the form $E_1 x \ldots x E_n$, where E_i is a Lebesgue measurable set on I_i, an interval of $\mathbb{R}$.

Then

$$\int_{E_1 x \ldots x E_n} f(\underset{\sim}{x})\, d\mu(\underset{\sim}{x}) = \int_{E_1} \ldots \int_{E_n} f(x_1, \ldots, x_n)\, dx_1 \ldots dx_n$$

has all the usual properties of multiple integrals.

Differentiation and Lebesgue integration

Consider

$$F(t) = F(a) + \int_a^t f(s)\, ds \qquad (1)$$

when the integral is the Lebesgue integral on $((a,b), \mathscr{M}, m)$.

We already know from Riemann integration theory that if f is continuous, $dF/dt = f$ on [a, b] and, conversely, if F is continuously differentiable (C^1) with $dF/dt = f$, then F satisfies (1). With the Lebesgue theory we can extend this fundamental theorem of calculus using the concept of absolute continuity, which is slightly weaker than continuous differentiability.

Definition 42. Let I be a compact (closed bounded) interval on $\mathbb{R}$, then F is absolutely continuous on I if for every $\epsilon > 0$ there is a $\delta > 0$ such that whenever $I_k = [a_k, b_k]$ are non-overlapping intervals in I with

$$\sum_{k=1}^{n} |b_k - a_k| \leq \delta, \qquad \sum_{k=1}^{n} |F(b_k) - F(a_k)| \leq \epsilon$$

F is absolutely continuous on an arbitrary interval if it is absolutely continuous on every compact subinterval.

For example, all C' functions are absolutely continuous and all functions which satisfy a Lipschitz condition on I:

$$|F(t) - F(s)| \leq K|t - s| \quad \text{for} \quad s, t \in I$$

Then the extended fundamental theorem of calculus is

$F: I \to \mathbb{R}$ is absolutely continuous iff F satisfies (1) for some integrable f on [a, b]. Then F' exists almost everywhere and F' = f almost everywhere. Two other extensions of the Riemann theory are:

1. Differentiation under the integral

If g(t) is Lebesgue integrable on (c, ∞), f(x, t) a measurable function of t for all x in (a, b) and $\partial f/\partial x\,(x, t)$ exists for all t in (c, ∞) with f, $\partial f/\partial x$ bounded on $(a, b) \times (c, \infty)$, then $F(x) = \int_{(c,\infty)} f(xt)\, g(t)\, dt$ is differentiable in (a, b) with

$$F'(x) = \int_{c}^{\infty} \frac{\partial f}{\partial x}(x, t)\, g(t)\, dt$$

2. Multiple integrals and change of order of integration

We have already noted that multiple integrals fall into our general theory of integration on $(\overline{X}, \mathscr{B}, \mu)$ and have the usual properties of integrals. The relation between these multiple integrals on $\mathbb{R}^n$ and successive Lebesgue integrals on $\mathbb{R}$ is stated in Fubini's theorem

(a) If f(x, y) is a measurable non-negative function on $\mathbb{R} \times \mathbb{R}$, then

$$\int_E dx \int_F f(x, y)\, dy = \iint_{E\times F} f(x, y)\, dx dy = \int_F dy \int_E f(x, y) dx$$

where E, F are Lebesgue measurable sets.

(b) If f(x, y) is a measurable function on $\mathbb{R} \times \mathbb{R}$ and any of the integrals is absolutely convergent, then

$$\int_E dx \int_F f(x, y)\, dy = \iint_{E\times F} f(x, y)\, dx dy = \int_F dy \int_E f(x, y) dx$$

$\left(\int_E f d\mu\right.$ is absolutely convergent if $\left.\int_E |f|\, d\mu \text{ exists}\right)$.

(Similar results hold for integrals of functions of n variables.)

The Lebesgue-Banach spaces

Our discussion of the Lebesgue integral was motivated by the fact that $\mathcal{C}[0,1]$ was not closed under the $\|\cdot\|_2$ norm. We now define spaces $L_p(E)$ which overcome this problem.

$$\mathcal{L}^p(E) = \left\{ \begin{array}{l} f: E \to \mathbb{R}, \text{ where } E \text{ is a finite Lebesgue measurable set} \\ \quad \text{in } \mathbb{R} \text{ and } \int_E |f|^p\, dt < \infty \end{array} \right\}$$

$1 \leq p < \infty$

Define

$$\|f\|_p = \left(\int_E |f(t)|^p dt \right)^{1/p}$$

Then by Minkowski's inequality

$$\left(\int_E |f+g|^p dt \right)^{1/p} \leq \left(\int_E |f|^p dt \right)^{1/p} + \left(\int_E |g|^p dt \right)^{1/p}$$

we see that $\|\cdot\|_p$ is a seminorm. It is not a norm, since $\|f\|_p = \|g\|_p$ does not imply that $f(t) \equiv g(t)$ at all points in E, but only "almost everywhere" (i.e. they can differ on a set of points of measure zero).

To define a normed linear space we need to consider equivalence classes as elements in $\mathcal{L}^p(E)$: $f = g$ iff $f(x) = g(x)$ for all x, except in a set of measure zero.

If we consider these equivalence classes as elements of a new space $L_p(E)$ then $\|\cdot\|_p$ is a norm, where equality in $L_p(E)$ means equal almost everywhere.

$L_p(E)$ is a Banach space (i.e. complete).

We also define $L_\infty(E)$ which is a generalization of bounded functions.

$$L_\infty(E) = \left\{ \begin{array}{l} f: E \to \mathbb{R}, \text{ where } f \text{ is a Lebesgue measurable function} \\ \qquad \text{on } E, \text{ which is bounded almost everywhere.} \end{array} \right\}$$

If we again consider equivalence classes of elements, then $L_\infty(E)$ becomes a normed linear space under the norm

$$\|f\|_\infty = \operatorname*{ess\,sup}_{t \in E} |f(t)|$$

where the ess sup is the infinum of sup g(t), where g ranges over all elements in the equivalence class for f,

i.e. ess sup $f(t) = \inf\{K : m\,\{t : f(t) > K\ = 0\}$

Example 34. Consider $f: [-1,1] \to \mathbb{R}$; $f(t) = \begin{cases} t^2 & -1 \leq t \leq 1 \\ 1 & t = \pm \frac{1}{2} \\ 1.5 & t = 0 \end{cases}$

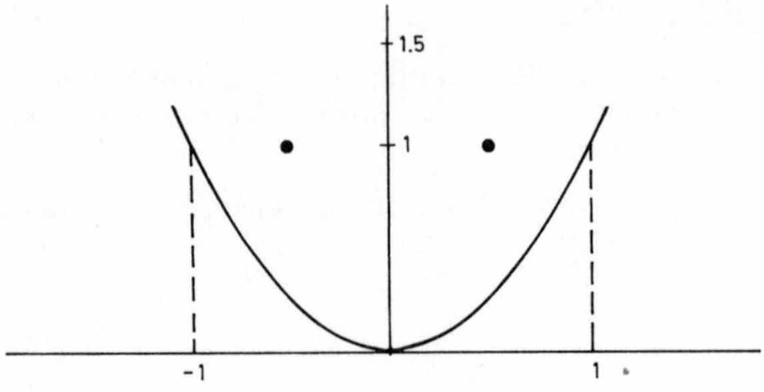

Then $\sup_{[-1,1]} f(x) = 2$, but $\operatorname{ess\,sup}_{[-1,1]} f(x) = 1$;

$L_\infty(E)$ is again a Banach space.

The Lebesgue spaces $L_p(E)$ are building blocks in analysis and so we summarize some useful properties:

1. $L_1 \supset L_2 \supset L_3 \supset \ldots$.
2. $\mathcal{D}_0^\infty$ is dense in $L_p(E)$; $1 \leq p < \infty$
   where $\mathcal{D}_0^\infty$ is the space of infinitely differentiable functions with compact support. (f has compact support if it is zero outside compact sets.)

3. Riesz representation theorem

   The dual space of $L_p(E)$ is $L_q(E)$; $1/p + 1/q = 1$.
   All continuous linear functionals G on $L_p(E)$ have the form

   $$G(f) = \int_E f g \, dt$$

where

   $g \in L_q(E)$ and $\|G\| = \|g\|_q$.

(This is essentially a consequence of the Hölder inequality.)

$$\int_E |fg| \, dx \leq \left(\int_E |f|^p \, dx \right)^{1/p} \left(\int_E |g|^q \, dx \right)^{1/q}; \quad 1/p + 1/q = 1; \; 1 \leq p < \infty.$$

(Note, however, that the dual of $L_\infty(E)$ is not $L_1(E)$.)

4. $L_2(E)$ is a Hilbert space.

Some extensions

1. We may similarly define $L_p(E)$, where E is a finite Lebesgue measurable set in $\mathbb{R}^n$ and f is a measurable real-valued function of n variables.

$L_p(E)$ is a Banach space under the norm

$$\|f\|_p = \left(\int_E |f(x_1 \ldots x_n)|^p \, dx_1 \ldots dx_n\right)^{1/p}; \quad 1 \leq p < \infty$$

and

$$\|f\|_\infty = \text{ess}_E \sup |f(x_1, \ldots x_n)|$$

2. We may also consider $L_p(E; \mathbb{R}^k)$, the space of vector-valued functions of n variables $f: E \subset \mathbb{R}^n \rightarrow \mathbb{R}^k$ under the norm

$$\|f\|_p = \left(\int_E |f(x_1, \ldots, x_n)|_k^p \, dx_1 \ldots dx_n\right)^{1/p}$$

where $|\cdot|_k$ is the Euclidean norm on $\mathbb{R}^k$.

3. A final example is the space of second-order stochastic processes on $[0, T]$, which is a Hilbert space under the inner product

$$\langle x, y \rangle = \int_0^T E\{x(t)\, y(t)\}\, dt$$

$$= \int_0^T \int_\Omega x(t, \omega)\, y(t, \omega)\, d\mu\, dt$$

4. HILBERT SPACES

Since a normed linear space is characterized by its norm and its algebraic structure, we often find that supposedly different spaces are equivalent both algebraically and topologically.

Definition 43. Two normed linear spaces $\underline{X}$ and $\underline{Y}$ are topologically isomorphic if there exists a $T \in L(\underline{X}, \underline{Y})$ such that T^{-1} exists and is continuous. T is called a topological isomorphism and we write

$$\underline{X} \cong \underline{Y}$$

Definition 44. If further, $\|Tx\|_y = \|x\|_x$, for all $x \in \underline{X}$, we say that they are isometrically isomorphic.

A useful necessary and sufficient condition for the topological isomorphism between two spaces $\underline{X}$ and $\underline{Y}$ is $\exists m, M > 0$, such that

$$m \|x\|_x \leq \|Tx\|_y \leq M \|x\|_x, \quad \forall x \in \underline{X}$$

Example 35. Let $\underline{X} = \ell^2_\infty$ and

$$Z = \{z : [0,1] \to \mathbb{R}, \; z(t) = a_0 + a_1 t; \; a_0, a_1 \in \mathbb{R}\}$$

under the norm

$$\|z\| = \max\{|a_0|, |a_1|\}.$$

Then $Tx = x_1 + x_2 t$ for $x = (x_1, x_2)$ is a linear map from $\underline{X}$ to Z and $\|T_x\| = \max\{|x_1|, |x_2|\} = \|x\|_\infty$ and T is bounded. Also $T^{-1} : z \to \underline{X}$ is given by $T^{-1}z = \{z(0), dz/dt|_{t=0}\}$. T^{-1} is linear and $\|T^{-1}z\| = \|z\|$.

So $\underline{X} \cong Z$ and even isometrically isomorphic. This is an example of a very important result, namely:

All real n-dimensional normed linear spaces are topologically isomorphic, i.e. all are equivalent to $\mathbb{R}^n$.

So we can find out all about real finite-dimensional normed linear spaces from $\mathbb{R}^n$.

Properties of $\mathbb{R}^n$

1. $\underset{\sim}{e}_i = \{0, \; , 0, 1, 0, \ldots\}$ forms a basis for $\mathbb{R}^n$,

   i.e. $\underset{\sim}{x} = \{x_1, \ldots, x_n\} = \sum_{i=1}^{n} x_i \underset{\sim}{e}_i$.

2. All norms are equivalent.

3. $\mathbb{R}^n$ is complete (i.e. a B-space) and all its subspaces are closed.

4. All linear maps $L : \mathbb{R}^n \to \underline{Y}$ are continuous for any normed linear space $\underline{Y}$.

5. A corollary of 4 is that the algebraic and topological duals of $\mathbb{R}^n$ coincide.

6. All linear transformations $T : \mathbb{R}^n \to \mathbb{R}^m$ may be represented as $m \times n$ matrices. This representation is not unique, but depends on the basis of $\mathbb{R}^n$ and $\mathbb{R}^m$. In fact T is represented by a class of similar matrices.

Example 36. Let $\underline{X}$ be the space of polynomials of degree ≤ 3,

i.e. $x(t) = \alpha_1 + \alpha_2 t + \alpha_3 t^2 + \alpha_4 t^3$

Let $\underline{Y}$ be the space of polynomials of degree ≤ 2;

Let D be the derivative operator on $\underline{X}$. The range of D is $\underline{Y}$ and $D : \underline{X} \to \underline{Y}$ is linear.

If we take $\{1, t, t^2, t^3\} = \{x_1, x_2, x_3, x_4\}$ as a basis for $\underline{X}$ and $\{1, t, t^2\} = \{y_1, y_2, y_3\}$ as a basis for $\underline{Y}$. then

$$Dx_1 = 0, \quad Dx_2 = y_1, \quad Dx_3 = 2y_2, \quad Dx_4 = 3y_3$$

$$\therefore \begin{bmatrix} \beta_1 \\ \beta_2 \\ \beta_3 \end{bmatrix} = \begin{bmatrix} 0 & 1 & 0 & 0 \\ 0 & 0 & 2 & 0 \\ 0 & 0 & 0 & 3 \end{bmatrix} \begin{bmatrix} \alpha_1 \\ \alpha_2 \\ \alpha_3 \\ \alpha_4 \end{bmatrix} \qquad \text{if } Dx = \beta_1 + \beta_2 t + \beta_3 t^2$$

and relative to these bases D has the representation

$$\begin{bmatrix} 0 & 1 & 0 & 0 \\ 0 & 0 & 2 & 0 \\ 0 & 0 & 0 & 3 \end{bmatrix}$$

If we change the basis of $\underline{X}$ to $\{1+t,\ t+t^2,\ t^2+t^3,\ 1+t^2\}$ then D has the new representation

$$\begin{bmatrix} 1 & 1 & 0 & 0 \\ 0 & 2 & 2 & 0 \\ 0 & 0 & 3 & 3 \end{bmatrix}$$

When we consider infinite-dimensional spaces, however, the situation is not as simple. However, there is an interesting subclass of infinite-dimensional spaces like ℓ_p which can be thought of as a generalization of $\mathbb{R}^n$ to $n = \infty$.

Definition 45. An inner product on a linear space is a bilinear function $\langle .,. \rangle : \underline{X} \times \underline{X} \to \mathbb{C}$, such that

1. $\langle \alpha x + \beta y, z \rangle = \alpha \langle x, z \rangle + \beta \langle y, z \rangle$ for scalars α, β.

2. $\overline{\langle x, y \rangle} = \langle y, x \rangle$

3. $\langle x, x \rangle \geqq 0$ and $\langle x, x \rangle = 0$ iff $x = 0$.

A linear space $\underline{X}$ with an inner product $\langle .,. \rangle$ is called an inner product space.

Example 37. $\mathbb{R}^n$ with $\langle x, y \rangle = \sum_{i=1}^{n} x_i y_i$

where

$$x = \{x_1, \ldots x_n\}, \quad y = \{y_1, \ldots y_n\}$$

Example 38. $L_2[a, b]$ with $\langle x, y \rangle = \int_a^b x(t)\, \overline{y(t)}\, dt$

Other properties of the inner product

1. $\langle x, y \rangle = 0 \quad \forall x \in \underline{X} \quad y = 0.$

2. $\langle x, \alpha y \rangle = \bar{\alpha} \langle x, y \rangle$

3. $\underline{X}$ can be made a normed linear space by defining $\|x\| = \sqrt{\langle x, x \rangle}$

4. Parallelogram law

$$\|x+y\|^2 + \|x-y\|^2 = 2\|x\|^2 + 2\|y\|^2$$

where $\|\cdot\|$ is the norm induced by the inner product.

5. Schwarz inequality $|\langle x, y \rangle| \leq \|x\| . \|y\|$.

Definition 46. A Hilbert space is an inner product space which is complete as a normed linear space under the induced norm, i.e. a Hilbert space is a special case of a B-space. When is a B-space a Hilbert space?

If its norm obeys the parallelogram law, then we can define an inner product

$$\langle x, y \rangle = \frac{1}{4}\left\{\|x+y\|^2 - \|x-y\|^2 + i\|x+iy\|^2 - i\|x-iy\|^2\right\}$$

Example 52. $L_2[a, b]$ is a Hilbert space with inner product

$$\langle x, y \rangle = \int_a^b x(t)\,\overline{y(t)}\,dt$$

and norm

$$\|x\| = \left(\int_a^b |x(t)|^2\, dt\right)^{1/2}$$

as before.

Let $V = \{y \in L_2[a, b]: y' \in L_2[a, b] \text{ and } y(a) = 0 = y(b)\}$. Then V is a linear subspace of $L_2[a, b]$ and in fact is a dense subspace. On V we can define a new inner product

$$\langle y, z \rangle_V = \langle y, z \rangle + \langle y', z' \rangle$$

V is a Hilbert space under $\langle .,. \rangle_V$, but it is not closed under $\langle .,. \rangle$.

Example 53. Let Ω be a closed subset in $\mathbb{R}^3$ and $\underline{X} = \mathcal{C}^2(\Omega)$, the space of complex-valued functions with continuous second partials in Ω.

Define $\nabla u = \left(\frac{\partial u}{\partial x_1}, \frac{\partial u}{\partial x_2}, \frac{\partial u}{\partial x_3}\right)$ and

$$\langle u, v \rangle = \int_\Omega \left[u\bar{v} + \frac{\partial u}{\partial x_1}\frac{\partial \bar{v}}{\partial x_1} + \frac{\partial u}{\partial x_2}\frac{\partial \bar{v}}{\partial x_2} + \frac{\partial u}{\partial x_3}\frac{\partial \bar{v}}{\partial x_3}\right] dx_1 dx_2 dx_3$$

where $x = (x_1, x_2, x_3)$.

Then $\langle .,. \rangle$ is an inner product on $\underline{X}$ and it induces the norm

$$\|u\| = \left(\int_\Omega |u|^2 + |\nabla u|^2 dx\right)^{1/2}$$

Example 54. Let Ω be an open subset of $\mathbb{R}^K$ and $u \in \mathcal{C}^n(\Omega)$. As in Ex.22 define the differential operator $D^\alpha u$:

$$D^\alpha u = \frac{\partial^{|\alpha|} u}{\partial x_1^{\alpha_1} \dots \partial x_k^{\alpha_k}}, \quad |\alpha| = n$$

and the inner product $\langle .,. \rangle_n$ on $\mathcal{C}^n(\Omega)$.

$$\langle u, v \rangle_n = \int_\Omega \sum_{|\alpha| \le n} D^\alpha u(x) \overline{D^\alpha v(x)} \, dx$$

$\langle .,. \rangle_n$ induces the norm $\|\cdot\|_{n,2}$

$$\|u\|_{n,2} = \left(\int_\Omega \sum_{|\alpha| \le n} |D^\alpha u(x)|^2 dx\right)^{1/2}$$

The completion of $\mathcal{C}^n(\Omega)$ with respect to this norm is a Hilbert space, denoted $H^n(\Omega)$, a Sobolev space, which is used in distribution theory in partial differential equations.

Another important Sobolev space is $H_0^n(\Omega)$, which is the completion of $\mathcal{C}_0^\infty(\Omega)$ under this (n, 2) norm. ($\mathcal{C}_0^\infty(\Omega)$ is the space of infinitely differentiable functions with compact support.)

We note that the other (n, p) norms cannot be obtained from inner products (see Ex.22).

An important finite-dimensional concept which generalizes Hilbert spaces is orthogonality:

Definition 47. x is orthogonal (perpendicular) to y, $x \perp y$, iff $\langle x, y \rangle = 0$.

If $x \perp y$, then the parallelogram law reduces to

$$\|x + y\|^2 = \|x\|^2 + \|y\|^2$$

which certainly looks familiar.

Definition 48. If M is a subspace of a Hilbert space, H, then the orthogonal complement

$$M^\perp = \{x \in H : \langle x, y \rangle = 0 \quad \forall y \in M\}$$

$M^\perp$ is a closed linear subspace of H and H can be decomposed uniquely as a direct sum,

$$H = \overline{M} \oplus M^{\perp}$$

also $M^{\perp\perp} = \overline{M}$.

If $y \in H$, then $y = y_M + y_{M^\perp}$, where $y_M \in \overline{M}$, $y_{M^\perp} \in M^\perp$ and $\|y\|^2 = \|y_{M^\perp}\|^2 + \|y_{M^\perp}\|^2$.

M induces an orthogonal projection π on H. $\pi : H \to \overline{M}$ where $\pi y = y_M$. π is linear and bounded and $\|\pi\| = 1$.

Example 39. $H = L_2[-a, a]$, $M = \{$set of all even functions: $x(t) = x(-t)\}$ M is a linear subspace of H.

Let $P : H \to H$ be $y = P_x$, where $y(t) + \frac{1}{2}(x(t) + x(-t))$. Then P is a projection of $H \to M$.

Definition 49. An orthonormal set in a Hilbert space H is a non-empty subset

$\{\phi_n\}$ of H $\ni \langle \phi_n, \phi_m \rangle = \delta_{nm}$

i.e. $\{\phi_n\}$ are mutually orthogonal unit vectors. Of course any mutually orthogonal set $\{u_n\}$ may be normalized by

$$\left\{ \phi_n = \frac{u_n}{\|u_n\|} \right\}$$

Example 40. ℓ_2^n. The set $\phi_k = \{0, \ldots, 0, 1, 0, \ldots\}$ is an orthonormal set. Similarly for ℓ_2.

Example 41. $\{\sin \pi nt\}$ is an orthonormal set in $L_2[0, 2]$.

Example 42. Any linearly independent set $\{x_1, x_2, \ldots\}$ in H can generate an orthonormal set $\{\phi_n\}$ by the Gram-Schmidt orthogonalization process

Let $\phi_1 = x_1 / \|x_1\|$

Let $\phi_2 = \dfrac{x_2 - \langle x_2, \phi_1 \rangle \phi_1}{\|x_2 - \langle x_2, \phi_1 \rangle \phi_1\|}$, so $\phi_2 \perp \phi_1$

Let $\phi_3 = \dfrac{x_3 - \langle x_3, \phi_1 \rangle \phi_1 - \langle x_3, \phi_2 \rangle \phi_2}{\|x_3 - \langle x_3, \phi_1 \rangle \phi_1 - \langle x_3, \phi_2 \rangle \phi_2\|}$, so $\phi_3 \perp \phi_2 \perp \phi_1$

Continue in the following way:

$$\phi_n = \frac{x_n - \langle x_n, \phi_1 \rangle \phi_1 - \langle x_n, \phi_2 \rangle \phi_2 - \langle \ldots \rangle - \langle x\ , \phi_{n-1} \rangle \phi_{n-1}}{\|x_n - \langle x_n, \phi_1 \rangle \phi_1 - \ldots \qquad - \langle x_n, \phi_{n-1} \rangle \phi_{n-1}\|}$$

Orthonormal sets satisfy an important inequality: Bessel's inequality.

If $\{\phi_1, \ldots, \phi_n\}$ is a finite orthonormal set in H and $x \in H$ then

$$\sum_{i=1}^{n} |\langle x, \phi_i \rangle|^2 \leq \|x\|^2$$

and

$$x - \sum_{i=1}^{n} \langle x, \phi_i \rangle \phi_i \perp \phi_j$$

<u>Proof</u>

$$0 \leq \|x - \sum_{i=1}^{n} \langle x, \phi_1 \rangle \phi_i \|^2$$

$$= \langle x - \sum_{i=1}^{n} \langle x, \phi_i \rangle \phi_i, \quad x - \sum_{i=1}^{n} \langle x, \phi_i \rangle \phi_i \rangle$$

$$= \langle x, x \rangle - 2 \sum_{i=1}^{n} \langle x, \phi_i \rangle \langle \overline{x, \phi_i} \rangle + \sum_{ij=1}^{n}\sum \langle \overline{x, \phi_j} \rangle \langle x, \phi_i \rangle \delta_{ij}$$

$$= \|x\|^2 - \sum_{i=1}^{n} |\langle x, \phi_i \rangle|^2$$

and

$$\langle x - \sum_{i=1}^{n} \langle x, \phi_i \rangle \phi_i, \phi_j \rangle = \langle x, \phi_j \rangle - \sum_{i=1}^{n} \langle x, \phi_i \rangle \langle \phi_i, \phi_j \rangle$$
$$= 0$$

This has a useful geometric interpretation:

$$\sum_{i=1}^{n} \langle x, \phi_i \rangle \phi_i$$

is the projection of x onto the subspace spanned by $\{\phi_1, \ldots, \phi_k\}$ and

$x - \sum_{i=1}^{n} \langle x, \phi_i \rangle \phi_i$ is the "error".

Bessel's inequality is also valid for $n = \infty$.

Earlier on, we said that a Hilbert space can be thought of as '$\mathbb{R}^\infty$' in some sense. Well just as for $\mathbb{R}^n$, where we can always express any element as a linear combination of n basis elements, we can do a similar thing for Hilbert spaces, except of course we need infinitely many basis elements and we need enough of them to include all the elements, i.e. a complete set.

Definition 50. An orthonormal set $\{\phi_n\}$ in a Hilbert space H is complete or maximal if $H = \overline{Sp\{\phi_i\}}$.

The following are equivalent conditions for $\{\phi_i\}$ to be complete:

1. $x \perp \phi_i \ \forall i \Rightarrow x = 0$

2. $x = \Sigma \langle x, \phi_i \rangle \phi_i \qquad \forall x \in H$

This is called the Fourier expansion for x

$\langle x, \phi_i \rangle$ are called the Fourier coefficients

3. Parseval's equation holds: $\|x\|^2 = \Sigma |\langle x, \phi_i \rangle|^2$

4. There is no vector $y \in H$ such that $\{\phi_i, y/\|y\|\}$ is an orthonormal set larger than $\{\phi_i\}$.

Example 43. Let $H = L_2[0, 2\pi]$. Then $\{\phi_n(x) = e^{inx}/\sqrt{2\pi}\}$ is an orthonormal set in $L_2[0, 2\pi]$.

$$c_i = \langle f, \phi_i \rangle = \frac{1}{\sqrt{2\pi}} \int_0^{2\pi} f(x)\, e^{-inx}\, dx$$

are the usual complex Fourier coefficients of f.

Bessel's inequality is

$$\sum_{-\infty}^{\infty} |c_i|^2 \leq \int_0^{2\pi} |f(x)|^2\, dx < \infty$$

One can show that this set is complete and equality holds and we can write

$$f(x) = \frac{1}{\sqrt{2\pi}} \sum_{-\infty}^{\infty} c_n e^{inx}$$

where equality is of course in the L_2-sense, not pointwise in general. Since $H = \overline{Sp\{\phi_i\}}$, we can see why the Riesz-Fischer theorem holds:

If $\{c_n\}$ is a complex sequence such that $\sum_{-\infty}^{\infty} |c_n|^2 < \infty$, then there is an $f \in L_2[0, 2\pi]$, whose Fourier coefficients are the c_n's.

It is just saying that $y = \sum_{-\infty}^{\infty} c_n \phi_n \in H$, since $\|y\|^2 = \sum_{-\infty}^{\infty} |c_n|^2 < \infty$.

Example 44. $H = L_2[0, 2\pi]$ over the real numbers. This is similar to the previous example, except the complete orthonormal basis is

$$\{1/\sqrt{2\pi}, \quad 1/\sqrt{\pi}\sin nt, \quad 1/\sqrt{\pi}\cos nt\}$$

and

$$f(t) = c_0/\sqrt{2} + \sum_{1}^{\infty} c_n \cos nt + \sum_{1}^{\infty} d_n \sin nt \text{ in the } L_2\text{-sense.}$$

Example 45. $H = L_2[-1, 1]$. The Legendre polynomials are an orthogonal set, where

$$P_n(x) = \frac{1}{2^n n!} \frac{d^n}{dx^n} (x^2 - 1)^n .$$

We have

$$\int_{-1}^{1} P_n(x) P_m(x)\, dx = \frac{2}{2n+1} \delta_{mn}$$

Also $\{P_n(x)\}$ is complete in $L_2[-1, 1]$ and so for arbitrary $f \in L_2[-1, 1]$ it has the expansion

$$f = \sum_{n=0}^{\infty} c_n P_n$$

where each polynomial $P_n(x)$ is of degree n and the partial sums converge to f in the L_2-norm, i.e.

$$\sum_{n=0}^{k} c_n P_n(x)$$

approximates f(x) in the mean square sense.

Example 46. $H = L_2(0, \infty)$. The Laguerre polynomials $L_n(x)$ are often defined as the polynomial solutions of $xy'' + (1-x)y' + ny = 0$. Then $\{e^{-x/2} L_n(x)\}$ forms a complete orthonormal set for H.

These last two examples are two of many such ways of expressing

$$f(x) = \sum_{n=0}^{\infty} c_n \phi_n(x)$$

as an expansion in terms of orthonormal functions. Often these arise as eigenfunction expansions of solutions of partial differential equations after using separation of variables. These also give us good approximations for f as a finite sum of terms, which are invaluable for numerical solutions. The advantage of using expansions in terms of orthonormal functions, as opposed to, say, finite polynomials, $f(x) = \sum_{n=0}^{k} a_n x^n$, is that, if you wish to increase your accuracy for f in the orthonormal expansion, you simply calculate one extra coefficient and add it on and repeat until the accuracy is sufficient. However, with approximation by polynomials, you need to recalculate all the coefficients for each new approximation.

5. LINEAR FUNCTIONALS, WEAK CONVERGENCE, WEAK COMPACTNESS

Definition 51. A linear functional on a normed linear space $\underline{X}$ is a linear map $f: X \to \mathbb{R}$ (or $\mathbb{C}$ if X is a complex vector space).

Definition 52. The algebraic dual of X, denoted X^a, is the linear vector space of all linear functionals on X.

If we consider topological properties as well, we can look at the space of all continuous linear functionals (or equivalently all bounded linear functionals).

Definition 53. The topological dual of X or the conjugate space X^* is the normed linear space of all bounded linear functionals on X with the norm

$$\|f\| = \sup_{x \in \underline{X}} \{ |f(x)| : \|x\| = 1 \}$$

We remark that X^* is always a B-space, even when X is not.

We illustrate this useful concept with the following examples:

Example 47. Consider $X = \ell_2^n$ and define $f_a(x) = a_1 x_1 + a_2 x_2 + \ldots + a_n x_n$ for a fixed $a \in \ell_2^n$, $a \neq 0$.

$f_a : X \to \mathbb{R}$ and is linear. It is also bounded since

$$\begin{aligned} |f_a(x)| &= |a_1 x_1 + a_2 x_2 + \ldots + a_n x_n| \\ &\leq \sqrt{a_1^2 + \ldots + a_n^2}\ \sqrt{x_1^2 + \ldots + x_n^2} \\ &= \|a\|_2\ \|x\|_2 \end{aligned}$$

applying the useful Hölder inequality

$$\left| \sum_{i=1}^{n} a_i b_i \right| \leq \left(\sum_{i=1}^{n} |a_i|^p \right)^{1/p} \left(\sum_{i=1}^{n} |b_i|^q \right)^{1/q}$$

where $1/p + 1/q = 1$, valid for n finite or infinite and $1 \leq p < \infty$. We also have that

$$\|f_a\| = \sup \left\{ |f_a(x)| : \|x\|_2 = 1 \right\}$$
$$\leq \|a\|_2$$

Letting $x = \frac{a}{\|a\|_2}$, $f_a\left(\frac{a}{\|a\|_2}\right) = \|a\|_2^2$ and hence we have the equality $\|f_a\| = \|a\|_2$. So every $a \in X$ defines an element f_a in the dual space; similarly for ℓ_2.

Example 48. $X = \mathcal{C}[a, b]$.

Then $I(x) = \int_a^b x(t)\, dt$ is a linear functional and $|I(x)| \leq \sup|x(t)| \cdot (b - a)$

with equality when $x \equiv$ constant. So I is a continuous linear functional with the norm $(b - a)$.

Example 49. $X = \mathcal{C}[a, b]$ and $I_y = \int_a^b y(t)\, x(t)\, dt$; I_y is a linear functional on X for integrable y.

$$|I_y(x)| \leq \max_{a \leq t \leq b} |x(t)| \int_a^b |y(t)|\, dt = \|x\| \int_a^b |y(t)|\, dt$$

So I_y is continuous and in fact $\|I_y\| = \int_a^b |y(t)|\, dt$.

Example 50. $X = \mathcal{C}[a, b]$ and $\delta_{t_0}\, x = x(t_0)$

δ_{t_0} is a linear functional on X with

$$|\delta_{t_0}\, x| \leq |x(t_0)| \leq \|x\|$$

and we have equality when x is a constant. So our familiar "delta function" which is often represented as $\int_a^b x(t)\, \delta(t - t_0) dt$ is not a function, but a continuous linear functional on $\mathcal{C}[a, b]$ with norm 1.

It is all very well to define the dual of a space and to give some simple examples, but we would like to know more in the general case. The following famous Hahn-Banach theorem ensures us of the existence of lots of continuous linear functionals.

Hahn-Banach theorem I

Every continuous linear functional $f: M \to \mathbb{R}$ defined on a linear subspace M of X, a normed linear space, can be extended to a continuous linear functional F on all of X with preservation of norm.

If we take the particular subspace to be $\{\alpha x_0\}$ for $x_0 \neq 0$ and α any scalar, then a linear functional is $f_0(y) = \alpha\|x_0\|$, where $y = \alpha x_0$ and f_0 is bounded with norm 1. The Hahn-Banach theorem says that there is an F_0 in X^* with norm 1.

The Hahn-Banach theorem also has the following geometric interpretation, which is useful in proving the existence of optimal control theory problems.

Hahn-Banach theorem II

Let X be a normed linear space, M a manifold of X and A a non-empty convex, open subset of X not intersecting M. Then there exists a closed hyperplane in X containing M and not intersecting A.

For many spaces there is a nice relationship between X and its conjugate space X^*.

Example 51. Consider ℓ_p^n and define $f \in \ell_p^{n*}$ by $f(x) = \sum_{i=1}^{n} f_i x_i$; $1 \leq p < \infty$. f is clearly linear, and

$$|f(x)| = |\sum f_i x_i|$$

$$\leq \left(\sum |f_i|^q\right)^{1/q}\left(\sum |x_i|^p\right)^{1/p}, \quad \text{where } 1/p + 1/q = 1 \text{ using Hölder's inequality}$$

$$= \|x\|_p \left(\sum |f_i|^q\right)^{1/q}$$

so f is bounded and $\|f\| \leq \left(\sum_{i=1}^{n} |f_i|^q\right)^{1/q}$

If we let

$$x_i = \begin{cases} \dfrac{|f_i|^q}{f_i} & \text{if } f_i \neq 0 \\ 0 & \text{if } f_i = 0 \end{cases}$$

then $\|x\|_p = \left(\sum_{i=1}^{n} |f_i|^q\right)^{1/p}$

and

$$|f(x)| = \sum_{i=1}^{n} |f_i|^q$$

$$= \|x\|_p \left(\sum |f_i|^q\right)^{1/q}, \quad \text{since } 1/p + 1/q = 0.$$

So we actually have

$$\|f\| = \left(\sum |f_i|^q\right)^{1/q}$$

$$= \|f\|_q$$

In fact all continuous linear functionals have this form and the dual of ℓ_p^n is ℓ_q^n, where $1/p + 1/q = 1$.

Similarly $\ell_p^* = \ell_q$.

Example 52. Consider ℓ_∞^n with $\|x\|_\infty = \max_{1 \leq i \leq n} |x_i|$

Let $f(x) = \sum_{i=1}^{n} f_i x_i$. Then f is a linear functional and

$$|f(x)| \leq \sum |f_i| |x_i|$$

$$\leq \|x\|_\infty \sum_{i=1}^{n} |f_i| = \|x\|_\infty \|f\|_1$$

For $x_i = \operatorname{sgn} f_i$, we get equality and so $\|f\| = \|f\|_1$. We can show that all continuous linear functionals must have this form, i.e. $\ell_\infty^{n*} = \ell_1^n$. Similarly $\ell_1^{n*} = \ell_\infty^n$ and $\ell_1^* = \ell_\infty$. However, $\ell_\infty^* \neq \ell_1$.

Example 53. There is an analogous result for the Lebesgue spaces:

$$L^p[a,b]^* = L^q[a,b], \text{ where } 1/p + 1/q = 1; \quad 1 < p < \infty$$

i.e. linear functionals F_g on $L^p[a,b]$ have the form:

$$F_g(f) = \int_a^b f(t)\, g(t)\, dt, \qquad \text{where } g \in L^q[a,b].$$

It is easily verified that $F_g \in L^p[a,b]^*$ since it is linear and Hölder's inequality for integrals is

$$\left| \int_a^b f(t)\, g(t)\, dt \right| \leq \left(\int_a^b |f(t)|^p\, dt \right)^{1/p} \left(\int_a^b |g(t)|^q\, dt \right)^{1/q}$$

$$= \|f\|_p \|g\|_q, \text{ where } 1/p + 1/q = 1$$

i.e. F_g is bounded and $\|F_g\| \leq \|g\|_q$. In fact $\|F_g\| = \|g\|_q$. These duality results do not hold for $p = \infty$.

We see that since $1/p + 1/q = 1$, if we take a second dual of $L^p[a,b]$, i.e. $(L^p[a,b]^*)^* = L^q[a,b]^* = L^p[a,b]$. In general, for a normed linear space, one has $X^{**} \supset X$ and if $X \cong X^{**}$ we call X reflexive.

Definition 54. X is reflective if its second conjugate is itself, i.e. $X^{**} \cong X$. This is a rather special property of spaces like $L^p[a,b]$, ℓ_p, ℓ_p^n and is not shared by many common spaces, for example $\mathscr{C}[a,b]$ is not reflexive. However, all Hilbert spaces are reflexive and in fact we can say more — they are self-dual spaces.

Suppose $y \in H$ and define $f_y : H \to \mathbb{C}$ by

$$f_y(x) = \langle x, y \rangle \quad \forall x \in H$$

f_y is linear and $|f_y(x)| \leq \|x\| \, \|y\|$.
So f_y is bounded and $\|f_y\| \leq \|y\|$.

In fact we can show that $\|f_y\| = \|y\|$, and that every linear functional $f \in H^*$ corresponds to some $y \in H$ in this way, i.e. there is a 1 - 1 correspondence between $y \in H$ and $f_y \in H^*$ and $\|f_y\| = \|y\|$. H^* can be considered as a Hilbert space too, by defining $\langle f_x, f_y \rangle = \langle y, x \rangle$. So there is an isometric isomorphism between H and H^* — usually we identify them as the same space. It follows that Hilbert spaces are necessarily reflexive.

As convergence is a kep concept in analysis, we shall now have a careful look at what one means by $x_n \to x$. In fact, depending on the context we can mean many very different things.

Example 54. Let $x_n(t) = \begin{cases} 1/\sqrt{t} & \text{on } [1/n, 1] \\ 0 & \text{elsewhere} \end{cases}$

Then $x_n(.) \in L_p[0,1]$, for all finite p, since $\int_0^1 |x_n(t)|^p \, dt < \infty$. Clearly $x_n(t) \to 1/\sqrt{t}$ pointwise. Now $1/\sqrt{t} \in L_1[0,1]$, but $1/\sqrt{t} \notin L_2[0,1]$. So x_n converges in the space $L_1[0,1]$, but not in $L_2[0,1]$.

Example 55. If f(t) is integrable on (0, 1), we can form its Fourier series

$$a_0 + \sum_1^\infty (a_n \cos 2\pi nt + b_n \sin 2\pi nt)$$

where

$$a_0 = \int_0^1 f(t)\,dt, \qquad a_n = 2\int_0^1 f(t) \cos 2\pi nt \, dt, \qquad b_n = 2\int_0^1 f(t) \sin 2\pi nt \, dt$$

Now, in general, the Fourier series of f does not converge to f pointwise, even if f is continuous. However, the Fourier series always converges in L_2 [0, 1], i.e.

$$\int_0^1 \left[a_0 + \sum_{n=1}^k a_n \cos 2\pi nt + b_n \sin 2\pi nt - f(t) \right]^2 dt \to 0 \quad \text{as } k \to \infty$$

$\forall f \in L_2$ [0, 1].

This type of convergence we have been discussing is often called:

Definition 55. Convergence in norm (or strong convergence)

$x_n \to x$ in X means $\|x - x_n\| \to 0$ as $n \to \infty$.

Sometimes a sequence or series does not converge in this sense, but it does tend to something in a weaker sense, and we find the following concept useful.

Definition 56. $x_n \to x$ weakly in X if $f(x_n) \to f(x)$ as $n \to \infty$ for all $f \in X^*$.

Convergence in norm implies weak convergence, because if $x_n \to x$, $f(x_n) \to f(x)$ for all continuous functionals, i.e. for all $f \in X^*$.

Example 57. ℓ_p^n. Consider a sequence $\{x^k\}$, where $x^k = \{x_1^k, \dots, x_n^k\}$. What does weak convergence mean?

A functional on ℓ_p^n is $f_i(x) = x_i$, corresponding to $f_i = (0, \dots, 1, 0, \dots)$. So

$$f_i(x^k) = x_i^k \to x_i \text{ as } k \to \infty$$

i.e. each component of x^k tends to each component of x. But

$$\|x^k - x\|_p = \left(\sum_{i=1}^{n} |x_i^k - x_i|^p\right)^{1/p}$$

$$\to 0 \text{ as } k \to \infty, \text{ since n is finite,}$$

i.e. in this case, $x^k \to x$ in norm also.

This is a particular property of finite dimensional spaces: weak convergence and strong convergence are equivalent. However, for infinite dimensional spaces, this is certainly not the case.

Example 58. Consider ℓ_p and the sequence $y^k = (0, ., 1, 0, ..)$ with a 1 in the k-th position.

Then for $f \in \ell_p^* = \ell_q$, $f(y^k) = f_k \to 0$ as $k \to \infty$, since

$$\sum_{k=1}^{\infty} |f_k|^q < \infty.$$

So y^k converges weakly to the zero element. But

$$\|y^k \sim 0\|_p = 1 \not\to 0 \quad \text{as } k \to \infty.$$

Example 59. $\mathscr{C}[a, b]$ with the sup norm.

Here convergence in norm is exactly uniform convergence:

$$\|x_n - x\| \to 0 \quad \text{as} \quad n \to \infty \quad \text{iff} \quad \max_{a \le t \le b} |x_n(t) - x(t)| \to 0 \text{ as } n \to \infty.$$

Weak convergence is equivalent to pointwise convergence and uniform boundedness in this space, i.e.

$x_n(t) \to x(t)$ for each t

$|x_n(t)| \leq k$ uniformly in n and t.

Weak convergence in a Hilbert space becomes a particularly simple idea: We recall that $x_n \to x$ weakly in H iff $f(x_n) \to f(x)$ $\forall f \in H^*$

i.e. iff $\langle x_n, y\rangle \to \langle x, y\rangle$ $\forall y \in H$

i.e. iff $\langle x_n - x, y\rangle \to 0$ as $n \to \infty$ $\forall y \in H$

Example 60. In $L_2[0,1]$, $x_n(t) \to x(t)$ weakly means that

$$\int_0^1 (x_n(t) - x(t))\, y(t)\, dt \to 0 \text{ as } n \to \infty \quad \forall y \in L_2[0,1].$$

There is yet another type of convergence in X-weak* convergence. We need to identify X as the dual of some space, say $X = Y^*$.

Definition 57. A sequence $\{y_n^*\}$ in Y^* is weak* convergent to y^* if

$y_n^*(y) \to y^*(y)$ $\forall y \in Y$.

We recall that $Y \subset Y^{**}$ and so elements of Y define linear functionals on Y^*, and so weak* convergence is like weak convergence, except you only use a subset of all possible linear functionals. Of course, if Y is reflexive, $Y \cong Y^{**}$, then weak convergence on Y^* is just weak convergence on Y^*. (This is true for Hilbert spaces).

Just as it is often useful to use weaker types of convergence in a space, it is useful to have weak concepts of compactness. For example in optimization problems, one often seeks to maximize a linear functional over some set. A fundamental question is whether or not the given functional attains its maximum on the set and the main result is that a continuous functional on a compact set K of a normed linear space X achieves its maximum on K. For finite-dimensional spaces K is compact iff it is closed and bounded, but unfortunately this is not true for infinite-dimensional spaces. In fact the unit ball $\{x \in X : \|x\| \leq 1\}$ is compact iff X is finite dimensional. However, one can prove that the unit ball is "weak* compact". We recall that strong convergence or convergence on norm refers to the strong topology or norm topology of X, i.e. X considered as a metric space under the metric induced by the norm. Similarly, weak convergence induces the weak topology on X and weak* convergence the weak* topology and one has two new concepts of compactness.

Definition 58. A set A in X is weakly compact if for all sequences $\{x_n\} \subset A$ there is a weakly convergent subsequence with limit point in A.

Definition 59. A set A in $X = Y^*$ is weak* compact if for all sequences $\{x_n\} \subset A$ there is a weak* convergent subsequence with limit point in A.

Weak compactness and weak* compactness are equivalent for reflexive spaces.

6. LINEAR OPERATORS

We are often concerned with transformations between spaces. A very large class is the class of those transformations between linear spaces which preserve the algebraic structure.

Definition 60. A linear transformation T of a linear space X to a linear space Y is $\ni T(\alpha x + \beta y) = \alpha Tx + \beta Ty$ for all $x, y \in X$ and for all scalars α, β.

Example 61. Consider the spring-mass system

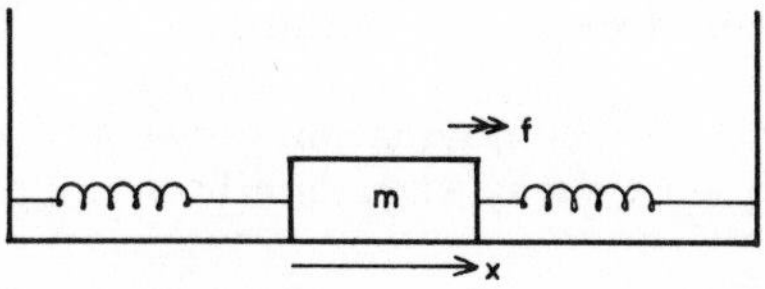

If we assume that the friction between the mass and the surface is $-b(dx/dt)$, the applied force is $f(t)$ and the combined restoring force of the springs is $-kx$, then if initially $x = 0$, $dx/dt = 0$, we have the usual equation of motion

$$f = m\frac{d^2x}{dt^2} + b\frac{dx}{dt} + kx \tag{i}$$

with solution

$$x(t) = \int_0^t h(t-s)\, f(s)\, ds \tag{ii}$$

where

$$h(r) = \frac{1}{m(\lambda_1 - \lambda_2)}\left(e^{\lambda_1 r} - e^{\lambda_2 r}\right)$$

and λ_1, λ_2 are the roots of $m\lambda^2 + b\lambda + k = 0$. Now if $f(.) \in \mathcal{C}[0,\infty]$, i.e. if it is a real continuous function, then so is x and so (ii) may be written $x = Lf$, where L is a linear transformation from $\mathcal{C}[0,\infty]$ to itself. Strictly speaking, L is an operator since $X = Y$. Similarly, (i) may be written $f = Tx$, where T again is a linear operator on $\mathcal{C}[0,\infty]$. L is an example of an integral operator and T a differential operator. Actually, since (i) and (ii) are equivalent, $L = T^{-1}$ or L is the inverse of T.

Definition 61. A map $F: X \to Y$ is invertible if there is a map $G: Y \to X$ such that GF and FG are the identity maps. G is called the inverse of F.

We know that the necessary and sufficient condition for F to be invertible is that F is 1 to 1 and $FX = Y$.

For linear maps, this reduces to the condition $Fx = 0$, only for $x = 0$.

From our example 61 we see that the solution of differential equations is one of finding inverses of certain transformations.

So far we have only considered algebraic properties of transformations, i.e. the preservation of the linear vector space structure. If X and Y are normed linear spaces, we may ask if a transformation preserves topological properties, for example

Definition 62. A transformation $T:X\to Y$ is continuous at $x_0 \in X$ if for every $\epsilon > 0$ there exists a $\delta > 0$ such that

$$\| x - x_0 \| < \delta \quad \text{implies that} \quad \| Tx - Tx_0 \| < \epsilon$$

(Note that this definition applies whether T is linear or not.)
An alternative form of the above definition is

Definition 62a. $T:X\to Y$, a transformation between two normed spaces, is continuous if $\{x_n\}$ being a Cauchy sequence in X implies that $\{Tx_n\}$ is Cauchy in Y.

Definition 63. $T:X\to Y$ is said to be bounded if $\| Tx\|_y \leq k\|x\|_x$ for some constant $k > 0$ and for all $x \in X$.

If T is bounded and linear then $\sup\{\| Tx\|_y : \|x\|_x = 1\}$ exists and is finite. We define this to be the norm of T, $\|T\|$. It satisfies all the properties of a norm, and we note the useful property:

$$\| Tx \|_y \leq \|T\| \; \|x\|_x \quad \text{for all} \quad x \in X.$$

For linear transformations you can show that the properties of continuity and boundedness are equivalent, and so we define

Definition 64. Let X, Y be normed linear spaces and $\mathscr{L}(X,Y) = \{T:X\to Y$ where T is bounded and linear$\}$. Under the norm defined above, $\mathscr{L}(X,Y)$ is a normed linear space.

For the special case, where $X = Y$, we write $\mathscr{L}(X)$ for the space of bounded linear operators on X. $\mathscr{L}(X)$ is not only closed under addition and scalar multiplication, but it is closed under operator multiplication or composition, i.e. $\mathscr{L}(X)$ is an algebra of operators.

For if $T_i \in \mathscr{L}(X)$, i.e. $T_i : X \to Y$ and is linear, it is readily seen that $T_1 T_2 : X \to Y$ and is also linear. The boundedness follows since

$$\begin{aligned} \| T_1 T_2 x \| &\leq \| T_1 \| \; \| T_2 x \| \\ &\leq \| T_1 \| \; \| T_2 \| \; \| x \| \end{aligned}$$

This also yields the important result $\| T_1 T_2 \| \leq \| T_1 \| \; \| T_2 \|$.

Of course not all linear operators are bounded, for example in Ex.61. L is bounded, whereas T is unbounded.

Example 62. Let $X = L_1[a,b]$, $Y = \mathscr{C}[a,b]$ and $T:X\to Y$ be given by

$$Tx(s) = \int_a^b k(s,t)\, x(t)\, dt$$

where $k(.,.) : [a,b] \times [a,b] \to \mathbb{R}$ is continuous in s and t. Then T is an integral operator, clearly linear, and

$$|Tx(s)| \leq \max_{0 \leq s,t \leq b} |k(s,t)| \, \|x\|_x$$

$$\|Tx(.)\|_y = \sup_{a \leq s \leq b} |Tx(s)| \leq \max_{0 \leq s,t \leq b} |k(s,t)| \, \|x\|_x$$

So $\quad T \in \mathscr{L}(X,Y)$ and $\|T\| \leq \max_{a \leq s,t \leq b} |k(s,t)|$

Example 63. $X = L_2[0,1]$ and $T: X \to X$ is given by $Tf = df/dt$.

T is linear, but cannot be bounded, as it is not even defined on all of X, but only on the subspace $\mathscr{D} = \{f \in X \ni f'(.) \in X\}$, called the domain of T. T is actually an example of a closed operator, which we shall define in the next section.

If, however, we take $Y = \mathscr{C}'[0,1]$, the space of all continuous functions with continuous first derivatives, then T is defined on Y and $T: Y \to X$ is bounded. This emphasizes the dependence of continuity and boundedness on the particular norm you are considering.

An interesting property of continuous linear transformation is used implicitly in the principle of superposition, a handy tool in differential equations. This depends on the mathematical theorem:

$L: X \to Y$ is a continuous linear transformation iff

$$L\left(\sum_{i=0}^{\infty} \alpha_i x_i\right) = \sum_{i=0}^{\infty} \alpha_i L x_i$$

for every convergent series

$$\sum_{i=0}^{\infty} \alpha_i x_i \in X$$

Again we emphasize that the continuity of L depends on the norms you choose for X and Y.

Example 64. Define $J: L_2[-\pi,\pi] \to L_2[-\pi,\pi]$ by $y = Jx$, where $y(t) = \int_0^t x(s)\,ds$. Then J is continuous and linear. Consider $\sum_{n=1}^{\infty} \frac{1}{n} \sin nt$ a convergent series in L_2, since $\int_{-\pi}^{\pi} \frac{1}{n^2} \sin^2 nt\,dt \to 0$ as $n \to \infty$. So by our above theorem,

$$J\left(\sum \frac{1}{n} \sin nt\right) = \int_0^t \sum_{n=1}^{\infty} \frac{1}{n} \sin ns \, ds$$

$$= \sum_{n=1}^{\infty} \int_0^t \frac{1}{n} \sin ns \, ds$$

$$= \sum_{n=1}^{\infty} \frac{1}{n^2} (\cos nt - 1)$$

Here we shall primarily be concerned with operators on Hilbert spaces as this case is similar to the theory of matrices.

Definition 65. Let $K \in \mathscr{L}(H)$, then the adjoint K^* is defined by

$$\langle Kx, y \rangle = \langle x, K^*y \rangle \quad \forall\, x, y \in H$$

That K^* always exists is easily seen by considering $\langle Kx, y\rangle = f(x)$; f is a linear functional on H and $|f(x)| \leqq \|Kx\| \, \|y\|$

$$\leqq \|K\| \, \|x\| \, \|y\|$$

$$\leqq \text{const} \, \|x\| \quad \text{for fixed } y$$

So $f \in H^*$ and so there is a $y^* \in H$, such that $f(x) = \langle x, y^* \rangle$, i.e. K induces a map $y \to y^*$ $\forall\, y \in H$ and we call this $y^* = K^*y$. The adjoint has the following easily verifiable properties:

1. $I^* = I$; $0^* = 0$

2. $(S+K)^* = S^* + K^*$

3. $(\alpha T)^* = \overline{\alpha}\, T^*$

4. $(ST)^* = T^*S^*$

5. $\|T^*\| = \|T\|$

We shall just prove the last property:

$$\|T^*y\|^2 = \langle T^*y, T^*y \rangle$$

$$= \langle TT^*y, y \rangle$$

$$\leqq \|TT^*y\| \, \|y\|$$

$$\leqq \|T\| \, \|T^*y\| \, \|y\|$$

$$\therefore \|T^*y\| \leqq \|T\| \, \|y\|$$

i.e. $\|T^*\| \leqq \|T\|$ and similarly $\|T\| \leqq \|T^*\|$

Example 65. $H = \ell_2^n$. Then $A \in \mathscr{L}(H)$ is representable by a matrix, and A^* is its transpose for the real scalars and the conjugate transpose for the complex case.

Example 66. $H = L_2[a,b]$ and $Tx(.) = \int_a^b k(.,s)\, x(s)\, ds$

$$\langle Tx, y \rangle = \int_a^b \int_a^b k(t,s)\, x(s)\, ds\, \overline{y}(t)\, dt$$

$$= \int_a^b x(s) \int_a^b k(t,s)\, \overline{y}(t)\, dt\, ds$$

$$= \int_a^b x(s) \overline{\int_a^b \overline{k(t,s)}\, y(t)\, dt}\, ds$$

and so

$$T^*y(.) = \int_a^b \overline{k(t,.)}\, y(t)\, dt$$

Example 69. Let $H = \left\{ \begin{array}{l} \text{space of functions } u : [0,T] \to \mathbb{R}^m \text{ with inner product} \\ \langle u, v \rangle = \int_0^T u'(s)\, v(s)\, ds \end{array} \right\}$

i.e. $H = L_2([0,T]; \mathbb{R}^m)$

($u(s)$ is a column vector and the prime denotes the row vector.)

If $B(t)$ is a positive symmetric $m \times m$ matrix, then we define the operator B by

$$(Bu)(t) = B(t)\, u(t) \quad \forall t \in T$$

Verify that $B^* = B$.

Note also that

$$\langle Bu, u \rangle = \int_0^T (B(t)u(t))'u(t)\, dt$$

$$= \int_0^T u(t)'B(t)\, u(t)\, dt$$

$$\geq 0 \quad \text{since } B(t) \text{ is a positive matrix.}$$

This leads us to two further definitions:

Definition 66. $A \in \mathscr{L}(H)$ is self-adjoint if $A^* = A$. (A self-adjoint operator has the property that $\langle Ax, x \rangle$ is always real.)

Definition 67. A positive operator $A \in \mathscr{L}(H)$ is a self-adjoint operator A such that $\langle Ax, x\rangle \geq 0 \;\; \forall x \in H$.

It is called strictly positive if $\langle Ax, x\rangle = 0$ only if $x = 0$. Of course the self-adjoint operators are generalizations of real symmetric matrices or Hermitian matrices, and positive operators correspond to the positive definite matrices with the property $x'Ax \geq 0$.

Self-adjoint and positive operators occur frequently in applications and they have several special properties which we shall list here as they are important in applications.

1. $$\|A\| = \sup_{\|x\|=1} |\langle Ax, x\rangle|$$

   or equivalently the smallest M such that $|\langle Ax, x\rangle| \leq M\|x\|^2$

2. A finer inequality is $m\|x\|^2 \leq \langle Ax, x\rangle \leq M\|x\|^2$

   If A is positive then $m \geq 0$, and if A is strictly positive, then $m>0$.
In the strictly positive case, the following also holds:

$$\frac{1}{M}\|x\|^2 \leq \langle A^{-1}x, x\rangle \leq \frac{1}{m}\|x\|^2 \qquad \forall x \in H$$

i.e. A^{-1} exists and is strictly positive.

3. Every strictly positive operator A has a unique strictly positive square root $A^{1/2}$.

4. An orthogonal projection is self-adjoint, since

$$\langle Px, x\rangle = \langle x_M, x_m + x_{m\perp}\rangle = \langle x_m, x_m\rangle$$

and

$$\langle x, Px\rangle = \langle x_M + x_{M\perp}, x_M\rangle = \langle x_m, x_m\rangle$$

Example 70. Consider the same space as Ex.69 (often denoted by $L_2([0,T]; \mathbb{R}^m)$ and let $W(t,\tau)$ be an $r \times m$ matrix-valued function on $[0,T] \times [0,T]$. Then the operator $\mathscr{W}$:

$$(\mathscr{W}u)(t) = \int_0^t W(t,\tau)\, u(\tau)\, d\tau$$

maps

$$H \to Y = L_2([0,T]; \mathbb{R}^r).$$

$$\langle \mathscr{W}u, v\rangle_y = \int_0^T \left(\int_0^t W(t,\tau)\, u(\tau)\, d\tau \right)' v(t)\, dt$$

$$= \int_0^T \int_0^t u(\tau)' W'(t,\tau)\, v(\tau)\, d\tau\, dt$$

$$= \int_0^T \left(\int_\tau^T u(\tau)'\, W(t,\tau)'\, v(t)\, dt \right) d\tau$$

changing the order of integration

$$= \int_0^T u(\tau)' \left(\int_\tau^T W(t,\tau)'\, v(t)\, dt \right) d\tau$$

$$= \langle u, \mathcal{W}^* v \rangle$$

where $\mathcal{W}^* v(t) = \int_t^T W(\tau, t)'\, v(\tau)\, d\tau$

We should note that although for simplicity we have limited the definitions of adjoint, self-adjointness and positivity to bounded linear operators, all these definitions can be extended to closed linear operators on a Banach space.

7. SPECTRAL THEORY

Most of the problems of linear algebraic equations, ordinary differential equations, integral equations and partial differential equations can be formulated as linear operator problems.

Example 67. The Fredholm integral equation is

$$\lambda f(t) - \int_a^b k(t,s)\, f(s)\, ds = g(t)$$

to be solved for the unknown function f(t), where $k(.,.) : [a,b] \times [a,b] \rightarrow \mathbb{R}$. $\int_a^b \int_a^b |k(t,s)|^2\, dt\, ds < \infty$. If we let $X = L_2[a,b]$ and define $K: X \rightarrow X$ by $K f(t) = \int_a^b k(t,s)\, f(s)\, ds$ then this is really an operator problem $(\lambda I - K) f = g$, where I is the identity operator and $f = (\lambda I - K)^{-1} g$, if $(\lambda I - K)^{-1}$ exists.

Example 68.

$$\frac{\partial^2 u}{\partial x^2} = \frac{\partial u}{\partial t} \quad \ldots\ldots \quad \left\{ \begin{array}{l} u(0,t) = 0 = u(1,t) \\ u(x,0) = u_0(x) \end{array} \right\}$$

This is the familiar heat equation. If we let $X = L_2[0,1]$, then we can define

$$A : X \to X \text{ by } Af = \frac{\partial^2 f}{\partial x^2} \quad \text{for} \quad f \in \mathscr{D}(A) = \left\{ g : g, \frac{\partial g}{\partial x}, \frac{\partial^2 g}{\partial x^2} \in L_2[0,1] \right\}$$

A is a closed linear operator on X with domain $\mathscr{D}(A)$ and the heat equation may be written

$$\frac{du}{dt} = Au(t)$$

In Eq.67, K is a linear bounded operator, but in Ex.66, A is linear but unbounded. In fact most differential operators are unbounded, although fortunately they do form a class with nice properties; the class of closed operators.

Definition 68. A linear operator $T : X \to X$ is closed if for all sequences $\{x_n\}$ in the domain of T, $\mathscr{D}(T)$ with $x_n \to x$ and $Tx_n \to y$ then $x \in \mathscr{D}(T)$ and $y = Tx$. Another way of saying this is to consider the product space $\mathscr{D}(T) \times R(T)$ with typical element (x, Tx). Then T is closed if $\mathscr{D}(T) \times R(T)$ is closed in $X \times X$. ($\mathscr{R}(T)$ is the range of T.)

A familiar example of a closed operator which occurs in physical problems is the Laplacian.

Example 69. Let Ω be the unit disc in $\mathbb{R}^2$, $\Omega = \{(x,y) : x^2 + y^2 \leq 1\}$ and consider the Laplace operator

$$\Delta u = \frac{\partial^2 u}{\partial x^2} + \frac{\partial^2 u}{\partial y^2} \quad \text{on } L_2(\Omega),$$

the space of Lebesgue integrable functions of two variables with finite norm

$$\|u\| = \left(\int_0^1 \int_0^1 u(x,y)^2 dxdy \right)^{1/2}$$

Let

$$\mathscr{D}(\Delta) = \{u \in L_2(\Omega) : u \text{ is } \mathscr{C}^2 \text{ and } \Delta u \in L_2(\Omega); \quad u = 0 \text{ on } \partial\Omega\}$$

($\partial\Omega$ is the boundary of the disc)

Then Δ is a closed operator on $L_2(\Omega)$ with the domain $\mathscr{D}(\Delta)$. The classical partial differential equation: $\Delta u = f$; $u = 0$ on $\partial\Omega$ has the solution

$$u(x,y) = \int_\Omega G(x,y;\xi,\eta) f(\xi,\eta) \, d\xi \, d\eta$$

where the Green's function is given by

$$G(x,y;\xi,\eta) = -\frac{1}{2\pi} \log \frac{\sigma_2}{\sigma_0 \sigma_3}$$

where the σ_i are the distances shown in the following figure:

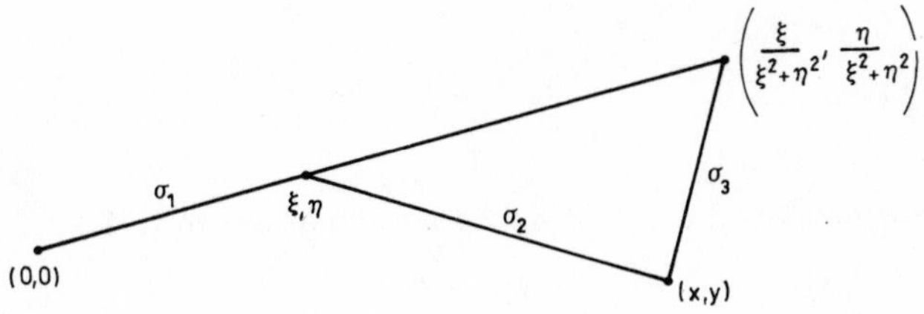

i.e. Δ has an inverse and $u = \Delta^{-1}f$. By inspection, Δ^{-1} is an integral operator and is linear and bounded on $L_2(\Omega)$. In a similar manner, other differential equations can be rephrased as problems of finding the inverse of a closed operator. Note that for a true solution to the problem, we need the inverse to be bounded. For example if $\Delta u = f$ is to have a solution for all $f \in L_2(\Omega)$, we need Δ^{-1} linear and bounded, i.e. defined on all $L_2(\Omega)$. Much of functional analysis is concerned with finding inverses of operators, but for now we shall just state the important result concerning linear bounded operators:

If X, Y are B-spaces and $L \in \mathscr{L}(X, Y)$ and L^{-1} exists in an algebraic sense, then it is bounded and linear. Now in finite dimensions the solution of linear equations depended on the eigenvalues of the matrices and although the alternative theorem is probably very familiar to you, we shall formulate the results in a Hilbert space context, under the heading

"Spectral theory in finite dimensions"

Let $X = \mathbb{R}^n$. Then $T \in \mathscr{L}(\mathbb{R}^n)$ has a matrix representation, which is not unique as it depends on the basis used for $\mathbb{R}^n$. However, each T is uniquely represented by an equivalence class of similar matrices (A and B are similar matrices if there exists a non-singular C, such that $A = C^{-1}BC$). Similar matrices have the same characteristic equations, and hence the same eigenvalues.

So for each $T \in \mathscr{L}(\mathbb{R}^n)$, we can define

1. The spectrum of T, $\sigma(T) = \{\text{set of eigenvalues of } T : \lambda_i\}$.

2. The eigenvectors of T corresponding to λ_i are x_i: $(T - \lambda_i I)x_i = 0$.

3. The eigenspace of T corresponding to λ_i is $M_i = Sp\{x : Tx = \lambda_i I\}$.

4. The projection operator P_i corresponding to M_i.

Using these definitions, we can decompose T as the sum,

$$T = \sum_{i=1}^{m} \lambda_i P_i$$

where λ_i are distinct eigenvalues and P_i is their projection operator. It can be shown that the P_i are pairwise orthogonal, i.e. $P_i P_j = 0$, $i \neq j$.

(You may remember this as the fact that eigenvectors of distinct eigenvalues are perpendicular.) So if we take powers of T, we have

$$T^k = \sum_{i=1}^{m} \lambda_i^k P_i$$

an easy calculation. Similarly, if f(t) is a polynomial in t, then

$$f(T) = \sum_{i=1}^{m} f(\lambda_i) P_i$$

The study of eigenvalues in matrices has particular significance for the algebraic linear equation

$$y = (\lambda I - T)x \quad \text{on } \mathbb{R}^n$$

The 'alternative' theorem states that it has a unique solution

$$x = (\lambda I - T)^{-1} y$$

(i.e. $(\lambda I - T)^{-1}$ exists as a matrix), i.e. provided λ is not an eigenvalue or provided $\lambda \notin \sigma(T)$.

This property that the existence of a solution of $y = (\lambda I - T)x$ depends on the spectrum of T generalizes to the infinite dimensional case.

Definition 69. Let T be a linear operator on a Banach space X, then

1. The resolvent of T, $\rho(T) = \{\lambda \in \mathbb{C} : (\lambda I - T)^{-1} \in \mathscr{L}(X)\}$

2. The spectrum of T, $\sigma(T) = \mathbb{C} - \rho(T)$

That is, for $\lambda \in \rho(T)$, the linear equation $y = (\lambda I - T)x$ has a unique solution $x = (\lambda I - T)^{-1} y$ for all $y \in X$. Note that $(\lambda I - T)x = y$ may now be an integral equation or a differential equation (see Exs 65 and 66).

An important subset of $\sigma(T)$ is the point spectrum:

$$P\sigma(T) = \{\lambda : (\lambda I - T) \text{ is not one to one}\}$$

In the finite-dimensional case, $\sigma(T) = P\sigma(T)$ and is a finite set, but in infinite dimensions, things are slightly more complicated. Certain operators, however, have similar properties to finite-dimensional operators (i.e. matrices). These are compact operators, a subclass of linear bounded operators, which have the special property that their spectrum is a point spectrum and has countably many elements,

i.e. $\sigma(T) = P\sigma(T) = \{\lambda_i;\ i = 1, 2, .. \infty\}$

This means then that $(\lambda I - T)x = y$ has a unique solution except for countably many λ_i, the $\lambda_i \in P\sigma(T)$.

<u>Definition 70.</u> <u>A compact operator T</u> is an operator $\in \mathscr{L}(X)$ which maps a bounded set A of X into a precompact set TA (i.e. TA has compact closure $\overline{TA}$).

An equivalent definition is a linear operator T which, for any bounded sequence $\{x_n\}$, $\{Tx_n\}$, has a convergent subsequence. All finite-dimensional operators are compact, as are linear operators with finite-dimensional range, since in a finite-dimensional space every closed bounded set is compact. Another example of compact operators are integral operators; K of Ex.65 is compact. Unfortunately, all linear operators are not compact and $\sigma(T)$ may be very complicated.

Example 70. $X = \mathscr{C}[a,b]$ and $T: x(t) \to \mu(t)\, x(t)$ where $\mu(t)$ is continuous. Consider

$$(T - \lambda I)\, x(t) = y(t)$$

$$\therefore (\mu(t) - \lambda)\, x(t) = y(t)$$

$$\therefore x(t) = \frac{y(t)}{\mu(t) - \lambda}, \quad \text{provided } \mu(t) - \lambda \neq 0$$

and

$$\sigma(T) = \{\lambda \ni \mu(t) - \lambda = 0, \text{ for } a \leq t \leq b\}$$

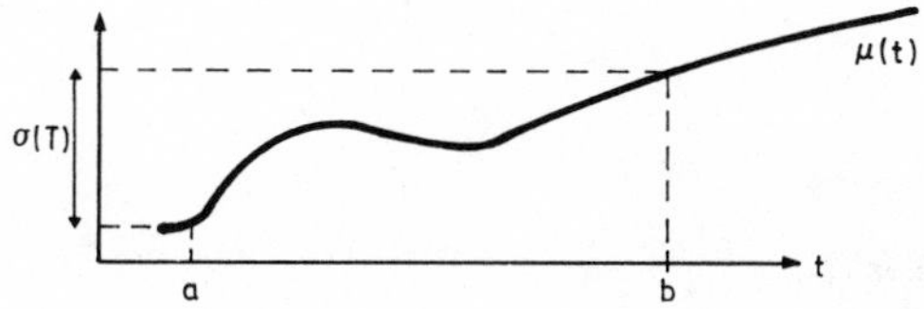

This is an example of a continuous spectrum.

Example 71. $X = \ell_1$, $Tx = \left\{\frac{x_1}{1}, \frac{x_2}{2}, \ldots, \frac{x_n}{n}, \ldots\right\}$

Consider

$$(\lambda I - T)x = y$$

Now

$$(\lambda I - T)x = \left\{\lambda x_1 - x_1, \lambda x_2 - \frac{x_2}{2}, \ldots\right\}$$

So $(\lambda I - T)$ is not 1 - 1 if $\lambda = 1/n$; $n = 1, 2, \ldots,$

i.e. $P\sigma(T) = \left\{\frac{1}{n};\quad n = 1, \ldots \infty\right\}$.

We can show that $0 \in P\sigma(T)$ also.

Example 72. Consider the bounded sequence cos x, cos 2x, cos 3x, ... and the integral map $T: Tx(t) = \int_0^t x(s)\,ds$, producing the sequence sin x, $\frac{1}{2}\sin 2x, \ldots, \frac{1}{n}\sin nx, \ldots$

Then this derived sequence is convergent, because T is a compact operator on $L_2(0,1)$.

Since a knowledge of the spectrum tells us when we can hope for a solution of $(\lambda I - T)x = y$, it is an important study in analysis. One very useful result for self-adjoint compact operators is that: $\sigma(T) \subset [-\|T\|, \|T\|]$, i.e. for $|\lambda| \geq \|T\|$, $(\lambda I - T)^{-1} \in \mathscr{L}(X)$ or equivalently, $(\lambda I - T)x = y$ has the unique solution $x = (\lambda I - T)_y^{-1}$. The proof is easy and an instructive exercise. First we consider the $\lambda = 1$ case: $(I-T)x = y$, $\|T\| < 1$ and let

$$B_k = \sum_{i=0}^{k} T^i$$

$\{B_k\}$ is Cauchy since

$$\| B_k - B_n \| \leq \sum_{k}^{n} \|T\|^i \to 0 \quad \text{as} \quad n, k \to \infty.$$

because $\|T\| < 1$. $\mathscr{L}(X)$ is complete and so $B = \sum_{i=0}^{\infty} T^i \in \mathscr{L}(X)$. But

$$B(I - T) = \sum_{i=0}^{\infty} T^i - \sum_{i=1}^{\infty} T^i = I$$

$$\therefore B = (I - T)^{-1} \in \mathscr{L}(X)$$

For $\lambda \neq 1$ we just consider $(\lambda I - \lambda)^{-1} = \lambda(I - (1/\lambda)T)^{-1}$ and $\|(1/\lambda)T\| < 1$ is equivalent to $\|T\| < \lambda$.

For certain classes of linear operators you can obtain a spectral decomposition for T. For example, if T is a compact, self-adjoint operator on a Hilbert space H, then

$$Tx = \sum_{k=1}^{\infty} \lambda_k \langle \lambda, e_k \rangle e_k$$

where $\lambda_k \in \sigma(T)$ and e_k are the corresponding eigenvectors. This looks very similar to the matrix case, however, for more general T, the best we can hope for is an integral decomposition of the type $Tx = \int \lambda\, dE(\lambda)x$.

Application to Fredholm integral equations

Consider

$$\lambda f(t) = \int_a^b k(t,s) f(s)\,ds = g(t)$$

the so-called Fredholm integral equation. Let K, X be as in Ex.67 and write it as

$$(\lambda I - K)f = g$$

It can be shown that if k(t, s) is integrable on [a, b]×[a, b] and

$$\int_a^b \int_a^b |k(t,s)|^2 \, dt\, ds < \infty,$$

then $K \in \mathscr{L}(X)$ and is compact. Now

$$\| k \| = \sup_f \left\{ \left(\int_a^b \left| \int_a^b k(t,s) f(s)\, ds \right|^2 dt \right)^{1/2} ; \int_a^b |f(s)|^2 \, ds = 1 \right\}$$

We know that if $\|K\| < |\lambda|$, then the equation has the unique solution

$$f = (\lambda I - K)^{-1} g$$

$$= \sum_{i=0}^{\infty} \left(\frac{K}{\lambda}\right)^i \frac{g}{\lambda}$$

One can verify that

$$K^n g = \int_a^b k_n(t,s)\, g(s)\, ds$$

where

$$k_n(t,s) = \int_a^b k_1(u,s)\, k_{n-1}(t,u)\, du$$

which gives an iterative method for finding the solution. As the series solution for f converges, a finite approximation will suffice.

If $\|K\| > |\lambda|$, we know that, except for countably many λ_i, the equation again has a unique solution. These are procedures for getting approximate solutions based on approximations on the kernel function k(t, s) but we shall only work out the case for $k(t,s) = \overline{k(s,t)}$, i.e. K is self-adjoint. In this case, K has the decomposition

$$K = \sum_{i=1}^{\infty} \lambda_i \langle \cdot, x_i \rangle x_i$$

where λ_k are its eigenvalues and x_k the corresponding eigenvectors.

We can now rewrite $\lambda f - Kf = g$, as

$$\lambda f - \sum_{i=1}^{\infty} \lambda_i \langle f, x_i \rangle x_i = g$$

$$\therefore \lambda \langle f, x_j \rangle - \lambda_j \langle f, x_j \rangle = \langle g, x_j \rangle, \quad \text{since } \langle x_i, x_j \rangle = \delta_{ij}$$

$$\therefore \langle f, x_j \rangle = \frac{\langle g, x_j \rangle}{\lambda - \lambda_j} \quad \text{for} \quad \lambda \neq \lambda_j$$

$$\therefore \lambda f = Kf + g$$

$$= \sum_{i=1}^{\infty} \lambda_i \langle f, x_i \rangle x_i + g$$

$$= \sum_{i=1}^{\infty} \lambda_i \frac{\langle g, x_i \rangle}{\lambda - \lambda_i} x_i + g$$

which solves the equation for f if $\lambda \notin \sigma(K)$.

$$\left(\langle g, x_i \rangle = \int_a^b g(t)\, x_i(t)\, dt \right)$$

Unfortunately, in practice many equations are represented by operators which are not compact or even bounded. These are the ordinary differential equations and partial differential equations, which in linear cases can be represented by closed operators.

Although closed operators are not as well behaved as compact or bounded operators, they also have nice properties and a knowledge of their spectrum also gives useful information. For example, if A is a closed linear operator on a Hilbert space H and for some λ, $(\lambda I - A)^{-1} \in \mathscr{L}(H)$ and is compact and self-adjoint, then A has a special eigenfunction decomposition:

$$Ax = \sum_i \mu_i \langle x, x_i \rangle x_i$$

for every x in the domain of A and $\{x_1, x_2, \ldots\}$ eigenvectors of A, and μ_n the eigenvalues of A.

Example 73. The heat equation of Ex.68

Assume that $u(x, t) = U(x)\, V(t)$

$$\therefore \frac{1}{V} \frac{dV}{dt} = \frac{1}{U} A U$$

Since the right-hand side depends only on x and the left-hand side only on t they must both equal a constant, λ, say

$$AU = \lambda U; \quad \frac{dV}{dt} = \lambda V$$

We can show that A has the property that $(\lambda_0 I - A)^{-1}$ is compact and self-adjoint for some λ_0,

i.e. $A = \sum \mu_n \langle \cdot, x_n \rangle x_n$, and $Ax_n = \mu_n x_n$

So a solution is $u(x,t) = e^{\mu_n t} x_n(x)$ and the general solution is $u(x,t) = \sum d_n e^{\mu_n t} x_n(x)$ where the d_n are determined by

$$u_0(x) = \sum d_n x_n(x)$$

i.e. $d_n = \langle u_0, x_n \rangle$

$$\therefore u(x,t) = \sum_{n=1}^{\infty} \langle u_0, x_n \rangle e^{\mu_n t} x_n(x)$$

Unfortunately, we cannot linger on this very important and vast area of application of functional analysis to differential equations, but merely refer the reader to Naylor and Sell or Dunford and Schwartz (see Bibliography).

8. CALCULUS IN B-SPACES

In this section, we shall be mainly concerned with developing a differential calculus for operators. The fundamental idea involved is the local approximation of operators by linear operators. Unfortunately, in order to understand this concept it will be necessary to "unlearn" the interpretation of the derivative of a real-valued function of a real variable. First of all, recall that for

$$f: \mathbb{R} \to \mathbb{R}$$

$$f'(x_0) = \lim_{t \to 0} \left[\frac{f(x_0 + t) - f(x_0)}{t} \right]$$

and that a good approximation to f(x) for x near x_0 is

$$L(x) = f(x_0) + f'(x_0)(x - x_0)$$

If we consider $f: \mathbb{R}^2 \to \mathbb{R}$, then the above expression for f' no longer makes sense because we cannot add the vector x_0 to the scalar t. However, since $f(x) = f(x_1, x_2)$, we can consider the partial derivatives

$$\frac{\partial f}{\partial x_1} = \lim_{t \to 0} \left[\frac{f(x_1 + t, x_2) - f(x_1, x_2)}{t} \right]$$

and

$$\frac{\partial f}{\partial x_2} = \lim_{t \to 0} \left[\frac{f(x_1, x_2 + t) - f(x_1, x_2)}{t} \right]$$

However, it is still not obvious how we should interpret

$$f'(x_0)(x - x_0) \text{ when } x - x_0 \in \mathbb{R}^2$$

For $\mathbb{R}'$, we have

$$f'(x_0)(x - x_0) = f'(x_0) \times (x - x_0)$$

and since the generalization of scalar multiplication is the inner product, we could take

$$f'(x_0)(x - x_0) = \langle \nabla f(x_0), x - x_0 \rangle$$

where

$$\nabla f = \left(\frac{\partial f}{\partial x_1}, \frac{\partial f}{\partial x_2} \right)$$

This is usually written

$$df = \frac{\partial f}{\partial x_1} dx_1 + \frac{\partial f}{\partial x_2} dx_2$$

But note that this definition depends on

(1) the inner product in $\mathbb{R}^2$

(2) a natural basis in $\mathbb{R}^2$

(3) the fact that f is real-valued.

Hence this approach will only generalize to functionals on $\mathbb{R}^n$.

Let us now return to $f: \mathbb{R} \to \mathbb{R}$ and

$$f'(x_0)(x - x_0) = \lim_{t \to 0} \left[\frac{f(x_0 + t) - f(x_0)}{t} (x - x_0) \right]$$

Write $x - x_0 = \eta$ and, replacing t by $t\eta$, we have

$$f'(x_0)\eta = \lim \left[\frac{f(x_0 + t\eta) - f(x_0)}{t} \right]$$

This latter interpretation makes sense for $f: X \to Y$, where X need only be a vector space and the space Y possesses some topological structure (so that the limit operation makes sense).

Definition 71. Gateaux derivative

Given x and η in X if

$$Df(x)\eta = \lim_{t\to 0}\left[\frac{f(x+t\eta)-f(x)}{t}\right]$$

exists, then f is called Gateaux differentiable at x in the direction of η.

We say f is Gateaux differentiable if it is Gateaux differentiable in every direction, and in this case the operator $Df(x): X \to Y$ which assigns to each $\eta \in X$ the vector $Df(x)(\eta) \in Y$ is called the Gateaux derivative at x.

Example 74. Consider $f: \mathbb{R}^n \to \mathbb{R}$ and the basis $e_i = (0,..,0,1,0,..)$ then

$$x = (x_1, x_2, \ldots x_n) = \sum_{i=1}^{n} x_i e_i$$

and

$$Df(x)\, e_i = \frac{\partial f}{\partial x_i}$$

Example 75. Let $f: \mathbb{R}^2 \to \mathbb{R}$ be given by

$$f(x) = \frac{x_1 x_2}{x_1^2 + x_2^2}\ , \quad x \neq 0; \qquad f(0) = 0$$

Then

$$Df(0)\eta = \lim_{t\to 0} \frac{1}{t}\left(\frac{\eta_1\eta_2}{\eta_1^2+\eta_2^2}\right)$$

exists iff

$$\eta = (\eta_1, 0) \text{ or } (0, \eta_2)$$

This example shows that the existence of the partial derivatives is not a sufficient condition for the Gateaux derivative to exist.

Example 76. Let $f: \mathbb{R}^2 \to \mathbb{R}$ be given by

$$f(x) = \frac{x_1 x_2^2}{x_1^2 + x_2^2}\ , \quad x \neq 0; \qquad f(0) = 0$$

Then

$$Df(0)\eta = \frac{\eta_1 \eta_2^2}{\eta_1^2 + \eta_2^2}$$

This example shows that the Gateaux derivative is not a linear operator.

Theorem: If the functional $f: X \to \mathbb{R}$ has a minimum or a maximum at $x \in X$ and $Df(x)$ exists, then $Df(x) = 0$.

Proof: If $\eta \in X$ is such that $Df(x)\eta > 0$, then for t sufficiently small $\frac{1}{t}(f(x+t\eta) - f(x)) > 0$. Consequently $f(x+t\eta) > f(x)$ if $t > 0$ and $f(x+t\eta) < f(x)$ if $t < 0$. A similar argument can be used if $Df(x)\eta < 0$.

Example 77. Let $\mathscr{C}[0,1]$ be the vector space of real-valued functions which are continuous on $[0,1]$. Consider $T: \mathscr{C}[0,1] \to \mathbb{R}$ defined by

$$T(y) = \int_0^1 [\tfrac{1}{2}(x+1)\, y(x)^2 - y(x)]dx$$

Then

$$DT(y)(\eta) = \int_0^1 [(x+1)y(x) - 1]\, \eta(x)\, dx$$

Hence for a minimum

$$DT(y)(\eta) = 0 \quad \forall \eta$$

Let

$$\eta(x) = (x+1)\, y(x) - 1; \text{ then we see}$$

$$(x+1)y(x) - 1 = 0$$

$$\therefore\ y(x) = 1/x+1.$$

The concept of the Gateaux derivative did not require any topology on the domain space, this can lead to "unusual" properties, for example: Consider $f: \mathbb{R}^2 \to \mathbb{R}$, where $f(x) = \dfrac{x_1^3}{x_2}$, $x \neq 0$; $f(0) = 0$, then $Df(0)(\eta) = 0\ \forall \eta \in X$. Hence $Df(0)$ exists and is a continuous linear operator. But f is not continuous at 0. In order to make sure that differentiable functions are continuous we now introduce the concept of a Fréchet derivative.

Definition 72. Fréchet derivative

Consider $f: X \to Y$, where both X and Y are normed linear spaces. Given $x \in X$; if a linear operator $df(x)$ exists which is continuous such that

$$\lim_{\|h\| \to 0} \left\{ \frac{\| f(x+h) - f(x) - df(x)(h) \|}{\| h \|} \right\} = 0$$

then f is said to be Fréchet differentiable at x and $df(x)(h)$ is said to be the Fréchet differential of f at x with increment h.

It is easy to see that if the Fréchet differential exists, then the Gateaux differential exists and the two are equal. Moreover, if f has a Fréchet differential at x, then f is continuous at x.

Example 78. Suppose $f: \mathbb{R}^n \to \mathbb{R}^m$ is Fréchet differentiable at x, then

$$df(x)(\eta) = \begin{pmatrix} \frac{\partial f_1}{\partial x_1} & \cdots\cdots\cdots & \frac{\partial f_1}{\partial x_n} \\ \\ \frac{\partial f_m}{\partial x_1} & \cdots\cdots\cdots & \frac{\partial f_m}{\partial x_n} \end{pmatrix} \begin{pmatrix} \eta_1 \\ \vdots \\ \eta_n \end{pmatrix}$$

If x is near x_0, i.e. $\|x - x_0\|$ is small, then $\|f(x) - f(x_0) - df(x_0)(x - x_0)\|$ is near zero. Here a good approximation to f(x) is

$$f(x) \simeq f(x_0) + df(x_0)(x - x_0)$$

So that the Fréchet derivative is essentially a linear approximation to $f(x + \Delta x) - f(x)$ for $\|\Delta x\|$ small.

We shall now see how these concepts can be applied to problems of approximation and optimization.

Newton's method

Consider the non-linear operator $P: X \to X$ and suppose we wish to find $x \in X$ such that $P(x) = 0$. Given $x_0 \in X$, let $\eta = x - x_0$. Then we require η so that $P(x_0 + \eta) = 0$, i.e. $P(x_0 + \eta) - P(x_0) = - P(x_0)$. But if x_0 is near x, a good linear approximation is

$$P(x_0 + \eta) - P(x_0) = dP(x_0)(\eta)$$

Hence $dP(x_0)(\eta) = -P(x_0)$ and if $dP(x_0)$ is invertible, $\eta = -(dP(x_0))^{-1} P(x_0)$ So that $x = x_0 - (dP(x_0))^{-1} P(x_0)$. This is readily recognized as Newton's method, so that the method makes use of the Fréchet derivative to replace a non-linear problem by a sequence of linear ones.

Euler-Lagrange equation

A classical problem in the calculus of variations is that of finding a function x on the interval [a, b], minimizing an integral of the form

$$J = \int_a^b f(x(t), x(t), t)\, dt$$

To specify this problem we must agree on the class of functions within which to seek the extremum. We assume that f is continuous in x, t and x and has continuous partial derivatives with respect to x and $\dot{x}$, and we seek a solution in the subspace of $\mathcal{C}^1[a, b]$ for which $x(a) = 0 = x(b)$.

A necessary condition for an extremum is that for all admissible h

$$DJ(x)(h) = 0$$

Now

$$DJ(x)(h) = \frac{d}{d\alpha}\int_a^b f(x+\alpha h,\ \dot{x}+\alpha\dot{h}, t)\, dt\Big|_{\alpha=0}$$

or

$$DJ(x)(h) = \int_a^b \left[\frac{\partial f}{\partial x}(x,\dot{x},t)\, h(t)\, dt + \frac{\partial f}{\partial \dot{x}}(x,\dot{x},t)\dot{h}(t)\right] dt$$

and it is easily verified that this differential is actually Fréchet. If we assume that a continuous partial derivative with respect to t exists, by integrating by parts, we can write

$$DJ(x)(h) = \int_a^b \left[\frac{\partial f}{\partial x} - \frac{d}{dt}\frac{\partial f}{\partial x}\right] h\, dt + \left[\frac{\partial f}{\partial \dot{x}}\, h\right]_a^b$$

Since $h(a) = 0 = h(b)$, for an extremum

$$\int_a^b \left[\frac{\partial f}{\partial x} - \frac{d}{dt}\frac{df}{d\dot{x}}\right] h\, dt = 0 \qquad \forall h \in \mathscr{C}^1[a,b] \text{ with } h(a) = 0 = h(b)$$

It can be shown that this implies

$$\frac{\partial f}{\partial x} - \frac{d}{dt}\frac{\partial f}{\partial \dot{x}} = 0$$

The Euler-Lagrange equation

Example 79. What is the lifetime plan of investment and expenditure that maximizes total enjoyment for a man having a fixed quantity of savings? We assume the man has no other income except that obtained through investment. His rate of enjoyment at a given time is a certain function V of r, his rate of expenditure. Thus we assume it is desired to maximize:

$$\int_0^T e^{-\beta t}\, V(r(t))\, dt$$

where the $e^{-\beta t}$ term reflects the notion that future enjoyment is counted less today.

If $x(t)$ is the total capital at time t, then

$$\dot{x}(t) = \alpha x(t) - r(t)$$

where α is the interest rate. Thus we maximize

$$\int_0^T e^{-\beta t}\, V(\alpha x(t) - \dot{x}(t))\, dt$$

subject to $x(0) = s$, $x(T) = 0$. For the Euler-Lagrange equation

$$\alpha e^{-\beta t} V'(\alpha x - \dot{x}) + \frac{d}{dt} e^{-\beta t} V'(\alpha x - \dot{x}) = 0$$

Hence

$$V'(r(t)) = V'(r(0))\, e^{(\beta - \alpha)t}$$

If $V(r) = 2\sqrt{r}$, then $r(t)$ turns out to be

$$r(t) = r(0)\, e^{2|\alpha - \beta| t} \quad \text{and} \quad x(t) = \left[x(0) - \frac{r(0)}{2\beta - \alpha}\right] e^{\alpha t} + \frac{r(0)}{2\beta - \alpha}\, e^{2|\alpha - \beta| t}$$

If $\alpha > \beta > \alpha/2$, from $x(T) = 0$, we have

$$r(0) = \frac{2\beta - \alpha}{1 - e^{-(2\beta - \alpha)T}} \cdot x(0)$$

The total capital grows initially and then decreases to zero.

High-order derivatives

Just as in ordinary calculus, it is possible to define higher-order derivatives by induction. Let us first take an example.

Example 80. For a particular case of $f: \mathbb{R}^2 \to \mathbb{R}$, take

$$f(x) = x_1^2 + x_1 x_2^2 + x_2^4$$

Then

$$\frac{f(x_0 + t\eta) - f(x_0)}{t} = 2x_{01}\eta_1 + \eta_1 x_{02}^2 + 2x_{01}x_{02}\eta_2 + 4x_{02}^3\eta_2$$

$$+ t\eta_1^2 + t x_{01}\eta_2^2 + 2t\eta_1 x_{02}\eta_2 + 6x_{02}^2\eta_2 + 0(t^2)$$

where

$$x_0 = \begin{pmatrix} x_{01} \\ x_{02} \end{pmatrix}$$

Now let

$$x_0 + t\eta = x, \quad \text{i.e.} \quad t\eta_1 = x_1 - x_{01}, \quad t\eta_2 = x_2 - x_{02}.$$

Then up to quadratic terms in $(x - x_0)$

$$f(x) - f(x_0) = 2x_{01}(x_1 - x_{01}) + (x_1 - x_{01})x_{02}^2 + 2x_{01}x_{02}(x_2 - x_{02})$$

$$+ 4x_{02}^3(x_2 - x_{02}) + (x_1 - x_{01})^2 + x_{01}(x_2 - x_{02})^2$$

$$+ 2x_{02}(x_1 - x_{01})(x_2 - x_{02}) + 6x_{02}(x_2 - x_{02})^2$$

$$= -(x - x_0)'2b + (x - x_0)'Q(x - x_0)$$

where

$$2b = -\begin{bmatrix} 2x_{01} + x_{02}^2 \\ 2x_{01}x_{02} + 4x_{02}^3 \end{bmatrix}$$

and

$$Q = \begin{bmatrix} 2 & 2x_{02} \\ 2x_{02} & 2x_{01} + 12x_{02}^2 \end{bmatrix}$$

If $f: \mathbb{R}^2 \to \mathbb{R}$ is twice Fréchet differentiable, then its first Fréchet derivative ∇f is a vector, $\nabla f = 2b$, say, and its second Fréchet derivative is a matrix, say Q, and if x is near x_0, a good approximation to f(x) is

$$f(x) = f(x_0) + \langle 2b, x - x_0\rangle + \langle (x - x_0), Q(x - x_0)\rangle$$

where the inner product is just the scalar product of vectors.

In the more general case $f: H \to \mathbb{R}$, where H is a Hilbert space, the same approximation is valid, where $2b \in H$ and $Q \in \mathscr{L}(H)$, which has application in

Iterative methods

If a functional is twice Fréchet differentiable on a Hilbert space H and has, say, a minimum at x_0 then near x_0 the behaviour of f(x) must be of the form

$$f(x) = f(x_0) + \langle 2b, x - x_0\rangle + \langle (x - x_0), Q(x - x_0)\rangle$$

So that many minimization problems can be examined by looking at the minimization of the functional

$$f(x) = \langle x, Qx\rangle - 2\langle x, b\rangle$$

where Q is a positive-definite self-adjoint operator on H. Let us examine two methods of minimizing this functional.

1. Steepest descent

It is easy to see that the minimum is given by x_0, where $Qx_0 = b$. Write $r = b - Qx$ and note that 2r is the negative gradient of f at the point x. Now consider the following iterative programme:

$$x_{n+1} = x_n + \alpha_n r_n$$

where $r_n = b - Qx_n$ and α_n is to be chosen to minimize $f(x_{n+1})$. Now

$$f(x_{n+1}) = \langle (x_n + \alpha_n r_n), Q(x_n + \alpha_n r_n)\rangle - 2\langle (x_n + \alpha_n r_n), b\rangle$$

$$= \alpha^2 \langle r_n, Qr_n\rangle - 2\alpha \| r_n \|^2 + \langle x_n, Qx_n\rangle - 2\langle x_n, b\rangle$$

which is minimized by

$$\alpha = \frac{\|r_n\|^2}{\langle r_n, Qr_n \rangle}$$

Hence steepest descent is

$$x_{n+1} = x_n + \frac{\|r_n\|^2}{\langle r_n, Qr_n \rangle} r_n$$

where $r_n = b - Qx_n$.

2. Conjugate gradient

In this method we try to reformulate the problem as a minimum norm problem. Again consider

$$f(x) = \langle x, Qx \rangle - 2\langle x, b \rangle$$

and introduce a new inner product

$$\langle x, y \rangle Q = \langle x, Qy \rangle.$$

Since $Qx_0 = b$, the problem is equivalent to minimizing

$$\langle x - x_0, \ Q(x - x_0) \rangle = \|x - x_0\|_Q^2$$

Suppose we generate a sequence of vectors that are orthogonal with respect to $\langle , \rangle_Q$. These are usually called "conjugate directions". Let x_0 be expanded in a Fourier series with respect to this sequence. If x_n is the coefficient of this expansion, then by the theories developed in section 4 $\|x_n - x_0\|_Q$ is minimized over the subspace spanned by the first n of the conjugate directions (say $p_1, p_2, \ldots, p_n$).

In order to compute the Fourier series of x_0, we have to compute

$$\langle x_0, p_i \rangle_Q = \langle p_i, Qx_0 \rangle = \langle p_i, b \rangle$$

i.e. no knowledge of x_0 is necessary.

The conjugate gradient method is given by the sequence

$$x_{n+1} = x_n + \alpha p_n$$

$$\alpha_n = \frac{\langle p_n, r_n \rangle}{\langle p_n, Qp_n \rangle} \ ; \quad r_n = b - Qx_n$$

Too see this, we have

$$y_{n+1} = y_n + \frac{\langle p_n, b - Qx_1 - Qy_n \rangle}{\langle p_n, Qp_n \rangle} p_n$$

$$= y_n + \frac{\langle p_n, y - y_n \rangle_Q}{\langle p_n, Qp_n \rangle} p_n$$

Since y_n is in the subspace spanned by $p_1, \ldots, p_{n-1}$ and since the p_i's are orthogonal with respect to $\langle\ ,\ \rangle_Q$

$$\langle p_n, y_n \rangle_Q = 0$$

and

$$y_{n+1} = y_n + \frac{\langle p_n, y_0 \rangle_Q}{\| p_n \|_Q^2} p_n$$

Thus

$$y_{n+1} = \sum_{k=1}^{n} \frac{\langle p_k, y_0 \rangle_Q}{\| p_k \|_Q^2} p_k$$

which is the nth partial sum of the Fourier expansion of y_0. It follows that $y_n \to y_0$ or $x_n \to x_0$. The orthogonality relation $\langle r_n, p_k \rangle = 0$ follows from the fact that $y_n - y_0 = x_n - x_0$ is orthogonal to the subspace generated by $p_1, \ldots p_{n-1}$.

9. PROBABILITY SPACES AND STOCHASTIC PROCESSES

Probability theory and statistics can mean very different things to the pure mathematician or the statistician. To the pure mathematician probability theory is a very special case of a measure space (Ω, P, μ) where $\mu(\Omega) = 1$, but to the statistician it means likelihoods of events, normal distribution, χ-distribution or ways of dealing with random phenomena. Here we shall provide the mathematical framework of probability theory as it fits in nicely with our functional analytic approach.

Definition 73. A probability space is a measure space (Ω, P, μ), where $\mu(\Omega) = 1$. Sets in P we call events, and μ a probability measure.

It is useful to compare the different terminology used in probability theory and in measure theory for essentially the same things:

probability space (Ω, P, μ)	measure space
sample point $\omega \in \Omega$	element in Ω
event $A \in P$	measurable set
sure event Ω	whole space
impossible event $\emptyset$	empty set
almost surely a.s. / with probability one w.p.1	almost everywhere
random variable	measurable function
expectation	integral

Let us consider some simple illustrative examples.

Example 81. Let Ω be the possible outcomes of tossing a die three times. Ω clearly has 6^3 sample points of the form (2, 1, 6) and examples of measurable sets in P or events are

A = {a 6 is turned up in at least one of the tosses}
B = {a 3 is turned up 3 times running}

Then you can see that $\mu(A) = 3/6 = 1/2$
and $\mu(B) = 1/6^3$

Example 82. Let Ω be the collection of all possible outcomes of flipping a coin 50 times. A typical sample point in Ω is $\omega = (H, H, T, \dots, H, T)$ a 50-tuple. Let p be the probability of getting H on a toss (allowing for 'unfair' coins). Then $q = 1 - p$ is the probability of getting T. Then it is well known that the probability that you get n heads and 50-n tails is

$$\frac{50!}{(50-n)!\,n!}\, p^n q^{50-n}$$

This is the binomial distribution.

Example 83. Let

$$f(x) = \begin{cases} \alpha e^{-\alpha x} & \text{for } x > 0 \\ 0 & \text{otherwise} \end{cases}$$

If $\Omega = \mathbb{R}$, then $\mu(A) = \int_A f(x)dx$ is a probability measure on $\mathbb{R}$. This is called the exponential distribution.

We have already said that a random variable is a measurable function $x : \Omega \to \mathbb{R}$. In applications you usually think of it as some quantity whose value you can never predict exactly, but you can predict the probability that it will have a certain value. Example 81, for instance, in the die question, one random variable is the number turned up on the second throw; you cannot say what it will be, but you do know that it will be any of the numbers 1 - 6 with equal probability 1/6.

Definition 74. A convenient way of expressing this information is by a probability distribution function of x, denoted by F(t) and defined by

$$F(t) = \mu\{\omega : x(\omega) \leq t\}$$

i.e. F(t) is a real-valued monotonically increasing function of one variable. $F(-\infty) = 0$, $F(\infty) = 1$, $0 \leq F(t) \leq 1$.

Example 84. Let Ω be the 36 possible outcomes of rolling 2 dice. Let x be the random variable which assigns to each outcome the total points of the 2 dice. Its distribution function is shown in the following figure:

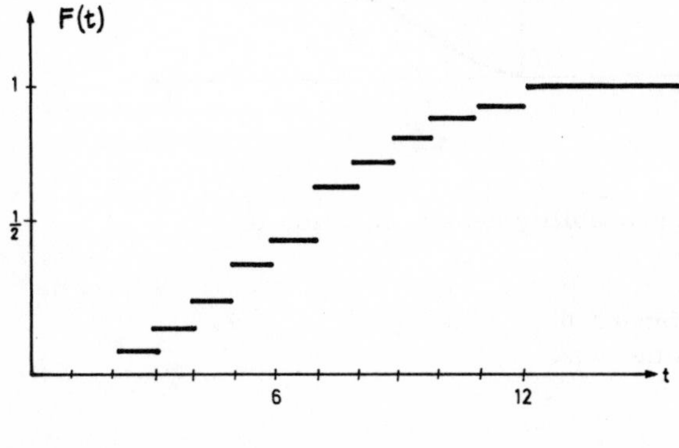

We note that

$$\mu\{t_1 \leq x(\omega) \leq t_2\} = F(t_2) - F(t_1).$$

Definition 75. If F is sufficiently smooth, there exists a function f(s), the probability density function such that

$$F(t) = \int_{-\infty}^{t} f(s)\, ds$$

where we may need to interpret the integration as an infinite summation in the case where x is a discrete random variable.

Discrete random variable x takes on countably many distinct values, for example in the die experiment.

Continuous random variable. The possible values of x are some interval in $\mathbb{R}$.

Example 85. $\Omega = \mathbb{R}$ and $\mu(A) = \int_A f(t)\, dt$, where

$$f(t) = \begin{cases} \alpha e^{-\alpha t} & \text{for } t > 0 \\ 0 & \text{otherwise} \end{cases}$$

Define the random variable $x(\omega) = \omega$. Then

$$F(t) = \mu\{\omega \leq t\} = \int_{-\infty}^{t} f(t)\, dt$$

$$= \begin{cases} 1 - e^{-\alpha t} & , \quad t > 0 \\ 0 & , \quad t \leq 0 \end{cases}$$

the exponential distribution function

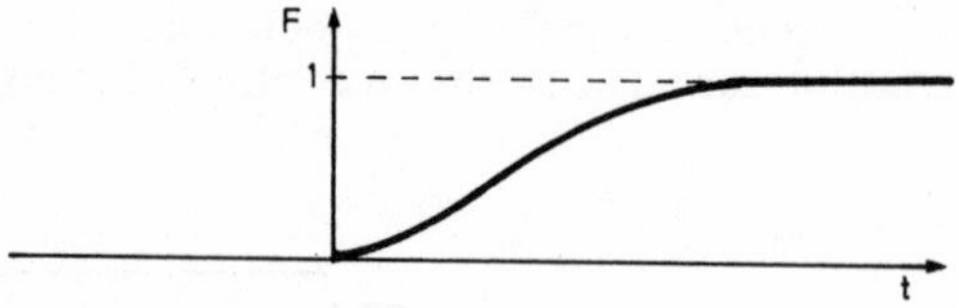

and the exponential probability density function is

$$f(t) = \begin{cases} \alpha e^{-\alpha t} & \text{for } t > 0 \\ 0 & \text{otherwise} \end{cases}$$

Joint distribution functions

If x and y are two random variables on the same probability space (Ω, P, μ), we define the joint probability distribution function

$$F(t, s) = \mu\{x(\omega) \leq t, \quad y(\omega) \leq s\}$$

Similarly you can define the joint distribution function for n random variables

$$F(t_1, \ldots t_n) = \mu\{x_i(\omega) \leq t_i; \quad i = 1, \ldots, n\}$$

The joint probability density function, if it exists, is given by

$$F(t_1, \ldots t_n) = \int_{-\infty}^{t_1} \ldots \int_{-\infty}^{t_n} f(t_1, \ldots t_n)\, dt_1 \ldots dt_n$$

Definition 76. Expectation

This mathematical definition is motivated by the notion of the mean or average value of a random event repeated infinitely many times.

$$E\{x\} = \int_{\Omega} x(\omega)\, d\mu(\omega)$$

provided of course x(.) is integrable or $x(.) \in L_1(\Omega, P, \mu)$.

$$L_1(\Omega, P, \mu) = \left\{ \begin{array}{l} \text{the equivalence class of measurable functions} \\ x : \Omega \to \mathbb{R} \ni \int_{\Omega} |x(\omega)|\, d\mu(\omega) < \infty \end{array} \right\}$$

We then say that x has finite expectation.

If $E\{|x|^2\} < \infty$ or equivalently $x(.) \in L_2(\Omega, P, \mu)$ we say that x has finite second moment $\int_{\Omega} x(\omega)^2\, d\mu(\omega)$, or that x is a second-order random variable.

$L_2(\Omega, P, \mu)$ is the space of second-order random variables.

Definition 77.

If $x \in L_2(\Omega, P, \mu)$ we can define the variance $\sigma^2(x) = E\{|x - Ex|^2\}$ which is a measure of the dispersion or spread of the values around the average value. $\sigma(x)$ is called the standard deviation of x.

Example 87. The normal distribution has the probability density function

$$f(t) = \frac{1}{\sqrt{2\pi}\,\sigma}\, e^{\frac{-(t-a)^2}{2\sigma^2}}$$

and

$$E\{x\} = a, \quad \sigma^2(x) = \sigma^2$$

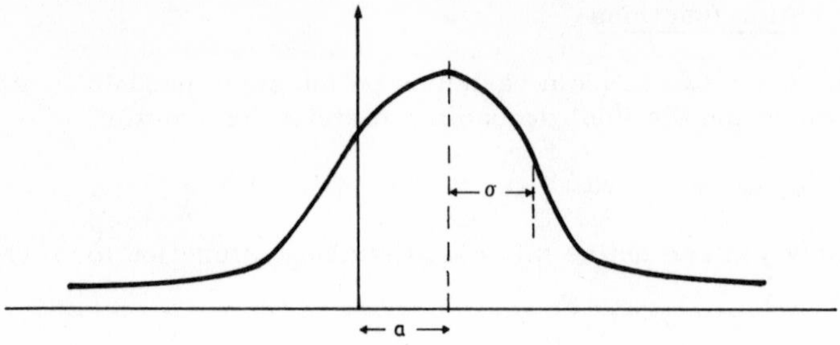

F(t) = probability that the outcome is ≦ t.

This space of second-order random variables $L_2(\Omega, P, \mu)$ is a Hilbert space with the inner product $\langle x, y \rangle = E\{xy\} = \int_\Omega x(\omega)\, y(\omega)\, d\mu(\omega)$.

Definition 78. We also define the covariance of $x, y \in L_2(\Omega, P, \mu)$

$$\text{cov}(x, y) = E\{(x - Ex)(y - Ey)\}$$

and the correlation coefficient

$$\rho(x, y) = \frac{\text{cov}(x, y)}{\sigma(x)\,\sigma(y)}$$

Stochastic independence

Intuitively this means that the outcomes of two events are not related. For example, if you toss a dice twice, the outcomes of the two tosses are independent. However, if you consider the event A = { the outcome of the first toss} and B = { the sum of the outcomes of the two tosses}, then clearly these are dependent events. This is formalized mathematically as follows:

Definition 79.

x and y are independent random variables if

$$\mu\{x(\omega) \leq \alpha, \quad y(\omega) \leq \beta\} = \mu\{x(\omega) \leq \alpha\}\, \mu\{y(\omega) \leq \beta\}$$

Similarly, $x_1, x_2, \ldots, x_n$ are mutually independent if

$$\mu\{x_i(\omega) \leq \alpha_i;\ i = 1 \ldots n\} = \prod_{i=1}^{n} \mu\{x_i(\omega) \leq \mu_i\}$$

An important result is that if $x, y \in L_2(\Omega, P, \mu)$ they are independent iff $E\{xy\} = E\{x\}\, E\{y\}$.

Definition 80. Stochastic processes

A stochastic process is a family of random variables x_t, t being a parameter.

Discrete stochastic process x_t; $t = 0, \pm 1, \pm 2, \ldots$

Continuous stochastic process $\{x_t\}$, $t \in$ an interval in $\mathbb{R}$.

For each t, x_t has a well-defined distribution function. An important class of random variables in applications are Gaussian random variables.

Definition 81. A random variable x is Gaussian if its probability density function is of the form

$$f(t) = \frac{1}{\sqrt{2\pi}\,\sigma} e^{-\frac{(t-a)^2}{2\sigma^2}}$$

i.e. it has a normal distribution.

Important properties of Gaussian random variables

1. If x and y are Gaussian, so is $\alpha x + \beta y$.

2. The probability properties of x are completely determined by its expectation and its standard deviation (i.e. by its first two moments only).

3. A weighted average of a sequence of n independent identically distributed random variables in $L_2(\Omega, P, \mu)$ tends to behave like a Gaussian random variable as $n \to \infty$ (Limit theorem), i.e. we can approximate a large number of independent random factors by a Gaussian law.

Definition 82. A Gaussian stochastic process is one for which each random variable x_t has a Gaussian distribution. The most general Gaussian process is completely specified by parameter functions $\mu(t)$ and $r(s,t)$, where $\mu(t)$ is the expectation function: $\mu(t) = E\{x_t\}$ and $r(s,t)$ is the covariance function: $r(s,t) = E\{x_t x_3\} - \mu(s)\,\mu(t)$. ($r(s,t) = r(t,s)$ and the matrix $[r(t_j, t_i)]$ is non-negative definite.)

The joint probability density function of $x_{t_1} \ldots . x_{t_n}$ is

$$f(s_1, \ldots, s_n) = \frac{|a_{ij}|^{1/2}}{(2\pi)^{n/2}} \exp\left\{-\frac{1}{2}\sum a_{ij}(s_j - \mu(t_j))(s_i - \mu(t_i))\right\}$$

where

$$(a_{ij}) = (r(t_i, t_j))^{-1} \text{ and its determinant is } |a_{ij}|.$$

A Gaussian stochastic process is a special case of second-order processes for which $x_t \in L_2(\Omega, P, \mu)$ for each t. For second-order processes we define $\mu(t) = E\{x_t\}$ and $r(s,t) = E\{x_t x_s\} - \mu(t)\,\mu(s)$, which, although they do not completely specify the process, they do provide very useful information.

An important subclass are wide-sense stationary processes where $r(t, s) = r(t - s)$, which are extensively used in engineering applications.

DEFINITIONS

In reading new material and in particular reading these rather concise notes it may be useful to have a list of the major definitions.

Definition 1. Linear vector space: A set of elements with a binary operation under which it forms a group and an associated scalar multiplication by the real or complex numbers, which is associative.

Definition 2. Linear subspace: if $\mathscr{V}$ is a linear vector space, then a subset S of $\mathscr{V}$ is a linear subspace if $x, y \in S \Rightarrow \alpha x + \beta y \in S$.

Definition 3. Affine subset: is the set $M = \{x : x = x_0 + c$, where $x_0 \in S$, some linear subspace, and c is a fixed element of $X\}$.

Definition 4. Convexity: A subset A of $\mathscr{V}$ is convex if $x, y \in A \Rightarrow \lambda x + (1 - \lambda) y \in A$, $\forall \lambda > 0 \leq \lambda \leq 1$.

Definition 5. Linear dependence: if $x_1, x_2, .., x_n \in X$ and $\exists \alpha_1 \ldots \alpha_n$ not all zero such that $\alpha_1 x_1 + \alpha_2 x_2 + \ldots + \alpha_n x_n = 0$, then $x_1, \ldots, x_n$ are linearly dependent.

Linear independence: No such $\alpha_1, \ldots, \alpha_n$ exist.

Definition 6. Dimension: If $x_1, \ldots, x_n$ are linearly independent and any vector in X can be represented as a linear combination of $x_1, \ldots, x_n$ then X is said to be of dimension n.

Definition 7. Basis: The set $x_1, \ldots, x_n$ of Definition 6 is called a basis for X.

Definition 8. Isomorphic: Vector spaces $\mathscr{V}$ and $\mathscr{W}$ are isomorphic if $\exists$ a bijective linear map $T : \mathscr{V} \to \mathscr{W}$.

Definition 9. A hyperplane is a maximal proper affine subset.

Definition 10. Metric space: A set X of elements $\{x, y, \ldots\}$ and a distance function $d(x, y)$ with the properties
(i) $d(x, y) \geq 0$
(ii) $d(x, y) = 0$ iff $x = y$
(iii) $d(x, y) = d(y, x)$
(iv) $d(x, y) \leq d(x, z) + d(z, y)$

$d(x, y)$ is a metric on X.

Definition 10a. Pseudo metric: Condition (ii) is replaced by $d(x, x) = 0$ ($d(x, y) = 0$ does not necessarily imply $x = y$).

Definition 11. Normed linear space X is a linear vector space with a norm on each element, i.e. to each $x \in X$ corresponds a positive number $\|x\|$

such that (i) $\|x\| = 0$ iff $x = 0$

(ii) $\|\alpha x\| = |\alpha| \, \|x\|$ for all scalars

(iii) $\|x+y\| \leq \|x\| + \|y\|$

Definition 11a. Seminorm $\|x\| = 0$ need not imply $x = 0$.

Definition 12. A subspace of a metric space (X, d) is (A, d) where $A \subset X$.

Definition 13. Product metric space $X \times Y = \{(x, y) : x \in X, \; y \in Y\}$, where (X, d_x), (Y, d_y) are metric spaces and the metric on $X \times Y$ is a suitable function of d_x, d_y, e.g. $d((x_1, y_1), (x_2, y_2)) = d_x(x_1, x_2) + d_y(y_1, y_2)$.

Definition 14. Continuity: $f : X \to Y$ a map between metric spaces (X, d_x) and (Y, d_y) is continuous at $x_0 \in X$ if given $\epsilon > 0$, $\exists$ a real number $\delta > 0$ such that $d_y(f(x), f(x_0)) < \epsilon$, whenever $d_x(x, x_0) < \delta$.

Definition 15. Uniform continuity: f is uniformly continuous if the above $\delta = \delta(\epsilon)$ is independent of the point x_0.

Definition 16. Convergence: A sequence $\{x_n\}$ in a metric space (X, d) converges to x_0 in (X, d) if $d(x_n, x_0) \to 0$ as $n \to \infty$.

Definition 17. Cauchy sequence: $\{x_n\}$ is Cauchy if $d(x_n, x_m) \to 0$ as $m, n \to \infty$.

Definition 18. Completeness: A metric space (X, d) is complete if each Cauchy sequence converges to a point in X.

Definition 19. Closed set: A set A in a metric space (X,d) is closed if it contains all its limit points.

Definition 20. Banach space: A complete normed linear space.

Definition 21. Dense: A linear subspace S of a metric space X is dense in X if its closure with respect to the metric $\supseteq X$.

Definition 22. Connected: A metric space is disconnected if it is the union of two open, non-empty, disjoint subsets. Otherwise it is connected.

Definition 23. Contraction mapping: $f : X \to X$, where (X, d) is a metric space, is a contraction mapping if $\exists \, k$, $0 \leq k \leq 1$ such that

$$d(f(x), f(y)) \leq k d(x, y) \quad \forall x, y \in X$$

Definition 24. Compact: A set A in a metric space is compact if every sequence in A contains a convergent subsequence with limit point in A.

Definition 25. Relatively compact: A set $A \subset X$ is relatively compact if its closure is compact.

Definition 26. Equicontinuity: A set $A \subset \mathscr{C}(X, Y)$, the space of continuous functions from (X, d_x), a compact metric space, to (Y, d_y), a complete metric space, is equicontinuous at $x_0 \in X$ if given $\epsilon > 0$ $\exists \delta > 0$ such that $|f(x) - f(x_0)| < \epsilon$ $\forall f \in A$, whenever $|x - x_0| < \delta$ and $x \in X$.

Definition 27. Open ball of a metric space (X, d)

$B_r(x_0) = \{x \in X;\ d(x, x_0) < r\}$, the open ball at x_0 of radius r. (Closed ball and sphere are similarly defined with $\leq r$ and $= r$, respectively.)

Definition 28. Local neighbourhood of a metric space (X, d). A subset N of X, such that $N = B_r(x_0)$ for some $r \neq 0$.

Definition 29. Equivalence of metrics. Metrics d_1 and d_2 on the space X are equivalent if

(a) $f: (X, d_1) \to (Y, d_3)$, an arbitrary metric space is continuous iff $f: (X, d_2) \to (Y, d_3)$ is continuous.

and

(b) A sequence $\{x_n\}$ converges to x_0 in (X, d_1) iff $\{x_n\}$ converges to x_0 in (X, d_2).

Definition 30. Open set. A set $A \subset (X, d)$ is open if it contains a local neighbourhood of each of its points.

Definition 31. Topology: The class of all open sets of (X, d) is the topology of (X, d).

Definition 32. Continuity: A map $f: (X, d_1) \to (Y, d_2)$ is continuous if the inverse image of each open set in (Y, d_2) is open in (X, d_1) (see Definition 14).

Definition 33. Convergence: A sequence $\{x_n\}$ in a metric space (X, d) converges to x_0 in X iff x_n is in every open set containing x_0 for sufficiently large n (see Definition 16).

Definition 34. Closed set: $A \subset X$ is closed if its complement is open (see Definition 19).

Definition 35. Separable metric space: (X, d) is separable if it contains a countable subset which is dense in X.

Definition 36. Measure space is a couple $(X, \mathscr{B})$ consisting of a set X and a σ-algebra $\mathscr{B}$ of subsets of X.
(A subset of X is measurable if it is in $\mathscr{B}$)

Definition 37. Measure μ on a measurable space $(X, \mathscr{B})$ is a non-negative set function defined for all sets in $\mathscr{B}$ with the properties

$$\mu(\emptyset) = 0$$

$\mu\left(\bigcup_{i=1}^{\infty} E_i\right) = \sum_{i=1}^{\infty} \mu E_i$ where E_i are disjoint sets in $\mathscr{B}$. $(X, \mathscr{B}, \mu)$ is then a measure space.

Definition 38. Complete measure space $(X, \mathscr{B}, \mu)$ is one which contains all subsets of sets of measure 0.

Definition 39. Measurable function $f: X \to \mathbb{R} \cup \{\infty\}$ is measurable if $\{x : f(x) < \alpha\} \in \mathscr{B}$ for all real α.

Definition 40. Simple function: $g(x) = \sum_{i=1}^{n} c_i \chi_{E_i}(x)$, where c_i are constants and χ_{E_i} is the characteristic function of $E_i \in \mathscr{B}$.

Definition 41. Integrable function $f: X \to \mathbb{R} \cup \{\infty\}$ is integrable if $\int_E f\, d\mu < \infty$.

Definition 42. Absolute continuity on I (compact interval in $\mathbb{R}$): For every $\epsilon > 0$, $\exists \delta > 0$ such that

$$\sum_{k=1}^{n} |F(b_k) - F(a_k)| < \epsilon$$

whenever $\sum_{k=1}^{n} |b_k - a_k| \leq \delta$, $\ldots b_{k-1} < a_k < b_k < a_{k+1} < \ldots$

For I, an arbitrary interval, F must be absolutely continuous on every compact subinterval.

Definition 43. Topological isomorphism of two normed linear spaces X, Y $\exists T \in \mathscr{L}(X, Y)$ such that $T^{-1} \in \mathscr{L}(Y, X)$; T is a topological isomorphism.

Definition 44. Isometric isomorphism:

$$\|Tx\|_y = \|x\|_x \quad \forall x \in X$$

Definition 45. Inner product is a bilinear function $\langle .,. \rangle : X \times X \to \mathbb{C}$ such that

1. $\langle \alpha x + \beta y, z \rangle = \alpha \langle x, z \rangle + \beta \langle y, z \rangle$ for scalars α, β.
2. $\overline{\langle x, y \rangle} = \langle y, x \rangle$
3. $\langle x, x \rangle = 0$

Definition 46. A Hilbert space is a complete normed linear space under the inner product norm $\|\langle .,. \rangle\|$.

Definition 47. Orthogonal. x is orthogonal to y iff $\langle x, y \rangle = 0$.

Definition 48. Orthogonal complement of $M \subset$ Hilbert space H is

$$M^{\perp} = \{ y \in H : \langle x, y \rangle = 0 \quad \forall x \in M\}$$

Definition 49. Orthonormal set in a Hilbert space H is a set $\{\phi_n\}$ such that $\langle \phi_n, \phi_m \rangle = \delta_{mn}$.

Definition 50. Complete orthonormal set when $H = \overline{Sp\{\phi_i\}}$.

Definition 51. Linear functional on a normed linear space X is a linear map $f : X \to R$.

Definition 52. Algebraic dual of X, X^a is a linear vector space of all linear functionals on X.

Definition 53. Topological dual of X, X^* is the normed linear space of all bounded linear functionals on X.

Definition 54. Reflexive space: $X^{**} \cong X$.

Definition 55. Convergence in norm: $x_n \to x$ in norm if

$$\|x_n - x\| \to 0 \quad \text{as} \quad n \to \infty.$$

Definition 56. Weak convergence $f(x_n) \to f(x)$ as $n \to \infty$ $\forall f \in X^*$.

Definition 57. Weak* convergence in $X^* : x_n^*(x) \to x^*(x)$ as $n \to \infty$, $\forall x \in X$.

Definition 58. Weak compactness: A set A in X is weakly compact if $\forall \{x_n\} \subset A$, there is a weakly convergent subsequence with limit point in A.

Definition 59. Weak* compactness: Replace weakly by weak* in Definition 58.

Definition 60. Linear transformation $T : X \to Y$, where X, Y are linear vector spaces is $\ni T(\alpha x + \beta y) = \alpha Tx + \beta Ty$. T is an operator if X = Y.

Definition 61. Invertible transformation $T : X \to Y$ is invertible if $\exists\, G : Y \to X \ni FG$ and GF are identity maps. G is the inverse of F.

Definition 62. Continuous transformation $T : X \to Y$ at x_0 if for every $\epsilon > 0$, $\exists\, \delta > 0$ such that $\|Tx - Tx_0\| < \epsilon$, whenever $\|x - x_0\| < \delta$.

Definition 63. Bounded transformation $T : X \to Y$ if $\|Tx\|_y \leq K \|x\|_x$, for some $K > 0$ and $\forall x \in X$.

Definition 64. $\mathcal{L}(X, Y) = \{ T : X \to Y$, where T is bounded and linear$\}$

$$\|T\| = \sup_{\|x\| \neq 0} \left\{ \frac{\|Tx\|}{\|x\|} \right\}$$

Definition 65. Adjoint T^* of $T \in \mathscr{L}(H)$, H Hilbert space, is given by

$$\langle Tx, y \rangle = \langle x, T^*y \rangle \quad \forall x, y \in H$$

Definition 66. Self-adjoint operator: $A \in \mathscr{L}(H)$, when $A = A^*$.

Definition 67. Positive operator: $A \in \mathscr{L}(H)$ if $\langle Ax, x \rangle \geq 0$. Strictly positive if $\langle Ax, x \rangle = 0$ only if $x = 0$.

Definition 68. Closed operator T on a Banach space X if for all sequences $\{x_n\}$ in the domain of T, $\mathscr{D}(T)$, with $x_n \to x$ and $Tx_n \to y$, then $x \in \mathscr{D}(T)$ and $y = Tx$.

Definition 69. Resolvent $\rho(T)$, spectrum $\sigma(T)$ of a linear operator T on a Banach space X

$$\rho(T) = \{\lambda \in \mathbb{C} : (\lambda I - T)^{-1} \in \mathscr{L}(X)\}$$

$$\sigma(T) = \mathbb{C} - \rho(T)$$

Definition 70. Compact operator $T \in \mathscr{L}(X)$ and T maps bounded sequences into sequences with convergent subsequences.

Definition 71. Gateaux derivative: Given x and η in X, if

$$Df(x)\eta = \lim_{t \to 0} \left[\frac{f(x + t\eta) - f(x)}{t} \right]$$

exists, then $Df(x)\eta$ is called the Gateaux derivative at x in the direction of η. If it exists in all directions η, $Df(x): X \to Y$ is called the Gateaux derivative at x.

Definition 72. Fréchet derivative: Consider $f: X \to Y$; X, Y normed linear spaces.

If $\exists\, df(x) \in \mathscr{L}(X, Y)$, such that

$$\lim_{\|h\| \to 0} \left\{ \frac{\| f(x+h) - f(x) - df(x)\,h \|}{\|h\|} \right\} = 0$$

then f is Fréchet differentiable at x and $df(x)(h)$ is the Fréchet differential at x with increment h.

Definition 73. Probability space is a finite measure space (Ω, P, μ) with $\mu(\Omega) = 1$.

Definition 74. Probability distribution function F of a random variable x is

$$F(t) = \mu\{\omega : x(\omega) \leq t\}$$

Definition 75. Probability density function f (if it exists) is given by

$$F(t) = \int_{-\infty}^{t} f(s)ds$$

Definition 76. Expectation of a random variable x in $L_1(\Omega, P, \mu)$ is

$$E\{x\} = \int_{\Omega} x(\omega)\, d\mu(\omega)$$

Definition 77. Variance of a random variable $x \in L_2(\Omega, P, \mu)$ is

$$\sigma^2(x) = E\{|x - E\{x\}|^2\}$$

Definition 78. Covariance function of two random variables $x, y \in L_2(\Omega, P, \mu)$ is

$$\text{cov}(x, y) = E\{(x - E\{x\})(y - E\{y\})\}$$

Definition 79. Independent random variables x, y if

$$\mu\{x(\omega) \leq \alpha,\ y(\omega) \leq \beta\} = \mu\{x(\omega) \leq \alpha\}\, \mu\{y(\omega) \leq \beta\}$$

Definition 80. Stochastic process is a family of random variables.

Definition 81. A Gaussian random variable has a probability density function

$$f(t) = \frac{1}{\sqrt{2\pi}\sigma} e^{-\frac{(t-a)^2}{2\sigma^2}}$$

Definition 82. A Gaussian stochastic process $\{x_t\}$ is one for which x_t is a Gaussian random variable.

BIBLIOGRAPHY

NAYLOR, A.W., SELL, G.R., Linear Operator Theory in Engineering and Science, Holt, Rinehart and Winston (1971).

TAYLOR, A.E., Introduction to Functional Analysis, Wiley, New York (1967).

SIMMONS, G.F., Introduction to Topology and Modern Analysis, McGraw-Hill (1963).

BACHMAN, G., NARICI, L., Functional Analysis, Academic Press, New York (1966).

YOSIDA, K., Functional Analysis, Springer, Berlin (1966).

KANTOROVICH, L.V., AKILOV, G.P., Functional Analysis in Normed Spaces, Moscow (1955).

DUNFORD, N., SCHWARTZ, J., Linear Operators I and II, Interscience Publ. (1963).

CONTROL THEORY AND APPLICATIONS

A.J. PRITCHARD
Control Theory Centre,
University of Warwick, Coventry,
United Kingdom

Abstract

CONTROL THEORY AND APPLICATIONS.

Most of the theories in control have been developed assuming a mathematical model in terms of differential equations. This part of the course will examine the existence, uniqueness, and regularity of both ordinary and partial, linear and non-linear differential equations.

A. ORDINARY DIFFERENTIAL EQUATIONS

1. Linear autonomous systems

The simplest systems in control theory are those described by equations of the form

$$\dot{x} = Ax, \quad x(0) = \overline{x} \tag{1.1}$$

where $x : [0,\tau] \to R^n$, $A \in \mathscr{L}(R^n, R^n)$ and is represented by an $n\times n$ matrix. x is assumed to be $C^1[0,\tau]$ so that the above equation makes sense.

Very many linear differential equations can be formulated in this manner. For instance, consider the damped harmonic oscillator

$$\ddot{x} + k\dot{x} + \omega^2 x = 0, \quad \dot{x}(0) = x_1, \quad x(0) = x_0$$

Introducing $\dot{x} = y$, we find

$$\dot{x} = y$$
$$\dot{y} = -ky - \omega^2 x$$

or

$$\begin{bmatrix} \dot{x} \\ \dot{y} \end{bmatrix} = \begin{bmatrix} 0 & 1 \\ -\omega^2 & -k \end{bmatrix} \begin{bmatrix} x \\ y \end{bmatrix}, \quad \begin{bmatrix} x(0) \\ y(0) \end{bmatrix} = \begin{bmatrix} x_0 \\ x_1 \end{bmatrix}$$

It is very useful to have an explicit representation of the solution of Eq.(1.1). To do this, we first introduce the normed linear space of $n\times n$ matrices $\mathscr{M}_n$

Definition

If A is a matrix of numbers (a_{ij}) with n rows and n columns we define the norm of A, $\|A\|$ by

$$\|A\| = \max\left\{\|Ax\|_{R^n},\ \|x\|_{R^n} = 1\right\}$$

where $\|\ \|_{R^n}$ is any norm on R^n. It is easy to show that

a) $\|Ax\|_{R^n} \leq \|A\|\ \|x\|_{R^n}$

b) $\|A+B\| \leq \|A\| + \|B\|$

c) $\|AB\| \leq \|A\|\ \|B\|$

We have, in fact, introduced many different spaces depending on the definition of $\|\ \|_{R^n}$. However, it is easy to see that all these spaces are topologically isomorphic (because of the equivalence of norms on R^n).

Definition The exponential

We define e^A by

$$e^A = I + A + \frac{A^2}{2!} + \frac{A^3}{3!} + \cdots + \frac{A^n}{n!} + \cdots$$

where I is the identity matrix and where we assume $\|A\| \leq k$ say. Then e^A is well defined since

$$\|e^A\| \leq 1 + k + \frac{k^2}{2!} + \cdots + \frac{k^n}{n!} + \cdots = e^k$$

and it is easy to show the following properties

a) $\frac{d}{dt} e^{At} = A e^{At} = e^{At} A$

b) $e^{-A} = (e^A)^{-1}$

c) $\det A = e^{\operatorname{tr} A}$ where tr A is the trace of A

d) However, $e^{A+B} \neq e^A e^B$ unless $AB = BA$

Solution of Eq.(1.1)

From the property (a) above it is obvious that

$$x(t) = e^{At}\,\overline{x}$$

is the required solution.

Let us illustrate this by considering the following example:

Example

$$\ddot{x} + x = 0, \quad \dot{x}(0) = x_1, \quad x(0) = x_0$$

Then

$$\begin{bmatrix} \dot{x} \\ \dot{y} \end{bmatrix} = \begin{bmatrix} 0 & 1 \\ -1 & 0 \end{bmatrix} \begin{bmatrix} x \\ y \end{bmatrix}, \quad \begin{bmatrix} x(0) \\ y(0) \end{bmatrix} = \begin{bmatrix} x_0 \\ x_1 \end{bmatrix}$$

To calculate e^A it is sometimes useful to use the characteristic equation

$$\det(\lambda I - A) = 0$$

In our case this becomes

$$\lambda^2 + 1 = 0$$

But we know that every matrix satisfies its own characteristic equation, so that

$$A^2 + I = 0$$

From this we obtain

$$A^{2n} = (-1)^n I$$

Hence

$$e^{At} = I[1 - t^2/2! + t^4/4! - \cdots] + A[t - t^3/3! + t^5/5! - \cdots]$$

$$= I \cos t + A \sin t$$

$$= \begin{bmatrix} \cos t & \sin t \\ -\sin t & \cos t \end{bmatrix}$$

Then

$$\begin{bmatrix} x(t) \\ y(t) \end{bmatrix} = \begin{bmatrix} \cos t & \sin t \\ -\sin t & \cos t \end{bmatrix} \begin{bmatrix} x_0 \\ x_1 \end{bmatrix} \qquad \text{as required.}$$

The inhomogeneous equation

Now consider

$$\dot{x} = Ax + f(t), \quad x(0) = \overline{x} \tag{1.2}$$

where for the moment we will assume that f is continuous. Then the solution is

$$x(t) = e^{At}\,\overline{x} + \int_0^t e^{A(t-s)} f(s)\,ds$$

We note that this solution is well defined, and verify that it is a solution of Eq.(1.2) by the direct calculation:

$$\dot{x}(t) = Ae^{At}\overline{x} + e^{A(t-t)} f(t) + \int_0^t A e^{A(t-s)} f(s)\, ds$$

$$= A x(t) + f(t)$$

In particular, if $f(t) = B u(t)$ where $u \in R^m$ is to be thought of as a control, and B is an $n \times m$ matrix:

$$x(t) = e^{At}\overline{x} + \int_0^t e^{A(t-s)} B u(s)\, ds$$

is the solution of

$$\dot{x} = Ax + Bu, \quad x(0) = \overline{x}$$

Of course, this requires that u(t) be continuous, and this is usually too strong a requirement for control problems. However, before we try to overcome this difficulty we really need to examine what we mean by a solution to a differential equation. This will be carried out in the next section, and then we shall return to the above problem.

2. Existence of solutions

Consider the system of non-linear ordinary differential equations

$$\dot{x} = f(t,x), \quad x(\tau) = \overline{x} \tag{2.1}$$

where x is a vector-valued function defined on an interval $I = [\tau, \tau + a]$ and f(t,x) is also vector-valued and defined on $I \times B = D$ where

$$B = \{x \in R^n, \ \|x - \overline{x}\| \leq b\}$$

We will assume that $f \in C^0(D)$ and define a solution x(t) if

a) $(t, x(t)) \in D, \ t \in I$

b) $\dot{x}(t) = f(t, x(t)), \ x(\tau) = \overline{x}$

If x is a solution on I then clearly $x \in C^1(I)$. Integrating Eq.(2.1), gives

$$x(t) = \overline{x} + \int_\tau^t f(s, x(s))\, ds \tag{2.2}$$

Obviously (because $f \in C^0(D)$), Eq.(2.1) will have a solution if and only if Eq.(2.2) has a solution. There are many different kinds of existence, and uniqueness theorems for Eq.(2.2). We shall describe the Picard-Lindeloff theorem because of its historical importance, and also because the proof enables the solution to be constructed.

Picard-Lindeloff theorem

Let $f \in C^0(D)$ and satisfy a Lipschitz condition

$$\| f(t,x) - f(t,y) \| \leq K \| x - y \|, \quad t \in I, \quad x,y \in B$$

Assume $\| f(t,x) \| \leq m$ for $(t,x) \in D$ and set $c = \min(a, b/m)$. Then for $\tau \leq t \leq \tau + c$ there exists a unique solution with

$$x(t) \in B$$

Proof

Consider the sequence of successive approximations

$$\begin{aligned} x_0(t) &= \bar{x} \\ x_{k+1}(t) &= \bar{x} + \int_\tau^t f(s, x_k(s))\, ds \quad k = 0,1,2\ldots \end{aligned} \tag{2.3}$$

We will show by induction that x_k exists on $[\tau, \tau+c]$, $x_k \in C^1$ and

$$\| x_{k+1} - \bar{x} \| \leq m(t - \tau) \quad k = 0,1,2 \ldots$$

Obviously, $\bar{x}$ satisfies these conditions. Assume x_k does the same, then $f(t, x_k(t))$ is defined and continuous on $[\tau, \tau+c]$. Hence from Eq.(2.3) x_{k+1} exists on $[\tau, \tau+c]$ and $x_{k+1} \in C^1$ and obviously

$$\| x_{k+1} - \bar{x} \| \leq \int_\tau^t \| f(s, x_k(s)) \|\, ds \leq m(t - \tau)$$

It now remains to show the convergence of x_k. Set

$$\Delta_k(t) = \| x_{k+1}(t) - x_k(t) \|, \quad t \in [\tau, \tau + c]$$

Then

$$\Delta_k(t) \leq \int_\tau^t \| f(s, x_k(s)) - f(s, x_{k-1}(s)) \|\, ds$$

Hence

$$\Delta_k(t) \leq K \int_\tau^t \| x_k(s) - x_{k-1}(s) \|\, ds = K \int_\tau^t \Delta_{k-1}(s)\, ds$$

But

$$\Delta_0(t) = \| x_1(t) - \bar{x} \| \leq m(t-\tau)$$

So, by induction,

$$\Delta_k(t) \leq mK^k (t-\tau)^{k+1}/(k+1)!$$

This shows that the series

$$\sum_{k=0}^{\infty} \Delta_k(t)$$

is majorized by the series $(m/K)(e^{Kc}-1)$ and hence the series is uniformly convergent on $[\tau, \tau+c]$. Thus the series

$$x_0(t) + \sum_{k=0}^{\infty} (x_{k+1}(t) - x_k(t))$$

is uniformly convergent and the partial sums

$$x_n(t) = x_0(t) + \sum_{k=0}^{n-1} (x_{k+1}(t) - x_k(t))$$

tend uniformly on $[\tau, \tau+c]$ to a continuous limit function x. We now need to show that x satisfies Eq.(2.2). Clearly,

$$\left\| \int_{\tau}^{t} [f(s,x(s)) - f(s,x_k(s))]\, ds \right\| \leq \int_{\tau}^{t} \| f(s,x(s)) - f(s,x_k(s)) \|\, ds$$

$$\leq K \int_{\tau}^{t} \| x(s) - x_k(s) \|\, ds$$

Now $\| x(s) - x_k(s) \| \to 0$ as $k \to \infty$ uniformly on $[\tau, \tau+c]$ and hence x(s) satisfies Eq.(2.2). To show the uniqueness of this solution, let us suppose there are two solutions x_1, x_2 on $[\tau, \tau+c]$. Then

$$x_1(t) - x_2(t) = \int_{\tau}^{t} [f(s,x_1(s)) - f(s,x_2(s))]\, ds$$

Hence

$$\| x_1(t) - x_2(t) \| \leq K \int_{\tau}^{t} \| x_1(s) - x_2(s) \|\, ds$$

by the Lipschitz condition. We conclude the proof by an application of the following lemma:

Gronwall's lemma

Let $a \in L^1(\tau, \tau+c)$, $a(t) \geq 0$, $\Delta \in L^\infty(\tau, \tau+c)$ and assume that b is absolutely continuous on $[\tau, \tau+c]$. If

$$\Delta(t) \leq b(t) + \int_\tau^t a(s)\Delta(s)\, ds$$

then

$$\Delta(t) \leq b(\tau) \exp \int_\tau^t a(s)\, ds + \int_\tau^t \dot{b}(s) \exp\left(\int_s^t a(\xi)\, d\xi\right) ds$$

Proof

Set

$$H(t) = \int_\tau^t a(s)\, \Delta(s)\, ds$$

Then

$$\Delta(t) \leq b(t) + H(t)$$

Multiplying by the integrating factor

$$\exp\left[-\int_\tau^t a(s)\, ds\right]$$

since

$\dot{H}(t) = a(t)\, \Delta(t)$ almost everywhere.

$$\frac{d}{dt}\left[H(t) \exp\left[-\int_\tau^t a(s)\, ds\right]\right] \leq a(t)\, b(t) \exp\left[-\int_\tau^t a(s)\, ds\right]$$

Integrating from τ to t since $H(\tau) = 0$, we obtain

$$H(t) \exp\left[-\int_\tau^t a(s)\, ds\right] \leq -\int_\tau^t b(s) \frac{d}{ds}\left[\exp\left[-\int_\tau^s a(\xi)\, d\xi\right]\right] ds$$

$$= b(\tau) - b(t) \exp\left[-\int_\tau^t a(s)\, ds\right] + \int_\tau^t \dot{b}(s) \exp\left[-\int_\tau^s a(\xi)\, d\xi\right] ds$$

Hence the lemma is proved.

To apply the lemma we set

$$\Delta(t) = \|x_1(t) - x_2(t)\|$$
$$b(t) = 0$$
$$a(t) = K$$

Hence

$$\|x_1(t) - x_2(t)\| \leqq 0$$

and so

$$x_1(t) = x_2(t)$$

This concludes the proof of the theorem.

3. Extension of the idea of a solution

The discussion in Section 2 required that f be continuous in the (t,x)-domain. This is very restrictive for control problems since if we consider the system

$$\dot{x} = f(t,u(t),x)$$

where u is a control, we do not wish to consider only continuous controls.

In proving the existence and uniqueness we wrote the differential equation in an equivalent integral form

$$x(t) = \bar{x} + \int_{\tau}^{t} f(s,x(s))\, ds \tag{3.1}$$

Now, Eq.(3.1) will make sense for a wider class of f than those in $C^0(D)$. In particular, we may ask whether or not an absolutely continuous function x defined on I satisfies the above integral equation. In this case, the differential equation

$$\dot{x} = f(t,x)$$

will only be satisfied almost everywhere (i.e. except on a set of Lebesgue measure zero). Defining the solution in this way, Carathéodory proved the following theorem:

Theorem

Let f be measurable in t for each fixed x and continuous in x for each fixed t. If there exists a Lebesgue integrable function m on the inverval $[\tau, \tau + a]$ such that

$$\|f(t,x)\| \leqq m(t) \qquad (t,x) \in D$$

then there exists a solution x(t) on some interval $[\tau, \tau+\beta]$ satisfying $x(\tau) = \bar{x}$.

Proof

Define M by

$$M(t) = 0 \qquad t < \tau$$

$$M(t) = \int_{\tau}^{t} m(s)\,ds \qquad t \in [\tau, \tau + a]$$

Then M is continuous, non-decreasing and $M(\tau) = 0$. Hence there exists β such that $(t,x) \in D$ for some interval $t \in [\tau, \tau + \beta]$ if $\|x - \bar{x}\| \leq M(t)$. The following iteration scheme is now introduced:

Set

$$x_j(t) = \bar{x} \qquad t \in [\tau, \tau + \beta/j]$$

$$x_j(t) = \bar{x} + \int_{\tau}^{t-\beta/j} f(s, x_j(s))\,ds, \quad t \in (\tau + \beta/j, \ \tau + \beta]$$

We now show that this scheme defines $x_j(t)$ as a continuous function on $[\tau, \tau+\beta]$. Clearly, $x_1(t) = \bar{x}$ is well defined. For any $j \geq 1$, the first formula defines x_j on $[\tau, \tau+\beta/j]$ and since $(t, \bar{x}) \in D$ for $t \in [\tau, \tau+\beta/j]$ the second formula defines x_j as a continuous function on $(\tau+\beta/j, \ \tau+2\beta/j]$. Furthermore, on this interval

$$\|x_j(t) - \bar{x}\| \leq M(t - \beta/j)$$

We can now define x_j on $(\tau+2\beta/j, \ \tau+3\beta/j]$ by the second formula, and so on. Note that for any two points t_1, t_2 we have

$$\|x_j(t_1) - x_j(t_2)\| \leq |M(t_1 - \beta/j) - M(t_2 - \beta/j)|$$

This implies that the set $\{x_j(t)\}$ is an equicontinuous, uniformly bound set on $[\tau, \tau+\beta]$. Hence, by Ascoli's theorem, $\exists$ a subsequence x_{j_k} which converges uniformly on $[\tau, \tau+\beta]$ to a continuous limit function

x as $k \to \infty$.

We now show that this limit function is a required solution. To do this, we shall apply the dominated-convergence theorem of Lebesgue. First note that

$$\|f(t, x_{j_k}(t))\| \leq M(t), \quad t \in [\tau, \tau+\beta]$$

and since f is continuous in x for fixed t:

$$f(t, x_{j_k}(t)) \to f(t, x(t)) \quad \text{as} \quad k \to \infty$$

Then by the dominated-convergence theorem

$$\lim_{k\to\infty} \int_{\tau}^{t} f(s, x_{j_k}(s))\, ds = \int_{\tau}^{t} f(s, x(s))\, ds$$

But

$$x_{j_k}(t) = \overline{x} + \int_{\tau}^{t} f(s, x_{j_k}(s))\, ds - \int_{t-\beta/j_k}^{t} f(s, x_{j_k}(s))\, ds$$

and we have

$$\left\| \int_{t-\beta/j_k}^{t} f(s, x_{j_k}(s))\, ds \right\| \to 0 \quad \text{as} \quad k\to\infty$$

Hence

$$x(t) = \overline{x} + \int_{\tau}^{t} f(s, x(s))\, ds$$

The conditions of the theorem only guarantee the existence of a solution. To obtain unique solutions, it is necessary to impose further conditions, for example Lipschitz conditions.

Another important problem is that of the continuation of solutions beyond $\tau+\beta$. We shall not consider this problem but refer those interested to the bibliography.

4. Linear systems

We now return to the problem

$$\dot{x} = Ax + Bu, \; x(0) = \overline{x}$$

where u is measurable on I and $\|u(t)\| \leq m(t)$, where m is Lebesgue integrable on I. We see that $f(t,x) = Ax + Bu$ is continuous in x for each t and measurable in t for each x. Moreover, for $(t,x) \in D$

$$\|f(t,x)\| \leq \|A\| \; \|x\| + \|B\| m(t)$$

f(t,x) satisfies a Lipschitz condition since

$$\|f(t,x_1) - f(t,x_2)\| = \|A(x_1 - x_2)\| \leq \|A\| \; \|x_1 - x_2\|$$

Hence all the requirements of Carathéodory's theorem are satisfied and $\exists$ a unique, absolutely continuous solution characterized by

$$x(t) = e^{At}\,\overline{x} + \int_0^t e^{A(t-s)} Bu(s)\, ds$$

Non-autonomous systems

We can generalize the problem to consider equations of the form

$$\dot{x} = A(t)x, \quad x(\tau) = \bar{x}$$

where we assume that A is measurable on I and

$$\|A(t)\| \leq m(t), \quad t \in I$$

where m is Lebesgue integrable, and uniformly bounded. Then by Carathéodory there exists a unique solution.

Now let us consider the problem for which $\bar{x} = e_i$ $i = 1,2,\dots n$, where the e_i are a basis for R^n. We might as well choose e_i to be the vector with zeros everywhere except in the i-th place where there is a one. We denote the solution of this equation by $\varphi_i(t,\tau)$. For any $\bar{x} \in R^n$ there is a unique expansion

$$\bar{x} = \sum_{i=1}^{n} c_i e_i$$

and by linearity the corresponding solution will be

$$x(t) = \sum_{i=1}^{n} c_i \varphi_i(t,\tau)$$

If we construct the matrix $\Phi(t,\tau)$ for which each column is the vector $\varphi_i(t,\tau)$ then

$$\dot{\Phi}(t,\tau) = A(t)\Phi(t,\tau)$$

Moreover, since

$$\varphi_i(\tau,\tau) = e_i, \quad \Phi(\tau,\tau) = I$$

We call $\Phi(t,\tau)$ the fundamental matrix.

The solution is given by

$$x(t) = \sum_{i=1}^{n} c_i \varphi_i(t,\tau) = \sum_{i=1}^{n} \Phi(t,\tau) c_i e_i$$

$$= \Phi(t,\tau)\bar{x}$$

Note 1

In the case $A(t) = A$ it is easy to verify that

$$\Phi(t,\tau) = e^{A(t-\tau)}$$

Note 2

In general,

$$\Phi(t,\tau) \neq \exp\left[\int_{\tau}^{t} A(s)\,ds\right]$$

Non-homogeneous system

For the controlled system

$$\dot{x} = A(t)x + Bu, \quad x(\tau) = \bar{x}$$

where

$$u \in L^2(I, R^m), \quad B \in \mathscr{L}(R^m, R^n)$$

It is easy to show that the solution is

$$x(t) = \Phi(t,\tau)\,\bar{x} + \int_{\tau}^{t} \Phi(t,s)\,Bu(s)\,ds$$

B. PARTIAL DIFFERENTIAL EQUATIONS

1. Introduction

The next section will be concerned with partial differential equations. There is a vast literature on this subject and it is obviously impossible to give anything more than a flavour of the subject in this paper. We shall, first of all, motivate the introduction of distributions and weak solutions, then go on to semigroups and mild solutions, and finally use monotone-operator theory to deduce results for non-linear partial differential equations.

A partial differential equation for a scalar function u is a relation of the form

$$F(x,y,\ \ldots,\ u_x, u_y \ldots,\ u_{xx}, u_{xy} \ldots) = 0$$

where

$$u_x = \frac{\partial u}{\partial x}, \quad u_{xy} = \frac{\partial^2 u}{\partial x \partial y} \quad \text{etc.}$$

It may happen that this equation is supplemented by constraints on u and its partial derivatives on the boundary Γ of the region Ω throughout which the independent variables x,y... vary. These constraints are called boundary conditions, and if one of the variables is identified as time the constraint associated with that variable is called an initial condition. The order of the partial differential equation is the order of the highest derivative occurring in F. We shall use, throughout this paper, three examples of partial differential equations which are of important physical significance. These examples are representatives of three classes of partial differential equations: parabolic, elliptic, hyperbolic, and we shall use them to illustrate the general results.

The equation governing the conduction of heat: parabolic

A good approximation to modelling the variation of the temperature T, in a rod of length ℓ, is given by

$$T_{t'} = k\, T_{x'x'}$$

This can be simplified by introducing $x'/\ell = x$, $t' = (\ell^2/k)t$ when

$$T_t = T_{xx}$$

The boundary conditions may take a variety of forms, e.g.

a) $T(0,t) = T(1,t) = h_1(t)$

or

b) $T_x\big|_{x=0} = T_x\big|_{x=1} = h_2(t)$

where h_1, h_2 are given functions of time.

The initial condition could be of the form

$$T(x,0) = T_0(x)$$

for some given function $T_0(x)$.

Laplace equation: elliptic

In R^2 this takes the form

$$\nabla^2\phi = \phi_{xx} + \phi_{yy} = 0$$

The equation represents many different phenomena ranging from the potential of some electric field to the stream function of a fluid flow. A variety of boundary conditions can be imposed depending on the particular physical situation. If, for example, the boundary is a circle C we could have

$$\phi\big|_C = f_1(C)$$

or

$$\phi_n\big|_C = f_2(C)$$

where ϕ_n is the derivative of ϕ in an outward direction normal to C and f_1, f_2 are two functions defined on C.

The wave equation: hyperbolic

Transverse vibrations of a taut string are governed by the wave equation

$$z_{tt} = z_{xx}$$

where z is the displacement from the x-axis. If the ends of the string are fixed, then the boundary conditions are

$$z(0,t) = z(1,t) = 0$$

The initial conditions could be

$$z(x,0) = g_1(x)$$

$$z_t(x,0) = g_2(x)$$

where g_1, g_2 are given functions.

A classical or strict solution to a partial differential equation (p.d.e.) is defined to be a function u(x,y,...) such that all the derivatives which appear in the p.d.e. exist, are continuous, and such that the p.d.e., boundary conditions, and initial conditions are all satisfied. It may be thought that with this definition of a solution the main emphasis should be focussed on deriving methods by which a solution can be obtained. However, the story is not quite so simple. If we consider the wave equation

$$z_{tt} = z_{xx}, \quad z(0,t) = z(1,t) = 0$$

$$z(x,0) = z_0(x)$$

$$z_t(x,0) = 0$$

then it is easily verified that the solution is

$$z(x,t) = \tfrac{1}{2}[z_0(x+t) + z_0(x-t)]$$

However, if $z_0(x)$ is given by

$$z_0(x) = x, \qquad 0 \leq x \leq 1/2$$

$$z_0(x) = 1-x, \qquad \tfrac{1}{2} \leq x \leq 1$$

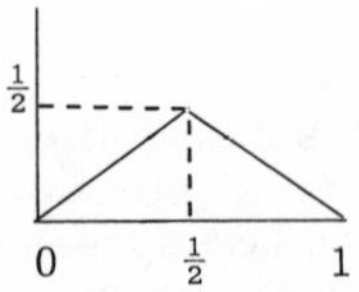

Then the above solution cannot be a classical solution since $z_0(x)$ is not differentiable at $x = \frac{1}{2}$. So what do we mean by a solution to this problem? It is obvious that we must widen the concept of a solution if we are going to allow initial conditions of the form given above.

Let us consider another example. An obvious computational method for determining a solution to the heat conduction equation is to approximate the equation by

$$\frac{T(x, t+\Delta t) - T(x, t)}{\Delta t} = T_{xx}(x,t)$$

If we assume that T(x,0) is a function of the form $z_0(x)$ (as given above), then the computation cannot even start because it is not possible to evaluate

$T_{xx}(x,0)$ at $x = 1/2$

It is obvious, therefore, that reasonable definitions of solutions must allow for these eccentricities, and must include a link between the spaces in which the initial and boundary conditions lie, and the space in which the solution is sought.

If we transform the wave equation by setting

$$y = x - t$$

$$\rho = x + t$$

we find

$$z_{\rho y} = 0 \quad \text{or } z_{y\rho} = 0$$

The equation $z_{\rho y} = 0$ is satisfied by any function of y only, but the expression $z_{y\rho}$ need not make sense for every such function. This is most peculiar and indicates the need for some generalized concept of a function.

We shall now show how the unnatural results of the above equation can be resolved by formulating a different concept of a solution, and this will lead to particular generalized functions — distributions.

If $z \in C^2(\Omega)$ (the space of twice continuously differentiable functions on Ω) where Ω is some bounded domain in R^2 and $f \in C^0(\Omega)$ then

$$z_{\rho y} = f \tag{1.1}$$

makes sense. Integrating by parts twice, we find

$$\iint_\Omega z\, \varphi_{y\rho}\, dy\, d\rho = \iint_\Omega f\varphi\, dy\, d\rho \tag{1.2}$$

for all

$$\varphi \in C_0^2(\Omega)$$

We define a <u>weak solution</u> to be a z such that formula (1.2) holds. If f is $C^0(\Omega)$ and $z \in C^2(\Omega)$ then the weak solution is a classical solution.

However, the concept of a weak solution allows us to consider a larger class of f. Note that since

$$\varphi_{\rho y} = \varphi_{y\rho}$$

we have

$$\iint_\Omega z\, \varphi_{\rho y}\, dy\, d\rho = \iint_\Omega z\, \varphi_{y\rho}\, dy\, d\rho = \iint_\Omega f\varphi\, dy\, d\rho$$

so that the weak solutions of

$$z_{\rho y} = f, \text{ and } z_{y\rho} = f$$

are the same.

One way of looking at the weak solution is to think of $z_{\rho y}$ as being represented by the linear form

$$\iint_{\Omega} z\, \varphi_{y\rho}\, dy\, d\rho$$

Then the study of differential operators leads to the study of continuous linear functionals on $C_0^\infty(\Omega)$.

2. Distribution theory

There are many different ways in which distributions can be defined. We shall choose to define them as elements of the dual of a certain space $D(\Omega)$, i.e. $D'(\Omega)$. So we shall first of all define and give some properties of the space.

The space $D(\Omega)$

Let K be any compact subset of $\Omega \subset R^m$ and let $D_K(\Omega)$ be the set of all functions $\phi \in C_0^\infty(\Omega)$ such that the supports of the ϕ's are in K. We define a family of seminorms on $D_K(\Omega)$ by

$$P_{K,n}(\phi) = \sup |D^\alpha \phi|$$

where the sup is taken over all $|\alpha| \leq n < \infty$ and all $x \in K$ and

$$D^\alpha \phi(x) = \frac{\partial^{\alpha_1 + \alpha_2 + \cdots + \alpha_n}}{\partial x_1^{\alpha_1} \partial x_2^{\alpha_2} \cdots \partial x_m^{\alpha_n}} \phi(x_1, x_2 \ldots x_m)$$

Then $D_K(\Omega)$ is a locally convex topological space whose open sets are determined via the fundamental system of neighbourhoods

$$V(K, 1/s, n) \qquad n = 0, 1, 2 \ldots \qquad s = 1, 2 \ldots$$

$$V(K, \epsilon, n) = \{\varphi \in D_K;\ P_{K,n}(\varphi) \leq \epsilon\}$$

If $K_1 \leq K_2$ then the topology of $D_{K_1}(\Omega)$ is the relative topology of $D_{K_1}(\Omega)$ as a subset of $D_{K_2}(\Omega)$. We define $D(\Omega)$ to be the "inductive limit" of the $D_K(\Omega)$'s as K ranges over all compact subsets of Ω.

Inductive limit

We say that a set is open in $D(\Omega)$ if and only if for every convex, balanced, and absorbing set $V \in D(\Omega)$ the intersection $V \cap D_K(\Omega)$ is an open set of $D_K(\Omega)$

containing the zero vector of $D_K(\Omega)$ for all K. The topology defined in this way is the "inductive limit" of the $D_K(\Omega)$.

A better grasp of these ideas can be obtained by understanding what is meant by convergence in $D(\Omega)$. It can be shown that

$$\lim_{r \to \infty} \phi_r = \phi$$

means that both of the following conditions hold:

a) There exists a compact $K \subset \Omega$ such that the support of ϕ_r, $r = 1,2 \ldots$ is in K

b) For all D^α, $D^\alpha \phi_r(x) \to D^\alpha \phi(x)$ as $r \to \infty$ uniformly on K.

Before we look at distributions in more detail, there are a number of useful technical devices which although we shall not use them are worth mentioning because of their importance in a thorough treatment of the subject.

First, we define

$$\varphi(x) = c\, f(\|x\|^2 - 1)$$

where

$$f(t) = \begin{cases} e^{1/t} & t < 0 \\ 0 & t \geq 0 \end{cases}$$

and c is a constant chosen so that

$$\int_{R^n} \varphi\, dx = 1$$

Then it is easy to show that $\varphi \in C_0^\infty(R^n)$.

Regularization of $z: z_\epsilon$

If z is an arbitrary integrable function, we define its regularization z_ϵ by

$$z_\epsilon = \int_{R^n} z(x - \epsilon y)\, \varphi(y)\, dy$$

Substituting $\epsilon y = y'$ we find

$$z_\epsilon = \epsilon^{-n} \int_{R^n} z(x - y')\, \varphi(y'/\epsilon)\, dy'$$

$$= \epsilon^{-n} \int_{R^n} z(y)\, \varphi[(x-y)/\epsilon]\, dy$$

Theorem

Let z be integrable and vanish outside a compact subset K of Ω. Then $z_\epsilon \in C_0^\infty(\Omega)$ if ϵ is smaller than the distance δ from K to Ω' (compliment of Ω). As $\epsilon \to 0$, $z_\epsilon \to z$ in $L_p, 1 \leq p < \infty$ and $z_\epsilon \to z$ uniformly if $z \in C_0^0(\Omega)$.

The proof of this theorem is immediate from the definition of z_ϵ and the representation

$$z_\epsilon(x) - z(x) = \int_{R^n} [z(x-\epsilon y) - z(x)]\,\varphi(y)\,dy$$

A direct result of this theorem is the so-called partition of unity.

Partition of unity

Let $\{K_i\}$ i = 1, 2... be open sets such that $\{K_i\}$ covers an open set $\Omega \subset R^n$. Then there exist functions α_i such that

a) $\alpha_i \geq 0, \quad \sum_{i=1}^{\infty} \alpha_i = 1$ in Ω.

b) $\alpha_i \in C^\infty$ and its support lies in some K_i.

c) Every compact set in Ω intersects only a finite number of the support of α_i.

Definition: Distributions

A distribution on Ω is an element T of $D'(\Omega)$, and will usually be denoted by $T(\varphi)$, $\varphi \in D(\Omega)$. Let us give some examples:

Example 1

Let f(x) be locally integrable, i.e.

$$\int_K |f(x)|\,dx < \infty \text{ for any compact } K \subset R^n$$

Then $T(\varphi) = \int_\Omega f(x)\,\varphi(x)\,dx$, $\varphi \in D(\Omega)$ defines a distribution on Ω which we usually denote by T_f.

Note

Two distributions T_{f_1}, T_{f_2} are equal if and only if $f_1 = f_2$ almost everywhere

Example 2. The Dirac delta function

Consider the distribution

$T(\varphi) = \varphi(0), \quad \varphi \in D(\Omega)$

We shall call this the Dirac delta function concentrated at the origin and write $\delta(\varphi)$. More loosely, this is written as $\delta(x)$. If instead we consider the distribution

$$T(\varphi) = \varphi(a), \quad \varphi \in D(\Omega)$$

we write

$$\delta_a(\varphi) \quad \text{or} \quad \delta(x-a)$$

Example 3

Both of the above examples are special cases of the following distributions. Let $d\mu$ be a measure on Ω, then

$$T(\varphi) = \int_\Omega \varphi d\mu, \quad \varphi \in D(\Omega)$$

is a distribution. If for example

$$d\mu = f dx, \quad f \in L_1^{loc}(\Omega)$$

then we obtain the distribution constructed in Example 1.

Example 4

If $T(\varphi)$ is a distribution, so is $T(f\varphi)$ for $f \in C^\infty(\Omega)$.

Example 5

If $T(\varphi)$ is a distribution, then so is $T(D^\alpha \varphi)$. A useful characterization of a distribution is: a linear functional T defined on $C_0^\infty(\Omega)$ is a distribution if and only if for every compact subset K of Ω there corresponds a positive constant c and integer k, so that

$$|T(\varphi)| \leq c P_{K,k} \quad \text{whenever} \quad \varphi \in D_K(\Omega)$$

We now make use of Example 5 to define the differentiation of a distribution.

Differentiation of a distribution

We define $(D^\alpha T)(\varphi)$ by

$$(D^\alpha T)(\varphi) = (-1)^{|\alpha|} T(D^\alpha \varphi), \quad \varphi \in D(\Omega)$$

Example 1

Consider the Heaviside function

$$H(x) = \begin{cases} 1 & x > 0 \\ 0 & x \leq 0 \end{cases}$$

Then

$$D\,T_H(\varphi) = -\int_{-\infty}^{\infty} H(x)\,\varphi'(x) = -\int_{0}^{\infty} \varphi'(x)\,dx = \varphi(0)$$

since φ has compact support. So symbolically

$$H'(x) = \delta(x)$$

Example 2

We can define the product of a function $f \in C^{\infty}$ and a distribution T as in Example 4.

$$fT(\varphi) = T(f\varphi), \quad \varphi \in D(\Omega)$$

Then the differentiation

$(DfT)(\varphi)$ is defined by

$$(DfT)(\varphi) = -fT(D\varphi) = -T(fD\varphi)$$

also

$$DfT(\varphi) = T(\varphi Df).$$

and

$$fDT(\varphi) = DT(f\varphi) = -T(D[f\varphi])$$

Now

$$D[f\varphi] = fD\varphi + \varphi Df$$

Therefore

$$(DfT)(\varphi) = DfT(\varphi) + fDT(\varphi)$$

Note It is not possible to define the multiplication of two arbitrary distributions.

3. Sobolev spaces

Definition $H^m(\Omega)$, where m is a positive integer, is the space of distributions $T(\varphi)$ such that $D^{\alpha}T \in L^2(\Omega)$ for all α, $|\alpha| \leq m$ provided with the norm

$$\|T\|_m = \Big(\sum_{|\alpha| \leq m} \|D^{\alpha} T\|^2_{L^2(\Omega)} \Big)^{1/2}$$

with the inner product

$$\langle T_1, T_2 \rangle_m = \sum_{|\alpha| \leq m} \langle D^\alpha T, D^\alpha T_2 \rangle_{L^2(\Omega)}$$

$H^m(\Omega)$ is a Hilbert space

Remark I

If $M > m$

$$H^M(\Omega) \subset H^m(\Omega) \subset L^2(\Omega) \underset{=}{\text{def}} H^0(\Omega)$$

Remark II

The delta function does not belong to any of the spaces $H^m(\Omega)$, $m \geq 0$. To see this we note

$$\|\delta\|_0 = \sup_\varphi \left\{ \frac{\langle \delta, \varphi \rangle_{L^2(\Omega)}}{\|\varphi\|_{L^2(\Omega)}} \right\} = \sup_\varphi \left\{ \frac{\varphi(0)}{\|\varphi\|_{L^2(\Omega)}} \right\}$$

It is always possible to choose a $\varphi \in D(\Omega)$ such that $\varphi(0) \geq k \|\varphi\|_{L^2(\Omega)}$ for any given k. Hence $\delta \notin H^0(\Omega)$ and so by Remark I cannot belong to $H^m(\Omega)$ for $m \geq 0$.

This remark indicates the need to consider more general spaces of distributions.

Tempered distributions

Let us consider the special case of $\Omega = R^n$. We define the Fourier Transform of $\varphi \in L^2(R^n)$ by $\hat{\varphi} = F\varphi$ where

$$\hat{\varphi}(y) = F\varphi = \frac{1}{(2\pi)^{n/2}} \int_{R^n} \exp[-i\langle x,y \rangle]\, \varphi(x)\, dx$$

where

$$\langle x,y \rangle = x_1 y_1 + x_2 y_2 + \ldots + x_n y_n$$

Then $\varphi \to \hat{\varphi}$ is an isomorphism of $L^2(R^n)$ onto $L^2(R^n)$ and

$$\varphi = \frac{1}{(2\pi)^{n/2}} \int_{R^n} \exp[i\langle x,y \rangle]\, \hat{\varphi}(x)\, dx$$

Definition: Tempered distributions

Let

$$S = \{\varphi : x^\alpha D^\beta \varphi \in L^2(R^n) \text{ for all } \alpha, \beta\}$$

Then with the seminorms

$$P_{\alpha,\beta}(\varphi) = \|x^\alpha D^\beta \varphi\|_{L^2(R^n)}$$

S is a Frechet space.

We define the tempered distribution as elements of

S' = dual of S with the strong topology

Note

For all $\varphi \in S$ and for all α

$$F(D^\alpha \varphi) = (iy)^\alpha F\varphi$$

and we can define the Fourier transform of $u \in S'$ by $\hat{u} = Fu$ where

$$\langle Fu,\varphi\rangle = \langle u,F\varphi\rangle \ \forall \ \varphi \in S$$

where $\langle , \rangle$ denotes the duality between S' and S.

Theorem

If $\Omega = R^n$, $H^m(R^n)$ can be defined by

$$H^m(R^n) = \left\{u : u \in S' \text{ and } (1 + \|y\|^2)^{m/2}\hat{u} \in L^2(R^n)\right\}$$

with

$$\|u\|_m = \|(1 + \|y\|^2)^{m/2}\hat{u}\|_{L^2(R^n)}$$

Proof

From Plancherel's theorem

$$\|D^\alpha u\|_{L^2(R^n)} = \|y^\alpha \hat{u}\|_{L^2(R^n)}$$

Hence

$$\|u\|_m = \int_{R^n} \sum_{|\alpha| \leq m} y^{2\alpha} |\hat{u}(y)|^2 \, dy$$

But there exist constants c_1, c_2 such that

$$c_1(1 + \|y\|^2)^m \leq \sum_{|\alpha| \leq m} y^{2\alpha} \leq c_2(1 + \|y\|^2)^m$$

which proves the theorem.

We note that the above definition of $H^m(R^n)$ does not require m to be a positive integer, and use the above to define $H^m(R^n)$ for all n positive and negative, m integer or not. It is then obvious that

$$H^{-s}(R^n) \supset H^0(R^n) \supset H^m(R^n), \quad -s < 0 < m$$

We can also show that

a) If $H^0(R^n)$ is identified with its dual

$$\{H^s(R^n)\}' = H^{-s}(R^n)$$

b) $D(R^n)$ is dense in $H^s(R^n)$ for all s.

The definition of the space $H^s(\Omega)$ is a far more complicated matter and does not fit into the above pattern in the sense that a) and b) are not true for general s. Although we will have cause to use results for these spaces the development is outside the scope of these lectures, and the interested reader is referred to the literature (e.g. the book by Lions and Magenes, see bibliography).

There is a particularly simple representation of the spaces $H^s(\Gamma)$ where Γ is the boundary of Ω and is assumed to be a C^∞-manifold.

If ψ_j are the eigenfunctions of the Laplace operator on Γ and $-\lambda_j$ the corresponding eigenvalues, so that

$$\Delta\psi_j + \lambda_j\psi_j = 0$$

where ψ_j are assumed to be orthonormalized in $H^0(\Gamma)$. If u is a distribution on Γ with u_j its Fourier coefficient relative to $\{\psi_j\}$

$$u = \sum_{j=1}^{\infty} u_j\psi_j$$

Then for all $s \in R$

$$H^s(\Gamma) = \left\{u : u \in D'(\Gamma), \sum_{j=1}^{\infty} \lambda_j^{2s} |u_j|^2 < \infty\right\}$$

Regularity

It is very important to know whether a weak solution to a partial differential equation is a classical solution. This quite often involves comparing $H^s(\Omega)$ spaces with $L^2(\Omega)$ and $C^k(\Omega)$ spaces. The important results in this area are contained in the following

Sobolev embedding theorem

We shall write $X \hookrightarrow Y$ to denote the continuous embedding of X in Y, i.e.

$$\|u\|_Y \leq K\|u\|_X \quad \text{for all} \quad u \in X$$

If

$$s > n/2 + k, \quad \Omega \subset R^n$$

Then

$$H^s(\Omega) \hookrightarrow C^k(\Omega)$$

We shall not prove this theorem but by a simple application of the Schwarz inequality indicate its validity.

For any $\varphi \in D(\Omega)$, $\Omega = (0,1)$

$$\varphi^2(x) = \left[\int_0^x \varphi_y(y)\,dy\right]^2 \leq \int_0^x 1^2 dv \int_0^x \varphi_y^2\,dy \leq \int_0^1 \varphi_y^2\,dy$$

Hence

$$\sup_{x\in\Omega} |\varphi(x)|^2 \leq \|D\varphi\|^2_{L^2(\Omega)}$$

So by continuity

$$\sup_{x\in\Omega} |u|^2 \leq \|u\|^2_{H^1(\Omega)}$$

for all u in the closure of $D(\Omega)$ in the H^1 norm. We shall denote this space by $H^1_0(\Omega)$ and we see

$$H'_0(\Omega) \hookrightarrow C^0(\Omega), \quad \Omega \subset R^1$$

An alternative definition of $H^s_0(\Omega)$ is the following

Definition $H^s_0(\Omega)$

Let Ω be a bounded open set $\subset R^n$ with a suitably smooth boundary Γ. Then $u \in H^s_0(\Omega)$ if and only if

a) $u \in H^s(\Omega)$

b) $\dfrac{\partial^j u}{\partial n_j} = 0$ on Γ, $0 \leq j < s - 1/2$

where $\partial^j/\partial n_j$ is the derivative of order j along the normal to Γ.

It can be shown that

(i) $D(\Omega)$ is not dense in $H^s(\Omega)$, $s > 1/2$
(ii) $H^s_0(\Omega)$ = closure of $D(\Omega)$ in $H^s(\Omega)$
(iii) $\{H^s_0(\Omega)\}' = H^{-s}(\Omega)$.

4. Application to partial differential equations

We shall begin by considering a simple class of elliptic equations

$$\left.\begin{aligned} \sum_{i,j=1}^{n} D^j(a_{ij} D^i u) &= -f \quad \text{in } \Omega \\ u &= 0 \quad \text{on } \Gamma \end{aligned}\right\} \qquad (4.1)$$

where

$$D^j = \frac{\partial}{\partial x_j}, \quad a_{ij} \in L^\infty(\Omega)$$

and we assume

$$\sum_{i,j=1}^{n} a_{ij}(x)\,\xi_i \xi_j \geq K \sum_{i=1}^{n} \xi_i^2$$

We define

$$a(u,\varphi) = \sum_{i,j=1}^{n} \int_\Omega a_{ij} D^i u\, D^j \varphi\, dx, \quad \langle f,\varphi \rangle = \int_\Omega f\varphi\, dx$$

for $\varphi \in D(\Omega)$. A weak solution of Eq.(4.1) is defined to be a solution of

$$a(u,\varphi) = \langle f,\varphi \rangle \qquad (4.2)$$

Because of the imposed conditions we obtain

$$a(\varphi,\varphi) \geq K \|\varphi\|^2_{H^1_0(\Omega)}$$

The proof of the existence now follows as a direct result of the Lax-Milgram theorem which states:

If $a(u,v)$ is a bounded bilinear form on a real Hilbert space X such that

$$a(v,v) \geq K\|v\|^2_X, \quad v \in X$$

For each $f \in X'$ there exists a unique $u \in X$ so that

$$a(u,v) = \langle f,v \rangle \text{ for all } v \in X$$

Of course,this reduces to the Riesz representation theorem when $a(u,v) = a(v,u)$ and so is an inner product on X.

We see therefore that for $f \in H^{-1}(\Omega)$, since $D(\Omega)$ is dense in $H^1_0(\Omega)$, there is a unique solution u of Eq.(4.2): $u \in H^1_0(\Omega)$.
If $f \in H^s(\Omega)$ it is possible to show that there exists $u \in H^{s+2}(\Omega) \cap H'_0(\Omega)$ such that

$$\nabla^2 u = f$$

In particular, if $\Omega \subset R^2$, $s = 1 + \epsilon$, $\epsilon > 0$, $u \in H^{3+\epsilon}(\Omega) \cap H^1_0(\Omega)$ and so by the Sobolev embedding theorem $u \in C^2(\Omega)$ and is therefore a classical solution. This indicates how it is possible to obtain classical solutions via weak solutions.

5. Evolution equations

We shall again consider a space $X \subset H$ where H is a real Hilbert space, then identifying H with its dual H', we have

$$X \subset H \subset X'$$

where X' is the dual of X.

We shall seek a solution of

$$\frac{dx}{dt} = A(t)x + f$$

$$x(0) = x_0, \; x_0 \in H$$

The precise spaces in which f and the solutions lie will be explained later. We define the bilinear form

$$a(t; \varphi, \psi) = -\langle A(t)\varphi, \psi\rangle$$

where $A(t)\varphi \in X'$ and $\langle \, , \rangle$ denotes the duality between X and X'. We assume

a) $a(t; \varphi, \psi)$ is measurable on $(0, \tau)$ and

$|a(t; \varphi, \psi)| \leq c \|\varphi\|_X \|\psi\|_X \quad \varphi, \psi \in X$, where $\| \; \|_X$ denotes the norm on X.

b) $a(t; \varphi, \varphi) \geq K\|\varphi\|^2$, $K > 0$ for all $\varphi \in X$, $t \in (0, \tau)$

so that

$$A(t) \in \mathscr{L}(L^2(0,\tau; X); \; L^2(0,\tau; X'))$$

i.e. if $f \in L^2(0,\tau; X)$, $A(t)f$ is the function

$$t \to A(t)f(t) \in X'$$

We know how to define df/dt for $f \in L^2(0,\tau; X)$. To do this we first define $f \in D'(0,\tau; X)$ by

$$D'(0,\tau; X) = \mathscr{L}(D(0,\tau); \; X)$$

the space of distribution on $(0, \tau)$ with values in X. Then if $f \in D'(0,\tau; X)$, $f(\varphi) \in X$ for all $\varphi \in D(0,\tau)$ and $\varphi \to f(\varphi)$ is a continuous map of $D(0,\tau) \to X$.

For $f \in L^2(0,\tau; X)$ we shall write

$$f(\varphi) = \int_0^\tau f(t)\varphi(t)\,dt$$

and identify $f \in L^2(0,\tau; X)$ with $f(\varphi) \in D'(0,\tau; X)$. The derivative df/dt is then defined by

$$\frac{df}{dt}(\varphi) = -f\left(\frac{d\varphi}{dt}\right)$$

Hence

$$\frac{df}{dt} \in D'(0,\tau; X)$$

We now introduce the space $W(0,\tau)$

$$W(0,\tau) = \left\{ f : f \in L^2(0,\tau; X),\ \frac{df}{dt} \in L^2(0,\tau; X') \right\}$$

endowed with the norm

$$\|f\|^2_{W(0,\tau)} = \int_0^\tau \|f(t)\|^2_X \, dt + \int_0^\tau \left\| \frac{df}{dt}(t) \right\|^2_{X'} dt$$

$W(0,\tau)$ is a Hilbert space. For more details the reader is referred to Lions and Magenes (see bibliography).

We seek a solution $x \in W(0,\tau)$ to the evolution equation

$$\frac{dx}{dt} = A(t)x + f$$
$$x(0) = x_0 \in H, \ f \in L^2(0,\tau; X') \qquad (5.1)$$

<u>Uniqueness</u>

Consider Eq.(5.1) with $f = 0$, $x_0 = 0$.

Then

$$a(t, x(t), x(t)) + \left\langle \frac{dx}{dt}(t), x(t) \right\rangle = 0$$

But on integrating by parts

$$\int_0^\tau \left\langle \frac{dx}{dt}(t), x(t) \right\rangle dt = \tfrac{1}{2} \|x(\tau)\|^2_H$$

Hence

$$K \int_0^\tau \|x(t)\|^2_X dt + \tfrac{1}{2} \|x(\tau)\|^2_H \leq 0$$

and so

$$x = 0$$

Existence

We shall assume that X is separable so that there exists a countable basis $e_1, e_2, \ldots, e_n, \ldots$ such that $e_1, \ldots e_n$ are linearly independent for all n and finite combinations are dense in X. Set

$$x_n(t) = \sum_{i=1}^{n} \alpha_{in}(t)\, e_i$$

and choose α_{in} so that

$$\left\langle \frac{d}{dt} x_n(t), e_j \right\rangle + a(t, x_n(t), e_j) = \langle f(t), e_j \rangle \tag{5.2}$$

$$1 \leq j \leq n$$

with $\alpha_{in}(0) = \alpha_{in0}$ and

$$\sum_{i=1}^{n} \alpha_{in0}\, e_i \to x_0 \text{ in H as } n \to \infty$$

The differential equations (5.2) are of the form

$$B_n \frac{dy_n}{dt} + A_n(t) y_n = f_n, \quad y_n(0) = \{\alpha_{in0}\}$$

where B_n is the matrix with elements $\langle e_i, e_j \rangle$,

$A_n(t)$ is the matrix with elements $a(t, e_i, e_j)$,

y_n is the vector with elements $\alpha_{in}(t)$ and $f_n(t)$ is the vector with elements $\langle f(t), e_j \rangle$. Since the e_j are linearly independent

$$\det B_n \neq 0$$

and so the above equations admit a unique solution.

We shall show that $x_n \to x$ (a solution of the original equation) as $n \to \infty$.

Multiplying Eq.(5.2) by $\alpha_{jn}(t)$ and summing over j, an easy computation yields

$$\frac{1}{2} \frac{d}{dt} \|x_n(t)\|^2 + a(t; x_n(t), x_n(t)) = \langle f(t), x_n(t) \rangle$$

Then using the lower bound on $a(t; x_n(t), x_n(t))$ and integrating

$$\|x_n(\tau)\|_H^2 + 2K \int_0^\tau \|x_n(t)\|_X^2 \, dt \leq \|x_n(0)\|_H^2 + 2 \int_0^\tau |\langle f(t), x_n(t) \rangle| \, dt$$

$$\leq \|x_n(0)\|_H^2 + 2\int_0^\tau \|f(t)\|_{X'} \|x_n(t)\|_X \, dt$$

by the Schwarz inequality.

But

$$2\int_0^\tau \|f(t)\|_{X'} \|x_n(t)\|_X \, dt \leq \beta \int_0^\tau \|x_n(t)\|_X^2 + \frac{1}{\beta}\int_0^\tau \|f(t)\|_{X'}^2 \, dt$$

and $\|x_n(0)\|_H^2 \leq a\|x_0\|_H^2$ for some a, $\beta > 0$.

Hence

$$\int_0^\tau \|x_n(t)\|_X^2 \, dt \leq c\left[\|x_0\|_H^2 + \int_0^\tau \|f(t)\|_{X'}^2 \, dt\right]$$

for some c. Thus x_n ranges over a bounded set in $L^2(0,\tau; X)$ and we may extract a subsequence so that

$$x_\rho \to z \text{ weakly in } L^2(0,\tau; X)$$

We shall now show that z is the desired solution.

Let i be fixed but arbitrary and let $\rho > j$, then multiplying Eq.(5.2) by $\varphi(t)$, where $n = \rho$ and $\varphi(t) \in C^1(0,\tau)$ $\varphi(\tau) = 0$, integrating on $(0,\tau)$ and setting $\varphi_i(t) = \varphi(t)\, e_i$, gives

$$\int_0^\tau \left[-\langle x_\rho(t), \varphi_i'(t)\rangle + a(t, x_\rho(t), \varphi_i(t))\right] dt = \int_0^\tau \langle f(t), \varphi_i(t)\rangle dt + \langle x_\rho(0), \varphi_i(0)\rangle$$

Since $x_\rho \to z$ weakly, we have

$$\int_0^\tau \left[-\langle z, \varphi_i'\rangle + a(t; z, \varphi_i)\right] dt = \int_0^\tau \langle f, \varphi_i\rangle dt + \langle x_0, \varphi_i(0)\rangle$$

Since the above is true for $\varphi \in C^1(0,\tau)$, $\varphi(\tau) = 0$ we may take $\varphi \in D(0,\tau)$. Then in the sense of distributions in $D'(0,\tau)$ we have

$$\frac{d}{dt}\langle z(t), e_j\rangle + a(t; z(t), e_j) = \langle f(t), e_j\rangle$$

But the e_j are dense in X so that

$$\frac{dz}{dt} = A(t)\, z + f$$

Hence

$$\frac{dz}{dt} \in L^2(0,\tau; X') \text{ and so } z \in W(0,\tau)$$

It is easy to verify that z satisfies the initial condition, so we have obtained the desired solution.

Example 1

Let $Q = \Omega \times (0,\tau)$ where Ω is an open bounded set in R^n with a smooth boundary Γ. Consider

$$\frac{du}{dt} = \sum_{i,j=1}^{n} \frac{\partial}{\partial x_j}\left(a_{ij}(x,t)\frac{\partial u}{\partial x_i}\right) + f(x,t)$$

where $u = 0$ on Γ, $u(x,0) = u_0(x)$

$$a_{ij} \in L^\infty(Q), \quad \sum_{i,j=1}^{n} a_{ij}\xi_i\xi_j \geq K\sum_{i=1}^{n}\xi_i^2, \quad \xi_i \in R^1$$

We take

$$X = H_0^1(\Omega), \quad \text{so that } X' = H^{-1}(\Omega)$$

$$H = H^0(\Omega) = L^2(\Omega)$$

Then for $\varphi, \psi \in H_0^1(\Omega)$

$$a(t,\varphi,\psi) = \sum_{i,j=1}^{n} \int_\Omega a_{ij}(x,t)\frac{\partial\varphi}{\partial x_i}\frac{\partial\psi}{\partial x_j}\,dx$$

so that all the conditions for the uniqueness and existence theorem are satisfied. Hence, for $f \in L^2(0,\tau, H^{-1}(\Omega))$, $u_0 \in L^2(\Omega)$ there exists a unique solution $u \in L^2(0,\tau;\ H'_0(\Omega))$.

Example 2

Consider the above example where f is to be thought of as a control. Let us associate with the problem a performance index C(f) and try to find the control which minimizes C(f). Before attempting to find criteria for optimality, it is necessary to consider whether or not the problem is well posed, and the existence results will play an important role. For example, if

$$C(f) = \int_0^\tau \left[\|u\|^2_{L^2(\Omega)} + \|f\|^2_{L^2(\Omega)}\right] dt$$

then we know the problem is well posed since $u \in L^2(0,\tau;\ H_0^1(\Omega))$. However, if

$$C(f) = \|u(\tau)\|^2_{H_0^1(\Omega)} + \int_0^\tau \left[\|u\|^2_{H_0^1(\Omega)} + \|f\|^2_{H^{-1}(\Omega)}\right] dt$$

the last two terms are well defined but it is not the case that $u(\tau)$ lies in $H_0^1(\Omega)$ and the problem is not well posed. This kind of analysis becomes even more important when the control forces lie on the boundary of the system; for example, see Lions and Magenes (see bibliography).

C. STRONGLY CONTINUOUS SEMI-GROUPS

1. Introduction

We have seen (Section A) that if $A \in \mathscr{L}(X)$, where X is a finite-dimensional Euclidean space, the solution of

$$\dot{x} = Ax + f, \quad x(0) = x_0 \tag{1.1}$$

is

$$x(t) = e^{At} x_0 + \int_0^t e^{A(t-s)} f(s)\, ds$$

for any integrable vector-valued function f. Here,

$$e^{At} = I + At + \frac{A^2t^2}{2!} + \ldots + \frac{A^nt^n}{n!} + \ldots$$

It is natural to attempt to define e^{At} for A unbounded on X, and then derive a similar expression for the solution of the evolution equation (1.1). The above definition is difficult to use for A unbounded since $D(A^n) \supset D(A^{n+1})$ and the domain becomes smaller as we consider higher powers of A. Moreover, there is the problem of convergence! Instead, we choose a different representation of the exponential function, motivated by the scalar formula

$$e^{At} = \lim_{n \to \infty} \left(I - \frac{t}{n} A\right)^{-n}$$

Of course, it is necessary to make sure that the limit exists, and for this we shall assume

a) A is a closed operator such that D(A) is dense in X.
b) $(\lambda I - A)^{-1}$ exists for $\lambda > 0$ and $\|(\lambda I - A)^{-1}\| \leq 1/\lambda$, $\lambda > 0$.

Note that (b) implies

$$\|(I - \rho A)^{-1}\| \leq 1, \quad \rho \geq 0$$

so that $(I - t/n\, A)^{-1}$ is bounded and may be iterated. Now set

$$V_n(t) = \left(I - \frac{t}{n} A\right)^{-n}, \quad t \geq 0, \; n = 1, 2 \ldots$$

Then

$$\| V_n(t) \| \leq 1$$

and $V_n(t)$ is holomorphic in $t > 0$ since we know the resolvent $(\lambda I - A)^{-1}$ is holomorphic for $\lambda > 0$. Hence

$$\frac{dV_n(t)}{dt} = A\left(1 - \frac{t}{n} A\right)^{-(n+1)}$$

$V_n(t)$ is not necessarily holomorphic at $t = 0$, but it is strongly continuous at $t = 0$. To see this, note that

$$\| V_1(t)x - x \| = t \| (1 - tA)^{-1} Ax \| \leq t \| Ax \|$$

Hence $V_1(t)x \to x$, as $t \downarrow 0$ for all $x \in D(A)$. $V_1(t)$ is uniformly bounded, and $D(A)$ is dense in X, and hence $V_1(t)\, x \to x$ for all $x \in X$. It is easy to show that the same must be true for the iterates $V_n(t)$, so that

$$V_n(t)\, x \to V_n(0)\, x = x, \quad t \downarrow 0$$

<u>Existence of the strong limit</u> $\lim_{n \to \infty} V_n(t)$

Write

$$V_n(t)\, x - V_m(t)\, x = \lim_{\epsilon \to 0} \int_{\epsilon}^{t-\epsilon} \frac{d}{ds} \left[V_m(t-s)\, V_n(s) \right] ds\, x$$

$$= \lim_{\epsilon \to \infty} \int_{\epsilon}^{t-\epsilon} \left[-A \left[1 - \frac{t-s}{m} A \right]^{-1} V_m(t-s)\, V_n(s)\, x \right.$$

$$\left. + A\left(1 - \frac{s}{n} A\right)^{-1} V_m(t-s)\, V_n(s)\, x \right] ds$$

$$= \lim_{\epsilon \to 0} \int_{\epsilon}^{t-\epsilon} \left(\frac{s}{n} - \frac{t-s}{m}\right) A^2 \left(1 - \frac{t-s}{m} A\right)^{-(m+1)} \left(1 - \frac{s}{n} A\right)^{-(n+1)} x\, ds$$

Hence

$$\| (V_n(t) - V_m(t))\, x \| \leq \| A^2 x \| \int_0^t \left(\frac{s}{n} + \frac{t-s}{m}\right) ds = \frac{t^2}{2}\left(\frac{1}{n} + \frac{1}{m}\right) \| A^2 x \|$$

Hence $V_n(t)\, x$ converges uniformly in t on any finite interval for all $x \in D(A^2)$.

Since $D(A^2)$ is dense in X, and $V_n(t)$ is uniformly bounded, $\lim_{n \to \infty} V_n(t)\, x$ exists for all $x \in X$, and we shall denote this limit by $T_t\, x$.

Properties of T_t

It is now necessary to show that T_t has the properties of the exponential function.

It is obvious that T_t is strongly continuous, and

$$\|T_t\| \leq 1, \quad T_0 = I \tag{1.2}$$

Now

$$\frac{dV_n(t)}{dt} = A\left(I - \frac{t}{n}A\right)^{-1} V_n(t) = V_n(t)\,A\left(I - \frac{t}{n}A\right)^{-1}$$

$$= AV_n(t)\left(I - \frac{t}{n}A\right)^{-1} \tag{1.3}$$

But

$$A\left(I - \frac{t}{n}A\right)^{-1} x = \left(I - \frac{t}{n}A\right)^{-1} Ax \to Ax$$

as $n \to \infty$ for all $x \in D(A)$. Therefore

$$V_n(t)\,A\left(I - \frac{t}{n}A\right)^{-1} x \to T_t Ax, \quad x \in D(A)$$

and since A is closed

$$AV_n(t)\left(I - \frac{t}{n}A\right)^{-1} x \to AT_t x$$

so that A commutes with T_t or $AT_t \supset T_t A$.

Integrating Eq.(1.3),

$$V_n(t)x - V_n(0)x = V_n(t)x - x = \int_0^t \left(I - \frac{t}{n}A\right)^{-(n+1)} Ax\,dt, \; x \in D(A)$$

But

$$\left(I - \frac{tA}{n}\right)^{-(n+1)} x = \left(I - \frac{tA}{n}\right)^{-1} V_n(t)\,x \to T_t x$$

uniformly for t in each finite interval.

Hence

$$T_t x - x = \int_0^t T_s A x\,ds, \quad x \in D(A)$$

so that

$$\frac{d}{dt}(T_t x) = T_t Ax = AT_t x, \quad x \in D(A)$$

Thus $x(t) = T_t x_0$ is the solution of

$$\dot{x} = Ax, \quad x(0) = x_0$$

if

$$x_0 \in D(A)$$

Utilizing this fact,it is not difficult to show that

$$T_{t+s} = T_t T_s \tag{1.4}$$

Such an operator T_t with the properties (1.2, 1.4) is called a strongly continuous contraction semi-group. It is possible to obtain other semi-groups for example, if we replace (b) by

$$\text{(i)} \quad \| (\lambda I - A)^{-k} \| \leq \frac{M}{\lambda^k}, \quad \lambda > 0, \ k = 1, 2 \ldots$$

Then it is possible to show that

$$\text{(ii)} \quad \begin{array}{l} \| T_t \| \leq M \\ \| (\lambda I - A)^{-k} \| \leq M(\lambda - \beta)^{-k} \quad \lambda > \beta, \ k = 1,2 \ldots \end{array}$$

then

$$\| T_t \| \leq Me^{\beta t}$$

In these cases we shall refer to strongly continuous semi-groups.

2. Solution of the inhomogeneous equation

We have seen that the solution of the homogeneous equation

$$\dot{x} = Ax, \quad x(0) = x_0$$

is

$$x(t) = T_t x_0 \quad \text{for} \quad x_0 \in D(A)$$

Consider now the equation

$$\dot{x} = Ax + f, \quad x(0) = x_0 \tag{2.1}$$

where f is assumed to be strongly continuous with values in X, and A generates a strongly continuous semi-group T_t.

Suppose that $x(t)$ is a solution, then

$$\frac{d}{ds}[T_{t-s}x(s)] = -T_{t-s}Ax(s) + T_{t-s}Ax(s) + T_{t-s}f(s)$$

Therefore

$$\frac{d}{ds}[T_{t-s}x(s)] = T_{t-s}f(s)$$

Integrating on $(0,t)$ we obtain

$$x(t) = T_t x_0 + \int_0^t T_{t-s} f(s)\, ds$$

Now one would hope that this solution is always the solution of Eqs (2.1) but this is not in general true. However, we can prove the following theorem:

Theorem

If A generates a strongly continuous semi-group T_t, and

a) $f(t)$ is continuously differentiable for $t \geqq 0$
b) $x_0 \in D(A)$

then

$$x(t) = T_t x_0 + \int_0^t T_{t-s} f(s)\, ds$$

is continuously differentiable and satisfies Eq.(2.1).

Proof

We need only show that

$$\int_0^t T_{t-s} f(s)\, ds$$

satisfies the differential equation and has initial value 0.

Let

$$V(t) = \int_0^t T_{t-s} f(s)\, ds$$

Then

$$V(t) = \int_0^t T_{t-s} \left[f(0) + \int_0^s f'(r)\, dr \right] ds$$

$$= \left[\int_0^t T_{t-s} ds \right] f(0) + \int_0^t \left[\int_r^t T_{t-s}\, ds \right] f'(r)\, dr \tag{2.2}$$

Now for $x \in D(A)$

$$\frac{d}{ds} [T_s x] = A T_s x$$

Hence

$$T_t x - T_r x = \int_r^t A T_s x\, ds$$

Since A is closed and D(A) is dense in X, we have

$$A \int_r^t T_s\, ds = T_t - T_r$$

or

$$A \int_r^t T_{t-s}\, ds = T_{t-r} - I$$

Using this result in the expression (2.2) ensures that $V(t) \in D(A)$ and

$$AV(t) = (T_t - I) f(0) + \int_0^t (T_{t-r} - I) f'(r)\, dr$$

$$= -f(t) + T_t f(0) + \int_0^t T_{t-r} f'(r)\, dr$$

Now

$$V(t) = \int_0^t T_s\, f(t-s)\, ds$$

Hence

$$\frac{dV(t)}{dt} = T_t\, f(0) + \int_0^t T_s\, f'(t-s)\, ds$$

Therefore

$$\frac{dV(t)}{dt} = AV(t) + f(t)$$

Other results of a different form more useful for control applications have appeared in the literature (see Balakrishnan).

3. Mild solution

If the conditions on f and x_0 are not satisfied, but sufficient conditions are imposed so that

$$x(t) = T_t x_0 + \int_0^t T_{t-s} f(s)\, ds \tag{3.1}$$

is well defined and strongly continuous, then the solution is said to be a "mild" solution (e.g. $f \in L^2(0,T;\ X)$). This concept of a solution is particularly important in control application since if f is regarded as a controller we may require it should only be piecewise continuous.

There is a theorem which gives a necessary and sufficient condition for an operator A to generate a strongly continuous semi-group:

Theorem

If X is a Banach space such that

a) D(A) is dense in X
b) The resolvent $(I - n^{-1}A)^{-1}$ exists such that

$$\| (I - n^{-1}A)^{-m} \| \leq C \qquad n = 1,2\ldots \qquad m = 1,2\ldots$$

then A generates a strongly continuous semi-group T_t. Unfortunately, these conditions are not easily verified. However, there are a variety of conditions which give sufficient conditions (Hille and Phillips, and Yosida, see bibliography).

D. NON-LINEAR SYSTEMS

There is a growing literature on these systems (Browder, Kato, see bibliography), and it is not intended to attempt to survey the main ideas in this paper. Instead we shall take a simple example which will illustrate the kind of approach which is frequently used. We shall follow closely the kind of analysis which led to the solution of the linear elliptic problem. First, we state an abstract theorem:

Theorem

Let T be a mapping (possibly non-linear) of the reflexive Banach space X into its dual X', which satisfies

a) T is continuous from lines in X to the weak topology in X'.
b) There exists c on R^1 with $\lim_{r \to \infty} c(r) = +\infty$ such that for all $x \in X$

$$\langle Tx, x \rangle \geqq c(\|x\|) \|x\|$$

where $\langle \, , \rangle$ denotes the duality between X, X' and $\| \; \|$ is the norm in X.

c) T is monotone, i.e. for all $x, y \in X$

$$\langle Tx - Ty, x - y \rangle \geqq 0$$

Then T maps X onto X'.

We shall not prove this theorem but show how a concrete example can be formulated in such a way so that the theorem can be used.

Example

Consider

$$Au = \sum_{|\alpha| \leq m} D^{\alpha} A_{\alpha}(x, u, Du, \ldots D^m u)$$

where A_{α} is a non-linear partial differential operator of order m on an open bounded set Ω. Assume

$$A_{\alpha} : \Omega \times R^m \to R^1$$

is continuous on R^m for fixed $x \in \Omega$ and measurable in x for fixed $\xi \in R^m$. We assume

$$A_{\alpha}(x, \xi) \leq c[1 + \|\xi\|_{R^m}]$$

and consider the non-linear equation

$$\begin{aligned} Au &= f \quad \text{on } \Omega \\ D^{\beta} u &= 0 \quad \text{on } \Gamma \quad |\beta| \leq m - 1 \end{aligned}$$

Set

$$a(u, \psi) = \sum_{|\alpha| \leq m} \langle (-1)^{\alpha} A_{\alpha}(x, u, Du, \ldots D^m u), D^{\alpha} \psi \rangle$$

and consider

$$a(u, \psi) = \langle f, \psi \rangle, \quad \psi \in D(\Omega)$$

with

$$u \in H_0^m(\Omega) = X$$

From the assumptions we have

$$|a(u, \psi)| \leq c(\|u\|)\,\|\psi\|_X,\ \psi \in D(\Omega)$$

This inequality will also hold for $\psi \in H_0^m(\Omega)$ since $D(\Omega)$ is dense in $H_0^m(\Omega)$, so that we may restate the problem

$$a(u,v) = \langle f,v \rangle, \quad v \in X$$

$$u \in X$$

We also note that a(u,v) is linear in v on X. Hence a(u,v) is a bounded linear functional on X, and so there exists a unique element $Tu \in X'$, such that

$$\langle Tu, v \rangle = a(u,v)\ v \in X$$

Similarly, we can impose conditions on f so that there exists a unique $\omega \in X'$ such that

$$\langle \omega, v \rangle = \langle f, v \rangle\ v \in X$$

Then the problem becomes

$$\langle Tu, v \rangle = \langle \omega, v \rangle \quad v \in X$$

i.e.

$$Tu = \omega$$

Now we may use the theorem to determine a solution. For more details see Browder (bibliography).

D. FUNCTIONAL DIFFERENTIAL EQUATIONS

1. General theory

In the previous sections, we have considered ordinary and partial differential equations which are examples of systems whose future behaviour depends on the present state and not on the past. However, there are many applications in control, mathematical biology, economics, etc., where the past does influence the future significantly. One class of such systems is described by differential delay equations or, more generally, by functional differential equations. For example, if we model a population by saying that the growth (or decay) is proportional to the number of people in the population between the ages of 15 and 45, then one model could be

$$\dot{N}(t) = k[N(t-15) - N(t-45)]$$

where N(t) is the number in the population at time t, and is a constant. We see here that the current rate of change of N(t) depends on the values of N(t), 15 and 45 units of time earlier.

A predator-prey model studied by Volterra is

$$\dot{N}_1(t) = \epsilon_1 - \gamma_1 N_2(t) - \int_{-r}^{0} F_1(-\theta)\, N_2(t+\theta)\, d\theta$$

$$\dot{N}_2(t) = -\epsilon_2 + \gamma_1 N_1(t) + \int_{-r}^{0} F_2(-\theta)\, N_1(t+\theta)\, d\theta$$

where $N_1(t)$ is the number of prey in the population at time t, and $N_2(t)$ the number of predators at time t, F_1, F_2 are given functions, and $\epsilon_1, \epsilon_2, \gamma_1$ are constants. Here, we see that the current rates of change depend on the past history through the integral terms. Several more examples may be found in the book of Hale.

The theory of existence and uniqueness of solutions of functional differential equations is very similar to that for ordinary differential equations except that it is necessary to consider solution segments

$$x(t+\theta) \quad -b \leqq \theta \leqq 0$$

over a time interval rather than just x(t). The following results are developed in much more detail in Hale's book (see bibliography).

Consider the functional differential equation of retarded type

$$\dot{x}(t) = f(t, x_t) \tag{1}$$

where $f: R \times C([-b,0]; R^n) \rightarrow R^n$, and

$$C([-b,0]; R^n) = \text{all continuous maps } \varphi: [-b,0] \rightarrow R^n \text{ with norm}$$

$$\|\varphi\| = \max_{[-b,0]} |\varphi(t)|_{R^n}$$

If $\sigma \in R$, $a \geqq 0$, and $x(\cdot) \in C([-b,0]; R^n)$ then for any $t \in [\sigma, \sigma + a]$, we let $x_t \in C([-b,0]; R^n)$ be defined by $x_t(\theta) = x(t+\theta)$, $-b \leqq \theta \leqq 0$. So $x_t(\theta)$ is the segment of the curve x(t), as indicated below

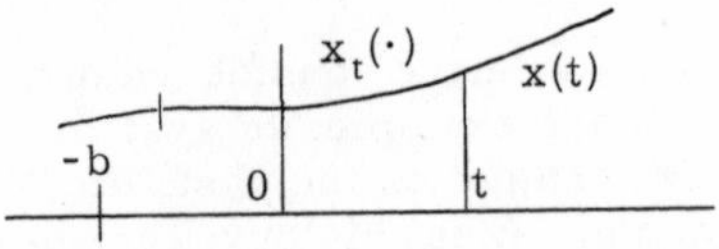

A function $x(\cdot)$ is said to be a solution of Eq.(1) if there exists $\sigma \in R$, $a > 0$ such that $x \in C([\sigma - b, \sigma + a]; R^n)$ and x(t) satisfies Eq.(1) on $[\sigma - b, \sigma + a)$. For a given $\sigma \in R$ and $h \in C([-b,0]; R^n)$ we say $x = x(\sigma, h)$ is a solution of Eq.(1) with initial value h at σ if there is an $a > 0$ such that $x(\sigma, h)$ is a solution of Eq.(1) on $[\sigma - b, \sigma + a)$ and $x_\sigma(\sigma, h) = h$.

Note

For the initial condition it is necessary to prescribe a function on $[-b,0]$.

One can prove the following existence and uniqueness theorem using methods similar to those used in Section A:

Theorem

a) Suppose D is an open set in $R \times C([-b,0], R^n)$ and $f: D \to R^n$ is continuous. Then if $(\sigma, h) \in D$ there exists a solution of Eq.(1) through (σ, h).

b) If f is also Lipschitzian in h on all compact subsets of D then the solution is unique.

Now let us see the implication of this theorem for the linear system

$$\dot{x}(t) = Ax(t) + Bx(t+\theta_1) + \int_{-b}^{0} C(\theta)x(t+\theta)\,d\theta \tag{2}$$

$$x_0 = h$$

Clearly, all the conditions of the theorem are satisfied and so there is a unique solution on $[0, \infty)$.

If we let $x_t(h)$ be the solution of Eq.(2) considered as a differential equation on $C([-b,0]; R^n)$ and write

$$x_t(h) = T(t)\,h$$

then T_t, $t \geq 0$ is the strongly continuous semi-group on $C([-b,0]; R^n)$ with infinitesimal generator $\mathscr{A}$ given by

$$\mathscr{A}\varphi(\theta) = \begin{cases} A\varphi(0) + B\varphi(\theta_1) + \int_{-b}^{0} C(\theta)\,\varphi(\theta)\,d\theta & \theta = 0 \\ \dfrac{d\varphi}{d\theta}(\theta) & \theta \neq 0 \end{cases}$$

with domain $D(a) = \{\varphi, \varphi' \in C([-b,0]; R^n)\}$.

Hence the equation may be written as an abstract evolution equation on $C([-b, 0]; R^n)$.

$$\dot{x}_t = \mathscr{A}x_t$$

$$x_0 = h$$

If we consider the inhomogeneous equation

$$\dot{x}_t = \mathscr{A}x_t + \tilde{f}(t)$$

$$x_0 = h$$

where $\tilde{f}$ is defined by

$$\tilde{f}(t)(\theta) = \left\{ \begin{matrix} 0 & \theta \neq 0 \\ f(t) & \theta = 0 \end{matrix} \right\}$$

then by analogy with Sections A and C, we may express the solution as

$$x_t = T(t)h + \int_0^t T(t-s)f(s)\,ds$$

at least for f(t) continuous.

2. Affine hereditary differential equations in M^2 spaces

It turns out that the abstract evolution equation formulation for linear delay systems is not the most appropriate for many control applications since $C([-b,0];\ R^n)$ is not a Hilbert space. However, it is possible to reformulate the system on a Hilbert space $M^2[-b,0;\ R^n]$ using the construction of Delfour and Mitter. Let us consider the system

$$\frac{dx}{dt}(t) = A\,x(t) + B\begin{matrix} x(t+\theta_1) \\ h(t+\theta_1)\end{matrix} + \int_{-b}^{0} C(\theta)\begin{matrix} x(t+\theta) \\ h(t+\theta)\end{matrix}\,d\theta \quad \begin{matrix} t+\theta \geq 0 \\ t+\theta < 0\end{matrix} \tag{3}$$

where $0 \leq t \leq T$, A, B are $n \times n$ matrices and $C \in L^\infty([-b,0], \mathscr{L}(R^n))$.

Definition M^2 space

Consider the space of functions

$\varphi : [-b,0] \to R^n$

with seminorm

$$\|\varphi\|_{M^2} = \left(|\varphi(0)|^2 + \int_{-b}^{0} |\varphi(\theta)|^2\,d\theta\right)^{\frac{1}{2}}$$

and $|\ \ |$ is the R^n norm.

We now define $M^2([-b,0];\ R^n)$ to be the quotient space of $L^2(-b,0;\ R^n)$ generated by the equivalence classes under $\|\ \ \|_{m^2}$, i.e. we say $\varphi_1 = \varphi_2$ in M^2 if $\|\varphi_1 - \varphi_2\|_{M^2} = 0$.

Thus M^2 is a function space of a point at $\theta = 0$ plus a curve

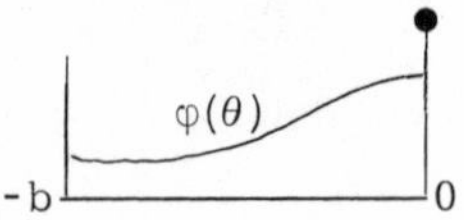

The point $\varphi(0)$ is uniquely specified, but the rest of the curve is only specified almost everywhere.

$M^2([-b,0];\ R^n)$ is a Hilbert space with inner product

$$\langle\varphi,\psi\rangle = \varphi'(0)\psi(0) + \int_{-b}^{0} \varphi'(s)\psi(s)\,ds$$

Note. In Hale's formulation, φ was continuous on $[-b, 0]$ and so was uniquely specified at all points.

Another way of looking at $M^2([-b,0], R^n)$ is to say that it is isometrically isomorphic to $R^n \times L^2(-b,0; R^n)$.

We now define another space $AC^2(t_0,t; R^n)$ to be the space of absolutely continuous maps $[t_0,t] \to R^n$ with derivative in $L^2(t_0,t_1; R^n)$ and norm

$$\|\varphi\|_{AC^2} = \left[|\varphi(t_0)|^2 + \int_{t_0}^{t} \left|\frac{d\varphi}{ds}\right|^2 ds \right]^{\frac{1}{2}}$$

Then we can write the homogeneous equation

$$\dot{\phi}(t) = \mathscr{A}\phi(t) \qquad (4)$$

$$\phi(0) = h$$

$\mathscr{A}$ is a closed linear operator on M^2 with domain AC^2 and is defined by

$$(\mathscr{A}h)(\theta) = \begin{cases} Ah(0) + Bh(\theta_1) + \int_{-b}^{0} C(\theta)h(\theta)\,d\theta & \theta = 0 \\ \dfrac{dh}{d\theta}(\theta) & \theta \neq 0 \end{cases} \qquad (5)$$

$\mathscr{A}$ is the infinitesimal generator of a strongly continuous semi-group $\Phi(t)$ on M^2 and Eq.(3) has the unique solution

$$\phi(t) = \Phi(t)h \qquad \text{for} \qquad h \in M^2$$

Consider now the inhomogeneous equation on

$$\phi(t) = \mathscr{A}\phi(t) + \tilde{f}(t) \qquad (6)$$

$$\phi(0) = h$$

where A is defined by expression (5), and $f \in L^2(0,\infty; R^n)$. Then Eq.(6) has the unique solution in

$$\phi(t) = \Phi(t)h + \int_0^t \Phi(t-s)\tilde{f}(s)\,ds$$

Thus we are able to allow for discontinuous inputs which are particularly important in control applications. Moreover, since M^2 is a Hilbert space we are able to obtain simple optimization results for the linear quadratic problem.

BIBLIOGRAPHY

YOSIDA, K., Functional Analysis, Springer Verlag (1968).

LIONS, J.L., MAGENES, E., Non-Homogeneous Boundary Value Problems, Springer Verlag (1972).

KATO, T., Perturbation Theory for Linear Operators, Springer Verlag (1966).

CARROLL, R.W., Abstract Methods in Partial Differential Equations, Harper and Row (1969).

CODDINGTON, E.A., LEVINSON, N., Theory of Ordinary Differential Equations, McGraw-Hill (1955).

LIONS, J.L., Optimal Control Systems Governed by Partial Differential Equations, Springer Verlag (1972).

HILLE, E., PHILLIPS, R., Functional Analysis and Semigroups, Am. Math. Soc. Colloq. 31 (1957).

BROWDER, F., Existence and Uniqueness Theorems for Solutions of Non-Linear Boundary Value Problems, Proc. Symp. Appl. Maths. A.M.S. (1965).

KATO, T., Non-Linear Evolution Equations in Banach Space, ibid.

HALE, J.K., Functional Differential Equations, Springer Verlag (1971).

DELFOUR, M.C., MITTER, S.K., Hereditary Differential Systems with Constant Delays I. General Case, J. Diff. Eqns 12 2 (1972) 213-235.

FINITE-DIMENSIONAL OPTIMIZATION

D.Q. MAYNE
Department of Computing and Control,
Imperial College of Science and Technology,
London, United Kingdom

Abstract

FINITE-DIMENSIONAL OPTIMIZATION.

Finite-dimensional optimization problems of the following types are considered in the paper: unconstrained optimization problem, inequality-constrained optimization problem, equally-constrained optimization problem, non-linear programming problem and optimization problems with special structures.

O. PRELIMINARIES

Optimization Problem

The finite-dimensional optimization problems discussed in these lectures have three ingredients: an objective function $f^0: R^n \to R$, an inequality constraint set Ω which is a subset of R^n, and an equality constraint set $\{x \in R^n \mid r(x) = 0\}$, where r maps R^n into R^k. The problems considered are:

P1. (Unconstrained optimization problem)

$$\min\{ f^0(x) \mid x \in R^n \}$$

which is shorthand for – find an $\hat{x}$ in the set R^n such that for all $x \in R^n$, $f^0(\hat{x}) \leq f^0(x)$.

P2. (Inequality-constrained optimization problem)

$$\min\{f^0(x) \mid x \in \Omega\}$$

P3. (Equality-constrained optimization problem)

$$\min\{f^0(x) \mid x \in R^n,\ r(x) = 0\}$$

P4. (Non-linear programming problem)

$$\min\{f^0(x) \mid x \in \Omega,\ r(x) = 0\}$$

Ω will generally be defined in terms of a function $f: R^n \to R^m$, that is:

1. $\Omega \triangleq \{x \in R^n \mid f(x) \leq 0\}$

(the notation $f(x) \leq z$, $f: R^n - R^m$, $z \in R^m$, denotes $f^i(x) \leq z^i$, $i = 1, 2, \ldots m$, where f^i, z^i denote the i-th component of f, z, respectively).

Optimization problems with special structure (for example, f^0, f, r, linear or quadratic), and, therefore, simpler to solve, will also be considered.

Conventions

The following conventions and notation will be used: R^n denotes the Euclidean space of ordered n-tuples of real numbers (R denotes R^1). $x \in R^n$ has components $x^1, x^2, \ldots x^n \in R$ ($x = (x^1, x^2, \ldots, x^n)$). x is treated as a column vector (n×1 matrix) in matrix operations. $\langle.,.\rangle$ denotes the scalar product in R^n, defined by $\langle x,y\rangle = x^T y = \sum_{i=1}^{n} x^i y^i$. $\|.\|$ denotes the norm in R^n, defined by $\|x\| = \langle x,x\rangle^{1/2}$. f or f(.) denotes a function. $f: A \to B$ denotes that the domain of f is A, and its codomain is B. Given a function $f: R^n \to R^m$, $f_x: R^n \to R^{m\times n}$ denotes the Jacobian matrix function whose ij-th element is $(\partial f_i/\partial x_j)$, $R^{m\times n}$ being the space of m×n real matrices. If $A \in R^{m\times n}$, $\|A\| \triangleq \max\{\|Ax\| \mid \|x\| \leq 1\}$. A^{-1} denotes the inverse of a matrix A, and A^T its transpose.

Symbols:

$\forall$	for all
$\exists$	there exists, $\nexists$ there does not exist
$\Rightarrow$	implies, $\not\Rightarrow$ does not imply
$\Leftrightarrow$	if and only if, is equivalent to
$\triangleq$	is defined by
$\ni$	such that
$\{x \mid P\}$	set of points having property P
$A + \bar{x}$	$\{x \mid x = a + x,\ a \in A\}$.
$\cup, \cap$	union, intersection
$A \subset B$	A is contained in B
$z \in A$, ($z \notin A$)	z belongs (does not belong) to A
$\emptyset$	the empty set
$Z = \{0,1,2,3\ldots\}$	the set of non-negative integers
$Z^+ = \{1,2,3\ldots\}$	the set of positive integers

$J^m = \{1,2,\ldots m\}$, $J_0^m = \{0,1,2,\ldots m\}$

Topology of R^n

Let $B_\epsilon(\bar{x}) \triangleq \{x \mid \|x - \bar{x}\| < \epsilon\}$. x is an interior point of $A \subset R^n$, if $\exists\, \epsilon > 0$, $\ni B_\epsilon(x) \subset A$, and a closure point if, $\forall \epsilon > 0$, $B_\epsilon(x) \cap A \neq \emptyset$. A is open if every point of A is an interior point, and closed if every closure point of A is in A. The closure $\bar{A}$ of A is the set of all closure points of A. The interior $\mathring{A}$ of A is the set of all interior points of A. (a,b) denotes the open interval $\{x \in R \mid a < x < b\}$. [a,b] denotes the closed interval $\{x \in R \mid a \leq x \leq b\}$.

Sequences

$\{x_i \in R^n\}$ denotes a sequence $x_1, x_2, x_3 \ldots$ in R^n. $\hat{x}$ is an accumulation point of $\{x_i\}$ if $\forall \varepsilon > 0$, $\forall n \in I$, $\exists n' \geq n \ni x^{n'} \in B_\epsilon(\hat{x})$. x^0 is a limit of the sequence $\{x_i\}$ (or the sequence $\{x_i\}$ converges to a limit x^0, written $x_i \to x^0$) if $\forall \varepsilon > 0$, $\exists n^0 \ni x_n \in B_\epsilon(x^0)$, $\forall n \geq n^0$. If $\hat{x}$ is an accumulative point of $\{x_i\}$, $\exists$ a subsequence of $\{x_i\}$ which converges to $\hat{x}$, that is, $\exists K \subset Z$ such that $x_i \to \hat{x}$ for $i \in K$. If $\hat{x}$ is a closure point of $A \subset R^n$, $\exists$ a sequence $\{x_i \in A\}$ having $\hat{x}$ as an accumulation point.

$A \subset R^n$ is bounded if $\exists \alpha \in R \ni \|x\| \leq \alpha$, $\forall x \in A$. $A \subset R^n$ is compact if it is closed and bounded.

2. Theorem

$A \subset R^n$ is compact $\Leftrightarrow$ every sequence $\{x_i \in A\}$ has an accumulation point in A.

3. Theorem (Cauchy convergence criterion)

$\{x_i \in R^n\}$ is convergent $\Leftrightarrow \{x_i\}$ is a Cauchy sequence, that is, $\forall \epsilon > 0$, $\exists n^0 \in I \ni \|x^m - x^n\| < \epsilon$, $\forall$ m, $n \geq n^0$.

Continuity

$f: R^n \to R^m$ is continuous at x_0 if $\forall \delta > 0$, $\exists \epsilon > 0 \ni x \in B_\epsilon(x_0) \Rightarrow f(x) \in B_\delta(f(x_0))$ (note B_ϵ is a ball in R^n, B_δ a ball in R^m). f is continuous on $A \subset R^n$ if it is continuous at all $x \in A$. $f: R^n \to R$ is lower semi-continuous at x_0 if $\forall \delta > 0$, $\exists \epsilon > 0 \ni x \in B_\epsilon(x_0) \Rightarrow f(x) > f(x_0) - \delta$. The definitions for lower semi-continuity on A, and upper semi-continuity (at x_0 or on A) are similar.

Infimum and minimum

$f: A \to R$ is bounded from below if $\exists \alpha \in R$ (α is a lower bound for f on A) $\ni f(x) \geq \alpha, \forall x \in A$. A lower bound $\bar{\alpha}$ (for f on A) is called the infimum for f on A ($\bar{\alpha} = \inf \{f(x) \mid x \in A\}$) if $\forall \epsilon > 0$, $\exists x \in A \ni f(x) < \bar{\alpha} + \epsilon$. Upper bound and supremum are similarly defined. If there exists an $\hat{x} \in A \ni f(x) \geq f(\hat{x})$ $\forall x \in A$, then $f(\hat{x})$ is called the minimum of f on A, and $\hat{x}$ the minimizer of f on A. Maximum is similarly defined.

Existence of minima

4. Theorem (existence)

If $f: A \to R$ is lower semi-continuous and $A \subset R^n$ is compact, then $\exists$ a solution $\hat{x} \in A$ to the problem $\min\{f(x) \mid x \in A\}$.

5. Examples

(i) f not continuous. $f: R \to R$, $f(0) = 1$, $f(x) = x$, $\forall x > 0$. $\inf\{f(x) \mid x \in [0,1]\} = 0$ but there does not exist a $\hat{x} \in [0,1]$ such that $f(\hat{x}) = 0$.

(ii) A not closed. $f: R \to R$, $x \mapsto x$; $A = (0,1]$
$\forall x \in A$, $\exists x' \in A \ni f(x') < f(x)$ (for example, $x' = x/2$). Again $\inf\{f(x) \mid x \in A\} = 0$, but $\hat{x}$ satisfying $f(\hat{x}) = 0$ does not lie in A.

(iii) A not bounded. $f: R \to R$, $x \mapsto x$; $A = R$. $\forall x \in A$, $\exists x' \in A \ni f(x') < f(x)$.

Differentiability

$f: R^n \to R$ is differentiable at $\bar{x}$, if $\exists$ a linear function $\tilde{f}: R^n \to R \ni \forall \delta > 0$, $\exists \epsilon > 0 \ni$:

6. $\| f(x^0) - \tilde{f}(x) \| < \delta \| x - \bar{x} \|$, $\forall x \in B_\epsilon(\bar{x})$ (equivalently $\| f(x) - \tilde{f}(x) \| = o(\| x - \bar{x} \|)$, where $o(\alpha)/\alpha \to 0$ as $\alpha \to 0$). Because $\tilde{f}$ is linear it may be expressed as:

$$\tilde{f}(x) = f(\bar{x}) + \langle \nabla f(\bar{x}), x - \bar{x} \rangle$$

The function $\nabla f: R^n \to R^n$ is called the gradient of f, and is regarded as a column vector in matrix operations.

7. Theorem

$f: R^n \to R$ is differentiable at $\bar{x}$, if all the partial derivatives $\partial f/\partial x^i$, $i = 1, \ldots n$, of f exist and are continuous at $\bar{x}$, in which case $\nabla f = f_x^T$, $f_x \triangleq (\partial f/\partial x_1, \ldots \partial f/\partial x_n)$.

$f: R^n \to R$ is twice differentiable at $\bar{x}$ if $\exists$ a quadratic function $\tilde{f}: R^n \to R$, $x \mapsto f(\bar{x}) + \langle \nabla f(\bar{x}), x - \bar{x} \rangle + \frac{1}{2} \langle x - \bar{x}, \nabla^2 f(\bar{x})(x - \bar{x}) \rangle \ni \forall \delta > 0 \; \exists \epsilon > 0 \ni$:

$$\| f(x) - \tilde{f}(x) \| < \delta \| x - \bar{x} \|^2, \quad \forall x \in B_\epsilon(\bar{x})$$

f is twice differentiable if the partial derivatives $\partial^2 f/\partial x_i \partial x_j$, $i, j = 1, 2, \ldots n$ exist and are continuous at $\bar{x}$, in which case $\nabla f = f_x^T$ and $\nabla^2 f = f_{xx}$, where $f_{xx}(\bar{x})$ is a $n \times n$ matrix, the Hessian of f at $\bar{x}$, whose ij-th element is $[f_{xx}(\bar{x})]_{ij} = \partial^2 f(\bar{x})/\partial x_i \partial x_j$, $i, j = 1, 2 \ldots n$.

If $A \in R^{n \times m}$, $\| A \| \triangleq \max \{ \| Ax \| \mid \| x \| \leq 1 \}$. $f: R^n \to R^m$ is differentiable at $\bar{x}$ if $\exists \tilde{f}: R^n \to R^m$ satisfying (6). If the partial derivatives $\partial f^i/\partial x^j$, $i = 1, 2, \ldots m$, $j = 1, 2, \ldots n$ exist and are continuous at $\bar{x}$, then $\tilde{f}: R^n \to R^m$ satisfying (6) exists and is defined by: $\tilde{f}(x) = f(\bar{x}) + f_x(\bar{x})(x - \bar{x})$; (the ij-th component of the Jacobian $f_x: R^n \to R^{m \times n}$ is $\partial f^i/\partial x^j$).

Mean-value theorems

7. Theorem

Let $f: R^n \to R$ be continuously differentiable (∇f exists and is continuous at all $x \in R^n$). Then $\forall x, h \in R^n$, $\forall \lambda \in R$, $\exists \xi \in [x, x + \lambda h]$, the line segment joining x and $x + \lambda h \ni$:

$$f(x + \lambda h) = f(x) + \lambda \langle \nabla f(\xi), h \rangle$$

8. Theorem

If $f: R^n \to R^m$ is continuously differentiable, then $\forall x, h \in R^n$, $\forall \lambda \in R^n$:

$$f(x + \lambda h) = f(x) + \left[\int_0^1 f_x(x + t.\lambda h) \, dt \right] \lambda h$$

Convex Sets

The set of points joining $x_1, x_2 \in R^n$ is denoted $[x_1, x_2] = \{x \in R^n \mid x = x_1 + \lambda(x_2 - x_1),\ \lambda \in [0,1]\}$, and is called a line segment. $(x_1, x_2) \triangleq \{x \in R^n \mid x = x_1 + \lambda(x_2 - x_1),\ \lambda \in (0,1)\}$.

A set $A \subset R^n$ is convex if, $\forall x_1, x_2 \in A$, the line segment $[x_1, x_2] \in A$. If $x \in A$, A convex, and $\exists$ no distinct $x_1, x_2 \in A \ni x \in (x_1, x_2)$, then x is an extreme point of A. The intersection of a finite or infinite family of convex sets in R^n is convex. If $x_1, x_2, \ldots x_m \in R^n$, then the convex hull of $x_1, x_2, \ldots x_m$, denoted co $(x_1, x_2, \ldots x_m)$ is the set $\left\{x \in R^n \mid x = \sum_{i=1}^{m} \lambda_i x_i,\ \lambda_i \geq 0 \text{ for } i = 1, 2, \ldots m,\ \sum_{i=1}^{m} \lambda_i = 1\right\}$.

If $x_0, x_1, \ldots x_n \in R^n$ are such that $(x_1 - x_0), (x_2 - x_0), \ldots (x_n - x_0)$ are linearly independent, then co $(x_0, x_1, \ldots x_n)$ is a simplex in R^n. A set C is a cone with vertex x_0 if $x \in C \Rightarrow x_0 + \lambda(x - x_0) \in C,\ \forall \lambda \geq 0$. The dimension of a convex set $C \subset R^n$ is the dimension of the smallest subspace $M \ni C \subset \bar{x} + M \triangleq L$ for some $\bar{x} \in R$. The relative interior of such a set, is rint $C \triangleq \{x \in L \mid \exists \epsilon > 0 \ni B_\epsilon(x) \cap C \subset L\}$. For example, if $a, b \in R^n$, int $[a, b] = \emptyset$, rint $[a, b] = (a, b)$.

Two sets are disjoint if they have no points in common. Two sets $A, B \subset R^n$ are separable (strictly separable) by a hyperplane $H_c(\alpha) \triangleq \{x \in R^n \mid \langle c, x \rangle = \alpha\}$ if $\exists c \in R^n$, $c \neq 0$, $\alpha \in R \ni$:

$$x \in A \Rightarrow \langle c, x \rangle \leq \alpha \quad (\langle c, x \rangle < \alpha)$$

$$x \in B \Rightarrow \langle c, x \rangle \geq \alpha \quad (\langle c, x \rangle > \alpha)$$

$H_c(\alpha)$ is said to separate A and B. Two sets may be separable but not disjoint, or disjoint but not separable. However:

9. Separation theorem

Let $A, B \subset R^n$ be convex and let A have dimension n. Then: A and B are separable $\Leftrightarrow$ int $A \cap$ rint $B = \phi$. Note, if $A, B \subset R^n$ are convex and disjoint, then A, B are separable.

The separation theorem may be used to prove a result, Farkas Lemma, much used in the sequel.

10. Farkas lemma

Let $A \in R^{m \times n}$, $b \in R^n$, and let the columns of A^T be $a_1, a_2, \ldots a_m \in R^n$. Then the following two statements are equivalent:

(i) $\{b\}$ and $S \triangleq \{x \in R^n \mid x = A^T y, y \geq 0\}$ are not disjoint $\left(\exists y \in R^k,\ y \geq 0 \ni A^T y = \sum_{i=1}^{m} a_i y_i = b\right)$

(ii) $\{b\}$ and S are not separable
$(\langle b, c \rangle \leq 0, \forall\ c \in R^n \ni \langle a_i, c \rangle \leq 0,\ i = 1, 2, \ldots m)$

Convex functions

Let $X \subset R^n$ be convex. $f: X \to R$ is convex if $x_1, x_2 \in X$, $\lambda \in [0,1] \Rightarrow$

$f(x_1 + \lambda(x_2 - x_1)) \leq f(x_1) + \lambda(f(x_2) - f(x_1))$.

f is concave if - f is convex.

$\hat{x}$ is a (strict) local minimizer of $f: R^n \to R$ on $X \in R^n$ if $\exists \epsilon > 0 \ni f(x) \geq f(\hat{x})$ $(f(x) > f(\hat{x}))$, $\forall x \in B_\epsilon(\hat{x}) \cap X$ $(\forall x \in B_\epsilon(\hat{x}) \cap (X \setminus \{\hat{x}\})$. $(X \setminus Y \triangleq \{x \in X \mid x \notin Y\})$

11. Theorem

If $f: X \to R$ is a convex function on a convex set $X \subset R^n$, then each local minimum of f on X is also the (global) minimum of f on X.

1. CONDITIONS OF OPTIMALITY

Assumption

In the sequel, $f^0: R^n \to R$, $f: R^n \to R^m$ and $r: R^n \to R^k$ will be assumed to be continuously differentiable.

1.1. Unconstrained optimization

1. Theorem

If $f^0(\hat{x})$ is a minimum (or local minimum) of f^0 on X, where X is an open subset of R^n, then:

$$\nabla f^0(\hat{x}) = 0$$

Proof

$\exists \epsilon > 0 \ni N_\epsilon(\hat{x}) \subset X$. Suppose $\nabla f^0(\hat{x}) \neq 0$ and let $h \triangleq \nabla f^0(\hat{x}) / \|\nabla f^0(\hat{x})\|$. By local optimality $\exists \epsilon_1 \in [0, \epsilon] \ni (\hat{x} - \alpha h) \in X$ and $f^0(\hat{x} - \alpha h) \geq f^0(\hat{x})$ $\forall \alpha \in [0, \epsilon_1]$. But:

$$f^0(\hat{x} - \alpha h) = f^0(\hat{x}) - \alpha \langle \nabla f(\hat{x}), h \rangle + o(\alpha)$$

$$= f^0(\hat{x}) - \alpha\, [c + o(\alpha)/\alpha], \quad c > 0$$

so that $\exists \alpha \in (0, \epsilon_1] \ni f^0(\hat{x} - \alpha h) \leq f^0(\hat{x}) - \alpha c/2$, a contradiction. Hence $\nabla f^0(\hat{x}) = 0$.

1.2. Inequality constraints

Consider P2 : $\min\{f^0(x) \mid x \in \Omega \subset R^n\}$ where $\Omega \triangleq \{x \in R^n \mid f(x) \leq 0\}$, where $f: R^n \to R^m$. $\forall x \in R^n$ let $I_0(x) \triangleq \{i \in \{1, 2, \ldots m\} \mid f^i(x) = 0\}$. The following condition of optimality is due to Zoutendijk.

1.   Theorem

If $\hat{x} \in \Omega \subset R^n$ is a solution to P2, then:

$$\theta(\hat{x}) \underset{=}{\triangle} \min_{h \in S} \max \{\langle \nabla f^0(\hat{x}), h \rangle; \langle \nabla f^i(\hat{x}), h \rangle, \ i \in I_0(\hat{x})\} = 0$$

where S is any subset of R^n containing the origin in its interior.
Proof

If $\theta(\hat{x}) = -\delta < 0$,   then $\exists \hat{h} \in S \ni$:

$$\langle \nabla f^i(\hat{x}), \hat{h} \rangle \le -\delta, \ \forall i \in \{0\} \cap I_0(\hat{x})$$

Hence $\exists \alpha > 0$ such that:

$$f^0(\hat{x} + \alpha h) \le f^0(\hat{x}) - \alpha \delta / 2$$

$$f^i(\hat{x} + \alpha h) \le -\alpha \delta / 2, \ \forall i \in I_0(\hat{x})$$

$$f^i(\hat{x} + \alpha h) \le 0, \ \forall i \in I_0^c(\hat{x})$$

where $I_0^c(x)$ denotes the complement of $I_0(x)$ in $J_m \underset{=}{\triangle} \{1, 2, \ldots m\}$, contradicting
the optimality of $\hat{x}$.

To proceed further, we have to introduce the concept of a 'linear
approximation' to Ω at $\hat{x}$. A convex cone is a linearization of
$(\Omega - \hat{x}) \underset{=}{\triangle} \{h \in R^n \mid h = x - \hat{x}, \ x \in \Omega\}$ if, $\forall$ sets $\{h^1, h^2, \ldots h^j\}$ of linearly independent
vectors in C, $\exists \epsilon > 0 \ni \text{co}\{0, \epsilon h^1, \epsilon h^2, \ldots \epsilon h^j\} \subset (\Omega - \hat{x})$. If $h \in C$,
$\exists \epsilon > 0 \ni \epsilon' h \in (\Omega - \hat{x}), \ \forall \epsilon' \in [0, \epsilon]$.

2.   Theorem

If either

A(i) $\{\nabla f^i(\hat{x}), \ i \in I(\hat{x})\}$

is a set of linearly independent vectors, or

A(ii) $\exists h \in R^n \supset \langle \nabla f^i(\hat{x}), h \rangle < 0, \ \forall i \in I_0(\hat{x})$

then $\widetilde{C} \underset{=}{\triangle} \{h \in R^n \mid \langle \nabla f^i(\hat{x}), h \rangle < 0, i \in I_0(\hat{x})\}$ is a linearization of $(\Omega - \hat{x})$, if
$\widetilde{C} \ne \phi$.
In fact, A(i) $\Rightarrow$ A(ii).

3.   Example

$$f^1(x) = (x^1 - 1)^3 + x^2, \ f^2(x) = -x^2, \ \hat{x} = (1, 0), \ I_0(\hat{x}) = \{1, 2\},$$

$\nabla f^1(\hat{x}) = (0, 1), \ \nabla f^2(\hat{x}) = (0, -1)$. $\widetilde{C}$ is not a linearization.

4. Theorem

Suppose $\hat{x} \in \Omega$ is a solution to P2, and C is a linearization of $(\Omega - \hat{x})$. Then:

$$\langle \nabla f^0(\hat{x}), h \rangle \geqq 0, \quad \forall h \in C$$

Proof

For, if not, $\exists h \in C, \ \delta > 0$

$$\langle \nabla f^0(\hat{x}), h \rangle < -\delta$$

Now, $\exists \epsilon_1 > 0 \ni \hat{x} + \epsilon h \in \Omega$, $\forall \epsilon \in [0, \epsilon_1]$, and $\exists \epsilon_2 \in [0, \epsilon_1] \ni f^0(\hat{x} + \epsilon h) < f^0(\hat{x})$, $\forall \epsilon \in [0, \epsilon_2]$. Thus $\forall \epsilon \in [0, \epsilon_2]$, $\hat{x} + \epsilon h \in \Omega$ and $f^0(x + \epsilon h) < f^0(x)$, contradicting the optimality of $\hat{x}$.

4a. Corollary. $\langle \nabla f^0(\hat{x}), h \rangle \geqq 0, \ \forall h \in \overline{C}$.

Proof

This follows from Theorem 4 and the continuity of the function $\langle \nabla f^0(x), \cdot \rangle : R^n \to R$.

5. Kuhn-Tucker Theorem

Suppose $\hat{x} \in \Omega$ is a solution to P2, and A(ii) of (1.2.2) holds. Then $\exists \{ \lambda^i \geqq 0, \ i \in I_0(\hat{x}) \} \ni$:

$$\nabla f^0(\hat{x}) + \sum_{i \in I_\epsilon(\hat{x})} \lambda^i \nabla f^i(\hat{x}) = 0$$

Proof

Since A(ii) holds we can replace C in (4) and (4a) by $\tilde{C}$, so that:

$$- \langle \nabla f^0(\hat{x}), h \rangle \leqq 0, \quad \forall h \in R^n \ni \langle \nabla f^i(\hat{x}), h \rangle \leqq 0, \quad \forall i \in I_0(\hat{x})$$

By Farkas' Lemma (0.10), $\exists \{ \lambda^i \geq 0 \mid i \in I_0(\hat{x}) \} \ni$:

$$\sum_{i \in I_0(x)} \lambda^i \nabla f^i(\hat{x}) = - \nabla f^0(\hat{x})$$

1.3. Equality and inequality constraints

Consider P3 : $\min \{ f^0(x) \mid f(x) \leqq 0, \ r(x) = 0 \}$. Let $g : R^n \to R^{k+1}$ be defined by

$$g(x) = (f^0(x), \ r(x))$$

g is continuously differentiable, its Jacobian being $g_x = \begin{bmatrix} f_x^0 \\ r_x \end{bmatrix}$. We consider first the relatively simple case when r is affine.

1. Theorem

If $\hat{x}$ is an optimal solution of P3, C is a linearization of $(\Omega - \hat{x})$ and r is affine, then $\exists \lambda = (\lambda^0, \lambda^1, \ldots \lambda^k) \in R^{k+1}$, $\lambda^0 \leqq 0$, $\lambda \neq 0$, $\ni$:

$$\langle \lambda, g_x(\hat{x})h \rangle \leqq 0, \quad \forall h \in \overline{C}$$

Proof

$$\text{Let } C_z \triangleq g_x(\hat{x})C$$

$$S \triangleq \{z \in R^{k+1} \mid z = -\beta(1, 0, \ldots 0), \ \beta > 0\}$$

C_z and S are convex cones.

We first prove that C_z and S are linearly separable. For, if they are not, C_z has dimension k+ 1, and all points of S are in the interior of C_z. Let $z^* \in S$; since $z^* \in \text{int}\, C_z$, $\exists h^* \in C \ni z^* = g_x(\hat{x})h^*$, that is, $(f_x^0(\hat{x})h^*, \ r_x(\hat{x})h^*) = -\beta^*(1, 0, \ldots 0)$ for some $\beta^* > 0$, so that:

$$f_x^0(\hat{x})h^* = -\beta^* < 0, \quad r_x(\hat{x})h^* = 0$$

Since $f_x^0(\hat{x})h^* = -\beta^*$ and C is a linearization, $\exists \epsilon > 0 \ni f^0(\hat{x} + \epsilon h^*) < f^0(\hat{x})$, $(\hat{x} + \epsilon h^*) \in \Omega$, and $r_x(\hat{x})\epsilon h^* = 0$. Since $r(\hat{x}) = 0$, and r is affine, the latter condition implies $r(\hat{x} + \epsilon h^*) = 0$. Hence $\hat{x}$ is not optimal, a contradiction, so that C_z and S are linearly separated.

Hence $\exists \lambda \in R^{k+1}$, $\lambda \neq 0$, $\ni$:

$$\langle \lambda, z \rangle \leqq 0, \quad \forall z \in C_z$$

$$\langle \lambda, z \rangle \geqq 0, \quad \forall z \in S$$

that is

$$\langle \lambda, g_x(\hat{x})h \rangle \leqq 0, \quad \forall h \in C$$

$$-\lambda^0 \geqq 0, \text{ that is } \lambda^0 \leqq 0$$

Since $\langle \lambda, \cdot \rangle$ is a continuous function, C may be replaced by $\overline{C}$, yielding the desired result.

We now show that the latter result can be expressed in the usual form:

2. Theorem

If $\hat{x}$ is an optimal solution of P3, A(ii) of (1.1.2) holds and r is affine, then $\exists \lambda \in R^{k+1}$, $\lambda \neq 0$, $\lambda^0 \leqq 0$, scalars $\mu^i \leqq 0$, $i \in I_0(\hat{x}) \ni$:

$$\lambda^0 \nabla f^0(\hat{x}) + \sum_{i=1}^{k} \lambda^i \nabla r^i(\hat{x}) + \sum_{i \in I_0(\hat{x})} \mu^i \nabla f^i(\hat{x}) = 0$$

Proof

From Theorem 1.3.1 replacing C by $\tilde{C}$

$$\langle g_x^T(\hat{x})\lambda, h\rangle \leq 0, \quad \forall h \in \tilde{\tilde{C}}$$

$$\Rightarrow \langle \lambda^0 \nabla f^0(\hat{x}) + \sum_{i=1}^{k} \lambda^i \nabla r^i(\hat{x}), h\rangle \leq 0, \quad \forall h \in R^n \ni \langle \nabla f^i(\hat{x}), h\rangle \leq 0, \quad \forall i \in I_0(\hat{x})$$

Applying Farkas Lemma, $\exists\{\eta^i \geq 0 \mid i \in I_0(\hat{x})\} \ni$

$$\lambda^0 \nabla f^0(\hat{x}) + \sum_{i=1}^{k} \lambda^i \nabla r^i(\hat{x}) = \sum_{i \in I_0(\hat{x})} \eta^i \nabla f^i(\hat{x})$$

The major task in the sequel is to remove the restriction 'r affine' in Theorems (1) and (2). This complicates the proof considerably, and use has to be made of a fixed-point theorem.

3. Theorem[1]

If for some $\epsilon > 0$, $\bar{x} \in R^n$, γ is a continuous map from $B_\epsilon(\bar{x})$ into $B_\epsilon(\bar{x})$, then γ has a fixed point, i.e. $\exists x \in B_\epsilon(\bar{x}) \ni$

$$\gamma(x) = x$$

4. Theorem

If $\hat{x}$ is an optimal solution of P3, C is a linearization of $(\Omega - \hat{x})$, then $\exists \lambda = (\lambda^0, \lambda^1, \ldots \lambda^k) \in R^{k+1}$, $\lambda^0 \leq 0$, $\lambda \neq 0 \ni$:

$$\langle \lambda, g_x(\hat{x})h\rangle \leq 0, \quad \forall h \in \bar{C}$$

Proof

Firstly we show that C_z and S are linearly separable. For if C_z and S are not linearly separable, then S lies in the interior of $C_z \Rightarrow \exists z^0 \in S$ $(z^0 = -\beta^0(1,0,\ldots0), \beta^0 > 0) \ni z^0 \in \text{int}\, C_z$, and $\exists$ simplex $\Sigma = \text{co}\{0, z^1, z^2, \ldots z^{k+1}\}$ $\ni z^0 \in \text{int}\,\Sigma$ and $\Sigma \subset C_z$. Since C_z has dimension $k+1$, $g_x(\hat{x})$ has maximal rank. $\exists\{h^i \in C \mid i = 1,2,\ldots k+1\} \ni$

5. $z^i = g_x(\hat{x})h^i, \quad i = 1, 2, \ldots k+1$

Also because C is a linearization of $(\Omega - \hat{x})$, z^i and, hence h^i, can be chosen so that

---

[1] In the following, $B_\epsilon(\bar{x}) \equiv \{x \mid \|x - \bar{x}\| \leq \epsilon\}$ denotes a closed ball.

6. $\text{co}\{0, h^1, \ldots h^{k+1}\} \subset \Omega$

Now, $\exists r > 0 \ni B_r(z^0) \in \Sigma$,

also $B_{\alpha r}(\alpha z^0) \in \Sigma, \quad \forall \alpha \in [0, 1]$.

Consider the map $\gamma^\alpha : B_{\alpha r}(\alpha z^0) \to R^{k+1}$ defined by:

7. $\gamma^\alpha(z) = z + \alpha z^0 - [g(\hat{x} + HZ^{-1}z) - g(\hat{x})]$

where the ith column of $H \in R^{(k+1)\times(k+1)}$ is h^i, and the ith column of $Z \in R^{(k+1)\times(k+1)}$ is z^i. (Z is invertible since Σ is a simplex and $Z = g_x(\hat{x})H$). If $h = HZ^{-1}z$, and $z \in B_{\alpha r}(\alpha z^0) \subset \Sigma \subset C_z$, then $h \in (\Omega - \hat{x})$, and $g_x(\hat{x})h = z$. Because g is continuously differentiable and $\|z\| \leq c$, for some $c < \infty$, $\forall z \in B_{\alpha r}(\alpha z^0)$, it follows that $\exists \alpha^0 \in (0, 1] \ni$

$$\|[g(\hat{x} + HZ^{-1}z) - g(\hat{x})] - g_x(\hat{x})HZ^{-1}z\|$$

$$= \|[g(\hat{x} + HZ^{-1}z) - g(\hat{x})] - z\| \leq \alpha r/2, \quad \forall z \in B_{\alpha^0 r}(\alpha^0 z).$$

Applying this to (7) yields:

$$\|\gamma^{\alpha^0}(z) - \alpha^0 z^0\| \leq \alpha^0 r/2, \quad \forall x \in B_{\alpha^0 r}(\alpha^0 z^0),$$

so that γ^{α^0} maps $B_{\alpha^0 r}(\alpha^0 z^0)$ into $B_{\alpha^0 r}(\alpha^0 z^0)$. Since γ^{α^0} is continuous, by Brouwer's fixed-point theorem, $\exists \bar{z} \in B_{\alpha^0 r}(\alpha^0 z^0) \ni$:

$$\gamma^{\alpha^0}(\bar{z}) = \bar{z}$$

that is

$$g(\hat{x} + \bar{y}) = g(\hat{x}) + \alpha^0 z^0$$

where

$$\bar{y} = HZ^{-1}\bar{z} \in (\Omega - \hat{x})$$

Hence

$$\bar{x} \triangleq \hat{x} + \bar{y} \in \Omega$$

and

$$f^0(\bar{x}) = f^0(\hat{x}) - \alpha^0\beta^0, \quad \alpha^0, \beta^0 > 0$$

$$r(\bar{x}) = 0$$

contradicting the optimality of $\hat{x}$.

Hence C_z and S are linearly separated. The rest of the proof is now the same as in the proof of Theorem 1.3.1.

5. Theorem

If $\hat{x}$ is the optimal solution of P3, and A(ii) of (1.1.2) holds, then $\exists\lambda \in R^{k+1}$, $\lambda^0 \leq 0$, $\lambda \neq 0$, scalars $\mu^i \leq 0$, $i \in I_0(\hat{x}) \ni$:

$$\lambda^0 \nabla f^0(x) + \sum_{i=1}^{k} \lambda^i \nabla r^i(\hat{x}) + \sum_{i \in I_0(\hat{x})} \mu^i \nabla f^i(\hat{x}) = 0$$

Proof

The proof is the same as that for Theorem (1.3.2), using the strengthened result Theorem 1.3.4 in place of theorem 1.3.1.

6. Corollary

If $\nabla r^i(\hat{x})$, $i = 1, 2, \ldots k$, $\nabla f^i(\hat{x})$, $i \in I_0(\hat{x})$, are linearly independent, then $\lambda^0 < 0$.

Proof

Assume $\lambda^0 = 0$. Since $\nabla r^i(\hat{x})$, $i = 1, 2, \ldots k$, $\nabla f^i(\hat{x})$, $i \in I_0(\hat{x})$ are linearly independent, $\exists h \in R^n$, $h \neq 0$, $\ni \langle \nabla r^i(\hat{x}), h \rangle = 0$, $i = 1, \ldots k$, $\langle \nabla f^i(\hat{x}), h \rangle < 0$, $\forall i \in I_0(\hat{x})$. From Theorem 1.3.5,

$$\sum_{i=1}^{k} \lambda^i \langle \nabla r^i(\hat{x}), h \rangle + \sum_{i \in I_0(\hat{x})} \mu^i \langle \nabla f^i(\hat{x}), h \rangle = 0$$

$$\Rightarrow \mu^i = 0, \quad \forall i \in I_0(\hat{x})$$

But $\exists h' \in R^n$, $h' \neq 0$, $\ni \langle \nabla r^i(\hat{x}), h' \rangle < 0$, $i = 1, \ldots k$, and this implies in turn that $\lambda^i = 0$, $i = 1, \ldots k$, that is, $\lambda = (\lambda^0, \lambda^1, \ldots \lambda^k) = 0$, a contradiction.

2. ALGORITHM MODELS – CONVERGENCE

Polak [1] gives three reasons for using simple algorithm models:

i) classification of computational methods;
ii) elucidation of essential features of algorithms guaranteeing convergence;
iii) providing a simple procedure for obtaining 'implementable' algorithms from 'conceptual' prototypes.

Each iteration of a conceptual algorithm may require an arbitrary number of arithmetical operations and function evaluations, while an implementable algorithm must require only a finite number. For example, many algorithms choose, at each iteration, a 'search direction' and then search along this direction; an algorithm which minimizes a function along this direction would be conceptual. We consider algorithms for determining points in $\Omega \subset R^n$ which are desirable. Conceptually, a point is desirable if it solves the optimization problem; more realistically, a point is desirable if it satisfies some necessary condition of optimality. The simplest algorithm for finding desirable points in a closed subset Ω of R^n employ two functions, $\bar{a} : \Omega \to \Omega$ for generating new points and $c : \Omega \to R$ for testing the desirability of a point.

1. Algorithm Model $\bar{a} : \Omega \to \Omega$, $c : \Omega \to R$

Step 0. Compute a $x_0 \in \Omega$. Set $i = 0$.

Step 1. Set $x_{i+1} = \bar{a}(x_i)$.

Step 2. If $c(x_{i+1}) \geq c(x_i)$ stop.

Else set $i = i + 1$ and go to step 1.

Under what conditions will this algorithm determine desirable points? The following theorem, due to Polak [1], shows that the essential condition is a kind of semi-continuity property of a.

2. Theorem

Suppose that:

\- (i) c is either continuous at all non-desirable points in Ω, or else c is bounded from below in Ω($\exists \bar{c} \in R \ni c(x) \geq \bar{c}$, $\forall x \in \Omega$)

(ii) $\forall x \in \Omega$ which is not desirable, $\exists \epsilon > 0$, $\delta > 0$ (possibly depending on x) $\ni$

$$c(\bar{a}(x')) - c(x') \leq -\delta < 0, \quad \forall x' \in B_\epsilon(x) \triangleq \{x' \in \Omega \mid \|x' - x\| \leq \epsilon\}$$

Then, either the sequence constructed by algorithm (1) is finite, and its next to last point is desirable, or it is infinite, and every accumulation point of $\{x_i\}$ is desirable.

Comment

There is an implicit test for desirability in algorithm (1). If $c(\bar{a}(x)) \geq c(x)$ then x is desirable.

Proof

If the sequence is finite, then, by step 4, $\exists k \ni c(x_{k+1}) \geq c(x_k)$, so that x_k is desirable. If the sequence is infinite and has an accumulation point x^*, it has a subsequence converging to x^* in Ω (that is $\exists K \subset Z \ni x_i \to x^*$ for $i \in K$). Assume that x^* is not desirable. Hence $\exists \epsilon > 0$, $\delta > 0$ and a $k \in K \ni$:

3. $x_i \in B_\epsilon(x^*)$, $c(x_{i+1}) - c(x_i) \leq -\delta$, $\forall i \geq k$, $i \in K$.

Hence, for any two consecutive points x_i, x_{i+j} of the subsequence K, $i \geq k$, we have:

$$c(x_{i+j}) - c(x_i) = [c(x_{i+j}) - c(x_{i+j-1})] + \ldots + [c(x_{i+1}) - c(x_i)]$$

$$\leqq [c(x_{i+1}) - c(x_i)]$$

$$\leqq -\delta$$

Hence, $\{c(x_i) \mid i \in K)\}$ is not a Cauchy sequence, and cannot converge. But, since c is either continuous, or bounded from below and $\{c(x_i) \mid i \in K\}$ is monotonically decreasing for $i \in K$, the sequence $\{c(x_i) \mid i \in K\}$ does converge, a contradiction. Hence x^* is desirable.

Comment

Note that the theorem does not state that accumulation points exist, only that if they do exist they are desirable. However, if Ω is compact, or if the set $\mathcal{L} \triangleq \{x \in \Omega \mid c(x) \leqq c(x_0)\}$ is compact (if $x_0 \in \mathcal{L}$ so does x_i, $\forall i \geqq 0$) then accumulation points in Ω do exist.

An algorithm can not always be expressed in terms of a function $\bar{a}: \Omega \to \Omega$. For example, once a search direction is chosen, the algorithm may select any one of many points satisfying some criterion. In such cases the function a should be replaced by a set-valued mapping $A: \Omega \to 2^\Omega$, where 2^Ω is the set of all subsets of Ω.

3. Algorithm Model $A: \Omega \to 2^\Omega$, $c: \Omega \to R$

Step 0. Compute a $x_0 \in \Omega$. Set $i = 0$.

Step 1. Compute a point $y \in A(x_i)$. Set $x_{i+1} = y$.

Step 2. If $c(x_{i+1}) \geqq c(x_i)$ stop.
Else set $i = i+1$ and go to step 1.

4. Theorem

Suppose that:

(i) c is either continuous or bounded from below in Ω.
(ii) $\forall x \in \Omega$, x not desirable, $\exists \epsilon > 0$, $\delta > 0 \ni$:

$$c(x'') - c(x') \leqq -\delta, \quad \forall x' \in B_\epsilon(x), \quad \forall x'' \in A(x')$$

Then the next to last point of any finite sequence, and every accumulation point of any infinite sequence generated by the algorithm is desirable.

These models, and this is one of their great virtues, can relatively easily be modified to cope with the approximations needed to make conceptual algorithms implementable. One method of modelling algorithms with approximations is to employ a set valued mapping $A: R^+ \times \Omega \to 2^\Omega$, $(\epsilon', x) \to y \in A(\epsilon', x)$, where ϵ' indicates the degree of approximation; as $\epsilon' \to 0$, $A(\epsilon', \cdot)$ tends, in some sense, to a 'conceptual' $\bar{A}$ satisfying Theorem 3. These comments are spelt out more precisely in Algorithm (5) and Theorem (6).

5. Algorithm Model $A : R^+ \times \Omega \to 2^\Omega$, $c : \Omega \to R$, $\epsilon_0 > 0$

Step 0. Compute $x_0 \in \Omega$. Set $i = 0$.

Step 1. Set $\epsilon = \epsilon_0$.

Step 2. Compute a $y \in A(\epsilon, x_i)$.

Step 3. If $c(y) - c(x_i) \leqq -\epsilon$, set $x_{i+1} = y$, set $i = i+1$ and go to step 1. Else set $\epsilon = \epsilon/2$ and go to step 2.

Comment

For simplicity the algorithm statement does not include a stop statement; for a version which does, see Polak [1], 1.3.26. Hence the algorithm either produces an infinite sequence $\{x_i\}$, or 'jams up' at some $x_s \in \Omega$, cycling between steps 2 and 3, halving ϵ at each cycle.

6. Theorem

Suppose that:

(i) c is either continuous or bounded from below in Ω.
(ii) $\forall x \in \Omega$, x non-desirable, $\exists \epsilon, \delta, \gamma > 0$

7. $c(x'') - c(x') \leqq -\delta$, $\forall x' \in B_\epsilon(x)$, $\forall x'' \in A(\epsilon', x')$, $\forall \epsilon' \in [0, \gamma]$

Then either the sequence $\{x^i\}$ jams up at x_s, where x_s is desirable (x is desirable if $c(x') - c(x) \geqq 0$, $\forall x' \in A(0, x)$), or the sequence is infinite, and all its accumulation points are desirable.

Proof

(i) First we show that the sequence cannot jam up at a non-desirable x_s. For if so, the algorithm generates a sequence $y_j \in A(\epsilon_0/2^j, x_s)$ (from step 3), $c(y_j) - c(x_s) > -\epsilon_0/2^j$. However, $\exists$ integer $J \ni \epsilon_0/2j \leqq \delta$ and $\epsilon_0/2^j \leqq \gamma$ (where $\delta = \delta(x_s) > 0$ and $\epsilon = \epsilon(x_s) > 0$ exist by assumption (ii)) $\forall j \geqq J$. Hence, from assumption (ii), $c(y_j) - c(x_s) \leqq -\delta \leqq -\epsilon_0/2^j$, $\forall j \geqq J$, which contradicts the statement above that $c(y_j) - c(x_s) > -\epsilon_0/2^j$, $\forall j \geqq 0$.

(ii) We next show that all accumulation points of an infinite sequence $\{x_i\}$ generated by the algorithm are desirable. For if x^* is a non-desirable accumulation point of $\{x_i\}$, then $\exists K \subset \{0, 1, 2, \ldots\} \ni x_i \to x^*$, for $i \in K$, and $\exists \epsilon, \delta, \gamma > 0$ satisfying (7) with $x = x^*$. Hence $\exists J_1 \ni x_i \in B_\epsilon(x^*)$, $\forall i \in K$, $i \geqq J_1$. Also, $\exists J \ni \epsilon_0/2^J \leqq \delta$, and $\epsilon_0/2^J \leqq \gamma$. From assumption (ii)

$$c(x_{i+1}) - c(x_i) \leqq -\delta \leqq -\epsilon_0/2^J$$

so that the test $c(x_{i+1}) - c(x_i) \leqq -\epsilon$ in step 3 of the algorithm is satisfied with $-\epsilon \leqq -\epsilon_0/2^J$. Hence, for any two consecutive points x_i, x_{i+j}, $i \geqq J_1$ in the convergent subsequence $\{x_i \mid i \in K\}$ satisfy

$$c(x_{i+j}) - c(x_i) = [c(x_{i+j} - c(x_{i+j-1}))] + \ldots [c(x_{i+1}) - c(x_i)]$$

$$\leq - \epsilon_0 / 2^J$$

Hence, the sequence $\{c(x_i) | i \in K\}$ is not Cauchy, but from assumption (i), since $\{x_i | i \in K\}$ converges, so does $\{c(x_i) | i \in K\}$, a contradiction.

3. UNCONSTRAINED OPTIMIZATION

In this section we shall describe several algorithms for P1, and examine their convergence properties using the algorithm models of Section 2. Since many algorithms choose at each iteration a search direction s_i, and then minimize, or approximately minimize, f^0 along s^i (that is, minimize the function $\theta : R^+ \to R$ defined by $\theta(\lambda) = f^0(x_i + \lambda s_i)$) we commence by describing several procedures for approximately minimizing a function $\theta : R^+ \to R$.

3.1. Algorithms for approximate minimization of $\theta : R^+ \to R$

1. Golden Section Search

Let θ: $R^+ \to R$ be convex. The golden-section algorithm consists of two parts. The first part determines an interval $[a_0, b_0]$ which brackets the minimum. The second part reduces the size of the interval bracketing the minimum to a pre-assigned value ϵ. Both parts rely on the convexity of θ. The first part follows. For a given $\rho > 0$, a sequence of points is calculated according to:

$$x_0 = 0$$

$$x_{i+1} = x_i + \rho$$

until the first point x_i is reached satisfying:

$$\theta(x_{i+1}) \geq \theta(x_i)$$

The minimum of θ will lie in the interval

$$I_0 = [a_0, b_0] \triangleq [x_{i-1}, x_{i+1}]$$

The second (golden-section part) is defined below: ($\forall j \geq 0$, $\ell_j \triangleq (b_j - a_j)$, $I_j \triangleq [a_j, b_j]$)

2. Golden Section Algorithm: $[a_0, b_0]$, $\epsilon > 0$, $F_1 = (3 - \sqrt{5})/\sqrt{2}$, $F_2 = 1 - F_1$

Step 0. Set $i = 0$.

Step 1. If $\ell_i \leq \epsilon$, set $\bar{x} = (a_i + b_i)/2$ and stop. Else proceed.

Step 2. Set $v_i = a_i + F_1 \ell_i$, $w_i = a_i + F_2 \ell_i$.

Step 3. If $\theta(v_i) < \theta(w_i)$, set $I_{i+1} \triangleq [a_{i+1}, b_{i+1}] = [a_i, w_i]$.

If $\theta(v_i) \geq \theta(w_i)$, set $I_{i+1} \triangleq [a_{i+1}, b_{i+1}] = [v_i, b_i]$.

Set $i = i+1$.

Note that $\ell_j = F_2^i \ell_0$. It can be shown that if $I_{i+1} \triangleq [a_{i+1}, b_{i+1}] = [a_i, w_i]$, then $w_{i+1} = v_i$ and if $I_{i+1} = [v_i, b_i]$, then $v_{i+1} = w_i$, reducing the computation required. If x^* denotes the solution to min $(\theta(x) \mid x \in R^+)$, then $|\bar{x} - x^*| \leq \epsilon/2$.

Recognizing the use of the algorithms presented in this section in the sequel, we now define $\theta : R^n \times R^+ \to R$ by:

3. $\theta(x, \lambda) \triangleq f^0(x + \lambda h(x)) - f^0(x)$

and a linear approximation $\tilde{\theta} : R^n \times R^+ \to R$ to θ by:

4. $\tilde{\theta}(x, \lambda) \triangleq \lambda \langle \nabla f^0(x), h(x) \rangle = \lambda \nabla_\lambda \theta(x, 0)$

It is assumed that

5. $\nabla_\lambda \theta(x, 0) = \langle \nabla f^0(x), h(x) \rangle < 0$

6. Armijo's algorithm: $\beta \in (0,1)$, $\rho > 0$

Step 0. Set $i = 0$, $\lambda_0 = \rho$.

Step 1. If $\theta(x, \lambda_i) \leq \frac{1}{2}\tilde{\theta}(x, \lambda_i)$, stop.

Else set $\lambda_{i+1} = \beta \lambda_i$ and go to step 1.

7. Comment

Figure 1 illustrates the situation.

8. Theorem

Let ∇f^0 and h be continuous, and $\langle \nabla f^0(x), h(x) \rangle = -\gamma < 0$. $\forall x \in R^n$ let $\lambda(x)$, denote the λ satisfying the stop condition in step 1, that is, $\lambda(x)$ is the largest λ of the form $\lambda = \beta^k \rho$ satisfying:

$$\theta(x, \lambda) \leq \frac{1}{2}\tilde{\theta}(x, \lambda)$$

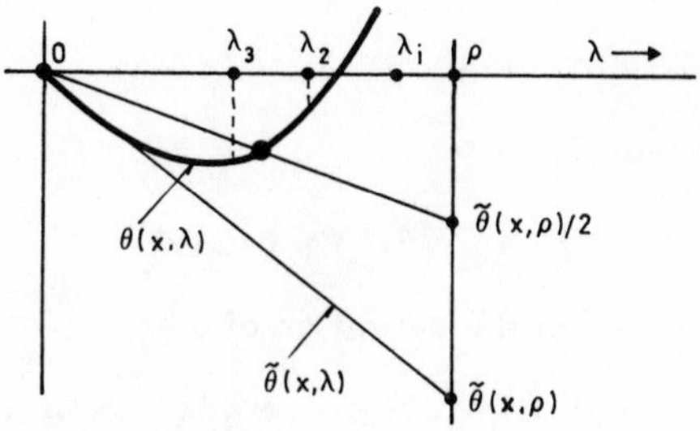

FIG. 1. Illustration of algorithm of approximate minimization.

Then, $\exists \epsilon > 0,\ \delta > 0\ (\delta = \gamma/2) \ni$:

$$\theta(x', \lambda(x')) \leq -\delta, \quad \forall x' \in B_\epsilon(x)$$

The proof is left to the reader as an exercise.

3.2. Algorithms for unconstrained minimization

Let us now return to the problem P1 : $\min\{f^0(x) \mid x \in R^n\}$. We assume that f^0 is continuously differentiable. The simplest algorithm is:

1. (Conceptual) Steepest Descent Algorithm

Step 0. Select a $x_0 \in R^n$. Set $i = 0$.

Step 1. Set $h(x_i) = -\nabla f^0(x_i)$. If $h(x_i) = 0$ stop.

Step 2. Compute smallest scalar $\lambda(x_i)$ which solves

$$f^0(x_i + \lambda(x_i)h(x_i)) = \min\{f^0(x_i + \lambda h(x_i)) \mid \lambda \geq 0\}$$

Step 3. Set $x_{i+1} = x_i + \lambda(x_i)\, h(x_i)$, set $i = i+1$ and go to step 1.

2. Theorem

Let $C(x_0) \triangleq \{x \mid f^0(x) \leq f^0(x_0)\}$ be bounded.

(i) If $\{x_i\}$ is finite, terminating at x_k, then x_k is desirable ($\nabla f^0(x_k) = 0$).

(ii) If $\{x_i\}$ is infinite, every accumulation point x^* is desirable.

Proof

(iii) Let $a : R^n \to R$ be defined by step 3 of (1); let $c \triangleq f^0$. c is continuous. Let θ, $\tilde{\theta}$ be defined by (4.1.3), (4.1.4). x^* not desirable $\Rightarrow \nabla_\lambda \theta(x^*, 0) = \langle \nabla f^0(x^*), h(x^*) \rangle = -\|\nabla f^0(x^*)\|^2 = -\gamma < 0$. Hence $\exists \lambda_1 > 0 \ni \theta(x^*, \lambda) < -\lambda\gamma/2$, $\forall \lambda \in [0, \lambda_1]$ so that $\lambda(x^*)$ (see step 2 of (1)) satisfies $\theta(x^*, \lambda(x^*)) \leq -\lambda_1\gamma/2$. Let $\phi : R^n \to R$ be defined by:

$$\phi(x') \triangleq f^0(x' + \lambda(x^*)h(x')) - f^0(x')$$

Clearly, ϕ is continuous, and satisfies:

$$\phi(x^*) = \theta(x^*, \lambda(x^*)) = -\lambda_1\gamma/2$$

Hence $\exists \epsilon > 0 \ni$

$$\phi(x') \leq -\lambda_1\gamma/4, \quad \forall x' \in B_\epsilon(x^*)$$

But, from the definition of ϕ and $\lambda(\cdot)$:

$$\theta(x', \lambda(x')) \leq \phi(x') \leq -\lambda_1\gamma/4, \quad \forall x' \in B_\epsilon(x^*)$$

But

$$\theta(x', \lambda(x')) = f^0(a(x')) - f^0(x')$$

Hence f^0 satisfies assumption (ii) of Theorem (2.2) (with $\delta = \lambda_1\gamma/4$), thus proving the result.

Algorithm 1 is conceptual since each iteration involves an optimization problem, (step 2) requiring an infinite number of steps. One method of making use of the golden section search of (3.1.1), (3.1.2) to replace step 2, together with a loop to reduce ϵ.

3. (Golden section) Steepest descent: $\epsilon_0 > 0$, $\rho > 0$

Step 0. Select $x_0 \in R^n$. Set $i = 0$.

Step 1. Set $\epsilon = \epsilon_0$.

Step 2. Let $h(x_i) = -\nabla f^0(x_i)$. If $h(x_i) = 0$ stop.

Step 3. Compute $\bar{\lambda}(x_i) \ni |\bar{\lambda}(x_i) - \lambda(x_i)| \leq \epsilon/2$, $\lambda(\cdot)$ defined in step 2 of (1), using (4.1.1), (4.1.2) on the function $\theta(x_i, \cdot)$.

Step 4. If $\theta(x_i, \bar{\lambda}(x_i)) \leq -\epsilon$, set $x_{i+1} = x_i + \bar{\lambda}(x_i)h(x_i)$, set $i = i+1$ and go to step 2. Else set $\epsilon = \epsilon/2$ and go to step 3.

Polak [1] shows that if f^0 is convex, $C(x_0)$ bounded, then any infinite sequence $\{x_i\}$ generated by (3) satisfies $f^0(x_i) \to \min\{f^0(x) | x \in R^n\}$.

4. (Armijo) Steepest descent: $\beta \in (0,1)$, $\rho > 0$.

Step 0. Select $x_0 \in R^n$, set $i = 0$.

Step 1. Set $h(x_i) = -\nabla f^0(x_i)$. If $h(x_i) = 0$, stop.

Step 2. Compute $\lambda(x_i)$ using Armijo's algorithm (4.1.6).

Step 3. Set $x_{i+1} = x_i + \lambda(x_i)h(x_i)$. Set $i = i+1$, and go to step 1.

That any accumulation point x^* of an infinite sequence generated by (5) is desirable ($h(x^*) = 0$) can be easily established, if $C(x_0)$ is bounded, using Theorem (3.1.8), identifying c with f^0 and defining $a: R^n \to R^n$ by step 3.

The convergence properties of algorithms (1), (3) and (4) are unaffected if h is defined by

$$h(x) = -D(x)\nabla f^0(x)$$

where $D: R^n \to R^{n\times m}$ is positive definite and continuous and $C(x_0)$ is bounded. If

$$D(x) = [f^0_{xx}(x)]^{-1}$$

then we have Newton-Raphson-type algorithms, for which ρ should be set equal to 1 in (3). If f^0 is quadratic, the Newton-Raphson version of algorithm (1) converges in 1 iteration.

3.3. Conjugate direction and conjugate gradient algorithms

To motivate the sequel we note that f^0 behaves like a quadratic function in the neighbourhood of any local (unconstrained) minimum x^*, in the sense that if f^0 is twice differentiable:

$$f^0(x^* + \delta x) = f^0(x^*) + \tfrac{1}{2}\langle \delta x,\, f^0_{xx}(x^*)\,\delta x\rangle + o(\|\delta x\|^2)$$

Hence, if an algorithm has poor behaviour when applied to a quadratic f^0, it will behave poorly when used for a general f^0. In this connection we note that the steepest descent algorithm, when f^0 is quadratic, and positive definite, that is:

1. $f^0(x) = a + \langle b, x\rangle + \tfrac{1}{2}\langle x, Cx\rangle$ where C is symmetric and positive definite, has the following property:

2. $f^0(x_j) - f^0(x^*) \leq \alpha^j\,[f(x_0) - f(x^*)]$ where

3. $\alpha \triangleq 1 - (\lambda_{min}|\lambda_{max})$ where λ_{min} and λ_{max} are the minimum and maximum eigenvalues of $C\,(\alpha \to 1$ as the 'condition number' $(\lambda_{min}/\lambda_{max}) \to 0)$. Note, in passing, that if f^0 satisfies 1, $f^0_x(x) \equiv Cx + b$, $f^0_{xx}(x) \equiv C$ and $x^* = -C^{-1}b$ in the unique local minimum (if C is p.d.). Also note that

$$f^0(x + \delta x) = f^0(x) + f^0_x(x)\,\delta x + \tfrac{1}{2}\langle \delta x, f^0_{xx}(x)\delta x\rangle$$

The Newton-Raphson algorithm satisfies

$$\begin{aligned} x_{i+1} &= x_i - \lambda(x_i)[f^0_{xx}(x_i)]^{-1}\nabla f^0(x) \\ &= x_i - \lambda(x_i)C^{-1}(Cx_i + b) \\ &= -C^{-1}b = x^*, \text{ if } \lambda(x_i) = 1 \end{aligned}$$

and so behaves excellently when f^0 is quadratic. However, the computation of f^0_{xx} is expensive. The algorithm described here tries to preserve the desirable 'second order' features of the Newton-Raphson algorithm at less expense. We assume, for simplicity, in the sequel that f^0 is quadratic and p.d.

4. Conjugate direction algorithm

This is of the form of the steepest descent algorithm (3.2.1) with the restriction $\lambda \geq 0$ removed from step 2 and step 1 replaced by: Step 1'. Set $h(x_i) = h_i$ where $\{h_i,\, i = 0, 1, 2, \ldots n-1\}$ are C-conjugate, that is:

5. $\langle h_i, Ch_j\rangle = 0\ \forall i \neq j,\ i, j \in 0, 1, 2, \ldots n-1$

which implies that $h_0, h_1, \ldots h_{n-1}$ are l.i. (linearly independent). Let, $\forall j = 1, 2, \ldots n$

6. $$M_j \triangleq \{x \mid x = \sum_{j=0}^{j-1} \alpha_i h_i,\ \alpha_i \in R\}$$

and

7. $$L_j \triangleq M_j + x_0$$

8. Fact. $\hat{x}$ solves $\min\{f^0(x) \mid x \in L_j\} \Leftrightarrow$

$$\nabla f^0(\hat{x}) \in {}^{\perp}M_j \ (\langle \nabla f^0(\hat{x}), h_i \rangle = 0,\ \forall i \in \{0, 1, 2, \ldots j-1\})$$

Proof

Left as an exercise to the reader.

9. Theorem

If $\{x_i \mid i = 1, 2 \ldots n\}$ is generated by the C.D. algorithm (and f^0 is quadratic and p.d.), then x_j minimizes f^0 on L_j (and hence x_n minimizes f^0 on R^n).

Proof.

x_1 minimizes f^0 on L_1. Assume x_j minimizes f^0 on L_j, that is $\langle \nabla f^0(x_j), h_i \rangle = 0$ for $i = 0, 1, \ldots j-1$. Now:

$$\nabla f^0(x_{j+1}) = \nabla f^0(x_j) + C(x_{j+1} - x_j)$$

$$= \nabla f^0(x_j) + \lambda_j C h_j$$

By virtue of the C-conjugate property:

$$\langle \nabla f^0(x_{j+1}), h_i \rangle = \lambda_j \langle h_i, C h_i \rangle$$

$$= 0, \text{ for } i = 0, 1, \ldots j-1$$

Also since x_{j+1} minimizes f^0 on $\{x_j + \lambda h_j \mid \lambda \in R\}$

$$\langle \nabla f^0(x_{j+1}), h_j \rangle = 0$$

that is, x_{j+1} minimizes f^0 on L_{j+1}, thus completing the proof by induction.

The following algorithm automatically generates C-conjugate directions ($g_i \triangleq \Delta f^0(x_i)$, $h_i \triangleq h(x_i)$, $\forall i \in Z$):

10. Conjugate gradient algorithm

Step 0. Select $x_0 \in R_n$, set $i = 0$, set $h_0 = g_0$.

Step 1. If $g_i = 0$, stop.

Step 2. Compute smallest scalar λ_i which solves

$$f^0(x_i + \lambda_i h_i) = \min\{f^0(x_i + \lambda h_i) \mid \lambda \geq 0\}$$

Step 3. Set $x_{i+1} = x_i + \lambda_i h_i$.

Set $\beta_{i+1} = \langle g_{i+1}, g_{i+1} \rangle / \langle g_i, g_i \rangle$

Set $h_{i+1} = -g_{i+1} + \beta_{i+1} h_i$

Set $i = i+1$ and go to step 1.

11. Lemma

Let f^0 be p.d. and quadratic, and assume that $h_0, h_1, \dots h_{j-1}$ are C-conjugate. Then:

(i) $\langle g_j, g_i \rangle = 0, \; i = 0, 1, \dots j-1$

(ii) $g_j \neq 0 \Rightarrow h_0, h_1, \dots h_j$ are C-conjugate.

Proof

(i) $(h_0, h_1, \dots h_{j-1})$ are C-conjugate $\Rightarrow$ (Theorem 9) x_j minimizes f^0 on $L_j \Rightarrow \langle g_j, h_i \rangle = 0, \; i = 0, \dots j-1$.
But for $i = 0, 1 \dots j-1$:

$$\langle g_j, h_i \rangle = \langle g_j, -g_i + \beta_i h_{i-j} \rangle$$

$$= -\langle g_j, g_i \rangle$$

(ii) $\langle h_j, Ch_i \rangle = \langle -g_j + \beta_j h_{j-1}, Ch_i \rangle$

$$= \langle -g_j, Ch_i \rangle, \; i = 0, 1, \dots j-2$$

But $g_i \neq 0 \Rightarrow \lambda_j \neq 0$ (prove) $\Rightarrow Ch_i = [g_{i+1} - g_i]/\lambda_i \Rightarrow$ (using i) $\langle h_j, Ch_i \rangle = 0$, $i = 0, 1, \dots j-2$.

$$\langle h_j, Ch_{j-1} \rangle = \langle -g_j + \beta_j h_{j-1}, (g_j - g_{j-1})/\lambda_{j-1} \rangle = 0$$

as can be shown using (i), the definition of λ_j, and the fact that λ_{j-1} satisfies:

$$0 = \langle g_j, h_{j-1} \rangle = \langle g_{j-1} + C\lambda_{j-1} h_{j-1}, h_{j-1} \rangle$$

$$\Rightarrow \lambda_{j-1} = \langle g_{j-1}, h_{j-1} \rangle / \langle h_{j-1}, Ch_{j-1} \rangle$$

12. Theorem

If f^0 is p.d. and quadratic, then algorithm (10) minimizes f^0 in at most n iterations.

Proof

h_0 and h_1 are C-conjugate (prove). By induction, from Lemma 11, $h_0, h_1, \ldots h_{n-1}$ are C-conjugate. The result follows from Theorem 9.

Polak-Ribiere have established [2] that a modified version of the algorithm ($\beta_{i+1} = \langle (g_{i+1} - g_i), g_{i+1} \rangle / \langle (g_i, g_i \rangle)$, which has the same behaviour when applied to quadratic functions (why?) has the property, when f^0 is strictly convex and twice continuously differentiable, that accumulation points of infinite sequences are desirable. This follows from the fact that $\exists \rho > 0 \ni - \langle g_i, h_i \rangle \geq \rho \, \| g_i \| \, \| h_i \|, \ \forall i.$

3.4. Pseudo-Newton-Raphson algorithms

There exists a class of algorithms, variously called pseudo- or quasi-Newton-Raphson, Variable Metric, Secant, which have the same or similar properties as the conjugate gradient algorithm when applied to quadratic functions, and in addition produce an estimate of the Hessian $\nabla^2 f^0$ or its inverse. The essential property of these methods is; that at iteration j, the estimate H_j of $(\nabla^2 f^0)^{-1}$ satisfies:

1. $\Delta x_i = H_j \Delta g_i, \ i = 0, \ldots j-1$

where $\Delta x_i \triangleq x_{i+1}$, $\Delta g_i \triangleq g_{i+1} - g_i$, $\forall i$. The search directions $\{h_j\}$ are generated according to:

2. $h_j = -H_j g_j$

Since, if f^0 is quadratic and p.d., $\Delta x_i = C^{-1} \Delta g_i, \ i = 0, \ldots j-1$, it can be seen that the restriction of H_j to CM_j — the linear subspace spanned by $\{\Delta g_0, \ldots \Delta g_{j-1}\}$ or $\{C\Delta x_0, \ldots C\Delta x_{j-1}\}$ — is equal to the restriction of C^{-1} to CM_j (we shall say $H_j = C^{-1}$ on CM_j). Assume H_j is non-singular. Then, if $h_j \in M_j$ from (2) and the non-singularity of H_j, $g_j \in CM_j$, so that $H_j g_j = C^{-1} g_j$. Since $x^* = x_j - C^{-1} g_j$, $x^* = x_j + h_j$ when $\lambda = 1$. Consequently, if $h_0, h_i, \ldots h_{n-1}$ are 1.i., $H_n = C^{-1}$, and $x_{n+1} = x_n + \lambda h_n = x^*$ for $\lambda = 1$, where x^* minimizes f^0 in R^n. Finally, it can be shown that if f^0 is quadratic and p.d., and H_i symmetric $\forall i$, the algorithm generates C-conjugate search directions, and hence minimizes f^0 in R^n in at most n iterations.

The most used algorithm for calculating $\{H_i\}$ is the Davidson-Fletcher-Powell (DFP) formula [3] H_0 is chosen to be symmetric, p.d. (for example, $H_0 = I$) and $\{H_i\}$ calculated according to:

3. $$H_{j+1} = H_j + \frac{\Delta x_j \Delta x_j^T}{\langle \Delta x_j, \Delta x_j \rangle} - \frac{(H_j \Delta g_j)(H_j \Delta g_j)^T}{\langle \Delta g_j, H_j \Delta g_j \rangle}$$

It can be shown that if f^0 is minimized exactly in each search direction, then, $\forall j \geq 0$

(1) $H_j \Delta x_i = 0, \ i = 0, 1, 2, \ldots j-1$

(2) H_{j+1} is p.d. symmetric.

This formula seems to be robust. Convergence of the resultant algorithm when applied to f^0, strictly convex and twice continuously differentiable, has been established by Powell (see Polak [1]).

Another similar formula is the rank 1 formula due to Sargent (see, e.g. Luenberger [3]):

$$4. \quad H_{j+1} = H_j + \frac{(\Delta x_j - H_j \Delta g_j)(\Delta x_j - H_j \Delta g_j)^T}{\langle \Delta x_j - H_j \Delta g_j, \Delta g_j \rangle}$$

if $\Delta x_j \neq H_j \Delta g_j$; otherwise $H_{j+1} = H_j$. Unlike the D.F.P. formula, this formula does not require exact minimizations of f^0 along search directions, and, therefore, lends itself to 'implementable' versions using, for example, Armijo's method. If we assume that H_j satisfies (1), that is, $H_j \Delta g_i = \Delta x_i$, $i = 0, 1 \ldots j-1$, then $H_{j+1} \Delta g_i = \Delta x_i$, $i = 0, 1, \ldots j$. If the algorithm does minimize f^0 along each search direction, and f^0 is quadratic and p.d., convergence in at most n iterations can be established.

4. RATE OF CONVERGENCE

Suppose $\{x_k\}$ is a sequence converging to x^*, and, for all non-negative integers k, let η_k^p denote $\|x_{k+1} - x^*\| / \|x_k - x^*\|^p$. $\{x_k\}$ is said to converge linearly if $\exists \beta \in (0,1) \ni$:

$$1. \quad \lim_{k \to \infty} \eta_k^1 = \beta$$

i.e. if $\exists k_1 \ni \|x_{k+1} - x^*\| \leq \beta \|x_k - x^*\| \ \forall k \geq k_1$. If $\{x_k\}$ converges linearly $\exists\, c \in R \ni \|x_k\| \leq c\beta^k$, $\forall k \geq k_1$, i.e. $\{x_k\}$ converges at least as fast as a geometric progression. β is called the convergence ratio. If $\beta = 0$ the convergence is said to be superlinear. The order of convergence is the supremum of $p \in R \ni \lim_{k\to\infty} \sup \eta_k^p < \infty$. ($\lim_{k\to\infty} \sup \eta_k^p = \lim_{k\to\infty} s_k$ where $s_k \overset{\Delta}{=} \sup\{\eta_j^p \mid j \geq k\}$). Convergence of any order greater than 1 is superlinear; however convergence of order 1 may be superlinear (if $\beta = 0$). As an example consider $\{x_k = a^k\}$, $a \in (0,1)$. $\eta_k^1 = a$, $\forall k \in Z$, and in fact $\{x_k\}$ converges linearly with convergence ratio $\beta = a$. If $x_k = a^{2^k}$, $\forall k \in Z$, $a \in (0,1)$, then $\eta_k^1 = a^{2^k}$, $\forall k \in Z$, and convergence is superlinear ($\beta = 0$). $\eta_k^2 = 1$, $\forall k \in Z$, and thus $\{x_k\}$ has order of Convergence 2.

These concepts enable us to compare the rate of convergence algorithms. Until now we have been merely concerned whether they converge or not. To illustrate the type of calculations required we consider two simple conceptual algorithms, steepest descent and Newton-Raphson, i.e. we shall consider algorithms of the form:

$$2. \quad a(x) = x + \lambda(x)\, h(x)$$

where $\lambda(x)$ is the minimizer of $f^0(x + \lambda h(x))$, $\lambda \in R^+$. As in section 3, $\theta(x, \lambda)$ denotes $f^0(x + \lambda h(x)) - f^0(x)$.

The steepest descent algorithm

For this algorithm $h(x) = -\nabla f^0(x)$. We assume that f^0 is twice continuously differentiable, and strictly convex in the neighbourhood of x^* which is assumed to be the limit point of a convergent sequence $\{x_i\}$. Hence $\exists \epsilon > 0,\ m > 0,\ M > 0,\ m \leq M \ni$:

3. $\langle h, \nabla^2 f^0(x) h \rangle \in [m\|h\|^2,\ M\|h\|^2]$

$\forall h \in R,\ \forall x \in B_\epsilon(x^*)$, and $\exists k_1 \ni x_i \in \mathring{B}_\epsilon(x^*),\ \forall i \geq k_1$. As a consequence of (1) we have:

$$\theta(x, \lambda) \leq \langle \nabla f^0(x), h(x) \rangle + M\|h(x)\|^2 \lambda^2 / 2$$

$\forall x, \lambda \ni x, x + \lambda h(x) \in \overset{0}{B}_\epsilon(x^*)$. Hence $\forall i \geq k$

4. $\theta(x_i, \lambda(x_i)) \leq -\|\nabla f^0(x_i)\|^2 / 2M.$

But we also have, $\forall i \geq k$:

5. $f^0(x_i) - f^0(x^*) \in [m\|x_i - x^*\|^2/2,\ M\|x_i - x^*\|^2/2]$

and

6. $\|\nabla f^0(x_i)\| \geq m\|x_i - x^*\|$

so that

$$\begin{aligned} f^0(x_{i+1}) - f^0(x_i) &= \theta(x_i, \lambda(x_i)) \\ &\leq -m^2\|x_i - x^*\|^2 / 2M \\ &\leq -(m/M)^2 [f^0(x_i) - f^0(x^*)] \end{aligned}$$

Hence

7. $f^0(x_{i+1}) - f^0(x^*) \leq [1 - (m/M)^2][f^0(x_i) - f^0(x^*)]$,

which establishes the linear convergence of $\{f^0(x_i)\}$ with convergence ratio $\beta = [1 - (m/M)^2]$. Convergence is possibly very slow as $\beta \to 1$ ($(m/M) \to 0$), since $f^0(x_i) - f^0(x^*) \leq \beta^i [f^0(x_{k_1}) - f^0(x^*)]/\beta^{k_1}$. Also $\forall i \geq k_1$:

$$\begin{aligned} \|x_i - x^*\|^2 &\leq (2/m)\,[f^0(x_i) - f^0(x^*)] \\ &\leq (2/m)\,[f^0(x_{k_1}) - f^0(x^*)]\,\beta^i / \beta^{k_1} \end{aligned}$$

so that $x_i \to x^*$ linearly with convergence ratio $\sqrt{\beta}$.

The above discussion can be generalized to deal with any algorithm of the type in (2), where $h(x)$ satisfies, for some $p \in (0, 1]$:

8. $\langle \nabla f^0(x), -h(x) \rangle \geq p \, \|\nabla f^0(x)\| \, \|h(x)\|$

$\forall x \in R^n$. (8) is clearly satisfied for the steepest descent algorithm ($-h(x) = \nabla f^0(x)$).

The Newton-Raphson algorithm

As before, we assume that $\{x_i\}$ converges to x^*. Let $H = \nabla^2 f^0$ be the Hessian of f^0. H and H^{-1} are assumed to exist and be continuous. Hence $\exists \epsilon, m, M \in R^+ \ni$:

9. $\|H(x) - H(y)\| \leq M, \quad \|H^{-1}(x)\| \leq 1/m$

$\forall x, y \; B_\epsilon(x^*)$. $\exists k_1 \ni x_i \in B_\epsilon(x^*) \; \forall i \geq k_1$. For the Newton-Raphson algorithm, $\forall x \in R^n$:

10. $a(x) = x - H^{-1}(x) \, g(x)$

and:

11. $\lambda(x) = 1$

where $g: R^n \to R^n$ denotes the gradient ∇f^0. Hence, $\forall i \geq 0$, $\{x_i\}$ satisfies:

12. $g(x_i) + H(x_i) \, [x_{i+1} - x_i] = 0$

Hence:

$$H(x_i)[x_{i+1} - x_i] = -g(x_i) + H(x_{i-1})[x_i - x_{i-1}] + g(x_{i-1})$$

so that:

$$x_{i+1} - x_i = -H^{-1}(x_i)\left[\int_0^1 [H(x_{i-1}) - H(x_{i-1} + s(x_i - x_{i-1}))] ds \, [x_i - x_{i-1}]\right.$$

Hence, $\forall i \geq k_1$.

13. $\|x_{i+1} - x_i\| \leq (M/2m) \|x_i - x_{i-1}\|^2 \triangleq \alpha \|x_i - x_{i-1}\|^2$

It follows from (13) that:

$$\lim_{i \to \infty} \|x_{i+1} - x^*\| / \|x_i - x^*\|^2 = \alpha$$

so that the order of convergence is two (and superlinear).

Superlinear convergence can also be established for the implementable version of this algorithm (employing Armijo), as well as for the conjugate gradient and quasi-Newton algorithms.

5. FEASIBLE DIRECTION ALGORITHMS FOR CONSTRAINED OPTIMIZATION PROBLEMS

We consider here the problem:

1. $\min\{f^0(x) \mid f^i(x) \leq 0,\ i = 1,2,\ldots m\}$

The feasible set Ω is defined by

2. $\Omega = \{x \mid f^i(x) \leq 0,\ i = 1,2,\ldots m\}$

$f^i: R^n \to R$, $i = 0,1,\ldots m$, are assumed to be continuously differentiable. An algorithm is said to be a method of feasible directions, if at each non-desirable point $x \in \Omega$, it generates a new point $x' \in \{x + \lambda h(x) \mid \lambda \in R^+\}$ by choosing $\lambda = \lambda(x) \ni x' = x + \lambda(x)\, h(x) \in \Omega$ and $f^0(x') < f^0(x)$. Hence such a method can only be employed if Ω has an interior (or relative interior if Ω is a subset of a linear manifold).

3. Definition. $\forall x_0, x \in \Omega$:

$$\mathcal{L}(x_0) \overset{\Delta}{=} \{x \in \Omega \mid f^0(x) \leq f^0(x_0)\}$$

$$B'_\epsilon(x) \overset{\Delta}{=} \{x' \in \Omega \mid \|x' - x\| \leq \epsilon\}$$

One of the simplest feasible direction algorithms is that due to Topkis and Veinott. Let $S \overset{\Delta}{=} \{h \in R^n \mid |h^i| \leq 1,\ i = 1,2 \ldots n\}$, $J_m \overset{\Delta}{=} \{1,2,\ldots m\}$.

$J_m^0 \overset{\Delta}{=} \{0,1,2,\ldots m\}$.

4. Feasible directions algorithm 1.

Step 0. Compute $x_0 \in \Omega$, set $i = 0$.

Step 1. Set $x = x_i$; compute $h(x)$ which solves:

$$\theta(x) = \min_{h \in S} \max\{\langle \nabla f^0(x), h\rangle;\ f^i(x) + \langle \nabla f^i(x), h\rangle,\ i \in J_m\}$$

If $\theta(x) = 0$ stop.

Step 2. Compute $\lambda(x)$, the smallest λ which solves:

$$f^0(x + \lambda(x)\, h(x)) = \min\{f^0(x + \lambda h(x)) \mid \lambda \in R^+,\ x + \lambda h(x) \in \Omega\}.$$

Step 3. Set $x_{i+1} = x + \lambda(x)\, h(x)$. Set $i = i + 1$.

Go to Step 1.

5. Comment.

$h(x)$ can be determined in Step 1 by solving the following linear program:

6. $\theta(x) = \min\{\sigma \mid -\sigma + \langle \nabla f^0(x), h\rangle \leq 0; \ -\sigma + f^i(x) + \langle \nabla f^i(x), h\rangle \leq 0$

$i = 1, 2 \ldots m; \ |h^i| \leq 1, \ i = 1, 2, \ldots n\}$

7. Theorem. Suppose that $\mathcal{L}(x_0)$ is compact and has an interior. Let $\{x_i\}$ be generated by algorithm (4). Then either $\{x_i\}$ is finite, ending at $x_{k+1} = x_k$ and $\theta(x_k) = 0$, or it is infinite, with every accumulation point x^* of $\{x_i\}$ satisfying $\theta(x^*) = 0$.

Proof. The finite case is trivially true. Since h(x) is not necessarily unique, suppose algorithm (4) defines a map $A : R^n \to 2^{R^n}$ ($x' \in A(x)$ if $x' = x + \lambda(x)\, h(x)$, for some h(x) solving (6). We note that S is compact, θ and f^i, $i = 0, 1, \ldots m$ are continuous and $\mathcal{L}(x_0)$ is compact. Assume x is not desirable, i.e. $\theta(x) = -v < 0$. The following results can then be established:

8. $\exists \epsilon_1 > 0 \ni \theta(x') \leq -v/2, \ \forall x' \in B_{\epsilon_1}(x)$

9. $\lambda \langle \nabla f^0(x'), h(x')\rangle \leq -\lambda v/2$ and

$f^i(x) + \langle \nabla f^i(x'), h(x')\rangle \leq -v/2, \ \forall x' \in B_{\epsilon_1}(x), \ \forall i \in J_m$

Also

$\exists \epsilon \in (0, \epsilon_1]$ and a $\lambda_1 > 0 \ni$

10. $|\langle \nabla f^i(x' + \lambda h), h\rangle - \langle \nabla f^i(x'), h\rangle| \leq v/4, \ \forall i \in J_m^0$

$\forall x' \in B_\epsilon(x), \ \forall \lambda \in [0, \lambda_1], \ \forall h \in S.$

Hence, using the mean-value theorem:

11. $f^0(x' + \lambda h(x')) - f^0(x') \leq -\lambda v/4$, and $f^i(x' + \lambda h(x')) \leq -\lambda v/4 \ \forall i \in J_m^0$,

$\forall x' \in B_\epsilon(x), \ \forall \lambda \in [0, \lambda_1]$

Since $x' \in B_\epsilon(x) \Rightarrow f^i(x') \leq 0, \ \forall i \in J_m$, it follows from (10) that $\lambda(x')$ (defined in Step 2 of (4)) satisfies:

12. $f^0(x'') - f^0(x') \leq -\delta$

13. $f^i(x'') \leq 0, \qquad \forall i \in J_m$

$\forall x' \in B_\epsilon(x), \ \delta = -\lambda_1 v/4, \ \forall x'' \in A(x').$

If we define $c \triangleq f^0$, we see that c is continuous, and that c, A satisfy the hypotheses of Theorem (2.2). Hence, any accumulation point x^* of an infinite sequence generated by (4) is desirable ($\theta(x^*) = 0$). Note, from Section 1, $\theta(x^*) = 0$ is a necessary condition of optimality of x^*.

Apart from the conceptual nature of Step 2 — which can be replaced by, e.g., Armijo's algorithm — the above algorithm suffers from the disadvantage of taking into account all the constraints all the time, thus increasing the dimensionality of the linear program (6). One might consider performing Step 1 (i.e. solving (6)) considering only the active constraints — those for which $f^j(x_i) = 0$. However, this strategy can lead to jamming, as shown by the example given by Wolfe [6]. Essentially, removal of constraint f^j from the linear problem because $f^j(x_i) \neq 0$, may result in $\langle \nabla f^j(x_i), h(x_i) \rangle$ being positive, and, if $|f^j(x_i)|$ is small, the resultant step size $\lambda(x_i)$ may become arbitrarily small, leading to jamming — convergence of $\{x_i\}$ to $x^* \ni \theta(x^*) \neq 0$. Zoutendijk overcame this phenomenon by taking into account ϵ-active constraints in choosing the search direction $h(x)$.

14. Definition. The ϵ-active constraint set at x is:

$$I_\epsilon(x) \triangleq \{i \in I_m \mid f^i(x) \geq -\epsilon\}$$

and

$$I^0_\epsilon(x) \triangleq \{0\} \cup I_\epsilon(x)$$

15. $\forall \epsilon \in R^+$, $\theta_\epsilon : R^n \to R$ and $h_\epsilon : R^n \to R^n$ are defined by:

$$\theta_\epsilon(x) = \min_{h \in S} \max_{i \in I^0_\epsilon(x)} \{\langle \nabla f^i(x), h \rangle\} = \max_{i \in I^0_\epsilon(x)} \{\nabla f^i(x), h_\epsilon(x)\}$$

Clearly, θ as defined in Step 1 of (4), satisfies $\theta = \theta_0$, and $h_0 = h$. The minimization problem in (15) can be easily written in the form of the linear program (6). The following results can be established fairly simply:

16. $\theta_\epsilon(x) \leq 0, \quad \forall x \in \Omega, \quad \forall \epsilon \in R^+$

17. $I_\epsilon(x) \subset I_{\epsilon'}(x), \quad \forall x \in \Omega, \quad \forall \epsilon, \epsilon' \in R^+ \ni \epsilon < \epsilon'$

18. $\theta_\epsilon(x) \leq \theta_{\epsilon'}(x), \quad \forall x \in \Omega, \quad \forall \epsilon, \epsilon' \in R^+ \ni \epsilon < \epsilon'$

19. $\forall \epsilon \in R^+, \quad \forall x \in \Omega, \quad \exists \epsilon_1 > 0 \ni I_\epsilon(x') \subset I_\epsilon(x), \quad \forall x' \in B'_{\epsilon_1}(x)$

20. $\forall \epsilon \in R^+, \quad \forall x \in \Omega, \quad \exists \epsilon_1 > 0 \ni I_{\epsilon'}(x) = I_{\epsilon'}(x), \quad \forall \epsilon' \in [0, \epsilon_1]$

The following algorithm is a modification by Polak [1] to an algorithm due to Zoutendijk.

21. Feasible directions algorithm 2: $\epsilon_0 > 0, \ \rho > 0$

Step 0. Compute a $x_0 \in \Omega$. Set $i = 0$.

Step 1. Set $\epsilon = \epsilon_0$.

Step 2. Compute a $h_\epsilon(x_i)$ which solves:

$$\theta_\epsilon(x_i) = \min_{h \in S} \max_{i \in I^0_\epsilon(x_i)} \{\langle \nabla f^i(x_i), h \rangle\}$$

Step 3. If $\theta_\epsilon(x) > -\epsilon$, set $\epsilon = \epsilon/2$ and go to Step 2.

Step 4. (Armijo) Compute the smallest integer k $\ni$

$$f^0(x_i + \rho h_\epsilon(x)/2^k) - f^0(x_i) \leq [\theta_\epsilon(x_i)\rho/2^k]/2$$

$$f^i(x_i + \rho h_\epsilon(x_i)/2^k) \leq 0, \ \forall i \in J_m$$

Step 5. Set $\lambda(x_i) = (\rho/2^k)$

Set $x_{i+1} = x_i + \lambda(x_i) h_\epsilon(x_i)$

Set $i = i + 1$

Go to Step 1.

22. Comment. For the purpose of analysis, no stop statement has been included. Hence, the algorithm either jams up at x_s, cycling between Steps 2 and 3, or produces an infinite sequence $\{x_i\}$.

23. Lemma. $\forall$ non-desirable $x \in \Omega$, $(\theta_0(x) \neq 0)$ $\exists j(x)$, $\epsilon(x) = \epsilon_0/2^{j(x)} > 0$,

$\ni \sigma_\epsilon(x) \leq -\epsilon(x)$.

Proof. $\exists \epsilon_1 \ni I_\epsilon(x) = I_{\epsilon_1}(x) \quad \forall \epsilon \in [0, \epsilon_1]$.

Hence, if $\theta_0(x) < 0$, then

$$\theta_\epsilon(x) = \theta_0(x) < 0, \quad \forall \epsilon \in [0, \epsilon_1]$$

Then $j(x)$ = smallest integer $\ni \epsilon(x) = \epsilon_0/2^{j(x)}$ satisfies

$$\epsilon(x) \leq \epsilon_1 \text{ and } \epsilon(x) \leq -\theta_0(x)$$

so that

$$\theta_\epsilon(x) = \theta_0(x) \leq -\epsilon(x), \quad \forall \epsilon \in [0, \epsilon_1]$$

Comment. Hence the map $A: \Omega \to 2^\Omega$, $D \triangleq \{x \in \Omega \mid \theta_0(x) = 0\}$ is well defined by

$$A(x) = \{x' \mid x' = x + \lambda(x) h_{\epsilon(x)}(x)\}, \ x \notin D$$

$$A(x) = x, \text{ if } x \in D.$$

24. Theorem

Let $\{x_i\}$ denote a sequence produced by (21).

(a) If $\{x_i\}$ jams up at x_s, then $\theta_0(x_s) = 0$.

(b) If $\{x_i\}$ is infinite then every accumulate point x^* satisfies $\theta_0(x^*) = 0$.

Outline of Proof

(a) is established in (23).

(b) Suppose that $x \in \Omega$ is non-desirable ($\theta_0(x) < 0$) so that

$\theta_{\epsilon(x)}(x) \leq -\epsilon(x) < 0$. From (19) $\exists \epsilon_1 > 0 \ni$:

25. $I_{\epsilon(x)}(x') \subset I_{\epsilon(x)}(x), \quad \forall x' \in B_{\epsilon_1}(x)$

Let $\psi : R^n \to R$ be defined by:

26. $\psi(x') = \min_{h \in S} \max_{i \in I^0_{\epsilon(x)}(x)} \langle \nabla f^i(x'), h \rangle$

Clearly ψ is continuous, $\psi(x) = \theta_{\epsilon(x)}(x)$ and $\theta_{\epsilon(x)}(x') \leq \psi(x')$. Hence $\exists \epsilon_2 \in [0, \epsilon_1] \ni$:

27.

$$\theta_{\epsilon(x)/2}(x') \leq \theta_{\epsilon(x)}(x') \leq \psi(x') \leq -\epsilon(x)/2, \quad \forall x' \in B_{\epsilon_2}(x)$$

(27) implies:

28.

$$\theta_{\epsilon(x')}(x') \leq \theta_{\epsilon(x)}(x)/2, \quad \forall x' \in B_{\epsilon_2}(x)$$

The rest of the proof proceeds as in the proof of 7, except for a minor modification due to use of the Armijo rule in Step 4.

For simplicity, we have restricted ourselves to the more direct feasible direction algorithms. For a discussion of a dual method of feasible directions, due to Pironneau and Polak, see Ref.[1].

6. PENALTY AND BARRIER FUNCTION METHODS

The basic idea behind penalty function methods is to convert constrained optimization problems into unconstrained problems, by discarding the constraints, and adding to f^0 a penalty function which penalizes constraint deviation. More precisely, the problem:

P1. $\min\{f^0(x) \mid x \in \Omega\}$

where $\Omega \subset R^n$ is closed (and may even have no interior if Ω represents equality constraints) is replaced by the following sequence of problems:

P2. $\min\{f^0(x) + p_i(x) \mid x \in R^n\}, \ i = 0, 1, 2, 3, \ldots$

whose solutions $\hat{x}_i \to \hat{x}$, the solution of (1), where $\forall i \in Z$, $p_i(x) = 0$ if and only if $x \in \Omega$. Note that x_i is not necessarily feasible, $\forall i \in z$.

In contrast, the Barrier Function method also replaces P1 by P2 but now, $\forall i \in Z$, p_i is defined only on the interior of $\overset{\circ}{\Omega}$ (assumed to be non-empty) and $p_i(x) \geq 0$, $\forall x \in \Omega$ and $p_i(x) \to \infty$ as $x \to \delta(\Omega)$, the boundary of Ω.

6.1. (Exterior) penalty function methods

1. Example

If $\Omega = \{x \mid f^i(x) \leq 0,\ i \in J_m\}$, then a useful sequence of penalty functions is given by:

2. $$p_i(x) = i\,P(x) = i \sum_{j=1}^{m} \max[0, f^j(x)]^2$$

Note, for each x, $p_j(x) > p_i(x)$ if $j > i$. In fact the defining properties of a sequence $\{p_i\}$ of penalty functions are:

3. For $i = 0, 1, 2, \ldots$: $p_i : R^n \to R$ continuous

$$p_i(x) = 0, \ \forall x \in \Omega$$

$$p_i(x) > 0, \ \forall x \notin \Omega$$

$$p_{i+1}(x) > p_i(x) \quad \forall x \notin \Omega$$

$$p_i(x) \to \infty \ \text{as} \ i \to \infty, \quad \forall x \notin \Omega$$

The sequence $\{p_i\}$ defined in (2) satisfies (3) if $\{f^j \mid j \in J_m\}$ are continuous. Let $\phi_i \triangleq f^0 + p_i : R^n \to R$ denote the cost function for P2, and let $\hat{x}_i$ minimize ϕ_i on R^n, and x^* minimize f^0 on Ω. If $\exists x' \ni \{x \mid f^0(x) \leq f(x')\} \triangleq \mathcal{L}(x')$ is compact, then x^* and $\{\hat{x}_i\}$ all exist and lie in $\mathcal{L}(x')$. The following result is a set of interesting and useful inequalities:

4. Lemma

$$\phi_0(\hat{x}_0) \leq \phi_1(\hat{x}_1) \leq \ldots f^0(x^*)$$

Proof

Since $p_i(x) \leq p_{i+1}(x)$, $\forall x \in R^n$

$$\phi_i(\hat{x}_i) \leq \phi_i(\hat{x}_{i+1}) \leq \phi_{i+1}(\hat{x}_{i+1}) \leq \phi_{i+1}(x^*), \ \forall i$$

Also since $x^* \in \Omega$, $p_i(x^*) = 0$, and

$$\phi_{i+1}(x^*) = f^0(x^*), \ \forall i$$

5. Theorem

Any accumulation point of $\{\hat{x}_i\}$ is a solution of P1. Proof (for the case when $\{p_i\}$ satisfy (2)).

Let $\{\hat{x}_i\}$ now denote a subsequence, converging to $\bar{x}$. Then $f^0(\hat{x}_i) \to f^0(\bar{x})$. Let $f^0(x^*)$ denote the minimum value of f^0 on Ω (x^* is a minimizer of f^0 on Ω. By Lemma (4) $\{\phi_i(\hat{x}_i)\}$ is a non-decreasing sequence bounded above by $f^0(x^*)$, so that $\phi_i(\hat{x}_i) \to \phi^*$, say, where $\phi^* \leq f^0(x^*)$. Hence

$$f^0(\hat{x}_i) \to f^0(\bar{x})$$

$$\phi_i(\hat{x}_i) \to \phi^* \leq f^0(\bar{x})$$

$$\phi_i(\hat{x}_i) = f^0(\hat{x}_i) + i\,P(\hat{x}_i)$$

Hence,

$$i\,P(\hat{x}_i) \to \phi^* - f^0(\bar{x}) \leq 0$$

Since $i \to \infty$, and $P(x) \geq 0$, $\forall x \in R^n$, this implies that $P(\hat{x}_i) \to 0 \Rightarrow P(\bar{x}) = 0$ $\Rightarrow \bar{x} \in \Omega$, that is, $\bar{x}$ is feasible.

Also: $f^0(\bar{x}) = \lim f^0(\hat{x}_i) \leq f^0(x^*)$

Hence, $\bar{x}$ is optimal for P1.

6.2. (Interior) barrier function methods

With the barrier function method, Ω must have an interior – problems with equality constraints cannot be handled. More precisely, Ω must be robust, that is, that $\bar{\mathring{\Omega}} = \Omega$ (the closure of the interior of Ω is equal to Ω). This rules out equality constraints, 'whiskers', isolated points, etc. Barrier function methods establish a barrier which prevents points generated by the algorithm leaving Ω:

1. Example

If $\Omega = \{x \mid f^i(x) \leq 0,\ i \in J_m\}$, then a useful sequence of penalty functions is given by:

2. $$p_i(x) = P(x)/i = \left[\sum_{j=1}^{m} (-1/f^j(x))\right]/i$$

The defining properties of a sequence $\{p_i\}$ of barrier functions are:

3. $0 < p_i(x) < p_{i+1}(x),\ \forall x \in \mathring{\Omega},\ \forall i \in Z.$

$p_i(x) \to 0$ as $i \to \infty$, $\forall x \in \mathring{\Omega}$, $\forall i \in Z.$

If

$$x_j \to x^* \in \delta(\Omega),\ p_i(x_j) \to \infty \ \text{as}\ j \to \infty,\ \forall i$$

As before, $\forall i \in Z$, let $\phi_i \triangleq f^0 + p_i$, let $\hat{x}_i$ denote a minimizer of ϕ_i on R^n, and x^* a minimizer of f^0 on $\mathring{\Omega}$ (because of the nature of p_i and ϕ_i, $\hat{x}_i$ is also the minimizer of ϕ_i on Ω.) It can be shown that:

4. $\phi_0(\hat{x}_0) \geq \phi_1(x_1) \geq \ldots \phi_i(\hat{x}_i) \geq \phi_{i+1}(\hat{x}_{i+1}) \ldots \geq f^0(x^*)$

5. Theorem

Any accumulation point $\bar{x}$ of $\{\hat{x}_i\}$ is a solution of P1.

There also exist mixed penalty-barrier function methods.

Penalty methods may be started at non-feasible x_0, and may be used for problems with equality and inequality constraints. The sequence $\{\hat{x}_i\}$ is, on the other hand, not necessarily feasible. Barrier methods, which require that $x_0 \in \overset{\circ}{\Omega}$, cannot be used for problems with equality constraints but produce feasible sequences. Both are conceptual, requiring minimization, $\forall i \in Z$, of ϕ_i. However, implementable versions have been developed (see Polak [1]). Both methods have the disadvantage that ϕ_i becomes 'nasty' as $i \to \infty$ in the sense that its Hessian becomes extremely ill-conditioned. For an interesting elaboration of this theme see Luenberger [3].

7. DUALITY

Consider the primal problem:

$$P: \quad \min\{f^0(x) \mid x \in \Omega\}$$

where $\Omega \triangleq \{x \in R^n \mid f(x) \leq 0\}$, which we can imbed in the family of problems ($\forall\, y \in R^m$):

$$P(y): \quad \min\{f^0(x) \mid x \in \Omega(y)\}$$

where

$$\Omega(y) \triangleq \{x \in R^n \mid f(x) \leq y\}$$

Let

$$\Gamma \triangleq \{y \in R^m \mid \Omega(y) \neq \emptyset\}$$

$\Gamma \subset R^m$ is convex provided $f: R^n \to R^m$ is convex.

For, if $y_1, y_2 \in \Gamma$, then $\Omega(y_1)$ and $\Omega(y_2)$ are not empty. If $x_1 \in \Omega(y_1)$ and $x_2 \in \Omega(y_2)$ and $\theta \in [0,1]$, then:

$$f(\theta x_1 + (1-\theta)x_2) \leq \theta f(x_1) + (1-\theta) f(x_2)$$

$$\leq \theta y_1 + (1-\theta) y_2$$

$$\Rightarrow \theta x_1 + (1-\theta)x_2 \in \Omega(\theta y_1 + (1-\theta)y_2) \neq \emptyset$$

$$\Rightarrow \theta y_1 + (1-\theta)y_2 \in \Gamma$$

We define $\omega: \Gamma \to R$ by:

1. Def. $\omega(y) \triangleq \inf\{f^0(x) \mid x \in \Omega(y)\}$

and the subset A, B of R^{m+1} by:

2. Def. $A \underline{\underline{\Delta}} \{(y^0, y) \in R \times R^m) \mid \exists x \in R^n \ni y^0 \geq f^0(x),\ y \geq f(x)\}$

$=$ region above the graph of Ω

3. Def. $\hat{y}^0 \underline{\underline{\Delta}}\ \omega(0) = \inf\{f^0(x) \mid x \in \Omega\}$

4. Def. $B \underline{\underline{\Delta}} \{(y^0, y) \in R^{m+1} \mid y^0 \leq \hat{y}^0,\ y \leq 0\}$

These definitions are illustrated in Fig.2; note, if $\hat{x}$ is a solution of P, $f^0(\hat{x}) = \hat{y}^0$.

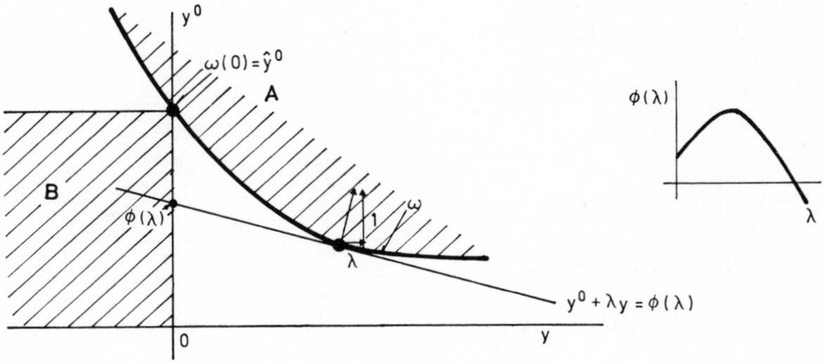

FIG.2. Definitions in the duality problem.

5. Lemma

(a) ω is non-increasing ($y_1 \geq y_2 \Rightarrow \omega(y_i) \leq \omega(y_2)$)

(b) If f^0, f are convex functions, ω is a convex function and A is a convex set.

The Lagrangian $L: R^n \times R^m \to R$ is defined by:

6. Def. $L(x, \lambda) \underline{\underline{\Delta}}\ f^0(x) + \langle \lambda, f(x) \rangle$

Associated with L is the unconstrained problem:

$P_\lambda : \min\{L(x, \lambda) \mid x \in R^n\}$

The dual function $\phi : R^m \to R$ is defined by:

7. Def. $\phi(\lambda) \underline{\underline{\Delta}} \inf\{L(x, \lambda) \mid x \in R^n\}$

and the dual problem is:

DP. $\max\{\phi(\lambda) \mid \lambda \geq 0\}$

8. Def. $\hat{\phi} \underline{\underline{\Delta}} \sup\{\phi(\lambda) \mid \lambda \geq 0\}$

Lemma 9 and 10 are true, whether or not Ω, f^0, f are convex,

9. Lemma

$\phi : R^{m+} \to R \cup \{-\infty\}$ is a concave function ($R^{m+} \triangleq \{\lambda \in R^m \mid \lambda \geqq 0\}$)

10. Lemma (weak duality)

If $x \in \Omega$, $\lambda \geqq 0$, then

$$f^0(x) \geqq \omega(0) \geqq \hat{\phi} \geqq \phi(\lambda)$$

Proof

$$f(x) \leqq 0, \quad \lambda \geqq 0 \Rightarrow \langle \lambda, f(x) \rangle \leqq 0, \text{ so}$$

$$f^0(x) \geqq L(x, \lambda), \quad \forall x \in \Omega, \ \lambda \geqq 0$$

Hence

$$\begin{aligned} f^0(x) &\geqq \inf\{f^0(x) \mid x \in \Omega\} = \omega(0) \\ &\geqq \inf\{L(x,\lambda) \mid x \in \Omega\} \\ &\geqq \inf\{L(x,\lambda) \mid x \in R^n\} = \phi(\lambda) \end{aligned}$$

that is

$$f^0(x) \geqq \omega(0) \geqq \phi(\lambda), \quad \forall x \in \Omega, \ \lambda \geqq 0$$

Taking infimum w.r.t. $\lambda \geqq 0$ yields the desired result.

The basic problem in duality theory is to determine conditions under which the solution to the primal and dual problems are equal, that is $\omega(0) = \hat{\phi}$.

11. Theorem (sufficiency)

Let $(\hat{x}, \hat{\lambda})$ satisfy the following optimality conditions:

(1) $\hat{x}$ is a solution of $P_{\hat{\lambda}}$
(2) $\hat{x} \in \Omega$, $\hat{\lambda} \geqq 0$
(3) $\hat{\lambda}^i = 0$ if $f^i(\hat{x}) < 0$ (i.e. $\langle \hat{\lambda}, f(\hat{x}) \rangle = 0$).

Then:

(i) $\hat{x}$ is a solution of P ($\omega(0) = f^0(\hat{x})$)
(ii) $\hat{\lambda}$ is a solution of DP ($\hat{\phi} = \phi(\hat{\lambda})$)
(iii) $\omega(0) = \hat{\phi}$ ($f^0(\hat{x}) = \phi(\hat{\lambda})$).

Proof

If

$$x \in \Omega = \Omega(0), \text{ then } \langle \hat{\lambda}, f(x) \rangle \leqq 0, \text{ and:}$$

$$f^0(x) \geqq f^0(x) + \langle \hat{\lambda}, f(x) \rangle = L(x, \hat{\lambda})$$

$$\geq \inf\{L(x,\hat{\lambda}) \mid x \in R^n\}$$

$$= L(\hat{x},\hat{\lambda}) \qquad \text{(by (1))}$$

$$= f^0(\hat{x}) \qquad \text{(by (3))}$$

i.e. $\hat{x}$ is a solution of P, and $f^0(\hat{x}) = \omega(0)$. Also $\phi(\hat{\lambda}) = L(\hat{x},\hat{\lambda})$

$$= f^0(\hat{x}) \qquad \text{(by (3))}$$

$$= \omega(0)$$

$\Rightarrow$ (by weak duality) $\omega(0) = \hat{\phi} = \phi(\hat{\lambda})$

i.e. $\hat{\lambda}$ is a solution of DP.

12. Theorem (strong duality)

Let f^0, f be differentiable convex functions, Ω convex, $0 \in \mathring{\Gamma}$ (i.e. $\exists\, x \in R^n \ni f(x) < 0$), $\omega(0) > -\infty$), then $\exists$ an optimal solution $\hat{\lambda}$ to DP and

$$\omega(0) = \phi(\hat{\lambda})$$

(If x is an optimal solution to P1, $\omega(0) = f^0(\hat{x})$.) Also, if $\hat{\lambda}$ is a solution for DP, then $\hat{x}$ is a solution for P $\Leftrightarrow$ $(\hat{x},\hat{\lambda})$ satisfy the optimality conditions of Theorem 11.

Hence, the solution $\hat{x}$ to P1, can be determined by finding $\bar{x}: R^{m+} \to R^n$ $\bar{x}(\lambda) \ni R^n$ minimizes $L(x,\lambda)$, and then finding λ^* which maximizes $\phi(\lambda) = L(\bar{x}(\lambda),\lambda)$ in R^{m+}. $\hat{x} = x(\lambda^*)$. However, this algorithm requires the (global) convexity of Ω, f^0 and f. It is possible, though, to obtain similar results of a local nature (local duality). Under suitable conditions (including f^0, f twice continuously differentiable) if P1 has a local minimum at $\hat{x}$, with associated Lagrange multiplier $\hat{\lambda}$, and if $\nabla_x^2 L(\hat{x},\hat{\lambda})$ is positive definite (local convexity) then $\hat{\lambda}$ is local maximum of the unconstrained dual problem DP, and $f^0(\hat{x}) = \phi(\hat{\lambda})$.

The above discussion holds, with minor modifications, for the case when $\Omega = \{x \in R^n \mid r(x) = 0\}$. For this case, observe that the two problems:

P1. $\min\{f^0(x) \mid r(x) = 0\}$

P2. $\min\{f^0(x) + c\|r(x)\|^2 \mid r(x) = 0\}$

are equivalent, for any $c > 0$. Also, that for c large enough:

$$\nabla_x^2\,[L(\hat{x},\hat{\lambda}) + c\|r(x)\|^2]$$

is positive definite. It can be shown that for every compact subset of R^n, there exists a $c < \infty$ such that if $\hat{x}$ is a local minimum of

$$\psi_c(x) = L(x,\lambda(x)) + c\|r(x)\|^2$$

where

$$\lambda(x) = (g_x(x)g_x^T(x))^{-1} g_x(x)\nabla f^0(x),$$

then $\hat{x}$ is a local minimum for P1, so that algorithms can be constructed with a rule for automatically increasing c, producing sequences whose accumulation points satisfy necessary conditions of optimality of the original problem.

ACKNOWLEDGEMENTS

These notes rely, in the main, on Polak [1], particularly in the exposition of the simple but powerful algorithm models of Chapter 2, and their application for establishing convergence of algorithms for unconstrained optimization (Chapter 3) and of algorithms of the feasible-directions type for constrained optimization. The discussion of conjugate-gradient and secant-type algorithms draws on the lecture notes of a colleague, Dr. J.C. Allwright. The discussion of conditions of optimality is based on Ref.[4], and the regrettably short treatment of duality is based on the treatment by Varaiya in Ref.[5].

REFERENCES

[1] POLAK, E., Computational Methods in Optimisation, Academic Press (1971).

[2] POLAK, E., RIBIERE, G., Note sur la convergence de méthodes de directions conjugées. Rev. Fr. Inform. Rech. Operation (16-R1) (1969) 35-43.

[3] LUENBERGER, D.G., Introduction to Linear and Nonlinear Programming, Addison-Wesley (1973).

[4] CANNON, M.D., CULLUM, C.D., POLAK, E., Theory of Optimal Control and Mathematical Programming, McGraw Hill (1970).

[5] VARAIYA, P.P., Notes on Optimization, Van Nostrand; REINHOLD: Notes on System Sciences (1972).

[6] WOLFE, P., "On the Convergence of Gradient Methods under Constraints", IBM Research Report, R.C. 1752, Yorktown Heights, New York (1967).

REACHABILITY OF SETS AND TIME- AND NORM-MINIMAL CONTROLS

A. MARZOLLO
Electrical Engineering Department,
University of Trieste
and
International Centre for Mechanical Sciences, Udine
Italy

Abstract

REACHABILITY OF SETS AND TIME- AND NORM- MINIMAL CONTROLS.

Standard continuous linear control systems as well as discrete time systems with "controller" and "anti-controller" representing disturbances are treated.

INTRODUCTION

This paper is divided into three parts; the first two parts concern standard continuous linear control systems, whereas the third part deals with discrete time systems whose state evolution is influenced both by a "controller" and an "anti-controller", which may represent disturbances (worst-case approach).

In Part I, functional-analysis methods are used to give necessary and sufficient conditions for the reachability of a given convex compact set in state space with norm-bounded controls, for different types of norms. In Part II, which again uses functional-analysis methods, these conditions are applied to obtaining both time-minimal and norm-minimal open loop controls, and a special case of Pontryagin's maximum principle is derived. These results are illustrated by an exercise at the end of Part II. Part III deals with the problem of finding the set in state space starting from which an initial state can be steered into a given target set, in the presence of an "anticontroller"; both the cases of bounded and unbounded controls and anticontrols are considered, and interesting relations are shown among possible situations depending on different information structures available to the controller.

Part I and II of these notes are extensions of ideas developed in Ref.[1].

PART I. CONTROLLABILITY WITH RESPECT TO GIVEN SETS, TIME INTERVAL AND BOUNDS ON THE NORM OF THE CONTROL

Let us consider a linear control system described by a linear differential equation of the type

$$\dot{x} = A(t)\,x(t) + B(t)\,u(t) \qquad \text{I(1)}$$

where the state vector $x(t)$ is in R^n, the control vector $u(t)$ is in R^m, and $A(t)$, $B(t)$ are matrices of appropriate dimensions.* Under the usual hypotheses on the functions in $A(t)$, $B(t)$, and on $u(t)$, if $x(t_0) = x_0$, the system I(1) has a unique solution

$$v_u(t) = V(t)\, x_0 + \int_{t_0}^{t_1} W(t_1, \tau)\, u(\tau)\, d\tau, \qquad t \geq t_0$$

where $V(t)$ is the solution of the matrix equation

$$\frac{d}{dt} V(t) = A(t)\, V(t), \qquad V(t_0) = I$$

and

$$W(t_1, \tau) = V(t_1) V^{-1}(\tau)\, B(\tau)$$

Let us equip u with a norm over the interval $K = [t_0, t_1]$, which, for the moment, will be an $L_{2,2}$ norm, for the sake of simplicity:

$$\|u\| = \left(\int_{t_0}^{t_1} \|u(\tau)\|\, d\tau \right)^{1/2}, \qquad \|u(\tau)\| = \left(\sum_i |u_i(\tau)|^2 \right)^{1/2}$$

(later on, we shall extend the considerations to more general $L_{r,p}$ norms, $1 \leq r \leq \infty$, $1 < p \leq \infty$). Let U^ρ be the class of admissible controls:

$$U^\rho = \{u : \|u\| \leq \rho, \quad \rho \geq 0\}$$

The problem we shall now consider is the following:

Problem I, 1

Given a non-empty convex compact set A in R^n; does an admissible control exist which "steers" x_0 at time t_0 into A at time t_1?

In other words, does an $u \in U^\rho$ exist such that $V_u(t_1) \in A$, or is the system I(1) "controllable with respect to x_0, A, U^ρ, K?

To solve Problem I,1 let us define the linear operator Λ from U^ρ into R^n:

$$\Lambda u = \int_{t_0}^{t_1} W(t_1, \tau)\, u(\tau)\, d\tau$$

and the "reachable set" R at time t_1

$$R = \{x : x = V(t_1)x_0 + \Lambda u, \ u \in U^\rho\} \qquad \text{I(2)}$$

Problem I, 1 clearly has a solution if and only if

$$R \cap A \neq \emptyset \qquad \text{I(3)}$$

The most natural way of verifying whether I(3) holds or not is to apply the separation theorem for convex compact sets, or, more precisely, its following corollary

Corollary I, 1

Two convex compact sets A and R of R^n have a non-empty intersection if and only if

$$\min_{r\in R} \langle x, r\rangle \leq \max_{a\in R} \langle x, a\rangle, \qquad \forall x \in R^n$$

From the convexity of U^ρ, the convexity of ΛU^ρ, hence of R, follows trivially. Its compactness may be obtained from classical theorem of functional analysis (see, e.g. Ref.[2]) by using the (weak) compactness of U^ρ and the (weak) continuity of Λ.

Using Corollary I, 1 and recalling I(2), we see that Problem I, 1 has a solution if and only if

$$\min_{u\in U^\rho} (\langle x, V(t_1)\, x_0\rangle + \langle x, \Lambda u\rangle) \leq \max_{a\in A} \langle x, a\rangle, \qquad \forall x \in R^n$$

or

$$\langle x, V(t_1)x_0\rangle - \max_{\|u\|\leq\rho} \langle x, \Lambda u\rangle \leq \max_{a\in A} \langle x, a\rangle, \qquad \forall x \in R^n$$

Defining Λ^* to be the adjoint operator of Λ, which takes R^n into U^*, we have

$$\langle x, V(t_1)x_0\rangle - \rho \max_{\|u\|\leq 1} \langle \Lambda^* x, u\rangle \leq \max_{a\in A} \langle x, a\rangle, \qquad \forall x \in R^n$$

or

$$\langle x, V(t_1)x_0\rangle - \rho \|\Lambda^* x\| \leq \max_{a\in A} \langle x, a\rangle, \qquad \forall x \in R^n \qquad \text{I(4)}$$

If A is the Euclidean sphere around x, of radius ϵ:

$$A = \{x : x = x_1 + y,\ \|y\| \leq \epsilon\}$$

I(4) becomes

$$\langle x, V(t_1)x_0 - x_1\rangle \leq \rho\|\Lambda^* x\| + \epsilon\|x\|, \qquad \forall x \in R^n$$

which is clearly equivalent to

$$\langle x, V(t_1)x_0\rangle \leq \rho\|\Lambda^* x\| + \epsilon, \qquad \forall x \in R^n,\ \|x\| = 1 \qquad \text{I(5)}$$

Representing $\|\Lambda^* x\|$ and noticing that, because of the arbitrariness of x, the first term of I(5) may be substituted by its absolute value, we have the following

Theorem I, 1

The system I(1) is "controllable with respect to x_0, $A = \{x, x = x_1 + y, \|y\| \leq \epsilon\}$, U^ρ, K " if and only if

$$|\langle x, V(t_1)x_0 - x_1 \rangle|^2 \leq \rho\left(\int_{t_0}^{t_1} \|W^*(t_1, \tau)x\|^2 \, d\tau\right)^{1/2} + \epsilon, \qquad \forall x, \quad \|x\| = 1 \quad I(6)$$

If $\epsilon = 0$, i.e. if $A = x_1$, condition I(6) may be easily verified by using standard techniques of linear algebra. Indeed, defining the vector Z of components z_i as follows:

$$z = V(t_1)x_0 - x_1$$

and taking the square of both terms of I(6), this inequality is equivalent to

$$\langle x, z \rangle \langle z, x \rangle \leq \rho^2 \left\langle x, \int_{t_0}^{t_1} W(t_1, \tau)\, W^*(t_1, \tau) d\tau\, x \right\rangle, \qquad \forall x \in R^n \qquad I(7)$$

and defining the $n \times n$ matrix $Z = z\rangle\langle z$ with elements z_i, z_j, the last inequality is equivalent to the following one:

$$\langle x, Zx \rangle \leq \rho^2 Y(t_1)x, \qquad \forall x \in R^n$$

where

$$Y(t_1) = \int_{t_0}^{t_1} W(t_1, \tau)\, W^*(t_1, \tau)\, d\tau \qquad I(8)$$

is the usual "controllability matrix" of system I(1). Defining the $n \times n$ matrix $C(t_1)$:

$$C(t_1) = Y(t_1) - \frac{1}{\rho^2} Z$$

from I(7) we have the following

Corollary I, 2

System I(1) is "controllable with respect to x_0, x_1, U^ρ, K" if and only if the matrix $C(t_1)$ is positive semi-definite.

The previous condition depends naturally on the system (through $Y(t_1)$), on x_0, x_1 (through Z), on ρ and on K (through $Y(t_1)$). It is easy to verify that letting x_0, x_1, ρ, K be arbitrary, we have the usual controllability condition:

Condition I,1

The state of the system I(1) may be carried from any x_0 at time t_0 into any x_1 at time t_1 by controls u of appropriately large norm if and only if the matrix $Y(t_1)$ defined in I(8) is positive definite.

Since the sufficiency is obvious, let us verify the necessity. Let $C(t_1)$ be positive semi-definite, i.e. let I(7) hold with arbitrary x_0, x_1 and appropriately large ρ. Since

$$\langle x, Y(t_1)x\rangle = \int_{t_0}^{t_1} \langle W^*(t_1, \tau)x, W^*(t_1, \tau)x\rangle \, d\tau$$

$$= \int_{t_0}^{t_1} \|W^*(t_1, \tau)x\|^2 \, d\tau \geq 0, \qquad \forall x \in R^n$$

$Y(t_1)$ is, at least, positive semi-definite. If it were singular, i.e. for some $x' \neq 0$, $Y(t)x' = 0$, from I(7) we would have $(\langle x', z\rangle)^2 \leq 0$, that is $\langle x', z\rangle = 0$ which is impossible for arbitrary x_0, x_1, hence for arbitrary z.

Until now, we considered U^ρ equipped with an L_{22}-norm. The previous results, appropriately adapted, would hold for more general norms, i.e. for $u \in L_{r,p}$:

$$\|u\|_{r,p} = \left(\int_{t_0}^{t_1} \|u(\tau)\|_r^p \, d\tau\right)^{1/p} \qquad 1 < p \leq \infty$$

where

$$\|u(\tau)\|_r = \left(\sum_i |u_i(\tau)|^r\right)^{1/r} \qquad 1 \leq r \leq \infty$$

The admissible class of controls would be

$$U^\rho_{r,p} = \{u : \|u\|_{r,p} \leq \rho\} \qquad \text{I(9)}$$

Instead of Theorem I,1 we would then have

Theorem I,1'

The system I(1) is "controllable with respect to x_0, A, $U^\rho_{r,p}$, $K_\|$ if and only if

$$|\langle x, x_1 - V(t_1)x_0\rangle| \leq \rho \left(\int_{t_0}^{t_1} \|W^*(t_1, \tau)x\|_s^q \, d\tau\right)^{1/q} + \epsilon, \qquad \forall x \in R^n \qquad \text{I(10)}$$

where

$$\frac{1}{s} + \frac{1}{r} = 1, \quad \frac{1}{q} + \frac{1}{p} = 1$$

For $r \neq 2$, $p \neq 2$, the verification of condition I(9) may not be reduced to the previous simple algebraic condition of Corollary I,2 but the following inequality should be verified

$$\max_{\|x\|\leq 1}\left[\langle x, x_1 - V(t_1)x_0\rangle - \rho\left(\int_{t_0}^{t_1}\|W^*(t_1,\tau)x\|_s^q\, d\tau\right)^{1/q} - \epsilon\right] \leq 0$$

which implies a maximization over the unit sphere $\|x\| = 1$, (or over the unit convex ball $\|x\| \leq 1$).

PART II. TIME-AND NORM-MINIMAL CONTROLS

In this part, we shall heavily rely on the results of Part I concerning the reachability of a given point x_1 at time t_1 from x_0 at time t_0 under controls u belonging to an admissible class $U^\rho_{r,p}$ defined in I(9), or (x_0, x_1, $U^\rho_{r,p}$, K) controllability, i.e. we shall use I(10) which we rewrite for convenience in the case $\epsilon = 0$:

$$|\langle x, x_1 - V(t_1)x_0\rangle| \leq \rho\left(\int_{t_0}^{t_1}\|W^*(t_1,\tau)x\|_s^q\, d\tau\right)^{1/q}, \quad \forall x \in R^n \qquad \text{II(1)}$$

We shall obtain results concerning the existence and the form of time optimal controls as well as of norm optimal controls, and shall derive a special case of Pontryagin's maximum principle. Throughout this chapter, we shall suppose $x_1 \neq V(t_1)x_0$, that is the function $u \equiv 0$ is not supposed to transfer x_0 into x_1. Let us first consider time optimal controls.

Theorem II, 1

If the system I(1) is controllable relative to points x_0 and x_1, time interval $K = [t_0, t_1]$ and the class $U^\rho_{r,p}$ of admissible controls, then there exists a least interval $\overline{K} = [t_0, \overline{t}]$, ($t_0 < \overline{t} \leq t_1$) in which it is controllable. In other words, if a system is (x_0, x_1, $U^\rho_{r,p}$, K) controllable, then there exists a time optimal control.

Proof

Let us consider the set $H = \{t, t > t_0\}$ over which the system is (x_0, x_1, $U^\rho_{r,p}$, K) controllable. This set is clearly non-empty. Let us define

$\overline{t} = \inf H$

We need to prove that the system I(1) is controllable relative to x_0, x_1, $U^\rho_{r,p}$ and $\overline{K} = [t_0, \overline{t}]$, i.e. the interval $\overline{K}$ is the minimal interval over which the system I(1) is controllable, which is equivalent to saying that $\overline{t}$ belongs to H. Indeed, suppose that $\overline{t}$ does not belong to H. Then, by theorem I, 1' there would exist a vector $\hat{x}$ in R^n such that

$$|\langle \hat{x}, x_1 - V(\overline{t})x_0\rangle| > \rho\left(\int_{t_0}^{\overline{t}}\|W^*(\overline{t},\tau)\hat{x}\|_s^q\, d\tau\right)^{1/q} \qquad \text{II(2)}$$

By the definition of $\overline{t}$, there exists a sequence $\{t_n\}$ with $\lim_{n\to\infty} t_n = \overline{t}$ and such that for each interval $K_n = [t_0, t_n]$ the system is controllable; and, therefore,

$$\left|\langle \hat{x}, x_1 - V(t_n)x_0\rangle\right| \leq \rho\left(\int_{t_0}^{t_n} \|W^*(t_n, \tau)\hat{x}\|_s^q \, d\tau\right)^{1/q} \qquad \text{II(3)}$$

or

$$\gamma(t_n) \leq \rho\delta(t_n)$$

with obvious meaning of $\gamma(t_n)$ and $\delta(t_n)$. Since $\gamma(t)$ and $\delta(t)$ are continuous functions, we have

$$\lim_{n\to\infty} \gamma(t_n) = \gamma(\overline{t}) = \gamma = \left|\langle \hat{x}, x_1 - V(\overline{t})x_0\rangle\right|$$

$$\lim_{n\to\infty} \rho\delta(t_n) = \rho\delta(\overline{t}) = \rho\delta = \rho\left(\int_{t_0}^{\overline{t}} \|W^*(\overline{t}, \tau)\hat{x}\|_s^q \, d\tau\right)^{1/q}$$

From inequality II(3),

$$\lim_{n\to\infty} \gamma(t_n) \leq \rho \lim_{n\to\infty} \delta(t_n)$$

that is

$$\gamma \leq \rho\delta$$

which would contradict II(2).

Theorem II, 2

If $\overline{K} = [t_0, \overline{t}]$ is the minimal time interval over which the system I(1) can be transferred from x_0 to x_1 with an admissible control belonging to $U^{\rho}_{r,p}$, then there exists a vector $x' \neq 0$, $x' \in R^n$ such that

$$\left|\langle x', x_1 - V(\overline{t})x_0\rangle\right| = \rho\left(\int_{t_0}^{\overline{t}} \|W^*(\overline{t}, \tau)x'\|_s^q \, d\tau\right)^{1/q} \qquad \text{II(4)}$$

Proof

Let the sequence $\{t_n\}$ converge to $\overline{t}$, with $t_n < \overline{t}$. For each t_n, there exists a vector x_n such that

$$\left|\langle x_n, x_1 - V(t_n)x_0\rangle\right| > \rho\left(\int_{t_0}^{t_n} \|W^*(t_n, \tau)x_n\|_s^q \, d\tau\right)^{1/q} \qquad \text{II(5)}$$

The inequality II(5) clearly holds if we replace x_n by $\lambda x'_n = x_n$, where $\|x'_n\| = 1$. In this way, all x'_n belong to the compact unit sphere of R^n, and, therefore, from the sequence $\{x'_n\}$ it is possible to extract a convergent subsequence $\{x'_{n_k}\}$ such that $\lim_{n_k \to \infty} x'_{n_k} = x'$, with $\|x'\| = 1$. Since both terms of inequality II(5) (with x'_n instead of x_n) depend continuously on t_n and on x'_n, we have

$$\lim_{\substack{x'_{n_k} \to x' \\ t_{n_k} \to \bar{t}}} |\langle x_n, x_1 - V(t_{n_k})x_0\rangle| = |\langle x', x_1 - V(\bar{t})x_0\rangle|$$

and

$$\lim_{\substack{x_{n_k} \to x' \\ t_{n_k} \to \bar{t}}} \left(\int_{t_0}^{t_n} \|W^*(t_{n_k},\tau)\, x'_{n_k}\|_s^q \, d\tau \right)^{1/q} = \int_{t_0}^{\bar{t}} \|W^*(\bar{t}, \tau)x'\|_s^q$$

Recalling inequality II(5), we have

$$|\langle x', x_1 - V(\bar{t})x_0\rangle| \geq \rho \left(\int_{t_0}^{\bar{t}} \|W^*(\bar{t}, \tau)x'\|_s^q \, d\tau \right)^{1/q}$$

Since the system is $(x_0, x_1, U^{\rho}_{r,p}, K)$ controllable, hence also II(1) holds for $x = x'$, the theorem is proved.

Let us now turn to the problem of the existence of the so called "minimal norm" control: given the system I(1), which is supposed to be $(x_0, x_1, U^{\rho}_{r,p}, K)$ controllable, does a control $\hat{u}$ exist such that $\|\hat{u}\|_{r,p} = \min_{u \in U^{\rho}_{r,p}} \|u\|_{r,p}$ and which transfers x_0 to x_1, within K?

The answer to this question is "yes", and the control u is called "minimal-norm" control. We are going to prove this fact in the following

Theorem II, 3

Let the system I(1) be $(x_0, x_1, U^{\rho}_{r,p}, K = [t_0, t_1])$ controllable and let $\bar{U} \subset U^{\rho}_{r,p}$ be the set of functions u which effect the transfer from x_0 to x_1 at time t_1, i.e. such that

$$\int_{t_0}^{t_1} W(t_1,\tau)\, u(\tau)\, d\tau = x_1 - V(t_1)\, x_0$$

Then there exists a function $\hat{u}$ such that

$$\|\hat{u}\|_{r,p} = \inf_{u \in \bar{U}} \|u\|_{r,p} = \hat{\rho} \qquad \text{II(6)}$$

Proof

Let us define $\hat{\rho}$ to be the $\inf_{u \in \bar{U}} \|u\|_{r,p}$. There exists, therefore, a sequence $\{\rho_n\}$, with $\rho_n > \hat{\rho}$ and such that $\lim_{n \to \infty} \rho_n = \hat{\rho}$. For each ρ_n the system is $(x_0, x_1, U_{r,p}^{\rho_n}, K)$ controllable; therefore

$$|\langle x, x_1 - V(t_1)x_0\rangle| \leq \rho_n\left(\int_{t_0}^{t_1} \|W^*(t_1, \tau)x\|_s^q \, d\tau\right)^{1/q}, \qquad \forall x \in R^n$$

This inequality is conserved as n tends to infinity, so we have

$$|\langle x, x_1 - V(t_1)x_0\rangle| \leq \hat{\rho}\left(\int_{t_0}^{t_1} \|W^*(t_1, \tau)x\|_s^q \, d\tau\right)^{1/q}, \qquad \forall x \in R^n \qquad \text{II(7)}$$

and, by Theorem I, 1', there exists a control function $\hat{u} \in \overline{U}$ with $\|\hat{u}\|_{r,p} \leq \hat{\rho}$. Since $\hat{u} \in \overline{U}$, and $\inf_{u \in \overline{U}} \|u\|_{r,p} = \hat{\rho}$, $\|\hat{u}\|_{r,p}$ must be equal to $\hat{\rho}$, and the infimum in II(1) is assumed. We now want to give an expression for $\hat{\rho}$ as a function of the parameters of the system I(1), and of x_0, x_1, t_0 and t_1. This expression is given in the following

Theorem II, 4

Using the definition of Theorem II(3) for $\overline{u}$, we have

$$\hat{\rho} = \inf_{u \in \overline{U}} \|u\|_{r,p} = \|\hat{u}\|_{r,p} = \sup_{\int_{t_0}^{t_1} \|W^*(t_1, \tau)x\|_s^q d\tau = 1} \langle x, x_1 - V(t_1)x_0\rangle \qquad \text{II(8)}$$

Proof

Since II(7) is satisfied, we have

$$\hat{\rho} \geq \sup_{x \in R^n} \left|\left\langle \frac{x}{\left(\int_{t_0}^{t_1} \|W^*(t_1, \tau)x\|_s^q \, d\tau\right)^{1/q}}, x_1 - V(t_1)x_0\right\rangle\right|$$

$$= \sup_{\left(\int_{t_0}^{t_1} \|W^*(t_1, \tau)x\|_s^q \, d\tau\right)^{1/q}} \langle x, x_1 - V(t_1)x_0\rangle = \rho' \qquad \text{II(9)}$$

Therefore, we only need to show that $\hat{\rho}$ cannot be larger than ρ'. Let us suppose the contrary. Then there exists a positive constant σ such that $\rho' < \sigma < \hat{\rho}$. Since $\sigma < \hat{\rho}$, by II(6), the system is not $(x_0, x_1, K, U_{r,p}^{\rho})$ controllable, and there exists a vector $x' \in R^n$:

$$|\langle x', x_1 - V(t_1)x_0\rangle| > \sigma\left(\int_{t_0}^{t_1} \|W^*(t_1, \tau)x'\|_s^q \, d\tau\right)^{1/q} \qquad \text{II(10)}$$

Since $\sigma > 0$, $|\langle x', x_1 - V(t_1)x_0\rangle|$ is positive and, therefore, since $\hat{u} \in \overline{U}$, hence $\hat{\rho} > 0$, by II(7) also $\left(\int_{t_0}^{t_1} \|W^*(t_1, \tau)x'\|_s^q \, d\tau\right)^{1/q}$ is positive, and we may therefore consider the quantity

$$\eta' = \frac{x'}{\left(\int_{t_0}^{t_1} \|W^*(t_1, \tau)x'\|_s^q \, d\tau\right)^{1/q}}$$

and by II(10) we have

$$|\langle \eta', x_1 - V(t_1)x_0\rangle| > \sigma > \rho'$$

which contradicts II(9).

Going back to consider the time optimal problem, let us suppose that $K = [t_0, t_1]$ is the minimal time interval over which the system I(1) can be transferred from x_0 at time t_0 to x_1 at time t_1 with controls whose norm does not exceed a given ρ. On the other hand, we know from Theorem II,4 and Corollary II,2 that, for the interval $\overline{K}$ considered given, the minimal norm controls have norm $\hat{\rho}$:

$$\hat{\rho} = \sup_{\left(\int_{t_0}^{\overline{t}} \|W^*(t_1, \tau)x\|_s^q \, d\tau\right)^{1/q} = 1} |\langle x, x_1 - V(t_1)x_0\rangle| \qquad \text{II(11)}$$

We may raise the following question: with reference to the same system, the same points, to ρ and hence $U^{\rho}_{r,p}$, can $\hat{\rho}$ be less than ρ?

We shall give a partial answer to this question in the following

Corollary II,3

If the system I(1) is controllable relative to any pair of points of R^n, to class $U^{\rho}_{r,p}$ of admissible controls and to the time interval $\overline{K} = [t_0, \overline{t}]$, where $\overline{K}$ is the minimal time interval over which the system is controllable relative to given x_0, x_1 and to $U^{\rho}_{r,p}$, with ρ given, then $\rho = \hat{\rho}$, where $\hat{\rho}$ is the minimal value of the norm of controls which effect the given transfer from x_0 to x_1, in $\overline{K}$.

Proof

Since $\overline{K} = [t_0, \overline{t}]$ is a time optimal interval, by Theorem II,2 there exists a vector x' such that II(4) holds. Since the system is controllable relative to any x_0, x_1, to $U^{\rho}_{r,p}$ and to $\overline{K} = [t_0, \overline{t}]$, by the already quoted Condition I,1, $Y(t_0, \overline{t}) = \int_{t_0}^{\overline{t}} W(\overline{t}, \tau)W^*(\overline{t}, \tau)d\tau$ is positive definite and, therefore, the integral $\left(\int_{t_0}^{\bar{t}} \|W(\bar{t}, \tau)x'\|_s^q \, d\tau\right)^{1/q}$ is positive. We may, therefore, write

$$\rho = \frac{|\langle x', x_1 - V(\bar{t})x_0\rangle|}{\left(\int_{t_0}^{\bar{t}} \|W^*(\bar{t}, \tau)x'\|_s^q \, d\tau\right)^{1/q}}$$

and recalling II(11), $\hat{\rho} \geq \rho$. Recalling that obviously $\hat{\rho} \leq \rho$, the corollary is proved. In what follows, we shall refer to time intervals $\bar{K} = [t_0, \bar{t}]$ which are minimal with respect to x_0, x_1 and to classes $U^{\rho}_{r,p}$ of admissible controls, $\hat{\rho}$ being also the minimal norm of controls transferring x_0 into x_1, in $\bar{K}$. Beside II(8), we shall also use

$$\hat{\rho} = \min_{u \in \bar{U}} \|u\|_{r,p} = \max_{\left(\int_{t_0}^{\bar{t}} \|W^*(\bar{t}, \tau)x\|_s^q \, d\tau\right)^{1/q} = 1} \langle x, x_1 - V(t_1)x_0\rangle = \frac{\langle \hat{x}, x_1 - V(\bar{t}_1)x_0\rangle}{\left(\int_{t_0}^{\bar{t}} \|W^*(\bar{t}, \tau)\hat{x}\|_s^q \, d\tau\right)^{1/q}} \qquad \text{II}(8')$$

We shall now see that the derivation of optimal controls (optimal in the sense just specified of being both time-minimal and norm-minimal) follows from the previous results and from a repeated use of Hölder's inequality. Let $\hat{u}$ be optimal with respect to x_0, x_1, $\hat{\rho}$, $\bar{K} = [t_0, \bar{t}]$. Since $\hat{u}$ transfers x_0 into x_1 in the interval $\bar{K} = [t_0, \bar{t}]$, we have

$$x_1 - V(\bar{t})x_0 = \int_{t_0}^{\bar{t}} W(\bar{t}, \tau)\hat{u}(\tau) d\tau \qquad \text{II}(12)$$

and since $\bar{K}$ is the minimal time interval, II(4) holds for some $\hat{x}$. Putting together II(4) and II(12) and taking the adjoint $W^*(\bar{t}, t)$ of $W(\bar{t}, t)$, we have

$$\hat{\rho}\left(\int_{t_0}^{\bar{t}} \|W^*(\bar{t}, \tau)\hat{x}\|_s^q \, d\tau\right)^{1/q} = \left|\int_{t_0}^{\bar{t}} \langle W^*(\bar{t}, \tau)\hat{x}, \hat{u}(\tau)\rangle d\tau\right|$$

Using Hölder's inequality in the finite-dimensional and infinite-dimensional case, we may write

$$\hat{\rho}\left(\int_{t_0}^{\bar{t}} \|W^*(\bar{t}, \tau)\hat{x}\|_s^q \, d\tau\right)^{1/q} = \left|\int_{t_0}^{\bar{t}} \langle W^*(\bar{t}, \tau)\hat{x}, \hat{u}(\tau)\rangle d\tau\right|$$

$$\leq \int_{t_0}^{\bar{t}} |\langle W^*(\bar{t}, \tau)\hat{x}, \hat{u}(\tau)\rangle| d\tau \leq \int_{t_0}^{\bar{t}} \|W^*(\bar{t}, \tau)\hat{x}\|_s \|u(\tau)\|_r \, d\tau$$

$$\leq \left(\int_{t_0}^{\bar{t}} \| \hat{u}(\tau) \|_r^p \, d\tau \right)^{1/p} \left(\int_{t_0}^{\bar{t}} \| W^*(\bar{t}, \tau)\hat{x} \|_s^q \, d\tau \right)^{1/q}$$

$$= \hat{\rho} \left(\int_{t_0}^{\bar{t}} \| W^*(\bar{t}, \tau)\hat{x} \|_s^q \, d\tau \right)^{1/q} = \hat{\rho} \| W^*\hat{x} \|_{s,q} \qquad \text{II(13)}$$

Since in II(13) the first and last terms are equal, we must have equality at each step. Recalling the conditions for Hölder's inequality to hold as an equality, we have almost everywhere in K:

$$\| \hat{u}(t) \|_r = C \| W^*(\bar{t}, t)\hat{x} \|_s^{q/p} \qquad \text{II(14)}$$

$$u_i(t) = K(t)[W^*(\bar{t}, t)\hat{x}]_i^{s/r} \qquad \text{II(15)}$$

and

$$\text{sign } u_i(t) = \text{sign}[W^*(\bar{t}, t)\hat{x}]_i^{s/r} \qquad \text{II(16)}$$

From II(8') and II(14) we may determine the constant C. Indeed,

$$\hat{\rho} = \| \hat{u} \|_{r,p} = C \left(\int_{t_0}^{\bar{t}} \| W^*(\bar{t}, \tau)\hat{x} \|_s^{\frac{q}{p} \cdot p} \, ds \right)^{1/p} = C(\| W^*\hat{x} \|_{s,q})^{q/p}$$

Hence,

$$C = \frac{\hat{\rho}}{(\| W^*\hat{x} \|_{s,q})^{q/p}} = \frac{\hat{\rho}}{(\| W^*\hat{x} \|_{s,q})^{q-1}}$$

Since (see II(4) or II(8')) the vector $\hat{x}$ is determined up to a multiplicative real constant, and since we suppose, as always, $x_1 - V(\bar{t})x_0 \neq 0$, we may determine $\hat{x}$ such that the scalar product $|\langle \hat{x}, x_1 - V(t_1)x_0 \rangle|$ is equal to one. In correspondence to such $\hat{x}$, we have from II(8')

$$\hat{\rho} = \frac{1}{\| W^*\hat{x} \|_{s,q}}$$

and, therefore,

$$C = \frac{1}{(\| W^*\hat{x} \|_{s,q})^q} \qquad \text{II(17)}$$

To determine the function K(t) let us take the finite-dimensional norm of $\hat{u}(t)$, using II(15),

$$\| u(t) \|_r = |K(t)| \left[\sum_i | W^*(\bar{t}, t)\hat{x}_i |^{\frac{s}{r} \cdot r} \right]^{\frac{1}{r}} = |K(t)| \, \| W^*(\bar{t}, t)\hat{x} \|_s^{\frac{s}{r}} \qquad \text{II(18)}$$

From II(15), II(17), and II(18) we have

$$|K(t)| = \frac{1}{\|W^*\hat{x}\|^q_{s,q}} \|W^*(\bar{t},t)\hat{x}\|^{\left(\frac{q}{p}-\frac{s}{r}\right)} \qquad \text{II(19)}$$

From II(15), II(16), II(19), we eventually have

$$\hat{u}_i(t) = \frac{1}{\|W^*\hat{x}\|_{s,q}} \|W^*(\bar{t},t)\hat{x}\|^{\left(\frac{q}{p}-\frac{s}{r}\right)} |W^*(\bar{t},t)\hat{x}|_i^{s-1} \operatorname{sign}(W^*(\bar{t},t)\hat{x})_i \qquad \text{II(20)}$$

which is the general expression for the optimal control when the norm of u is the general norm

$$\|u\|_{r,p} = \left(\int_{t_0}^{\bar{t}} \left[\sum_i |u_i(\tau)|^r\right]^{p/r} d\tau\right)^{1/p}$$

$\frac{1}{r}+\frac{1}{s} = 1, \frac{1}{p}+\frac{1}{q} = 1$, and $\hat{x}$ is given by II(8), II(8') with the additional constraint $|\hat{x}, x_1 - V(\bar{t})x_0| = 1$. Many particular cases of the general formula II(29) are interesting. For example, if $u \in L_{2,2}$ and its norm becomes

$$\|u\|_{2,2} = \left(\int_{t_0}^{\bar{t}} \sum_{i=1}^{m} |u_i(\tau)|^2 d\tau\right)^{1/2}$$

so as to have the intuitive meaning of "energy", the optimal control function is the following:

$$\hat{u}_i(t) = \frac{1}{\int_{t_0}^{\bar{t}} \sum_i |W^*(t,\tau)\hat{x}|_i^2 d\tau} (W^*(\bar{t},t)\hat{x})_i$$

if $u \in L_{\infty,\infty}$ and its norm is

$$\|u\|_{\infty,\infty} = \operatorname*{ess\,sup}_{t\in\bar{K}} \left[\max_{i=1\dots m} |u_i(t)|\right]$$

the optimal control function is

$$\hat{u}_i(t) = \frac{1}{\int_{t_0}^{\bar{t}} \sum_{i=1}^{m} |W^*(\bar{t},\tau)\hat{x}|_i d\tau} \operatorname{sign}[W^*(\bar{t},t)\hat{x}]_i$$

In general, the type of norm to be chosen is given by physical considerations, as is obvious. Let us consider the case $u \in L_{r,\infty}$, or

$$\|u\|_{r,\infty} = \operatorname*{ess\,sup}_{t \in \overline{K}} \left[\sum_{i=1}^{m} |u_i(t)|^r \right]^{1/r} ; \quad (1 \leq r \leq \infty)$$

and let us see how it is now possible to derive the well-known Pontryagin's maximum principles for this case.
Let us consider the system I(1) with $u \equiv 0$:

$$\dot{x} = A(t)x \qquad \text{II(21)}$$

and its adjoint equation

$$\dot{\psi} = - A^*(t)\psi \qquad \text{II(22)}$$

Let the fundamental matrix solution $V(t)$ of II(21) be non-singular at $\overline{t}$, hence invertible almost everywhere. Calling $\Psi(t)$ the fundamental matrix solution of II(22) with $\Psi^*(\overline{t}) = V^{-1}(\overline{t})$, we immediately see that

$$\Psi^*(t) = V^{-1}(t)$$

(Indeed, $(d/dt)[\Psi^*(t)\,V(t)] = -[A^*(t)\Psi(t)]^*\,V(t) + \Psi^*(t)\,A(t)\,V(t) = 0$, hence $\Psi^*(t)\,V(t) = I$, where I is the identity matrix). As a consequence, we may write the general vector solution of II(22) in the form

$$w(t) = \Psi(t)\,\Psi^{-1}(\overline{t})\hat{x} = V^{-1^*}(t)V^*(\overline{t})\hat{x} \qquad \text{II(23)}$$

where $\hat{x} = w(\overline{t})$. The optimal control law u makes all inequalities of II(13) equalities, as we have seen. In particular,

$$|\langle W^*(\overline{t},t)\hat{x},\, \hat{u}(t)\rangle| = \|W^*(\overline{t},t)\hat{x}\|_s \,\|\hat{u}(t)\|_r$$

a.e. in $\overline{K}$ or, equivalently,

$$|\langle V^{-1^*}(t)\,V^*(\overline{t})\hat{x},\, B(t)\hat{u}(t)\rangle| = |\langle B^*(t)V^{-1^*}(t)V^*(\overline{t})\hat{x},\, \hat{u}(t)\rangle|$$

$$= \|B^*(t)\,V^{-1^*}(t)V^*(\overline{t})\hat{x}\|_s \,\|\hat{u}(t)\|_r$$

and since for every u such that

$$\operatorname*{ess\,sup}_{t \in \overline{K}} \left(\sum_{i=1}^{m} |u_i(t)|^r \right)^{1/r} \leq \rho$$

$$|\langle V^{-1^*}(t)V^*(\overline{t})\hat{x},\, B(t)u(t)\rangle| \leq \|B^*(t)V^{-1^*}(t)V^*(\overline{t})\hat{x}\|_s \,\|u(t)\|_r \qquad \text{II(24a)}$$

we may conclude that for every t the function $\hat{u}(t)$ makes the quantity $|\langle V^{-1^*}(t)V^*(\overline{t})\hat{x},\, B(t)u(t)\rangle|$ maximal with respect to all $u(t)$ such that $\|u\|_{r,\infty} \leq \rho$ i.e. to all admissible controls. Since, by II(23), $w(t) = V^{-1^*}(t)V^*(\overline{t})\hat{x}$ is the solution of the adjoint equation $\dot{w}(t) = -A^*(t)\,w(t)$ satisfying $w(\overline{t}) = \hat{x}$, we may say that the optimal control $\hat{u}(t)$ maximizes for every t, a.e. and with respect to all admissible controls, the scalar product $|\langle w(t), B(t)\,u(t)\rangle|$, where w(t) is an appropriate solution of the quoted adjoint vector equation. This is precisely Pontryagin's maximum principle for this case.

Let us now go back to the general optimal control II(20) and notice that the only quantity whose computation is not explicit in it is the vector $\hat{x}$, which corresponds to

$$\hat{\rho} = \max_{|\langle x, x_1 - V(t_1)x_0\rangle| = 1} \frac{1}{\left(\int_{t_0}^{\bar{t}} \| W^*(\bar{t}, \tau)x\|_s^q \, d\tau\right)^{1/q}} \quad \left(\frac{1}{p} + \frac{1}{q} = 1; \frac{1}{r} + \frac{1}{s} = 1\right) \qquad \text{II(24b)}$$

For a general type of norm, this computation is a rather difficult problem which we are not going to treat here.

We shall only show how the computation of $\hat{x}$ becomes an easy task when $u \in L_{2,2}$, hence u belongs to a Hilbert space. From II(20) we see that in this case the form of the optimal control is

$$\hat{u}(t) = \frac{W^*(\bar{t}, t)\hat{x}}{\int_{t_0}^{\bar{t}} \langle W^*(\bar{t}, \tau)\hat{x}, W^*(\bar{t}, \tau)\hat{x}\rangle \, d\tau} = \hat{\rho}^2 W^*(\bar{t}, t)\hat{x} \qquad \text{II(25)}$$

Since $\hat{u}$ transfer x_0 into x_1 at $\bar{t}$, we have, putting again $z(\bar{t}) = x_1 - V(\bar{t})x_0$,

$$\int_{t_0}^{\bar{t}} W(\bar{t}, \tau)u(\tau)d\tau = z(\bar{t})$$

which, by II(24), becomes

$$\hat{\rho}^2 \int_{t_0}^{\bar{t}} W(\bar{t}, \tau)W^*(\bar{t}, \tau)\hat{x} \, d\tau = z(\bar{t})$$

or, with the usual position for $Y(\bar{t})$, the determination of $\hat{x}$ reduces to the solution of the linear algebraic equation

$$\hat{\rho}^2 \, Y(\bar{t})\hat{x} = z(\bar{t}) \qquad \text{II(26)}$$

with (from II(24))

$$\hat{\rho}^2 = \frac{1}{\langle \hat{x}, Y(\bar{t})\hat{x}\rangle} \qquad \text{II(27)}$$

Let us distinguish two cases: in the first one, the system I(1) is controllable at time $\bar{t}$ relative to any pair of points of R^n. In this case, $Y(\bar{t})$ is non-singular, therefore invertible, and Eq.II(26) has the unique solution

$$\hat{x} = \frac{1}{\hat{\rho}^2} Y^{-1}(\bar{t})\, z(\bar{t})$$

with

$$\hat{\rho}^2 = \langle Y^{-1}(\bar{t})\, z(\bar{t}), z(\bar{t})\rangle$$

and from II(25), the expression of the optimal control law $\hat{u}$, is

$$\hat{u}(t) = B^*(t)\, V^{-1*}(t) V^*(\bar{t}) Y^{-1}(\bar{t})\, z(\bar{t}) \text{ for } t \in [t_0, \bar{t}] \qquad \text{II(28)}$$

In the second case, the matrix $Y(\bar{t})$ is singular, but the system is still controllable with respect to x_0, x_1, $K = [t_0, \bar{t}]$, ρ, and therefore the matrix $C(\bar{t}) = Y(\bar{t}) - (1/\hat{\rho}^2)Z(\bar{t})$ is positive semi-definite. We shall see that it is still possible to find an explicit expression of the optimal control law. To do that, we need the following

Theorem II, 5

Let $Ra[Y(\bar{t})]$ be the range of matrix $Y(\bar{t})$. Then, if $C(\bar{t})$ is positive semi-definite, the vector $z(\bar{t})$ belongs to $Ra[Y(\bar{t})]$.

Proof

Let $N[Y(\bar{t})]$ be the null space of $Y(\bar{t})$, that is the set of vectors x such that $Y(\bar{t})x = 0$. (The dimension of the null space $N[Y(\bar{t})]$ is equal to the multiplicity of 0 as an eigenvalue of $Y(\bar{t})$). Since $Y(\bar{t})$ is real and symmetric, its eigenvectors span the space R^n, and

$$R^n = Ra[Y(\bar{t})] \oplus N[Y(\bar{t})]$$

Let P_N be the projection operator of R^n into $N[Y(\bar{t})]$, and consider the vector $P_N\, z(\bar{t})$. Since $C(\bar{t})$ is positive semi-definite, we have

$$\langle C(\bar{t}) P_N\, z(\bar{t}), P_N\, z(\bar{t})\rangle \geq 0$$

and, by definition of $N[Y(\bar{t})]$,

$$\langle C(\bar{t})\, P_N z(\bar{t}), P_N z(\bar{t})\rangle = -\frac{1}{\hat{\rho}^2}\, (P_N\, z(\bar{t}), z(\bar{t})\rangle \leq 0$$

and therefore $P_N\, z(\bar{t})$ is orthogonal to $z(\bar{t})$, which means that $z(\bar{t})$ does not have components in $N[Y(\bar{t})]$ and belongs to $Ra[Y(\bar{t})]$, as we wanted to prove.

Let now ℓ be the dimension of $N[Y(\bar{t})]$, so that $n - \ell$ is the dimension of $Ra[Y(\bar{t})]$. The vector $z(\bar{t})$ can then be written as $z(\bar{t}) = \sum_{i=\ell+1}^{n} \mathfrak{z}_i\, d_i$, where d_i are the eigenvectors of $Y(\bar{t})$ corresponding to the non-null eigenvalues of $Y(\bar{t})$. Since $\hat{x}$ can be written as $\hat{x} = \sum_{i=1}^{n} \zeta_i\, d_i$, where $d_1, \ldots, d_\ell$ are the vectors spanning $N[Y(\bar{t})]$, equation II(26) becomes

$$\hat{\rho}^2 Y(\bar{t}) \sum_{i=1}^{n} \zeta_1 d_i = \sum_{i=\ell+1}^{n} \mathfrak{z}_i d_i$$

$$\hat{\rho}^2 \sum_{i=\ell+1}^{n} \zeta_i \lambda_i d_i = \sum_{i=\ell+1}^{n} \mathfrak{z}_i d_i$$

where λ_i are the non-null (and real, since $Y(\bar{t})$ is real and symmetric) eigenvalues of $Y(\bar{t})$. The solution of II(26) is, therefore,

$$\hat{x} = \frac{1}{\hat{\rho}^2}\left(\sum_{i=1}^{\ell} \zeta_i d_i + \sum_{i=\ell+1}^{n} \frac{\mathfrak{z}_i}{\lambda_i} d_i\right)$$

and, from II(25), we see that the optimal control law is

$$\hat{u}(t) = B^*(\bar{t}) V^{-1*}(t) V^*(\bar{t}) \left(\sum_{i=1}^{\ell} \zeta_i d_i + \sum_{i=\ell+1}^{n} \mathfrak{z}_i d_i\right) \qquad \text{II(29)}$$

where $\zeta_1, \ldots, \zeta_n$ are arbitrary constants. We may check that the values of $\zeta_1, \ldots, \zeta_n$ do not affect the norm $\hat{\rho}$ simply by computing $\hat{\rho}$, indeed,

$$1 = |\langle \hat{x}, z(\bar{t})\rangle| = \frac{1}{\hat{\rho}^2}\left|\left\langle \sum_{i=1}^{\ell} \mathfrak{z}_i d_i + \sum_{i=\ell+1}^{n} \frac{\mathfrak{z}_i}{\lambda_i} d_i, \sum_{i=\ell+1}^{n} \mathfrak{z}_i d_i \right\rangle\right| = \frac{1}{\hat{\rho}^2} \sum_{i=\ell+1}^{n} \frac{\mathfrak{z}_i^2}{\lambda_i}$$

from which

$$\hat{\rho} = \left(\sum_{i=\ell+1}^{n} \frac{\mathfrak{z}_i^2}{\lambda_i}\right)^{1/2} \qquad \text{II(30)}$$

Formula II(30), with $\ell = 0$, is, of course, valid also when $Y(\bar{t})$ is non-singular.

<u>Exercise II, 1</u>

Given the system

$$\dot{x} = -\frac{1}{4}\begin{Vmatrix} 2 & 20 \\ 5 & 2 \end{Vmatrix} x + \begin{Vmatrix} 2 & 4 \\ 1 & 2 \end{Vmatrix} u(t) \qquad x_0 = \begin{vmatrix} 2 \\ -3 \end{vmatrix} \qquad x_1 = \begin{vmatrix} 8 \\ -4 \end{vmatrix} \qquad \text{E(1)}$$

and admissible controls u such that

$$\int_0^T \|u(s)\|^2 ds \le 3$$

where $\|u(t)\|$ is the Euclidean norm of u(t).

a) Prove that it is controllable relative to x_0, x_1 to the interval $K = [0, T]$ for some $T > 0$, and to the given class of admissible controls;

b) determine the minimal time $\bar{t}$ such that the system is controllable relative to the same quantities and construct a corresponding time optimal control;

c) verify whether there exists only one time-optimal control;

d) verify whether there exist admissible controls which transfer x_0 in x_1, and which are time-optimal but not norm minimal;

e) if the class of admissible controls is $U = \{u: \sup_{t \in [0, T]} \|u(t)\| \leq 3\}$, with respect to which T is the system controllable?

<u>Solution</u>

The eigenvalues of the matrix $A = -\frac{1}{4}\begin{Vmatrix} 2 & 20 \\ 5 & 2 \end{Vmatrix}$ are $\lambda_1 = 2$, $\lambda_2 = -3$.

The diagonal matrix $A' = S^{-1}AS$, with $S = \frac{1}{4}\begin{Vmatrix} 2 & 2 \\ -1 & 1 \end{Vmatrix}$ and $S^{-1} = \begin{Vmatrix} 1 & -2 \\ 1 & 2 \end{Vmatrix}$

is, therefore, $A' = \begin{Vmatrix} 2 & 0 \\ 0 & -3 \end{Vmatrix}$, and the fundamental matrix solution V(t) of

E(1) with $V(0) = I$ is

$$V(t) = \frac{1}{4}\begin{Vmatrix} 2 & 2 \\ -1 & 1 \end{Vmatrix}\begin{Vmatrix} e^{2t} & 0 \\ 0 & e^{-3t} \end{Vmatrix}\begin{Vmatrix} 1 & -2 \\ 1 & 2 \end{Vmatrix}$$

Therefore

$$W(T, t) = W(T - t) = V(T)V^{-1}(t)B(t) = \frac{1}{4}\begin{Vmatrix} 2 & 2 \\ -1 & 1 \end{Vmatrix}\begin{Vmatrix} e^{2(T-t)} & 0 \\ -1 & e^{-3(T-t)} \end{Vmatrix}$$

$$\times \begin{Vmatrix} 1 & -2 \\ 1 & 2 \end{Vmatrix}\begin{Vmatrix} 2 & 4 \\ 1 & 2 \end{Vmatrix} = e^{-3(T-t)}\begin{Vmatrix} 2 & 2 \\ -1 & 1 \end{Vmatrix}\begin{Vmatrix} 0 & 0 \\ 1 & 2 \end{Vmatrix} \qquad E(2)$$

and the matrix product $W(T, t)\, W^*(T, t)$ is

$$W(T - t)\, W^*(T - t) = e^{-6(T-t)}\begin{Vmatrix} 2 & 2 \\ -1 & 1 \end{Vmatrix}\begin{Vmatrix} 0 & 0 \\ 0 & 5 \end{Vmatrix}\begin{Vmatrix} 2 & -1 \\ 2 & 1 \end{Vmatrix} \qquad E(3)$$

and, therefore,

$$Y(T) = \int_0^T W(T-s)\, W^*(T-s)\,ds = \frac{1-e^{-6T}}{6} \begin{Vmatrix} 2 & 2 \\ -1 & 1 \end{Vmatrix} \begin{Vmatrix} 0 & 0 \\ 0 & 5 \end{Vmatrix} \begin{Vmatrix} 2 & -1 \\ 2 & 1 \end{Vmatrix}$$

$$= 5\,\frac{1-e^{-6T}}{6} \begin{Vmatrix} 2 & 2 \\ -1 & 1 \end{Vmatrix} \begin{Vmatrix} 0 & 0 \\ 0 & 1 \end{Vmatrix} \begin{Vmatrix} 2 & -1 \\ 2 & 1 \end{Vmatrix} \qquad E(4)$$

We notice that since $Y(T)$ has 0 as eigenvalue it is not positive definite; hence the system is not controllable relative to any pair of points x_0, x_1.

From the previous position $z(T) = x_1 - V(T)x_0$ we have

$$z(T) = \begin{vmatrix} 8 \\ -4 \end{vmatrix} - \frac{1}{4} \begin{Vmatrix} 2 & 2 \\ -1 & 1 \end{Vmatrix} \begin{Vmatrix} e^{2T} & 0 \\ 0 & e^{-3T} \end{Vmatrix} \begin{Vmatrix} 1 & -2 \\ 1 & 2 \end{Vmatrix} \begin{vmatrix} 2 \\ -3 \end{vmatrix}$$

$$= \frac{1}{4} \begin{Vmatrix} 2 & 2 \\ -1 & 1 \end{Vmatrix} \left[\begin{Vmatrix} 1 & -2 \\ 1 & 2 \end{Vmatrix} \begin{vmatrix} 8 \\ -4 \end{vmatrix} - \begin{Vmatrix} e^{2T} & 0 \\ 0 & e^{-3T} \end{Vmatrix} \begin{vmatrix} 8 \\ -4 \end{vmatrix} \right] \qquad E(5)$$

and,therefore,

$$Z(T) = z(T)\, z^*(T) = \begin{Vmatrix} 2 & 2 \\ -1 & 1 \end{Vmatrix} \begin{Vmatrix} 1-e^{2T} & -2 \\ 1 & 2-e^{-3T} \end{Vmatrix} \begin{vmatrix} 2 \\ -1 \end{vmatrix}$$

$$\times \begin{vmatrix} 2 & -1 \end{vmatrix} \begin{Vmatrix} 1-e^{2T} & 1 \\ -2 & 2-e^{-3T} \end{Vmatrix} \begin{Vmatrix} 2 & -1 \\ 2 & 1 \end{Vmatrix}$$

From the definition of $C(t)$ and from E(4), E(5) we have

$$C(T) = Y(T) - \frac{1}{\rho^2} Z(T)$$

$$= \begin{Vmatrix} 2 & 2 \\ -1 & 1 \end{Vmatrix} \left[\begin{Vmatrix} 0 & 0 \\ 0 & \frac{5}{6}(1-e^{-6T}) \end{Vmatrix} - \frac{1}{\rho^2} \begin{Vmatrix} \frac{1}{16}(16-8e^{2T})^2 & c_1(T) \\ c_2(T) & e^{-6T} \end{Vmatrix} \right] \begin{Vmatrix} 2 & -1 \\ 2 & 1 \end{Vmatrix}$$

E(6)

where $c_1(T)$ and $c_2(T)$ are functions obtainable from E(4), but which are not relevant for our purposes.

From E(6) we see that C(T) has the same eigenvalues as the matrix

$$R(T) = \begin{Vmatrix} -\frac{1}{\rho^2}\frac{1}{16}(16-8e^{2T})^2 & -\frac{1}{\rho^2}C_1(T) \\ -\frac{1}{\rho^2}C_2(T) & \frac{5}{6}(1-e^{-6T})-\frac{1}{\rho^2}e^{-6T} \end{Vmatrix}$$

and, therefore, C(T) has the same character as R(T) as far as positive or negative semi-definiteness is concerned.

By inspecting R(T), we easily see that a necessary condition for it to be positive semi-definite is that $16-8e^{2T}$ be zero, and this is verified only for $T = \frac{1}{2}\ln 2 = t^*$. So this is the only candidate for the system to be controllabl It turns out that the system is controllable relative to the given x_0, x_1, $\rho = \sqrt{3}$ and the time interval $[0, t^*]$. To prove that, let us first see that the equation

$$Y(t^*)x = z(t^*) \qquad E(7)$$

has a solution. From E(4), E(5), Eq.E(7) with $t^* = \frac{1}{2}\ln 2$ is

$$\frac{35}{48}\begin{Vmatrix} 2 & 2 \\ -1 & 1 \end{Vmatrix}\begin{Vmatrix} 0 & 0 \\ 0 & 1 \end{Vmatrix}\begin{Vmatrix} 2 & -1 \\ 2 & 1 \end{Vmatrix}\hat{x} = \begin{Vmatrix} 2 & 2 \\ -1 & 1 \end{Vmatrix}\begin{vmatrix} 0 \\ 2^{-3/2} \end{vmatrix} \qquad E(8)$$

Since $\begin{Vmatrix} 2 & 2 \\ -1 & 1 \end{Vmatrix}$ is non-singular, we may pre-multiply both sides of E(8) by its inverse, obtaining

$$\frac{35}{12}\begin{Vmatrix} 0 & 0 \\ 0 & 1 \end{Vmatrix}\begin{Vmatrix} 2 & -1 \\ 2 & 1 \end{Vmatrix}\hat{x} = \begin{vmatrix} 0 \\ 4\cdot 2^{-3/2} \end{vmatrix}$$

or,

$$(2\hat{x}_1 + \hat{x}_2) = \frac{12}{35}\cdot 4\cdot 2^{-3/2} \qquad E(9)$$

We know that any solution $\hat{x}$ of E(9) is such that the control $\hat{u}(t) = B^*(t) \times V^{-1*}(t)\, V(t^*)\hat{x}$ takes x_0 into x_1, at time t^*. Actually, E(9) has infinitely many solutions, all with this property and all corresponding to norm optimal controls. Among these solutions let us take one of the form $\hat{x} = \lambda\begin{bmatrix} 1 \\ 2 \end{bmatrix}$, so that from E(9) we have $\lambda = \frac{12}{35}\, 2^{-3/2}$.

The corresponding norm-minimal control is

$$\hat{u}(t) = \frac{6}{35}\begin{bmatrix} 1 \\ 2 \end{bmatrix} e^{3t} \qquad 0 \leq t \leq \tfrac{1}{2}\ln 2 \qquad \text{E(10)}$$

and its squared norm is

$$\hat{\rho}^2 = \left(\frac{6}{35}\right)^2 5 \int_0^{\frac{1}{2}\ln 2} e^{6s}\, ds = \frac{6}{35} \qquad \text{E(11)}$$

Since $\hat{\rho} = \sqrt{\frac{6}{35}} < 3$, the control $\hat{u}$ is admissible and we have proved that the given system is controllable relative to $\begin{bmatrix} 2 \\ -3 \end{bmatrix}, \begin{bmatrix} 8 \\ -4 \end{bmatrix}$, time interval $[0, \frac{1}{2}\ln 2]$ and $\rho = \sqrt{3}$.

Of course, since $t^* = \frac{1}{2}\ln 2$ is the only time for which the system is controllable, then it is also the minimal time, and the control $\hat{u}(t)$ given by E(10) is not only norm-minimal, but also time-optimal.

Since in E(11) we saw that $\hat{\rho} < \rho = \sqrt{3}$, we expect that, besides all infinitely many norm-minimal controls which effect the same transfer in the same time, there are other controls which do the same job, being also admissible but not norm-minimal. Indeed, let us add to the norm-minimal control $\hat{u}(t)$ given by E(10) any other control $u^1(t)$ such that

$$\int_0^{\frac{1}{2}\ln 2} V(\tfrac{1}{2}\ln 2)V^{-1}(s)B(s)u^1(s)ds = 0 \qquad \text{E(12)}$$

Taking $\hat{u}(s) = \begin{bmatrix} K_1 \\ K_2 \end{bmatrix}$, with K_1 and K_2 constants, we have from E(2)

$$\int_0^{\frac{1}{2}\ln 2} V(\tfrac{1}{2}\ln 2)V^{-1}(s)\, B(s)\begin{bmatrix} K_1 \\ K_2 \end{bmatrix} ds = C \begin{Vmatrix} 2 & 4 \\ 1 & 2 \end{Vmatrix} \begin{vmatrix} K_1 \\ K_2 \end{vmatrix} = C \begin{bmatrix} 2K_1 + 4K_2 \\ K_1 - 2K_2 \end{bmatrix}$$

with C constant, and therefore all controls of the type $u^1(t) = \begin{bmatrix} K_1 \\ -\frac{1}{2}K_1 \end{bmatrix}$ satisfy E(12). Therefore, all infinitely many controls u of the type

$$u(t) = \hat{u}(t) + \begin{bmatrix} K_1 \\ -\frac{1}{2}K_1 \end{bmatrix}$$

with $|K_1| > 0$ limited by the constraint $\int_0^{\frac{1}{2}\ln 2} \|u(s)\| ds < 3$ are admissible and transfer x_0 in x_1 in the time interval $[0, \frac{1}{2}\ln 2]$, and are therefore time optimal but not norm minimal.

Passing now to the last question e), let us verify the general controllability condition I(11) in this specific case.

For the system to be controllable at time T, we must have

$$|\langle x, z(T)\rangle| \leq \sqrt{3}\int_0^T \|W^*(T,s)x\|\,ds$$

for every $x \in R^n$.

From E(2), E(5) we must therefore have

$$\left\langle \begin{Vmatrix} 2 & -1 \\ 2 & 1 \end{Vmatrix} \begin{vmatrix} x_1 \\ x_2 \end{vmatrix}, \begin{vmatrix} 4 \cdot 2e^{2T} \\ e^{-3T} \end{vmatrix} \right\rangle$$

$$\leq \sqrt{3}\int_0^T e^{-3(T-s)} \left\| \begin{Vmatrix} 0 & 1 \\ 0 & 2 \end{Vmatrix} \begin{Vmatrix} 2 & -1 \\ 2 & 1 \end{Vmatrix} \begin{vmatrix} x_1 \\ x_2 \end{vmatrix} \right\| ds, \qquad \forall x_1, \forall x_2 \qquad \text{E(13)}$$

Since the matrix $\begin{Vmatrix} 2 & -1 \\ 2 & 1 \end{Vmatrix}$ is non-singular, the vector

$$\begin{vmatrix} y_1 \\ y_2 \end{vmatrix} = \begin{Vmatrix} 2 & -1 \\ 2 & 1 \end{Vmatrix} \begin{vmatrix} x_1 \\ x_2 \end{vmatrix}$$

spans the space R^2 as well as $\begin{vmatrix} x_1 \\ x_2 \end{vmatrix}$ and, therefore, we may write

$$\left| \left\langle \begin{vmatrix} y_1 \\ y_2 \end{vmatrix}, \begin{vmatrix} 4 - 2e^{2T} \\ e^{-3T} \end{vmatrix} \right\rangle \right| \leq \sqrt{3}\int_0^T e^{-3(T-s)} \left\| \begin{Vmatrix} 0 & 1 \\ 0 & 2 \end{Vmatrix} \begin{vmatrix} y_1 \\ y_2 \end{vmatrix} \right\| ds,$$

$$\forall y_1, \forall y_2$$

or

$$|(4 - 2e^{2T})y_1 + e^{-3T} y_2| \leq \frac{\sqrt{5}}{\sqrt{3}}(1 - e^{-3T})|y_2|, \qquad \forall y_1, \forall y_2 \qquad \text{E(14)}$$

as a condition of controllability instead of E(13).

A necessary condition for E(14) to hold is that the coefficient of y_1 be zero, i.e. $T = \frac{1}{2}\ln 2$. For such a value of T, E(14) becomes

$$\frac{2^{-3/2}}{1 - 2^{-3/2}}|y_2| \leq \sqrt{\frac{5}{3}}|y_2|, \qquad \forall y_2$$

which is satisfied.

Therefore we may conclude that the system E(1) is controllable with respect to x_0, x_1, class of admissible controls $U = \{u : \sup_{t \in [0, \frac{1}{2}\ln 2]} \|u(t)\| \leq 3\}$, only for the time interval $K = [0, \frac{1}{2}\ln 2]$.

PART III. REACHABILITY OF SETS IN THE PRESENCE OF DISTURBANCES

Given a control system subject to partially unknown perturbations, a problem which arises naturally is to find the set X_0 of initial states x_0 which may be transferred "for sure" into a given target set X_N by an admissible control.

Many practical examples of such a problem may be found in various fields of system sciences, of operational research and management science, when a "worst case" or conservative philosophy is to be adopted, that is when the aim is to reach a goal for sure, not to maximize the expectation of a given event: an intuitive case is, for example, the problem of determining the zone starting from which a given vehicle can reach for sure a target in a noisy environment. This kind of problem, as well as the determination of the best control in the worst possible conditions, has been formally treated only recently, either in itself or in the framework of differential game theory, in which it may be embedded if the partially unknown disturbances or environmental situations are conservatively treated as the adversary of game theory.

The algorithms which are presently available for the solution of the problem above, even for linear systems with admissible controls and disturbances belonging to convex compact sets, are generally quite difficult to be implemented and it seems that further research is needed in order to furnish flexible algorithms for facing practical situations.

The purpose here is to sketch the key points of the methods of solution proposed until now, for different information structures available both to the controller and to the adversary disturbances, and to show some interesting simplifications which are the consequence of taking a subspace as a target set and any control and disturbance as admissible; for this case, it is possible to state some interesting results in a compact form, as we shall do.

As an example of the problem mentioned, let us consider the following linear discrete time system

$$x_{K+1} = Ax_K + Bu_K + Cw_K \qquad (k = 0, 1, ...) \qquad \text{III(1)}$$

where the states x_K belong to R^n, controls u_K and disturbances w_K belong to closed compact sets U of R^m and W of R^p respectively and the target is a given closed compact set X_N of R^n (many types of extensions, like the one to time-varying systems with sets U and W varying also with time instants are possible but inessential to our purposes).

We formulate the problem of reachability of X_N from X_0 under disturbances by giving two among the main possible information structures.

<u>Problem III,1</u>

Determine the set X_0^1 of initial states x_0 which may be transferred into X_N at the time instant N by an admissible sequence $u_0(x_0), u_1(x_1), \dots u_{N-1}(x_{N-1})$ of controls, for any admissible sequence $(w_0, w_1, \dots w_{N-1})$ of disturbances. Clearly, for each of its "moves" u_i the controller may take into account his perfect information about the present state x_i, but is ignorant about the "move" w_i of the disturbances.

Problem III, 2

Determine the set X_0^2 of initial states x_0 which may be transferred into X_N at time instant N by an admissible sequence $u_0(x_0, w_0), \ldots, u_{N-1}(x_{N-1}, w_{N-1})$ of controls for any admissible sequence $w_0, \ldots, w_{N-1}$ of disturbances. Obviously, the information available to the controller is in this case "larger" than in the previous case, since in each step he knows the "move" of the disturbances.

Even for these simple examples of the general problem, with the clearly defined information structures mentioned, the used methods for finding X_0^1 or X_0^2 have required the use of the so-called operation of geometric difference of sets[1] and either the use of separation theorems for convex compact sets [3], or the use of support functions to describe sets and set-inclusion [4, 5] or some ellipsoid-type or polyhedrical-type approximation of sets [6], which have the advantage of describing sets with a finite number of parameters but can give only sufficient conditions for a point x_0 to be transferrable into X_N, that is can give only subsets of X_0^1 or X_0^2.

When the instant of time at which reachability occurs is of interest, game-theoretical approaches to the above kind of problems are interesting, but computational results are rather involved indeed [7, 8].

Referring to problem III, 1 and using Eq.III(1), we see that at the (N-1)-th step the state x_{N-1} is "transferrable" into X_N if there exists $u_{N-1}(x_{N-1}) \in U$ such that

$$Ax_{N-1} + Bu_{N-1} + Cw_{N-1} \in X_N$$

for every $w_{N-1} \in W$, that is iff there exists $u_{N-1}(x_{N-1}) \in U$ such that

$$Ax_{N-1} + Bu_{N-1} + CW \in X_N$$

or

$$Ax_{N-1} + Bu_{N-1} \in (X_N \stackrel{*}{-} CW)$$

that is iff

$$Ax_{N-1} \in (X_N \stackrel{*}{-} CW) \stackrel{*}{-} BU$$

Therefore the set X_{N-1} of states x_{N-1} which may be transferred into X_N in one step satisfies the following equation:

$$Y_{N-1}^1 = AX_{N-1}^1 = (X_N \stackrel{*}{-} CW) - BU$$

and, similarly, the set X_{N-1} $(i = 1, \ldots, N)$ of states which may be transferred into X_N in i steps satisfies

$$Y_{N-i}^1 = A^i X_{N-1}^1 = (Y_{N-i+1}^1 \stackrel{*}{-} A^{i-1} CW) - A^{i-1} BU \qquad (i = 1, \ldots, N) \qquad \text{III(2)}$$

[1] Given two sets S and T, the geometric difference $Z = S \stackrel{*}{-} T$ is defined as $Z = \{z : z + T \subset S\}$. For its properties, see Ref.[9].

Equation III(2), together with the obvious equality

$$Y_N^1 = X_N \qquad \text{III(3)}$$

is a recursive algorithm for building $Y_{N-1}, \ldots, Y_0$ from U, W, X_N and, therefore, X_0^1 which is characterized by

$$A^N X_0^1 = Y_0^1 \qquad \text{III(4)}$$

Proceeding in an analogous manner for Problem III, 2, and taking into account the different information structure, we have that in this case x_{N-1} may be transferred into X_N in one step iff for every $w_{N-1} \in W$ there exists $u_{N-1}(x_{N-1}, w_{N-1}) \in U$ such that

$$Ax_{N-1} + Bu_{N-1} + Cw_{N-1} \in X_N$$

that is iff

$$Ax_{N-1} + CW \in X_N - BU$$

Therefore, the set X_{N-1}^2 of states x_{N-1} of this type is given by

$$Y_{N-1}^2 = AX_{N-1}^2 = (X_N - BU) \stackrel{*}{-} CW$$

and similarly the sets X_{N-i}^2 $(i = 1, \ldots, N)$ of states which may be transferred into X_N in i steps by admissible controls and for every disturbance satisfy

$$Y_{N-i}^2 = A^i X_{N-i}^2 = (Y_{N-i+1}^2 - A^{i-1}BU) \stackrel{*}{-} A^{i-1}CW, \qquad (i = 1, \ldots, N) \qquad \text{III(2')}$$

We have again found a recursive algorithm which, starting again backward from

$$Y_N^2 = X_N \qquad \text{III(3')}$$

gives the following characterization of X_0^2:

$$A^N X_0^2 = Y_0^2 \qquad \text{III(4')}$$

As is apparent from III(2), (3), (4) or from III(2'), (3'), (4'), the determination of X_0^1 or of X_0^2 essentially involves the computation of N geometric differences and N additions of sets. As we saw in Part I, the analogue of Problem III, 1 and Problem III, 2 for continuous systems, considering only open loop controls would lead to a similar computation for only one step, but the matrices $A^{i-1}B$ and $A^{i-1}C$ $(i = 0, \ldots, N-1)$ would be substituted by linear integral operators from sets U and W of control and disturbance functions into R^n.

Going back to consider Eqs III(2), (3), (4) or III(2'), (3'), (4'), we see that the key difficulty is the description of sets resulting from operations on sets.

A natural tool to be used for this purpose is the one given by support functions (see, e.g. [5]), when the sets X_N, U, W, and therefore also all other sets involved, are convex and closed. Support functions are linear with respect to the addition of such a kind of sets; unfortunately, the operation of geometric difference of sets does not possess such a property: if h is the support function of the set X : $h_X(p) = \sup_{x \in X} \langle p, x \rangle$, $\forall p \in X^*$, where X is the dual of the space X, then for every set S, T, we have

$$h_{S \underline{*} T}(\vec{p}) \leq h_S(\vec{p}) - h_T(\vec{p}) \qquad \text{III(5)}$$

It is therefore important to find conditions on the sets involved in III(2), (3), (4) (or III(2'), (3'), (4')) for III(5) to be held as an equality (we suppose, for simplicity, that the geometric differences are never the empty set). In such a case, the construction of the sets $Y^1_{N-1}, \ldots, Y^1_0$ from Y^1_N and of $Y^2_{N-1}, \ldots, Y^2_N$ would be straightforward and we would also have the interesting consequence that these two sequences of sets would coincide if $Y^1_N = Y^2_N$; therefore X^1_0 would also coincide with X^2_0 and the difference in the information structures of Problems 1 and 2 would not have any effect.

Defining $S_T = (S \underline{*} T) + S$ as the "regular part" of S with respect to T, we define the set S to be regular with respect to T iff $S_T = S$. It is easy to prove that III(5) holds as an equality iff S is regular with respect to T; indeed, from the regularity it follows that

$$h_{S \underline{*} T}(p) + h_T(p) = h_S(p), \qquad \forall p$$

that is

$$h_{S \underline{*} T}(p) = h_S(p) + h_T(p), \qquad \forall p \qquad \text{III(6)}$$

and the regularity follows from III(6) for any couple of sets S and T completely described by the support functions.

These last remarks seem an important reason for research work on conditions easy to be expressed on couples of sets (representing, respectively target sets and reachable regions) in order that the first one be regular with respect to the second one (see, e.g. Ref. [9]).

A case in which the regularity conditions are trivially satisfied is when the sets under consideration are linear subspaces. Sets of this kind have, furthermore, the following absorption property:

$$S \underline{*} T = \begin{cases} S \text{ if } S \supset T \\ \{\emptyset\} \text{ if } S \not\supset T \end{cases} \qquad \text{III(7)}$$

An interesting consequence of this property applied to our Problems III, 1 and III, 2 when U and W are R^m and R^r, respectively, and the target set X_N is a linear subspace M of R^n. In this case we can define the orthogonal complement N of M in R^n: $R^n = N \oplus M$ and the projection operator π of R^n onto N. We have

$$\pi x \in N, \quad \forall x \in R^n, \quad \pi x = \{\emptyset\}, \quad \forall x \in M$$

The target set X_N is therefore characterized by

$$M_N = \pi X_N = \pi M = \{\emptyset\} \qquad \text{III(8)}$$

Referring to Problem III,1 and using Eq.III(1) we have therefore, for x to be transferrable into M in one step

$$\pi(Ax_{N-1} + Bu_{N-1} + Cw_{N-1}) = M_N = \{\emptyset\}, \qquad \forall w_{N-1} \in W$$

that is

$$\pi A X^1_{N-1} = (M_N \stackrel{*}{-} CW) + BU = M^1_{N-1}$$

and using Eq.III(1)

$$M^1_{N-i} = \pi A^i X^1_{N-i} = (M^1_{N-i+1} \stackrel{*}{-} \pi A^{i-1} CW) - \pi A^{i-1} BU \qquad (i = 1, \ldots, N)$$

which gives an iterative algorithm for the construction of sets X^1_{N-i} of states which may be transferred into X_N in i steps, until

$$M^1_0 = A^N X_0 = (M^1_1 \stackrel{*}{-} A^{N-1}\ CW) - \pi A^{N-1}\ BU$$

which characterizes X^1_0.

Recalling $U = R^m$ and $W = R^r$, and defining

$$P_i = \pi\,(\text{span}\ A^{i-1} B) \qquad Q_i = \pi\,(\text{span}\ A^{i-1} C) \qquad \text{III(9)}$$

we have

$$M^1_{N-i} = \pi A^i X_{N-i} = (M^1_{N-i+1} \stackrel{*}{-} Q_i) + P_i \qquad (i = 1, \ldots, N) \qquad \text{III(10)}$$

that is, using the absorption property of geometrical difference for subspaces

$$M^1_{N-i} = \pi A^i X^1_{N-i} = M_{N-i+1} + P_i \quad \text{if} \quad M^1_{N-i+1} \supset P_i$$

$$M_{N-i} = \{\emptyset\} \quad \text{if} \quad M^1_{N-i+1} \not\supset P_i \qquad \text{III(11)}$$

From III(11) we see that M_{i-1} is not empty iff

$$Q_1 = \{O\};\ \text{and} \quad P_1 \supset Q_2, \qquad (P_1 + P_2) \supset Q_3 \ \ldots \sum_{j=1}^{i=1} P_j \supset Q_i \qquad \text{III(12)}$$

Proceeding in an analogous way for Problem III,2 in this case, we have again III(8) and

$$M^2_{N-i} = \pi A^i X^2_{N-i} = (M^2_{N-i+1} + P_i) \stackrel{*}{-} Q_i \qquad \text{III(10')}$$

until

$$M_0^2 = \pi A^N X_0^2 = (M_1^2 + P_N) \stackrel{*}{-} Q_N$$

Using again the absorption property we have

$$M_{N-i}^2 = M_{N-i+1}^2 + P_i \quad \text{if} \quad M_{N-i+1}^2 + P_i \supset Q_i$$

$$M_{N-1} = \{\emptyset\} \quad \text{if} \quad M_{N-i+1}^2 + P_i \not\supset Q_i \qquad \text{III(11')}$$

From III(11') we see that M_{N-i}^2 is not empty iff

$$P_1 \supset Q_1 \qquad P_1 + P_2 \supset Q_2 \; \ldots\ldots \sum_{j=1}^{i} P_j \supset Q_i \qquad \text{III(12')}$$

As is intuitively comprehensible, condition III(12) for the existence of some x_0 which can be brought into X_N in i steps (for example, in i = N steps) is stricter than condition III(12'), which corresponds to an information structure more favourable for the controller. Nevertheless, it is very interesting to notice that when III(12) is satisfied, then $M_0^1 = \sum_{j=1}^{N} P_j = M_0^2$.

We may summarize the preceding results in the following

Theorem III, 1

The set X_0^1 solution of Problem III, 1 for $U = R^m$, $W = R^r$ and X_N given by a subspace M of R^n is not empty iff Eqs III(12) are satisfied and is characterized by

$$A^N X_0^1 = M_0^1$$

where M_0^1 is given by the recursive algorithm III(11) starting from $M_N^1 = 0$. Analogously, the set X_0^2 is not empty iff Eqs III(12') are satisfied, and are characterized by

$$\pi A^N X_0^2 = M_0^2$$

where M_0^2 is given by the recursive algorithm III(11') starting from $M_N^2 = \{0\}$. Furthermore, if X_0^1 is not empty, then $X_0^1 = X_0^2$, with

$$M_0^1 = M_0^2 = \sum_{j=1}^{N} P_j$$

For the particular system

$$x_{k+1} = Ax_k + Bu_k + w_k \qquad (k = 0, 1, \ldots) \qquad \text{III(13)}$$

we have the following

Corollary

Considering Problem III, 2, if X_0^2 is not empty for some N, then
(a) any $x_0 \in X_0^2$ can be brought into M in an arbitrary number of steps,
(b) $X_0^2 = R^n$.

Proof

Since III(12') is satisfied with $Q_1 = \pi I = L$, we have $P_1 \supset Q_1 = L$; on the other hand $\sum_{j=1}^{i} P_j \supset \sum_{j=1}^{i-1} P_j \dots \supset P_1$; therefore, since all P_i are in L, $\sum_{j=1}^{i} P_j = L$ for any i, that is $M_0^2 = M_i^2 = \dots M_N^2 = L$. Recalling $M_i^2 = \pi A^i X_0^2$, part (a) is proved. Part (b) follows from the characterization of X_0^2:

$$X_0^2 = \{x : \pi A^N x \in M_0\}$$

from the definition of π and from $M_0 = L$.

We may observe that in our case, since C = I, $\sum_{j=1}^{i-1} P_j \supset \dots \supset P_1 \supset Q_1 = \pi R^n$, that is π span $[B, AB, \dots\dots, A^{N-1} B] \supset \pi R^n$. This means that the system was controllable (in the classical sense), at least in the subspace L.

REFERENCES

[1] ANTOSIEWICZ, H., Linear Control Systems, Arch. Rat. Mech. Anal. 12 (1963) 313-24.
[2] DUNFORD, N., SCHWARZ, J.T., Linear Operators, Part I, Interscience (1964).
[3] MARZOLLO, A., An algorithm for the determination of ζ-controllability conditions in the presence of noise with unknown statistics, Automation and Remote Control (February 1972).
[4] WITSENHAUSEN, H., "A MinMax Control Problem for Sampled Linear Systems", IEEE AC (february 1968).
[5] WITSENHAUSEN, H., "Sets of possible states given perturbed observations", IEEE AC (October 1968).
[6] GLOVER, J.D., SCHWEPPE, F.C., "Control of Linear Dynamic Systems with Set-constrained Disturbances", IEEE AC, october 1971.
[7] PONTRYAGIN, L., Linear Differential Games, I, II, Soviet Math. Doklady, 8 3 (1967); 4 (1967).
[8] BORGEST, W., VARAIYA, P., "Target Function Approach to Linear Pursuit Problems", IEEE AC (October 1971).
[9] MARZOLLO, A., PASCOLETTI, A., Computational Procedures and Properties of the Geometrical Difference of Sets", J. Math. Anal. Appl. (to appear).

EXISTENCE THEORY IN OPTIMAL CONTROL

C. OLECH
Institute of Mathematics,
Polish Academy of Sciences,
Warsaw, Poland

Abstract

EXISTENCE THEORY IN OPTIMAL CONTROL.

This paper treats the existence problem in two main cases. One case is that of linear systems when existence is based on closedness or compactness of the reachable set and the other, non-linear case refers to a situation where for the existence of optimal solutions closedness of the set of admissible solutions is needed. Some results from convex analysis are included in the paper.

INTRODUCTION

Perron paradox. A necessary condition for N to be the largest positive integer is that N = 1. Indeed if $N \neq 1$ then $N^2 > N$. So N is not the largest integer, contrary to the definition; thus, N = 1.

We have proved the theorem: "If N is the largest integer, then N = 1."

There is nothing wrong with this statement. It is correct, except that the assumption is never satisfied. From such an assumption everything follows. This example, though it looks like a joke, has an important implication, i.e. necessary conditions in optimization may be useless if we do not know that the solution we are talking about exists, since, if it does not exist, one may derive a wrong conclusion from correct necessary condition.

A simple example of such a situation is the following functional:

$$\int_0^1 (1+\dot{x}^2)^{1/4}\, dt \tag{0.1}$$

if we want to minimize it over all functions of class C^1 on [0, 1] and satisfying the boundary condition x(0) = 0, x(1) = 1. This is a simple problem of calculus of variations. It is well known that the solution of the problem, if it does exist, has to satisfy the Euler equation which in this case is simply

$$\dot{x}(1+\dot{x}^2)^{-3/4} = \text{constant}$$

Therefore, any solution of the Euler equation of class C^1 has to be linear ($\dot{x}(t)$ = constant) and the one which satisfies the boundary condition is x(t) = t. The value of the functional (0.1) for x(t) = t is $2^{1/4}$ and it is easy to see that it is not minimal. The infimum in this case is equal to 1 and the optimal solution does not exist. Without checking the existence of the minimum we seek, we may be led to a wrong conclusion, from the necessary condition.

Consider another example. In the set of all functions x(t) of class C^1 on [-1, 1] such that

$$x(-1) = x(1) = 0 \quad \text{and} \quad |\dot{x}(t)| \leq 1$$

we wish to minimize the functional

$$I(x) = \int_{-1}^{+1} t\dot{x}(t)dt$$

The solution of this problem as stated does not exist. Indeed, it is clear that the integrand is estimated from below by $-|t|$. Thus, $I(x) \geq -1$ and equal to -1 iff $x(t) = 1$ for $t \leq 0$ and -1 for $t \geq 0$ while for any $x(t)$ of class C^1 the set $\{t \; |\dot{x}(t)| < 1\}$ has positive measure and, therefore, $I(x) > -1$. The example shows, and this is rather typical, that, to ensure the existence of the minimum, we sometimes have to enlarge or complete the set on which we wish to minimize the functional in question. In fact, the above problem has no solution in the class of C^1, but it has a solution in the class of absolutely continuous functions.

Hilbert said: "Every problem of the calculus of variation has a solution provided the word 'solution' is suitably understood". The meaning of this statement is as follows: If we look for a minimum of a functional and the infimum is finite, then we always have a sequence approaching the infimum – the so-called minimizing sequence. The question of the existence of a minimum reduces to the problem of when we can produce the optimal solution out of the minimizing sequence. Often this is possible if we are able to prove that the minimizing sequence converges. But, even if this is possible, then not always in such a way that the limit belongs to the set on which we originally wished to minimize the functional. The last example is such a case. Indeed, one can prove that if $x_n(t)$ satisfies the boundary condition, is continuously differentiable and such that $I(x_n) \to -1$ as $n \to \infty$, then $x_n(t)$ converges uniformly on $[-1, 1]$ to $x_0(t) = t + 1$ for $t \leq 0$ and $-t + 1$ for $t \geq 0$ while

$$\int_{-1}^{+1} |\dot{x}_n(t) - \dot{x}_0(t)|dt$$

converges to zero but the $\dot{x}_n(t)$ do not converge uniformly to $\dot{x}_n(t)$. Hence, x_n is not convergent in C^1 space but is convergent in the space AC of absolutely continuous functions. It is, therefore, natural to adjust the notion of "solution" and to treat this problem in the space of absolutely continuous functions rather than in C^1. This situation is typical and, for this reason, in control problems usually the control function is assumed to be measurable. Even in the case where the optimal solution is more regular, it is convenient to introduce this enlargement of the space in order to prove that the optimal solution exists. For the similar reason of completion, the Sobolev spaces were invented and became so successful in the theory of partial differential equations (weak solutions).

In this paper, we discuss the existence of optimal solutions first to linear problems and then to a general non-linear problem. In the first case, compactness and convexity of the so-called reachable set is instrumental, while in the second case the closedness of the set of admissible solutions is important. In both cases, convex analysis is used extensively.

In Section 1, an existence theorem for time-optimal linear problems with constant coefficients is stated. Section 2 contains a detailed discussion of the integral of set-valued functions. Also, several propositions concerning convex sets are stated and proved in this section. Section 3, besides a proof of the theorem stated in Section 1, also contains some other applications of the integration of set-valued functions. Among these, we discuss the so-called "bang-bang" principle.

The existence theorem given in Section 4 is a typical example of an extension of the direct method well-known from the calculus of variations to optimal control problems. The direct method originated by Hilbert was developed by Tonelli, McShane and Nagumo for problems in calculus of variations. Roughly speaking, it consists in establishing that the minimizing sequence converges (compactness), that the limit is in the set in which we minimize the functional in question (closedness or completeness) and that the value of the functional at the limit is not greater than the infimum (lower semicontinuity). For optimal control problems, it has been extended in the last decade, mainly by Cesari. Theorem 4.1 stated and proved in Section 4 is an example of possible existence theorems along these lines.

The important difference (and perhaps the only one, too) between classical results from calculus of variations and those concerned with the existence of a solution of optimal control problems lies in regularity assumption. To cover problems of interest, one is led to consider problems with as weak regularity assumption as possible. For this purpose, also a theory of measurable set-valued mappings, selector theorems, etc. is used in an essential way. Mathematical control theory, among others, gave an impact to the recent development of this theory. These questions will also be briefly discussed in Sections 4 and 5.

In Section 5, we discuss the necessity of the convexity assumption in the existence theorem given in Section 4. Actually, convexity is essential for lower semicontinuity or lower closure, and in Section 5 we give a result (Theorem 5.1), which shows that, in certain special cases, convexity is necessary and sufficient for weak lower semicontinuity of an integral functional. Also, as an example, another result on lower semicontinuity is stated there, but without detailed proof.

It would be impossible to cover the whole subject in such a short paper, which is, therefore, far from complete. The choice of material presented was highly influenced by the author's own contribution to the subject.

A selected list of references is included, and, at the end of the paper, we also give some comments concerning the literature.

The author wishes to acknowledge the help of Miss Barbara Kaśkosz in preparing this paper.

1. TIME-OPTIMAL LINEAR CONTROL PROBLEM

Consider the linear control system

$$\dot{x} = Ax + Bu \tag{1.1}$$

where $x \in R^n$, $u \in R^m$, A is an $n \times n$ matrix and B is an $n \times m$ matrix. Let $U \subset R^m$ be a given fixed set. Let Ω be a set of functions u taking values from U, called admissible control functions. In this section, we take as admissible control functions which are piecewise constant.

A solution of Eq. (1.1) is of the form

$$x(t) = X(t, t_0)\left(x_0 + \int_{t_0}^{t} X^{-1}(s, t_0)\, Bu(s)\, ds\right) \tag{1.2}$$

where $X(t, t_0)$ is the so-called fundamental matrix of the homogeneous linear differential equation corresponding to Eq. (1.1). That is, we have

$$\frac{d}{dt} X(t, t_0) = AX(t, t_0) \quad \text{and} \quad X(t, t_0) = E \tag{1.3}$$

where E is the identity $n \times n$ matrix. In the case of A constant, $X(t, t_0) = \exp(A(t-t_0))$.

A solution of Eq. (1.1) is unique if the initial value x_0 at time $t = t_0$ and the control function $u = u(t)$ are fixed. Thus, we shall use the notation $x(t; t_0, x_0, u)$ for solution (1.2) of Eq. (1.1) to indicate this relation.

Time-optimal control problem. Given an initial state x_0 at time t_0, we wish to transfer it by means of an admissible control function to a target point x_1, and we want to do this in the shortest time possible, i.e. we seek an admissible u_* and a time t_* such that $x(t_*; t_0, x_0, u_*) = x_1$ while $x(t; t_0, x_0, u) \neq x_1$ for each $u \in \Omega$ if $t < t_*$. We shall call t_* and u_* optimal time and optimal control, respectively.

For the above optimal problem, the following existence theorem holds (due to LaSalle):

Theorem 1.1. If A, B in Eq. (1.1) are constant, U is a compact polyhedron, Ω the set of piecewise constant control functions taking values in U and $x_1 = x(t; t_0, x_0, u)$ for some $u \in \Omega$ and $t > t_0$, then the time-optimal solution exists.

In Section 3, we shall present a result which would contain this result as a special case. Now, we wish to call the reader's attention to the fact that Theorem 1.1 is, in general, not true if A and B are not constant and U is not the polyhedron. In other words, the piecewise constant control function is, in general, too narrow a class for the optimum to be attained.

For the existence of a time-optimal solution, the properties of the reachable set are decisive. By reachable set of a control system we mean the set of all states which can be reached in time t by means of admissible control functions from a fixed initial condition. In the case of Eq. (1.1), it is given by

$$\mathscr{R}(t) = \{x(t; t_0, x_0, u) \mid u \in \Omega\} \tag{1.4}$$

The properties of $\mathscr{R}(t)$ which make the above theorem true are

$$\mathscr{R}(t) \text{ is convex and compact for each } t \tag{1.5}$$

$$\mathscr{R}(t) \text{ is Hausdorff continuous in } t \tag{1.6}$$

The latter means that $h(\mathscr{R}(t), \mathscr{R}(s))$ tends to 0 if $|s-t|$ tends to 0 where $h(P,S) = \max(\sup_{p \in P} d(p,S), \sup_{s \in S} d(s,P))$ is the Hausdorff distance between two sets.

For the existence of optimal solution only compactness in (1.5) is needed. Convexity is decisive in deriving necessary and sufficient conditions for optimality. In the next section a proof of (1.5) and (1.6) will be given in a more general setting.

2. REACHABLE SET AND INTEGRATION OF SET-VALUED FUNCTIONS

It follows from Eqs (1.2) and (1.4) that

$$\mathscr{R}(t) = X(t;t_0)(x_0 + S(t))$$

where

$$S(t) = \left\{ b \mid b = \int_{t_0}^{t} X^{-1}(t, t_0)\, Bu(t)\, dt, \quad u(t) \in U \right\}$$

Putting

$$P(t) = \{ p \mid p = X^{-1}(t, t_0)\, Bu, \; u \in U \}$$

we have

$$S(t) = \left\{ \int_{t_0}^{t} v(t)\, dt \mid v(t) \in P(t) \right\} \tag{2.1}$$

The above set, if v is an arbitrary, but integrable function, is called the integral of set-valued function P(t). It is clear that properties of $\mathscr{R}(t)$ like closedness, convexity and continuity hold iff the same properties hold for S(t) given by (2.1). Thus, in what follows we shall discuss the basic properties of the integral of the set-valued mapping P. For simplicity, we shall consider P to be defined on [0, 1] with values P(t) being subsets of R^n. A point-valued function v from [0, 1] into R^n is a selection of P (or a selector) if $v(t) \in P(t)$. We shall require the latter to hold almost everywhere (a. e.) since we shall deal with measurable or integrable selections of P. By K_P we denote the set of all integrable selections of P. Thus the integral of P is the set

$$\begin{aligned} I(P) = \int_0^1 P(\tau)\, d\tau &= \left\{ \int_0^1 v(t)\, dt \mid v(t) \in P(t) \text{ a.e. in } [0,1] \right\} \\ &= \left\{ \int_0^1 v(t)\, dt \mid v \in K_P \right\} \end{aligned} \tag{2.2}$$

The aim of this section is to prove that I(P) is convex and to prove some other properties of I(P). The main results are stated at the end of the section.

Before, we recall some elementary facts concerning the extremal structure of convex sets.

Let S be a convex set. A convex subset $E \subset S$ is called an extremal face of S if for each s, $p \in S$ and any $0 < \lambda < 1$, condition $\lambda s+(1-\lambda) p \in E$ implies that both s and p belong to E. If an extremal face reduces to a single point $\{e\}$ then e is called an extreme point of S. The set of all extreme points of S is called the profile of S and we denote if by ext(S).

Our nearest goal is to give a characterization of an extremal face throu[gh] some order relations. By C we denote a convex cone in R^n and we shall writ[e] $x \leq_C y$ iff $y - x \in C$. Because of the convexity of C, this relation is transitive. Below we shall only deal with convex cones which satisfy the relation

$$C \cup (-C) = R^n \tag{2.3}$$

This implies that any two points x, y of R^n are comparable, that is either $x \leq_C y$ or $y \leq_C x$. Both inequalities hold iff $x - y \in C \cap (-C) = M_C$. Clearly, M is a subspace of R^n. We shall write $y = \max_C S$ iff $y \in S$ and $x \leq_C y$ for each $x \in S$. If S is closed and bounded then $\max_C S$ exists and it is unique if $M_C = \{0\}$.

Two following propositions about cones satisfying (2.3) will be useful:

(α) For each convex cone C satisfying (2.3) there is an orthogonal sequence $a_1, \ldots, a_k$ such that $M_C = \{x \mid \langle x, a_i\rangle = 0, \ i = 1, \ldots, k\}$ and $x \in C$ iff the first non-zero member of the sequence $\{\langle x, a_i\rangle\}$ is positive.

Indeed, suppose $z = (x+y)/2 \in M$ where $x, y \in C$. But $-x = -2z+y$ and $-2z \in C$, thus $-x \in C$. Similarly, $-y \in C$. Therefore both x and y belong to M. This shows that $C\setminus M$ is convex (M is an extremal face of C). Thus $-C$ and $C\setminus M$ are convex and disjoined. We can separate them. Therefore, there is $a_1 \neq 0$ such that $\langle x, a_1\rangle \geq 0$ for each $x \in C$ and equal zero if $x \in M$. Putting $C_1 = \{x \in C \mid \langle x, a_1\rangle = 0\}$, we see that $C_1 \cup (-C_1) = \{x \mid \langle x, a_1\rangle = 0\}$ and either $C_1 = M$ then $k = 1$ and we are finished or there is $a_2 \perp a_1$ and $a_2 \neq 0$ such that $\langle x, a_2\rangle \geq 0$ for each $x \in C_1$ and $\langle x, a_2\rangle = 0$ if $x \in M$. An easy induction argument proves (α).

The dimension of a convex set S is the dimension of the smallest linear manifold containing S (the linear manifold spanned by S). We say that a cone C satisfying (2.3) is spanned by S if $S \subset C$ and if for each C_1 satisfying (2.3) and containing S the inclusion $C_1 \subset C$ implies that $C_1 = C$. In general, there are more than one cone spanned by a set S. We shall now prove the following:

(β) For any set S there exists a cone C satisfying (2.3) and spanned by S.

Indeed, let C_α, $\alpha \in A$ be a family of cones satisfying (2.3) and linearly ordered by inclusion; $S \subset C_\alpha$ for each α. Put $C = \bigcap_{\alpha \in A} C_\alpha$. Manifestly $S \subset C$. If there were $x \in R^n$, $\alpha, \beta \in A$ such that $x \notin C_\alpha$, $x \notin C_\beta$ then $C_\alpha \subset C_\beta$ or $C_\beta \subset C_\alpha$ and $x \notin C_\alpha \cup (-C_\alpha)$ or $x \notin C_\beta \cup (-C_\beta)$. Thus $(-C) \cup C = R^n$. The Kuratowski-Zorn lemma completes the argument.

Notice that if C is spanned by convex set S then M_C is the linear span of $S_1 = \{x \in S \mid$ there is $\alpha > 0$ that $-\alpha x \in S\}$. S_1 is the extremal face of S containing zero in the relative interior.

One more notation: If $a \in R^n$ then we put

$$S_a = \{x \in S \mid \langle x, a\rangle = \max_{s \in S} \langle s, a\rangle\}$$

and, inductively,

$$S_{a_1,\ldots,a_k} = (S_{a_1,\ldots,a_{k-1}})_{a_k}$$

We are now able to state the announced characterization of extremal faces to a convex set.

(γ) The following conditions are equivalent for a subset E of a convex set S:

(i) E is an extremal face of S
(ii) There is a convex cone C satisfying (2.3) such that $E = \{y \mid y = \max_C S\}$
(iii) There is an orthogonal sequence $a_1, \ldots, a_k$ such that $E = S_{a_1,\ldots,a_k}$, where $n-k \leq \dim E$.

Proof. If (i) holds then take as C a cone spanned by $x_0 - S$ (see item (β)), where x_0 is any point from relative interior of E. By definition of C, $x_0 = \max_C S$ and since $x_0 \in$ relative int E, $E \in x_0 + M_C$ thus for each $y \in E$, $y = \max_C S$. On the other hand if $y = \max_C S$ then $y \in x_0 + M_C$ and the interior of the segment y, x_0 has non empty intersection with E. But E is an extremal face, thus $y \in E$. Therefore (i) implies (ii). Assume now (ii). Let $a_1, \ldots a_k$ be the orthogonal sequence corresponding by (α) to the cone C. Notice that from (ii) it follows that $E = S \cap (x_0 + M_C)$, where x_0 is any point from E. This together with (α) applied to $x_0 - S$ shows that $E = S_{a_1,\ldots,a_k}$. Manifestly $\dim E \leq \dim M_C = n-k$. Thus (iii) follows from (ii). Assume now (iii) and let $z = \lambda x + (1-\lambda)y \in E = S_{a_1,\ldots,a_k}$ while $x, y \in S_{a_1,\ldots,a_i}$, $i < k$, and $0 < \lambda < 1$. Then $\langle z, a_j \rangle = \langle x, a_j \rangle = \langle y, a_j \rangle$ for $j = 1, \ldots, i$ and $\langle z, a_{i+1} \rangle \geq \max \langle x, a_{i+1} \rangle, \langle y, a_{i+1} \rangle)$. But since $0 < \lambda < 1$, and $\langle z, a_{i+1} \rangle = \lambda \langle x, a_{i+1} \rangle + (1-\lambda) \langle y, a_{i+1} \rangle$ the latter inequality implies that $\langle z, a_{i+1} \rangle = \langle x, a_{i+1} \rangle = \langle y, a_{i+1} \rangle$ which means that $x, y \in S_{a_1,\ldots,a_{i+1}}$. Thus, the induction argument implies that $x, y \in S_{a_1,\ldots,a_k} = E$, hence we proved that E is an extremal face if S. Therefore, (iii) $\Rightarrow$ (i) which completes the proof of (γ).

A special case of interest of (γ) is when $E = \{e\}$ and e is an extreme point of S. Then, C in (ii) can be chosen so that $C \cap (-C) = \{0\}$. In this case, the order $\leq_C$ is called lexicographical. We have the following corollary of (γ):

(γ') The following conditions are equivalent for a point e of a convex set S.

(i) e is an extreme point of S.
(ii) There is a lexicographical order $\leq$ in R^n such that $x \leq e$ for each $x \in S$.
(iii) There is an orthogonal basis $a_1, \ldots, a_n$ in R^n such that $\{e\} = S_{a_1,\ldots,a_n}$.

Example: Consider the set $\Delta = \left\{x \in R^n \mid \sum_{i=1}^{n} x_i = 1,\ x_i \geq 0\right\}$

i.e. an (n-1)-dimensional simplex in R^n. It is convex, compact; $e \in \Delta$ is an extreme point of Δ if at least one co-ordinate of e is equal to one (then the remaining have to be equal zero); extremal faces are subsets of all points in Δ with certain fixed co-ordinates equal to zero. Notice that each $x \in \Delta$ can be written as a sum

$$x = \sum_{i=1}^{n} \lambda_i e_i,\ \lambda_i \geq 0,\quad \sum_{i=1}^{n} \lambda_i = 1$$

where $\{e_1, \ldots, e_n\}$ is the profile of Δ.

Let P be a subset of R^n. The smallest convex set containing P is called the convex hull of P and it is denoted by coP. Correspondingly, a closed convex hull of P, denoted by clco P, is the smallest closed convex set containing P. Since the intersection of convex and/or closed sets is convex and/or closed thus both co P and clco P are well defined and clco P is the closure of co P.

The basic result concerning convex hulls is the following:

(δ) Carathéodory theorem: If $P \subset R^n$ then

$$\text{co } P = \bigcup_{\{p_0, \dots, p_n\} \subset P} \left\{ \sum_{i=0}^{n} \lambda_i p_i \mid \lambda_i \geqq 0, \ \sum_{i=0}^{n} \lambda_i = 1 \right\}$$

The proof of this theorem can be found in many textbooks. However, for convenience of the reader we shall also prove it here. Denoting the right-hand side of the last relation by D, we clearly see that $D \subset \text{co } P$, thus only convexity of D should be established to prove (δ). Let x, y be two points of D, that is

$$x = \sum_{i=0}^{n} \lambda_i p_i, \ \ y = \sum_{i=0}^{n} \mu_i q_i, \ \ \lambda_i \geqq 0, \ \ \mu_i \geqq 0, \ \ \sum_{i=0}^{n} \lambda_i = \sum_{i=0}^{n} \mu_i = 1, \ p_i \in P, \ q_i \in P$$

Let $z = \alpha x + (1-\alpha) y$, $0 < \alpha < 1$. Clearly, the set

$$B = \left\{ \sum_{i=0}^{n} \gamma_i p_i + \sum_{i=n+1}^{2n+1} \gamma_i q_{i-n-1} \mid \sum_{i=0}^{2n+1} \gamma_i = 1, \ \gamma_i \geqq 0 \right\}$$

is convex and contains z. Let

$$A = \left\{ (\gamma_i) \mid \sum_{i=0}^{n} \gamma_i p_i + \sum_{i=n+1}^{2n+1} \gamma_i q_{i-n-1} = z, \ \sum_{i=0}^{2n+1} \gamma_i = 1, \ \gamma_i \geqq 0 \right\}$$

It is clear that A is a convex, closed subset of the (2n+1)-dimensional simplex Δ_{2n+1} and its dimension is, at least, n+1. $A = M \cap \Delta_{2n+1}$ where M is linear manifold spanned by A. Let e be an extreme point of A; $e = \{\gamma_i(e)\}_{0 \leq i \leq 2n+1}$. We claim that, at least, n+1 out of 2n+2 co-ordinates of e are zeros. In fact, let F be the smallest extremal face of Δ_{2n+1} containing e. Then $F \cap M = \{e\}$. Therefore, the dimension of F is at most n+1, thus $\gamma_i(e) = 0$ for, at least, (n+1) "i's". That implies that

$$z = \sum_{i=0}^{n} \gamma_i(e)\, p_i + \sum_{i=n+1}^{2n+1} \gamma_i(e)\, q_{i-n-1}$$

is, in fact, a convex combination of (n+1) points from P, thus $z \in D$, which completes the proof of (δ).

The next results will give an answer to the question of how small a subset of a convex closed set can be in order that the original set can be reconstructed by taking a convex hull. A reasonable result of this sort can only be obtained for closed convex subsets.

(ϵ) Assume that $S \subset R^n$ is closed convex and does not contain a line and let P be a subset of S composed of all extreme points and all extremal rays of S. Then S = co P. On the other hand, if S = co P then P contains the profile of S and for each extremal ray E of S the intersection $E \cap P$ is unbounded.

Proof by induction with respect to the dimension of S. If dim S = 1 then (ϵ) clearly holds. Assume that (ϵ) holds if the dimension of the set in question is smaller than n and assume dim S = n. Each point x of the boundary of S is in co P because it belongs to an extremal face E of S and dim E < n. But both the profile and extremal rays of E are contained in P, thus we may apply (ϵ). Since S does not contain a line then there is a hyperplane H with the property that $(x+H) \cap S$ is bounded for each x. Thus if $x \in \operatorname{int} S$ then it can be represented as a convex combination of points from the boundary of S thus also $x \in \operatorname{co} P$. The second part follows from an easy observation that $S \setminus E$ is convex if E is an extremal face. Thus if E is an extremal ray or an extreme point and E is not equal to $\operatorname{co}(E \cap P)$ where $P \subset S$ then co P is not equal S.

Thus if co P = S and E = {e}, $e \in \operatorname{ext}(S)$, then $e \in P$ and if E is an extremal ray then $\operatorname{co}(E \cap P) = E$ which is the case only if $E \cap P$ is unbounded and contains the extreme point of E.

As an immediate corollary of (ϵ) we have the following result.

(ϵ') If S is convex and either compact or closed and without extremal rays, then $S = \operatorname{co}(\operatorname{ext} S)$. If S = co P then $\operatorname{ext} S \subset P$.

The next result will concern the closed convex hull of arbitrary sets (not necessarily bounded or closed).

(ζ) If $P \subset R^n$ and S = clco P, then the profile ext S of S is contained in cl P.

From the inclusion $P \subset \operatorname{cl} P \subset \operatorname{clco} P = S$ and (δ) we obtain the equality clco P = clco(cl P). Thus for bounded P, (ζ) follows from (ϵ'). Notice also that if P is compact then S = co P. Suppose now that P is closed and that $e \in \operatorname{ext}(\operatorname{clco} P)$. By ($\delta$) we have

$$e = \lim_k \sum_{j=0}^{n} \lambda_j^k x_j^k, \text{ where } \lambda_j^k \geqq 0, \ \sum_{j=0}^{n} \lambda_j^k = 1$$

for each k and $x_j^k \in P$. If e were not in P then there would exist $\epsilon > 0$ such that $|e - x_j^k| \geqq \epsilon > 0$ for each j and k. In this case we could represent e as a limit of

$$\sum_{j=0}^{n} \mu_j^k z_j^k, \text{ where } \mu_j^k \geqq 0, \ \sum_{j=0}^{n} \mu_j^k = 1, \ z_j^k \in \operatorname{co} P$$

and $|e - z_j^k| = \epsilon$.

Without any loss of generality, we may assume that both $\mu_j^k \to \mu_j$ and $z_j^k \to z_j \in \operatorname{clco} P$. Hence $e = \sum_{j=0}^{n} \mu_j z_j$ and $|e - z_j| = \epsilon > 0$, which contradicts the assumption that e is an extreme point of S. Thus $e \in P$ and (ζ) is proved.

We begin the study of the properties of the integrals (2.2) with the following:

Lemma 1. Let $S = \text{clco } I(P)$ and let $e \in S$ be an extreme point of S. Then to each $\epsilon > 0$ there is $\delta = \delta(\epsilon) > 0$ such that if $|I(u)-e| < \delta$ and $|I(v)-e| < \delta$ for any pair $u, v \in K_p$ then $\|u-v\| = \int_0^1 |v(t)-u(t)|dt \leq \epsilon$.

Proof. Let $\epsilon > 0$ be arbitrary. Put $\eta = \epsilon/4\sqrt{n}$ and let $B(e, \eta)$ be the closed ball centred at e and of radius η. There is an open halfspace H such that $e \in H \cap S \subset B(e, \eta)$. Indeed, consider the set Q beeing the intersection of S with the boundary of $B(e, \eta)$. Q is compact hence also $\text{co}\, Q \subset B(e, \eta) \cap S$ is compact. Manifestly e does not belong to co Q. Thus we can separate e from co Q; that is there is an open halfspace H, such that $H \cap (S \setminus B(e, \eta)) = \emptyset$ because Q is equal $S \cap \partial B(e, \eta)$. Now take δ small enough that

$$S \cap B(e, \delta) \subset H \cap S \subset B(e, \eta) \tag{2.4}$$

and assume $u, v \in K_p$ are such that $|I(u)-e| < \delta$ and $|I(v)-e| < \delta$. Let $A \subset [0, 1$ be an arbitrary measurable set and put $w_1 = u + \chi_A(v-u)$ and $w_2 = v + \chi_A(u-v)$, where χ_A is the characteristic function of the set A. Of course both w_1 and w_2 belong to K_p and we have

$$I(w_1) = I(u) + \int_A (v(\tau)-u(\tau))\, d\tau \text{ and } I(w_2) = I(v) - \int_A (v(\tau)-u(\tau))\, d\tau$$

This implies that, at least, one of those two points belongs to $H \cap S$ and therefore by (2.4)

$$\left|\int_A (v(\tau)-u(\tau))\, d\tau\right| \leq 2\eta.$$

The latter holds for each $A \subset [0, 1]$, therefore we have

$$\int_0^1 |v_i(\tau)-u_i(\tau)|\, d\tau \leq 4\eta$$

where v_i is the i-th co-ordinate of v and consequently,

$$\|u-v\| = \int_0^1 |v(\tau)-u(\tau)|\, d\tau \leq 4\sqrt{n}\,\eta = \epsilon$$

which completes the proof.

As a consequence of Lemma 1, we shall state two corollaries.

Corollary 1. Assume P(t) is closed for each t. Let C be a convex cone such that $C \cup (-C) = R^n$ and $F = \{y \,|\, y = \max_C x \text{ if } x \text{ clco } I(P)\}$ be a compact extremal face of clco I(P). Then

(i) The set $K_P^C = \{v \in K_P \mid I(v) \in F\}$ is not empty.

(ii) $F = \text{co}\, I(K_P^C) = \text{co}\left\{\int_0^1 v(\tau)d\tau \mid v \in K_P^C\right\}$

(iii) $v \in K_P^C$ iff $v \in K_P$ and $u(t) \leq_C v(t)$ a.e. in $[0,1]$ for each $u \in K_P$

Proof. Since F is compact there is an extreme point e of clco I(P) belonging to F. Thus by (ζ) there is $\{u_i\} \subset K_P$ such that $I(u_i) \to e$ and by Lemma 1 $\{u_i\}$ is convergent in L_1 norm to a function v, hence without any loss of generality we may assume that $u_i(t) \to v(t)$ pointwise. Thus $v \in K_P$ and of course $I(v) = e \in F$ hence (i) holds and $I(K_P^C)$ contains the profile of F, therefore (ii) holds as well. To prove (iii) let us fixe $v \in K_P^C$ and $u \in K_P$ be arbitrary. Put $w(t) = \max_C(u(t), v(t))$. We have the inequality $v(t) \leq_C w(t)$ and therefore $I(v) \leq_C I(w)$. Therefore $I(w) \in F$. That means that $w(t)-v(t) \in C$ and

$$\int_0^1 (w(t)-v(t))\,dt \in C \cap (-C)$$

which is the case only if $w(t)-v(t) \in C \cap (-C)$ a.e. in $[0,1]$. The latter set is a subspace thus $v(t)-w(t) \in C \cap (-C)$ also. Hence $u(t) \leq_C v(t)$ a.e. in $[0,1]$. On the other hand, if the latter inequality holds for each $u \in K_P$ and $v \in K_P$ then $I(v) = \max_{u \in K_P} {}_C I(u)$. By (i) there is $w \in K_P$ such that $I(w) \in F$. $I(w) \leq_C I(v)$ as well as $I(v) \leq_C I(w)$. Hence $I(w)-I(v) \in C \cap (-C)$ which implies that $I(v) \in F$ and completes the proof.

The following is a specification of Corollary 1 to the case $F = \{e\}$.

Corollary 2. Under the assumption of Corollary 1 for each extreme point e of clco I(P) and each convex cone C such that $C \cap (-C) = \{0\}$, $C \cup (-C) = R^n$ and clco $I(P) \subset e-C$ there exists a unique $v \in K_P$ such that $i(v) = e$ and for each $u \in K_P$ $u(t) \leq_C v(t)$ a.e. in $[0,1]$.

Lemma 1 as well as both corollaries hold if we integrate on $[0,t]$ instead of $[0,1]$.

Put

$$I_t(P) = \int_0^t P(\tau)d\tau = \left\{\int_0^t v(\tau)d\tau \mid v \in K_P\right\}$$

and

$$F_t = \text{co}\left\{\int_0^t v(\tau)d\tau \mid v \in K_P^C\right\}$$

We shall prove now the following Lemma 2. Let P(t) be closed and F, C and K_P^C be as in Corollary 1. Then

(i) F_t is Hausdorff continuous in t.

(ii) F_t is a compact extremal face of clco $I_t(P)$ and $F_t = \{y \mid y = \max_C \text{clco}\, I_t(P)\}$.

(iii) F_t can be represented as a sum $\varphi(t)+F_t'$, where φ is continuous and $F_t' \subset M_C = C \cap (-C)$ for each t.

Proof. From Corollary 1, (iii) it follows that if u, $v \in K_P^C$ then both $u(t)-v(t) \in C$ and $v(t)-u(t) \in C$. Thus if we fix $u_0 \in K_P^C$ then each $w \in K_P^C$ is of the form $w = u_0+v$ where $v(t) \in M_C = C \cap (-C)$ a.e. in [0, 1]. This proves (iii) of Lemma 2.

Let $F = \{y \,|\, y = \max_C \text{clco}\, I_t(P)\}$. It is clear that if $s_n \in \text{co}\, I_t(P)$ then

$$s_n + \int_t^1 u_0(\tau)d\tau \in \text{co}\, I(P)$$

and if the first sequence tends to a point of F_t then the second tends to a point of F. This shows that F_t is compact because F is compact. Therefore, by Corollary 1,

$$F_t = \text{co}\left\{\int_0^t v(\tau)d\tau \,\middle|\, u(\tau) \leq_C v(\tau),\ 0 \leq \tau \leq t \text{ for each } u \in K_P\right\}$$

It is clear that each v in the set above is a truncation of an element of K_P^C. Thus,

$$\widetilde{F} = \text{co}\left\{\int_0^t v(\tau)d\tau \,\middle|\, v \in K_P^C\right\} = F_t$$

Hence (ii) is also proved. Part (i) follows from the inequality $|v(t)| \leq \lambda(t)$ for each $v \in K_P^C$, where $\lambda \in L_1$. Such a λ exists. Indeed, take any $b \in M_C$ different than zero. Apply Corollary 1 to $\widetilde{C} = (C \setminus M_C) \cup \{a \in M_C \,|\, \langle a, b\rangle \geq 0\}$, $K_P^{\widetilde{C}} \subset K_P^C$ and $K_P^{\widetilde{C}} = \{v \in K_P^C \,|\, \langle v(t), b\rangle = \beta(t) \geq \langle u(t), b\rangle$ for each $u \in K_P^C\}$. Note that β is integrable. This shows that for each $b \neq 0$, $b \in M_C$, $\langle v(t), b\rangle$ is uniformly bounded by an integrable function if $v \in K_P^C$, which together with (iii) proves the existence of a uniform integrable bound for K_P^C and completes the proof of Lemma 2.

Definition: We shall call $v \in K_P$ an extremal element of K_P or extremal selector of P if there is a lexicographical order; that is a cone C such that $C \cap (-C) = \{0\}$ and $C \cup (-C) = R^n$ such that

$$u(t) \leq_C v(t) \qquad \text{a.e. in } t \tag{2.5}$$

$v \in K_P$ is piecewise extremal if there is a partition $t_0 = 0 < t_1 < \ldots < t_k = 1$ of the interval [0, 1] such that v(t) is equal on each of subintervals $[t_i, t_{i+1}]$ to an extremal element of K_P.

Lemma 3. If P(t) closed and F is compact extremal face of clco I(P) then for each $x \in F$ there is a piecewise extremal selector v_x of P such that $x = \int_0^1 v_x(t)dt$. Moreover, the number of subintervals on which v_x is extremal can be made not greater than the dimension of F plus one. In particular, $F \subset I(P)$.

Proof. We shall prove Lemma 3 by induction with respect to the dimension of F. If dim F = 0, that is F = {e}, where e is an extreme point of clco I(P), then by Corollary 2 there is an extremal v such that I(v) = e. Suppose now that Lemma 3 holds for dim F < k and let dim F = k. Let $x_0 \in F$ be arbitrary. Take any extremal selector v_0 in K_P^C (C is the cone corresponding to F) and consider function

$$x(t) = x_0 - \int_t^1 v_0(\tau)d\tau \tag{2.6}$$

By Lemma 2 (iii), $x(t)-\varphi(t) \in M_C$ for each t, therefore, $x(t) \in \varphi(t) + M_C \supset F_t$. But $x(1) \in F$, and by Lemma 2, (i), F_t is continuous, thus there is $0 \leq s \leq 1$ such that $x(t) \in F_t$ if $s \leq t \leq 1$ and if $s > 0$ then for a sequence $0 \leq t_i < s$, $t_i \to s$ $x(t_i) \notin F_{t_i}$. Hence x(s) belongs to the relative boundary of F_s, thus to an extremal face of dimension m < k. Therefore $x(s) = \int_0^s v_1(\tau)d\tau$, where v_1 is piecewise extremal with m+1 pieces at the most. Putting $v(t) = v_1(t)$ if $t \leq s$ and $v_0(t)$ if $t \geq s$ we see that v is piecewise extremal with k+1 pieces at the most and by (2.6)

$$x_0 = \int_0^s v_1(\tau)d\tau + \int_s^1 v_0(\tau)d\tau = \int_0^1 v(\tau)d\tau$$

which was to be proved.
Remark 1. Notice that if x_0 is not an extreme point of F then we have a choice of v_0 in the proof above and, therefore, such x_0 can be reached by, at least, two different piecewise extremal elements.

From Lemma 2 we have also the following

Corollary 3. The integral I(P) of P is convex.

Indeed, if v_1, $v_2 \in K_P$ then $P_1(t) = \{v_1(t), v_2(t)\} \subset P(t)$ is closed and clco $I(P_1)$ is compact thus by Lemma 2 $I(v_i) \in \text{clco}\, I(P_1) \subset I(P_1) \subset I(P)$, i = 1, 2, which proves convexity of I(P).

Remark 2. Notice the very mild assumption for P(t) in obtaining these results. Actually, what was relevant that far was the following property of set $K \subset L_1$:

(P) if u, $v \in K$ and A is measurable then $u\chi_A + v\chi_{[0,1]\setminus A} \in K$

Manifestly, if $K = K_P$ that is K is the set of measurable selections of a set-valued map then property (P) holds. This only property was needed in the proof of Lemma 1. For the remaining lemmas, we needed the closedness of K_P in L_1-strong topology which is the case if P(t) is closed for each t.

We are in a position to state the basic results of this section:

Theorem 2.1. If P is a map from [0, 1] into closed subsets of R^n and $\int_0^1 P(t)dt$ is bounded then

$$S(t) = \int_0^t P(t)dt \qquad 0 < t \leq 1$$

is convex and compact, and it is Hausdorff continuous in t. Moreover if K_* is the class of piecewise extremal selectors of P(t) then

$$S(t) = \left\{ \int_0^t v(t)dt \,\middle|\, v \in K_* \right\} \tag{2.7}$$

Proof. $S = \text{clco} \int_0^t P(t)dt$ is compact thus Lemmas 2 and 3 applied to S give the Theorem.

Theorem 2.2. (the unbounded case) If P is a map from [0, 1] into closed subsets of R^n then $\int_0^1 P(t)dt$ is convex and each compact extremal face of $\text{cl} \int_0^1 P(t)dt$ is contained in $\int_0^1 P(t)dt$. In particular, if $\text{cl} \int_0^1 P(t)dt$ does not contain any extremal ray then $\int_0^1 P(t)dt$ is closed. In the latter case, (2.7) also holds.

Proof. Convexity of $\int_0^1 P(t)dt$ is given by Corollary 3. The remaining parts follow from Lemma 3 and proposition (ϵ').

Remark 3. Theorem 2.1 is an extension of well known Liapunov theorem concerning the range of vector-valued measure, stating that the range of finite and non-atomic vector measure is closed and convex. A special case of this is when the measure is given by $\mu(A) = \int_A f(t)dt$ and the general case essentially can be reduced to this case. Without any loss of generality, we may assume that the measure μ is given on the interval [0, 1]; then we see that

$$\{\mu(A) \mid A \text{ measurable subset of } [0,1]\} = \int_0^1 P(t)dt$$

where P(t) is composed of two points: f(t) and 0. On the other hand, the Liapunov theorem could be used to prove the convexity part of Theorem 2.1 and to some extent also the compactness part. However, the second part of Theorem 2.1 cannot be obtained from the Liapunov theorem. This part of Theorem 1 is the reason for a rather detailed analysis of the integral of set-valued functions presented here and perhaps for a rather long proof of Theorem 2.1.

3. SOME APPLICATIONS; EXISTENCE; BANG-BANG PRINCIPLE

As a first application of the integral of set-valued function we shall give a proof of Theorem 1.1. For the sake of simplicity, we assume that $t_0 = 0$ in this theorem, and, we shall write X(t) for X(t;0). Put

$$P(t) = \{p \mid p = X^{-1}(t)Bu,\ u \in U\} \tag{3.1}$$

$$S(t) = \int_0^t P(t)dt \tag{3.2}$$

$$\tilde{\mathscr{R}}(t) = X(t)\,(x_0 + S(t)) \tag{3.3}$$

It is clear that the reachable set for (1.1) defined by (1.4) is contained in $\tilde{\mathscr{R}}(t)$. By Theorem 2.1, however, we have

$$\tilde{\mathscr{R}}(t) = \left\{ X(t)\left(x_0 + \int_0^t v(t)dt\right) \mid v \in K_P \text{ piecewise extremal} \right\} \tag{3.4}$$

By definition, v is an extremal selector of P(t) if there is a lexicographical order $\leqq_C$ such that (2.5) holds. Take $w(t) = \max_C P(t)$. It is easy to see that $w(t) = \max_C\{X^{-1}(t)Bu_i \mid 1 \leqq i \leqq k\}$ where u_i are vertices of U.

Each of the functions $X^{-1}(t)Bu_i$ is analytic. If we restrict ourselves to only these vertices of U for which $Bu_j \neq Bu_i$ if $i \neq j$, then function u(t) taking values from $\{u_1, \ldots, u_k\}$ and such that $X^{-1}(t)Bu \leqq_C X^{-1}(t)Bu(t)$ for each $t > 0$ and $u \in U$, is uniquely defined and piecewise constant. Indeed, notice that $C \cap (-C) = \{0\}$ and, therefore, by (α), there is an orthogonal sequence $a_1, \ldots, a_n$ such that $x \in C$ iff the first non-zero member of the sequence $\langle a_i, x\rangle$ is positive. Thus for each t, $u(t) = u_{j(t)}$, $1 \leqq j(t) \leqq k$ and $\langle a_1, X^{-1}(t)Bu_{j(t)}\rangle = \max\{\langle a_1, X^{-1}(t)Bu_j\rangle \mid 1 \leqq j \leqq k$. This condition either defines u(t) uniquely everywhere except for a finite number of values of t or the maximum is attained at $u_{j_1}, \ldots, u_{j_s}$ on an interval. The latter, because of the analyticity of $X^{-1}(t)B$, is only in the case where $\langle a_1, X^{-1}(t)Bu_{j_1}\rangle \equiv \langle a_1, X^{-1}(t)Bu_{j_2}\rangle \equiv \ldots \equiv \langle a_1, X^{-1}(t)Bu_{j_s}\rangle$ equal to $\max \langle a_1, X^{-1}(t)Bu_j\rangle$ on an interval $[\alpha, \beta]$, $\alpha < \beta$. If this is the case, then $\langle a_2, X^{-1}(t)Bu_{j(t)}\rangle = \max\limits_{1 \leq k \leq s} \langle a_2, X^{-1}(t)Bu_{j_k}\rangle$. Applying the same argument again, we can prove, by induction, that u(t) is piecewise constant. Therefore, if v(t) is a piecewise extremal selector of P(t) given by (3.1), then there exists a u(t) which is piecewise constant and takes values from the set of vertices of U such that $v(t) = X^{-1}(t)Bu(t)$. Thus, by Theorem 2.1,

$$\mathscr{R}(t) = \tilde{\mathscr{R}}(t)$$

where $\mathscr{R}(t)$ is the reachable set corresponding to piecewise constant control function and $\mathscr{R}(t)$ is convex, compact and continuous in t. The existence of a time-optimal solution follows from the fact that the set

$$\{t \mid x_1 \in \mathscr{R}(t)\} \tag{3.5}$$

is closed and contained in $[0, +\infty]$. Indeed, by continuity of $\mathscr{R}(t)$, the set $\{(x, t) \mid x \in \mathscr{R}(t)\}$ = graph $\mathscr{R}$ is closed. Hence the set (3.5) is closed and there is $t_* = \min\{t \mid x_1 \in \mathscr{R}(t)\}$ which completes the proof of Theorem 1.1.

Notice that continuity was used only through the relation that the graph of $\mathscr{R}$ is closed. The latter property is less than continuity and it is called upper semicontinuity in Kuratowski sense of set-valued function. There is another definition of upper semicontinuity (u. s. c.) of set-valued maps: A set-valued map P into subsets of R^m is u. s. c. if

$$P^- G = \{t \mid P(t) \cap G \neq \emptyset\}$$

is closed for each closed subset G of R^m.

In the case of P(t) compact and bounded uniformly (i. e. $P(t) \subset P_0$ for each t, P_0 – bounded), the above two definitions are equivalent.

Notice also that we did not use the convexity of $\mathscr{R}(t)$. This property and continuity allows the conclusion that if t_* is the optimal time then $x_1 \in \partial\mathscr{R}(t_*)$ which in turn leads to necessary conditions for optimal control.

Theorem 1.1 can be easily extended to a more general situation. Namely to the case when the control system is of the form

$$\dot{x} = A(t)x + f(t, u) \tag{3.6}$$

where A(t) is an integrable matrix-valued function and f satisfies the Carathéodory conditions:

(C) f(t, u) is measurable in t for each fixed u, continuous in u for each fixed t

However, in this case the admissible control function u will be measurable selections of a given set-valued map U from $[t_0, T]$ into subsets of R^m and such that f(t, u(t)) is integrable. Under those assumptions, for each admissible u the solution of (3. 6) satisfying given initial condition $x(t_0) = x_0$ is uniquely defined and the reachable set is contained in

$$\mathscr{R}(t) = X(t)\left(x_0 + \int_0^t P(t)dt\right) \tag{3.7}$$

where $P(t) = X^{-1}(t)f(t, U(t))$. To have the opposite inclusion, we need to check that if $v(t) \in P(t)$ and is integrable then there is an admissible u such that $v(t) = X^{-1}(t)f(t, u(t))$ on the interval in question. A positive answer to this question requires some properties of U. For this purpose we recall the notion of measurable set-valued mapping.

<u>Definition.</u> We say that a set-valued map P from [0, 1] into subsets of R^n is measurable if the set

$$P^- G = \{t \mid P(t) \cap G \neq \emptyset\}$$

is measurable for each $G \subset R^n$ closed.

Notice that upper semicontinuity implies measurability.

The following result is referred to often in the literature as Filippov lemma and gives condition under which (3. 7) holds.

<u>Proposition 3.1.</u> If f satisfies Carathéodory condition (C) and U(t) is closed and measurable in t, then for any measurable v such that $v(t) \in f(t, U(t))$ there is a measurable selector u of U(t) such that $v(t) = f(t, u(t))$.

We omit the proof of this proposition but only explain two main steps in it. Firstly, one proves that the set-valued map $V(t) = \{u \in U(t) \mid f(t, u) = v(t)\}$ is measurable. Secondly, one applies a selection theorem which states that measurable closed set-valued mapping admits measurable selection.

The next proposition suggests one way selection theorem can be obtained:

Proposition 3.2. Let P(t) be closed and measurable in t on [0, 1] and let the cone C induces a lexicographical order in R^n. Then $v(t, C) = \max_C P(t)$ is finite on a measurable set on which it is measurable and v(t, C) is integrable on [0, 1] if and only if $\max_C \mathrm{cl} \int_0^1 P(t)dt$ is finite.

This proposition gives both; existence of measurable selectors as well as characterization of extremal elements of K_P if P is measurable. The proof of Proposition 3.2 is by induction argument. In fact, it is enough to prove that $P_a(t)$ for fixed $a \in R^n$ is measurable if P(t) is measurable.

To state an analogue of Theorem 1.1 for system (3.6) we would like to introduce the notion of an extremal solution of (3.6). Namely, x(t) on $[t_0, t]$ is an extremal solution of (3.6) if $x(t_1)$ is an extreme point of the reachable set $\mathscr{R}(t)$ corresponding to the initial condition $x(t_0) = x_0$. Now Proposition 3.1 and Theorem 2.1 implies:

Theorem 3.1. If f satisfies assumption (C), f(t, U(t)) and U(t) are closed and U is measurable in t and we assume that for each $u(t) \in U(t)$ measurable, f(t, u(t)) is integrable, then:

(a) the reachable set (3.7) for system (3.6) is compact and convex and depends continuously on t.

(b) the problem of passing from x_0 at time t_0 into a closed subset Z of $[t_0, +\infty) \times R^n$ in a minimal time has an optimal solution $x_*(t)$. Moreover, there is also a time-optimal solution of the above problem, which is piecewise extremal. $x_*(t)$ is unique iff $x_*(t_*)$ (t_* the optimal time) is an extreme point of $\mathscr{R}(t_*)$.

Notice the following characterization of extremal solutions: x(t) is an extremal solution of (3.6) on the interval [a, b] iff for each $t_1, t_2 \in [a, b]$, $t_1 < t_2$, and any other solution $\tilde{x}(t)$ of (3.6) the condition holds:

$$\tilde{x}(t_i) = x(t_i),\ i = 1, 2, \text{ implies that } x(t) = x(t) \text{ for } t \in [t_1, t_2]$$

For system (1.1) and $U = \{(u_1, \ldots, u_m)/ |u_i| \leq 1,\ i = 1, \ldots, m\}$ the time-optimal control function u_* has the property that $|u_{*i}(t)| = 1$ for each t. This is so because the extremal control functions have that property. Thus the equality $\mathscr{R}(t) = \tilde{\mathscr{R}}(t)$ which we have proved, holds also if for admissible controls Ω we take all piecewise constants u(t) such that $|u_i(t)| = 1$ for each i. This can, in other words, be expressed as follows: each point which can be reached in time t from the initial state x_0 at time t_0 by means of a measurable control function $(u_1(t), \ldots, u_n(t))$, $|u_i(t)| \leq 1$ a.e. in $[t_0, t]$, can also be reached, in the same time, by a solution corresponding to a control function u(t) whose co-ordinates only take on the values +1 or -1. This phenomenon of jumping from one extremity to another has been given a colorful name of "bang-bang" principle. A generalization of this principle follows from Theorem 2.1 and will be stated in the next theorem. Before we introduce the following property (compare with property (P)), we shall say that a class K of functions of a real variable is closed with respect to "piecewise" operation if

(P^*) for each $u_1, \ldots, u_k \in K$ and $t_0 = 0 < t_1 \ldots < t_k = 1$ the function

$$u = \sum \chi[t_{i-1}, t_i)\, u_i \in K$$

Theorem 3.2. Under the assumptions of Theorem 3.1, any point which can be reached from a fixed initial condition in a fixed time by an admissible solution of (3.6) can be also reached – and in the same time – by a piecewise extremal solution of (3.6). Moreover, if K denotes the class of all admissible solutions of (3.6) with fixed initial condition (t_0, x_0), K_* is the class of all piecewise extremal solutions of (3.6) and $\widetilde{K} \subset K$ has the property (P^*) and for each $t_1 > t_0$ and any $x \in K$ there is $\widetilde{x} \in \widetilde{K}$ such that $x(t_1) = \widetilde{x}(t_1)$ then $K_* \subset \widetilde{K}$.

Proof. The first part follows from Theorem 2.1. To prove the second part, we notice that any subset $\widetilde{K} \subset K$ with the property that for each $x \in K$ there is $\widetilde{x} \in \widetilde{K}$ that $x(t_1) = \widetilde{x}(t_1)$ has to contain an extremal solution of (3.6) because, if $e = x(t_1)$ is an extreme point of the reachable set $\mathscr{R}(t_1)$, then x(t) is the only solution of (3.6) and e can be reached at time t_1. This, together with (P^*), implies $K_* \subset \widetilde{K}$.

Theorem 3.2 is, in a sense, the best version possible of the "bang-bang" principle. Notice that uniqueness of admissible solutions leading to extreme points of $\mathscr{R}(t)$ does not necessarily imply uniqueness of the corresponding control functions. Notice also that extremal control functions u(t) are such that f(t, u(t)) takes values from the profile of clco f(t, U(t)), similarly to the linear case discussed above, where extremal (and also optimal, in some case control functions take on values from the set of vertices of the cube U, thus from the profile of U. However, even in this simple case it may happen that to an extremal solution there correspond two different control functions.

As another application of Theorems 2.1 and 2.2, let us consider the following more general problem: Suppose that, in the class of admissible solutions of (3.6) with fixed initial condition, we wish to find a solution such that $(t_1, x(t_1; x_0, t_0, u)) \in Z$ for some $t_1 > t_0$ and that the functional

$$\int_{t_0}^{t_1} (\langle a^0(t), x(t)\rangle + f^0(t, u(t))\, dt + \varphi(t_1, x(t_1))$$

attains its minimum. This is a Bolza type of problem. Consider the reachable set $\mathscr{R}(t)$ for an extended system

$$y = \widetilde{A}(t)\, y + F(t, u), \quad y(t_0) = (x_0, 0) \tag{3.8}$$

where $y = (x, x^0) \in R^{n+1}$, $\widetilde{A}(t)$ is $(n+1) \times (n+1)$ matrix of the form:

$$A(t) = \begin{vmatrix} A(t), & 0 \\ a^0(t), & 0 \end{vmatrix}$$

and $F(t, u) = (f(t, u)\ f^0(t, u))$.

Under similar assumptions as in Theorem 3.1, we can prove that the reachable set is compact and continuous. The problem reduces to finding a minimum of the function $\Phi(t, y) = \varphi(t, x) + x^0$, on the intersection of the graph of the reachable set $\mathscr{R}(t)$ with Z, and one sees that a solution exists if φ is such that $\varphi(t, x) \to +\infty$ if $t \to +\infty$ or if we, in addition, require in the problem that t_1 be bounded from above by a constant. We leave to the reader the exact statement of an appropriate theorem and its proof.

So far, we considered the bounded case, i.e. the case where the reachable set of the corresponding integral of a set-valued function is bounded and, thus, compact. This is not the case in the example mentioned in the introduction, i.e. the problem: minimize $\int_0^1 (1+\dot{x}^2)^{1/4}\,dt$ in the class of all absolutely continuous functions satisfying the boundary condition $x(0) = 0$, $x(1) = 1$. This problem is also a special case of above control problem with $A = 0$, $a^0 = 0$, $f(t, u) \equiv u$ and $f^0(t, u) = (1+u^2)^{1/4}$. The reachable set is simply the integral

$$\mathscr{R}(t) = \int_0^t P(t)dt$$

where $P(t) = \text{const} = \text{graph}\,(1+u^2)^{1/4}$. This integral is unbounded and not closed. In fact, one can check that $\int_0^t P(t)dt = t\,\text{co}P$ if P constant and, therefore, in our case $\mathscr{R}(t)$ is a halfspace $x^0 > t$ with one point of the boundary $(0, t)$ included. Thus, the solution of the problem exists only if the end condition is the same as the initial condition.

Let us dwell a little bit more on this kind of problem, i.e. the problem of minimizing the functional

$$\min \int_0^1 f(t, \dot{x}(t))dt \tag{3.9}$$

in the class of absolutely continuous functions on $[0, 1]$ satisfaying the boundary condition $x(0) = a$, $x(1) = b$. This is a special case of the classical Lagrange problem from the calculus of variation (the integral does not depend on $x(t)$). Put

$$P(t) = \text{graph}\, f(t, \cdot) = \{(x, y) \mid y = f(t, x),\ x \in R^n\}$$

and $Q(t) = \text{epigraph}\, f(t, \cdot) = \{(x, y) \mid y \geq f(t, x),\ x \in R^n\}$

As to f, we make the assumption that both $P(t)$ and $Q(t)$ are closed and measurable in t. The solution of the problem exists if and only if the intersection of $\int_0^1 Q(t)dt$ with the line $\{(x, y) \mid x = b,\ y \text{ arbitrary}\}$ is closed from below. A sufficient condition for the existence of a solution of this problem for any finite value b can be deduced from Theorem 2.2. A sufficient condition is that either the closure of the integral does not have proper unbounded extremal faces or that it does not contain a ray different from that the parallel to the positive y-axis. Indeed, consider the intersection $\left(\mathrm{cl}\int_0^1 Q(t)dt\right) \cap \{(x, y) \mid x = b\} = \{(x, y) \mid x = b,\ y \geq \beta\}$ (only the case β is finite is of interest). The optimal solution exists iff $(b, \beta) \in \int_0^1 Q(t)dt$. But (b, β) belongs to an extremal face F of $\mathrm{cl}\int_0^1 Q(t)dt$ which is transversal to direction of y-axis. Thus if F were not contained in $\int_0^1 Q(t)dt$ then F as well as $\mathrm{cl}\int_0^1 Q(t)dt$ would contain an extremal ray not parallel to the positive y-axi which contradicts the assumption.

For both of these cases, we shall give analytic conditions in terms of the so-called conjugate function to f.

First, however, we shall discuss the support function to a convex set. If $S \subset R^n$, then

$$\varphi_S(p) = \sup_{s \in S} \langle p, s \rangle$$

is called the support function of S. Note that φ can assume $+\infty$ values. The following properties of φ are well known and easy to check:

(i) (convexity) $\varphi_S(\lambda p_1 + (1-\lambda)p_2) \leq \lambda\varphi_S(p_1) + (1-\lambda)\ \varphi_S(p_2)$ if $\lambda \in [0, 1]$

(ii) (homogeneity) $\varphi_S(\lambda p) = \lambda\varphi_S(p)$ if $\lambda > 0$

(iii) (l. s. c.) $\liminf_{p \to p_0} \varphi_S(p) \geq \varphi_S(p_0)$

Denote by H(p, a) the halfspace $\{s \mid \langle s, p\rangle \leq a\}$. $H(p, \varphi_S(p))$ is called the support halfspace of S and we have:

$$\bigcap_p H(p, \varphi_S(p)) = \mathrm{clco}\ S$$

On the other hand, if $\varphi(p)$ satisfies (i), (ii) and (iii) then $\varphi = \varphi_S$ where $S = \bigcap_p H(p, \varphi(p))$.

An important property of the integral $I(P) = \int_0^1 P(t)dt$ is the following relation:

$$\varphi_{I(P)}(p) = \int_0^1 \varphi_{P(t)}(p)dt \tag{3.1C}$$

This is a consequence of Lemma 1 of the previous section. In this case, we integrate the epigraph Q(t) then the integral $S = \int_0^1 Q(t)dt$ has the property that if $(x, y) \in S$ then $(x, z) \in S$ for each $z > y$. But S is convex, thus for each x we can define

$$g(x) = \inf\{y \mid (x, y) \in S\}$$

if the latter set is not empty and $+\infty$ otherwise. One can check that g(x) is convex and if Q is closed then g is l. s. c. In this case, the support function $\varphi_S(p)$ of S is $+\infty$ if the last co-ordinate of p is positive and

$$\varphi_S(p, -1) = \sup_x(-g(x) + \langle p, x\rangle) = g^*(p)$$

where g^* is the so-called conjugate function of g. It is clear that g^* is convex and l. s. c. We notice also that g^{**} = cl g where cl g denotes the so-called closure of g, that is the largest convex and l. s. c. function bounded by g. From (3.10), we have

$$g^*(p) = \int_0^1 f^*(t, p)dt \tag{3.11}$$

where f^* is the conjugate function to $f(t, \cdot)$.

By the effective domain of g we mean the set

$$A = \{p \mid g^*(p) < +\infty\}$$

The set A is convex and we have the following proposition:

(η) Let Q be the epigraph of a convex and l. s. c. function $g: R^n \to R \cup \{+\infty\}$. Then p belongs to the interior of the effective domain of conjugate function g^* iff the set

$$G(p) = \{(x, g(x)) \mid -g(x) + \langle x, p\rangle = g^*(p)\}$$

is non empty and compact.

Proof. G(p) is convex and closed because g is l. s. c. If G(p) is unbounded then it contains a half-line, i. e. there exists $x_0, y_0 \in R^n$, $y_0 \neq 0$ such that

$$-g(x_0 + \lambda y_0) + \langle x_0 + \lambda y_0, p\rangle = g^*(p) \qquad \text{for each } \lambda \geq 0$$

It is clear that for q such that $\langle q, y_0\rangle > 0$ we have

$$\sup(-g(x_0 + \lambda y_0) + \langle x_0 + \lambda y_0, p + \epsilon q\rangle) = +\infty$$

if only $\epsilon > 0$. Thus, $g^*(p + \epsilon q) = +\infty$ if $\epsilon > 0$, hence p does not belong to the interior of A. On the other hand, suppose that $G(p_0)$ is compact for some p_0 fixed and let $r > 0$ be such that

$$F = \{(x, y) \mid |x|^2 + |y|^2 = r, \ -y + \langle x, p_0\rangle = g^*(p_0)\}$$

is disjoint from $G(p_0)$ as well as from Q. Let $\epsilon = \min_{z \in F} d(z, Q)$ and let $z_0 = (x_0, g(x_0))$ be a point from $G(p_0)$. There is $\delta > 0$ such that $d(z, D_p) \leq \epsilon/2$ for each $z \in F$ and any p such $|p - p_0| < \delta$ and $D_p = \{(x, y) \mid -y + \langle x, p\rangle = -g(x_0) + \langle p, x_0\rangle\}$. If $D_p \cap Q$ were unbounded, then because of the convexity of $D_p \cap Q$, we would have a contradiction with the definition of r in F. Thus, for each p such that $|p - p_0| < \delta$, $S = \{(q, r) \in Q \mid -r + \langle q, p\rangle \geq -g(x_0) + \langle p, x_0\rangle\}$

is compact and since $G(p) = \{(x, y) \in S \mid -y + \langle x, p\rangle = \max_{(q,r)\in S} (-r + \langle q, p\rangle\}$, therefore G(p) is not empty and compact.

We may now summarize what we have proved in the following

Theorem 3.3. Suppose that $f: [0, 1] \times R^n \to R \cup \{+\infty\}$ is measurable in t and lower semicontinuous in u. If the set

$$\{p \mid f^*(t, p) \text{ is integrable}\}$$

is open, then the integral $\int_0^1 P(t)dt$ is closed, where

$$P(t) = \text{epigraph } f(t, \cdot)$$

In particular, there exists $\min \int_0^1 f(t, \dot{x}(t))dt$ over all x(t) absolutely continuou[s] $x: [0, 1] \to R^n$ such that $x(0) = a$, $x(1) = b$.

A particular case of this situation is that where $f^*(t, p)$ is integrable for each p. In this case, we say that f satisfies the growth condition. Notice that this assumption means that

$$-f(t, u) + \langle p, u\rangle \leq \psi_p(t) \tag{3.12}$$

where $\psi_p \in L_1$ or equivalently that to each $\epsilon > 0$ there is an integrable functio[n] $\psi_\epsilon(t)$ such that

$$\epsilon f(t, u) \geq |u| + \psi_\epsilon(t) \tag{3.13}$$

From (3.13) it follows that, roughly speaking, f grows, with respect to u, faster than any linear function. A special case of the growth condition is tha[t] where $f(t, u) \geq \Phi(|u|)$ and $\Phi(s)/s \to +\infty$ as $s \to \infty$. This corresponds to (3.12) or (3.13) with function ψ_p or ψ_ϵ bounded or simply constant. Thus, it is a uniform growth condition. If the growth condition holds, then the only ray contained in the integral $\int_0^1 P(t)dt$ is that parallel to the positive y-axis. This is the direction in which all P(t) are unbounded.

The last theorem can be generalized to a more complicated situation when we wish to minimize

$$I(x, u) = \int_0^1 (\langle a^0(t), x(t)\rangle + f^0(t, u(t))dt$$

under the conditions

$$\dot{x} = A(t)x + f(t, u), \quad u \in U(t)$$

$$x(0) = a, \quad x(1) = b$$

The solution of this problem exists if f and f^0 satisfy a Carathéodory condition i.e. f, f^0 are measurable in t and continuous in u, U(t) is closed and measurable in t, F(t, U(t)) is closed and for each $\epsilon > 0$ there is $\psi_\epsilon \in L_1$ such that

$$\epsilon f^0(t, u) \geq |f(t, u)| + \psi_\epsilon(t) \quad \text{if} \quad u \in U(t)$$

F, as before, stands for (f, f^0).

4. A NON-LINEAR OPTIMAL PROBLEM

In contrast to the case considered up so far, for a non-linear control system, the reachable set may be not closed, even if the set of admissible values of the control parameter is compact. Let us consider the following example:

$$\dot{x} = 1/(1+y^2), \quad \dot{y} = u/(1+y^2), \quad u = 1 \text{ or } u = -1 \tag{4.1}$$

Notice that dy/dx = u, and it is clear how the trajectories of Eq.(4.1) look like. However, the reachable set of Eq.(4.1) is not closed and the time-optimal problem of passing from (0, 0) to (1, 0) has no solution. Indeed, the infimum is 1 because for each $\epsilon > 0$ there is a solution of Eq.(4.1) such that $|y(t)| \leq \epsilon$, $x(t_\epsilon) = 1$, $y(t_\epsilon) = 0$. But we have:

$$\int_0^{t_\epsilon} \dot{x}(t)dt = 1 = \int_0^{t_\epsilon} \frac{1}{1+y^2}\, dt \geq \frac{1}{1+\epsilon^2}\, t_\epsilon, \quad \text{thus } t_\epsilon \leq 1+\epsilon^2$$

On the other hand, x travels with a speed less or equal one, thus the optimal time is bounded from below by 1. Therefore, the optimal time, if it exists, has to be equal one. But that would be possible if and only if $\dot{x}(t) \equiv 1$ hence if $y(t) \equiv 0$, which is impossible.

Another example of the sort is the following Lagrange problem: Minimize

$$I(x) = \int_0^1 (1+x^2)(|\dot{x}^2 - 1| + 1)\, dt \tag{4.2}$$

over the class of all absolutely continuous functions satisfying the boundary condition:

$$x(0) = 0, \quad x(1) = 0 \tag{4.3}$$

The infimum of I(x) over x satisfying (4.3) is 1. Indeed, taking a polygonal line x_n with slope +1 or -1 we see that

$$I(x_n) = \int_0^1 (1+x_n^2(t))dt$$

and it is clear that we can do it so that $|x_n| \leq \epsilon$ for arbitrary $\epsilon > 0$. On the other hand $I(x) \geq 1$ for each x. But the value 1 is not attained because that would be possible only if $(1+x^2)(|1-\dot{x}^2|+1) = 1$. Thus if $x(t) \equiv 0$ and $\dot{x}(t) \equiv 1$. A contradiction.

In both those cases, the lack of convexity is the reason for the non-existence of an optimal solution. If, in the first case, we convexify the right-hand side, i.e. we take the convex hull of the set $\{(1/(1+y^2),\ u/1+y^2)) \mid |u| = 1\}$ which comes to allowing u to be $|u| \leq 1$, or, in the second case, replace $f(x, u) = (1+x^2)(|1-u^2|+1)$ by $\operatorname{co} f(x, u) = (1+x^2)(\max(0, u^2-1)+1)$ which is the largest function convex in u and bounded from above by f, then the optimal solution would exist. Notice that in both cases we could define sequences $x_n(t)$ of admissible trajectories such that the value of cost function on x_n tends to the infimum. Such sequence is called a minimizing sequence. In both cases, the minimizing sequence is uniformly convergent but in the first case the limit function is not an admissible trajectory and in the second the functional evaluated at the limit is greater than the infimum. In the following theorem we give a condition guaranteeing that neither of the above effects holds.

Problem 1. Minimize the functional

$$I(x, u) = \int_0^{t_1} f^0(t, x(t), u(t))dt \tag{4.4}$$

over the class Ω of admissible pairs (x, u) such that x is an absolutely continuous function, u-measurable and

$$\begin{cases} \dot{x}(t) = f(t, x(t), u(t)) & \text{a.e. in } [0, t_1] \\ x(0) = a \quad x(t_1) \in \Phi(t_1) \\ u(t) \in U(t) & \text{- a given set-valued function} \end{cases} \tag{4.5}$$

where Φ is continuous function from $[0, T]$ into R^n, and t_1 is not fixed but $t_1 \leq T$.

<u>Theorem 4.1.</u> (existence theorem) Assume that f^0, f are continuous in x, u for fixed t, measurable in t for x, u fixed; U(t) is closed and measurable in t; the set

$$Q(t, x) = \{(q, r) \mid q = f(t, x, u),\ r \geq f^0(t, x, u),\ u \in U(t)\} \tag{4.6}$$

is convex and closed for each (t, x); for each t fixed Q is u.s.c. in x in Kuratowski sense; and the growth condition holds, that is

$$\sup_{(q, r) \in Q(t, x)} (-r + \langle p, q \rangle) = \varphi(t, p, x) \leq \psi(t, p) \text{ for each } x \tag{4.7}$$

where ψ is integrable in t, for each $p \in R^n$.

Under these assumptions there exists optimal solution of the problem (4.4) - (4.5); that is there are x_*, u_*, t_* which satisfy (4.5) and such that

$$\int_0^{t_*} f^0(t, x_*(t), u_*(t))dt \leq \int_0^{t_1} f^0(t, x(t), u(t))dt$$

for any other admissible pair (x, u).

Proof. The proof of this theorem will be split in a few steps. First we notice that because of (4.7) of I(x, u) is finite. Therefore we have a sequence $x_n(t)$, $u_n(t)$, t_1^n of admissible pairs such that

$$I(x_n, u_n) = \int_0^{t_1^n} f^0(t, x_n(t), u_n(t))dt \to \alpha = \underset{(x,u) \in \Omega}{\text{infimum}}\, I(x, u)$$

Put $q_n(t) = \dot{x}_n(t) = f(t, x_n(t), u_n(t))$, $r_n(t) = f^0(t, x_n(t), u_n(t))$. By (4.6) we have

$$(q_n(t), r_n(t)) \in Q(t, x_n(t)) \quad \text{and} \quad \int_0^{t_1^n} r_n(t)dt \to \alpha \tag{4.8}$$

We assume that x_n and u_n are defined on the interval [0, T] and thus also (4.8) holds on [0, T]. This is not much of restriction. It will allow us to avoid certain technical difficulties. By (4.8) we get

$$\int_0^T r_n(t)dt \leq M < +\infty \tag{4.9}$$

We shall prove first that both q_n and r_n are bounded in L_1-norm. $r_n(t)$ is bounded because of (4.9) and the inequality below obtained from (4.9) and (4.7) for p = 0: $-r_n(t) \leq \psi(t, 0)$. Similarly, by (4.7) and (4.8)

$$\langle p, q_n(t) \rangle \leq r_n(t) + \psi(t, p)$$

which holds for each p and which implies that $|q_n(t)| \leq K(r_n(t) + \psi(t))$ where K is constant, ψ is integrable and independent on n. Thus there is M such that $\|q_n\|_{L_1} + \|r_n\|_{L_1} \leq M$ for each n. This implies that $\{(q_n, r_n)\}$ is precompact in the weak* topology of the conjugate space C* of the space C of continuous functions from $[0, T] \to R^{n+1}$. Without any loss of generality we may assume that (q_n, r_n) is convergent in that sense. Thus there is a measure μ taking values in R^n and a scalar measure ν, such that

$$\int_0^T \langle \xi(t), q_n(t) \rangle dt + \int_0^T r_n(t)\, \eta(t)dt \to \int_0^T \langle \xi(t), d\mu(t) \rangle + \int_0^T \eta(t)d\nu(t) \tag{4.10}$$

for each ξ: $[0, T] \to R^n$ and η: $[0, T] \to R$ continuous.

From (4.8) and (4.7) we have the inequality

$$\langle p, q_n(t) \rangle - r_n(t) - \psi(t, p) \leq 0 \tag{4.11}$$

Denote by μ_a, μ_s and ν_a, ν_s absolutely continuous and singular part of μ and ν, respectively. The left-hand side of (4.11) is also converging weak*, and the limit is because of the inequality (4.11) a non-positive measure. Thus both absolutly continuous part and singular part of the limit are non-positive, too. Therefore from (4.10) and (4.11) we obtained inequalities:

$$\langle p, \frac{d\mu_a}{dt}(t)\rangle - \frac{d\nu_a}{dt}(t) - \psi(t,p) \leq 0 \tag{4.12}$$

$$\langle p, \mu_s(A)\rangle - \nu_s(A) \leq 0 \quad \text{for each A measurable} \tag{4.13}$$

Since (4.13) holds for each p thus $\mu_s(A) = 0$ for each A. Therefore

$$\int_0^T \langle \xi(t), q_n(t)\rangle dt \to \int_0^T \langle \xi(t), q_0(t)\rangle dt$$

where $q_0(t) = \frac{d\mu_a}{dt}(t)$, for each continuous ξ which implies:

$$x_n(t) = a + \int_0^t q_n(t)dt \to a + \int_0^t q_0(t)dt = x_0(t) \tag{4.14}$$

We shall prove now that

$$\limsup_n \varphi(t,p,x_n(t)) \leq \varphi(t,p,x_0(t)) \quad \text{a.e. in } [0,T] \tag{4.15}$$

Without any loss of generality, we may assume that in (4.7) $\psi(t,p) = \sup_x \varphi(t,p$
Then $\psi(t,p)$ is convex in p and there is a set $N \subset [0,T]$ of measure zero such that if $t \notin N$ then $\psi(t,p)$ is finite for each p. Inequality (4.15) holds for $t \in [0,T]\setminus N$. Indeed, fixe $t \in [0,T]\setminus N$, then $Q(t,x) \subset P(t)$, where P(t) is the epigraph of the conjugate function $\psi^*(t,\cdot)$ to $\psi(t,\cdot)$. We have the inclusion

$$Q(t,x) \cap \{(q,r) \mid -r + \langle q,p\rangle \geq \varphi(t,p,x_0(t))\}$$
$$\times \subset P(t) \cap \{(q,r) \mid -r + \langle q,p\rangle \geq \varphi(t,p,x_0(t))\}$$

and by (η) the latter set is compact. Thus if there is $(q_n, r_n) \in Q(t, x_n(t))$ and such that $\varphi(t,p,x_n(t)) = -r_n + \langle p, q_n\rangle \geq \varphi(t,p,x_0(t))$ then (q_n, r_n) containes a convergent subsequence and we may assume as well that (q_n, r_n) converges. Then the limit (q_0, r_0), since $x_n(t) \to x_0(t)$, belongs to $Q(t, x_0(t))$ because of u.s.c. of Q with respect to x. Thus we have (4.15).

It is clear from (4.15) and (4.7) that putting

$$\psi_n(t) = \max(\varphi(t,p,x_0(t)), \sup_{i \geq n} \varphi(t,p,x_i(t)))$$

we have $\lim \psi_n(t) = \varphi(t,p,x_0(t))$ and $\psi_n(t) \leq \psi(t,p)$, $n = 1, 2, \ldots$ This leads us to a possibility of replacing, in inequality (4.12), $\psi(t,p)$ by ψ_n for arbitrary n hence also by $\varphi(t,p,x_0(t))$. We have then

$$\langle p, q_0(t)\rangle - r_0(t) \leq \varphi(t,p,x_0(t)) \quad t \in [0,T]\setminus N(p) \tag{4.16}$$

It follows from (4.16) that for each denumerable set $\{p_j\}$ there is a set N of measure zero such that

$$-r_0(t) + \langle p, q_0(t)\rangle \leq \varphi(t, p, x_0(t)) \text{ for } p \in \{p_j\} \text{ and } t \in [0, T] \setminus N \qquad (4.16')$$

Since both sides of Eq. (4.16') are continuous in p, if $\{p_j\}$ is dense in R^n, then (4.16') holds for each p and $t \in [0, T] \setminus N$. Hence we conclude that

$$(q_0(t), r_0(t)) \in Q(t, x_0(t)) \quad \text{a.e. in } [0, T] \qquad (4.17)$$

It follows from (4.11) that for each $\epsilon > 0$ there is δ such that $\int_A |q_n(t)| dt < \epsilon$ if the measure of A is less than δ, which implies, in particular, that the convergence in (4.14) is uniform and that $q_n \to q_0$ weakly in L_1. From this follows that there is t^* equal to the limit of t_1^n or of a subsequence of it such that $x_0(t^*) = \Phi(t^*)$. The regularity assumption on f and f^0 are enough to conclude that there is measurable $u_0(t)$ (Proposition 3.1) such that

$$q_0(t) = \dot{x}_0(t) = f(t, x_0(t), u_0(t)) \quad \text{and} \quad r_0(t) \geq f^0(t, x_0(t), u_0(t))$$

and $u_0(t) \in U(t)$. Obviously, $x_0(0) = a$ and, therefore, $x_0(t)$, $u_0(t)$ satisfies (4.5), hence is admissible. Now by (4.10) and (4.13) for each $\epsilon > 0$ there is $\delta > 0$ such that if φ: $[0, T] \to [0, 1]$ is continuous and $\varphi(t) = 1$ if $t \leq t^* + \delta/2$ and $\varphi(t) = 0$ if $t > t^* + \delta$ then

$$\left| \int_0^t r_0(t)dt - \int_0^T \varphi(t) r_0(t)dt \right| < \epsilon/4$$

$$\left| \int_0^{t_1} r_n(t)dt - \int_0^T \varphi(t) r_n(t)dt \right| < \epsilon/4 \qquad \text{for} \quad n \geq N(\epsilon)$$

$$\int_0^T \varphi(t) r_n(t)dt - \int_0^T \varphi(t) r_0(t)dt \geq -\epsilon/4 \quad \text{for} \quad n \geq N(\epsilon)$$

$$\left| \int_0^{t_1^n} r_n(t)dt - \alpha \right| < \epsilon/4 \qquad \text{for} \quad n \geq N(\epsilon)$$

The above inequalities imply

$$\int_0^{t^*} r_0(t)dt \leq \alpha + \epsilon$$

But ϵ is arbitrary, therefore we have

$$I(x_0, u_0) = \int_0^t f^0(t, x_0(t), u_0(t))dt \leq \int_0^t r_0(t)dt$$

which shows that x_0, u_0, t^* is optimal solution and completes the proof of the theorem.

Theorem 4.1 can be extended, without much additional difficulty, to the case of multiple integrals.

Problem 2. Suppose we wish to minimize the functional

$$I(x, u) = \int_G f^0(t, x(t), u(t))dt \quad \text{for} \quad (x, u) \in \Omega$$

where G is a bounded domain in R^m, $x: G \to R^k$ and $u: G \to R^n$. The set Ω of admissible pairs (x, u) is defined by the following conditions:

x belongs to the Sobolev space $H_1^1(G, R^k)$, u is measurable and $u(t) \in U(t)$ a.e. in G, where U is given set-valued mapping,

$$\nabla x(t) = f(t, x(t), u(t)) \quad \text{a.e. in } G$$

and the boundary value of x are fixed in the sense that $x \in x_0 + H_{10}^1(G, R^k)$, where $x_0 \in H_1^1$ is fixed and H_{10}^1 is the subspace of H_1^1 which is obtained by closing in H_1^1 the set of C^∞-function with compact support contained in G.

Theorem 4.2. Assume that f^0, f are continuous in x, u for fixed t, measurable in t for fixed x, u, U(t) is closed and measurable in t, the set

$$Q(t, x) = \{(q, r) \mid q = f(t, x, u), \; r \geq f^0(t, x, u), \; u \in U(t)\}$$

is convex and closed for each (t, x) and for each t fixed Q is u.s.c. in x in Kuratowski sense;

$$\sup_{(q, r) \in Q(t, x)} (-r + \langle p, q\rangle) = \varphi(t, p, x) \leq \psi(t, p) \text{ for each } x,$$

where ψ is integrable in t for each fixed $p \in R^{km}$.
Under these assumptions there exists an optimal solution (x_0, u_0) of the problem 2.

The proof is analogous and we just point out those places where there is a difference.
As in the previous case we put

$$q_n(t) = \nabla x_n(t) = f(t, x_n(t), u_n(t))$$

$$r_n(t) = f^0(t, x_n(t), u_n(t))$$

where (x_n, u_n) is the minimizing sequence. In exactly the same manner we can prove that there is a subsequence of (q_n, r_n) (for simplicity still denoted by (q_n, r_n)) such that $q_n \to q_0$ weakly in L_1 and $\int_G r_n(t)dt \to \alpha \geq \int_G r_0(t)dt$.

The space H_{10}^1 has the property that $x_n \in H_{10}^1$ and ∇x_n converges weakly in L_1 then x_n converges strongly to a function x_0 in H_{10}^1 and ∇x_n converges weakly to ∇x_0. So we have the limit function, we see that $q_0 = \nabla x_0$ and we conclude that

$$(q_0(t), r_0(t)) \in Q(t, x_0(t))$$

The remaining part goes exactly in the same way or even more simply since here the integration is always over a fixed set G while in the previous case we allow the right end of the integration interval to be variable.

5. LOWER CLOSURE AND LOWER SEMICONTINUITY

The optimal problem which we discussed in the previous section can have the following "control-free" formulation:
Let Q be a set-valued map defined on $[0, T] \times R^n$ into subsets of R^{n+1}. We assume that $Q(t, x)$ has the property that if $(q, q_0) \in Q(t, x)$ (q_0 is the last co-ordinate of a point from Q) and $r > q_0$ then $(q, r) \in Q(t, x)$. Let Ω be a certain set of pairs (x, w), called admissible pairs, both being integrable functions from $[0, T]$ into R^n.
Problem 3:

$$\text{minimize} \left\{ \int_0^T v(t)dt \,\middle|\, (w(t), v(t)) \in Q(t, x(t)),\ (x, w) \in \Omega \right\}$$

This problem under the assumptions of Theorem 4.1 concerning f^0, f and the set-valued map U is equivalent to the optimal-control problem considered in the previous section in the sense that if we define Q(t, x) by (4.6) and an integrable $(w(t), v(t)) \in Q(t, x(t))$ then there is u(t) measurable such that $u(t) \in U(t)$ and $w(t) = f(t, x(t), u(t))$ and $v(t) \geq f^0(t, x(t), u(t))$. On the other hand, the above problem can be equivalently expressed as the Lagrange problem if we consider Q(t, x) to be the epigraph of a function $g(t, x, \cdot)$. More precisely

$$g(t, x, u) = \inf\{v \mid (u, v) \in Q(t, x)\} \tag{5.1}$$

if the latter set is non-empty and $+\infty$ otherwise.

If Q(t, x) is closed then it is the epigraph of $g(t, x, \cdot)$ defined by (5.1).
Problem 4.

$$\text{Minimize} \int_0^T g(t, x(t), u(t))dt$$

in a given class Ω of admissible pairs (x(t), u(t)). Under some mild regularity condition concerning Q, Problem 4 is equivalent to Problem 3 in the sense that a solution of Problem 4 is the solution 3 with the same Ω and vice versa. The assumption sufficient for this is that g(t, x(t), u(t)) is measurable for each x, u measurable and that the infimum in (5.1) is attained. The latter is the case if Q(t, x) is closed while the first assumption holds if Q is $\mathscr{L} \times \mathscr{B}$ measurable in the sense that the graph of Q is a $\mathscr{L} \times \mathscr{B}$-measurable subset of $[0, T] \times R^n \times R^{n+1}$ where $\mathscr{L}$ is the Lebesgue δ-field on $[0, T]$ and $\mathscr{B}$ is the Borel δ-field in $R^n \times R^{n+1}$.

For the existence of optimal solution the following two conditions are usually established:

(C_1) The set

$$D = \left\{(x, u, a) \,\middle|\, (u(t), v(t)) \in Q(t, x(t)) \text{ and } a = \int_0^T v(t)dt\right\}$$

is closed.

(C_2) There is a such that the set:

$$\Omega_a = \left\{(x, u) \,\middle|\, \Omega \,\middle|\, \text{there is v such that } (u(t), v(t)) \in Q(t, x(t)) \text{ on } [0, T] \text{ and } \int_0^T v(t)dt \leq a\right\}$$

is non-empty and compact.

Both conditions were established in the course of the proof of Theorem 4.1 with strong sequential convergence for x, and weak sequential convergence for u. It should be emphasized that the topology here is not given by a problem and any topology is good provided both conditions C_1 and C_2 are established. In fact, different topologies have been used. This shows also that one can have a great variety of existence theorems, even for the same type of problem. Condition C_1 is equivalent to the so-called lower closure property of the orientor field: $(u(t), v(t)) \in Q(t, x(t))$; that is the property that if $u_n \to u_0$, $x_n \to x_0$, $\int_0^1 v_n(t)dt \to a$, and $(u_n(t), v_n(t)) \in Q(t, x_n(t))$ then there is v_0 such that $\int_0^T v_0(t)dt \leq a$ and $(u_0(t), v_0(t)) \in Q(t, x_0(t))$. Condition C_1 when translated in terms of function g means lower semicontinuity of the functional $I(x, u) = \int_0^T g(t, x(t), u(t))dt$. Indeed, if a sequence or a generalized sequence (x_α, u_α) converges to (x_0, u_0) then by C_1 $(x_0, u_0, \liminf_\alpha I(x_\alpha, u_\alpha)) \in D$. But because of (5.1) the infimum of $\{a \,|\, (x_0, u_0, a) \in D\}$ equals $\int_0^T g(t, x_0(t), u_0(t))dt = I(x_0, u_0)$, thus $I(x_0, u_0) \leq \liminf_\alpha I(x_\alpha, u_\alpha)$ and vice versa. Condition C_2 gives us compactness of the minimizing sequence. The growth condition is responsible for this condition in Theorem 4.1. Both conditions C_1 and C_2 plus closedness of Ω imply existence of a optimal solution to Problem 3. Of course, in a specific existence theorem, a topology or a mode of convergence in x-space and u-space has to be chosen and conditions C_1 and C_2 have to be established with respect to those topologies. In choosing topologies in x-space and u-space, we are faced with a conflict between C_1 and C_2. The weaker the topology is which we choose the more probable is it that C_2 would hold, while, for condition C_1, the situation is opposite. As in the previous section, we shall here consider strong L_1-topology for x and L_1-weak sequencial convergence for u. There are two types of existence theorems following the above pattern: One is valid when condition C_2 holds, for reasons independent on the structure of Q. For example, if we beforehand know the topology in which Ω is compact, then the only assumptions we need to impose on Q should be such that C_1 holds with respect to this particular topology. The other type is such that the properties of Q which we seek should imply both C_1 and C_2. Theorems 4.1 and 4.2 are of this type. There, the growth condition gives C_2 while the convexity assumption is essential (and the growth condition is very helpful) to obtain C_1.

We shall prove that convexity of Q(t, x) is, in a sense, necessary for C_1 to hold. To be more precise, we have the following:

Theorem 5.1. Let Q(t) be a set-valued map defined on the interval [0, 1] with values being subsets of R^{n+1} space with the property that if $(q, q_0) \in Q(t)$ and $r > q_0$ then $(q, r) \in Q(t)$, too (q_0 is the last co-ordinate of point in R^{n+1}).

Assume that the graph of Q, that is the set $\{(t, v) \mid t \in [0, 1], v \in Q(t)\}$ is $\mathscr{L} \times \mathscr{B}$-measurable subset of $[0, 1] \times R^{n+1}$. Then the following conditions are equivalent:

(i) $\inf \left\{ \int_0^1 v(t)dt \mid (u_0(t), v(t)) \in Q(t) \text{ a.e. in } [0, 1] \right\}$ is greater than $-\infty$ for each u_0, and $(u(t), v(t)) \in Q(t)$ has l.c.p. with respect to weak convergence in L_1.

(ii) The set $K = \{(u, v) \mid u \in L_1([0, 1], R^n), v \in L_1([0, 1], R), (u(t), v(t)) \in Q(t)\}$ is convex and closed in weak topology of L_1 and there is $\varphi \in L_\infty$ and $\psi \in L_1$ such that for each $(u, v) \in K$

$$-v(t) + \langle u(t), \varphi(t) \rangle \leq \psi(t) \quad \text{a.e. in } [0, 1] \tag{5.2}$$

(iii) Q(t) is closed and convex a.e. in [0, 1] and

$$Q(t) = \bigcap_{\varphi \in \Phi} \{(q, q_0) \mid -q_0 + \langle q, \varphi(t) \rangle \leq \psi_\varphi(t)\} \quad \text{a.e. in } [0, 1] \tag{5.3}$$

where $\Phi \subset L_\infty([0, 1], R^n)$ is not empty and denumerable and $\psi_\varphi \in L_1$ for each φ.

(iv) $Q(t) = \{(q, q_0) \mid q_0 \geq f(t, q)\}$ where $f: [0, 1] \times R^n \to R \cup \{+\infty\}$ is $\mathscr{L} \times \mathscr{B}$-measurable, lower semicontinuous and convex in u for each fixed t and the functional

$$I(u) = \int_0^1 f(t, u(t))dt$$

is well defined on L_1 with values from $R \cup \{+\infty\}$ and weakly lower semicontinuous.

Proof. (i) ⇒ (ii). The l.c.p. of $(u(t), v(t)) \in Q(t)$ with respect to weak topology implies, in particular, that the set K in (ii) is weakly closed. In fact, if $\{u_\alpha, v_\alpha\}$ converges weakly to (u_0, v_0) and $(u_0, v_0) \notin K$ then without any loss of generality we may assume that

$$\int_0^1 v_0(t)dt < \alpha = \inf\left\{\int_0^1 v(t)dt \,\middle|\, (u_0, v) \in K\right\} \tag{5.4}$$

Indeed, let $A = \{t \mid (u_0(t), v_0(t)) \in Q(t)\}$ and v_1 be such that $(u_0, v_1) \in K$ and $\int_0^1 v_1(t)dt - \alpha < \epsilon$. Then, putting $(\tilde{u}_\alpha(t), \tilde{v}_\alpha(t)) = (u_0(t), v_1(t))$ if $t \in A$ and $(u_\alpha(t), v_\alpha(t))$ otherwise we obtain a sequence weakly convergent to (u_0, v_ϵ), where $v_\epsilon(t) = v_1(t)$ if $t \in A$ and $v_0(t)$ if $t \notin A$. If ϵ is small enough (5.4) holds for v_ϵ because $\int_{[0,1]\setminus A} (v_1(t) - v_0(t))dt > 0$. But (5.4) contradicts the lower closure property.

The closedness of K, together with the fact that K is the set of integrable selectors of a set-valued function, implies that K is convex. Indeed, let $w_1, w_2 \in K$ and put $w_\lambda = \lambda w_1 + (1-\lambda)w_2$, $\lambda \in (0, 1)$. Take any sequence $\varphi_1, \ldots, \varphi_k$ of L_∞ functions and consider the set:

$$B = \left\{\left(\int_0^1 \langle w, \varphi_1\rangle dt, \ldots, \int_0^1 \langle w, \varphi_k\rangle dt\right) \,\middle|\, w = w_1 \chi_A + w_2 \chi_{[0,1]\setminus A}; \text{A measurabl}\right.$$

By theorem 2, 1 it is compact and convex and it does contain points

$$a_i = \left(\int_0^1 \langle w_i, \varphi_1\rangle dt, \ldots, \int_0^1 \langle w_i, \varphi_k\rangle dt\right), \quad i = 1, 2$$

Therefore, it contains also $\lambda a_1 + (1-\lambda)a_2 = \left(\int_0^1 \langle w_\lambda, \varphi_1\rangle dt, \ldots, \int_0^1 \langle w_\lambda, \varphi_k\rangle dt\right)$ which proves that in any weak neighbourhood of w_λ there is $w = w_1\chi_A + w_2\chi_{[0,1]\setminus A} \in K$ which implies that $w_\lambda \in K$ if K is weakly closed, and if K is weakly sequentially closed $w_\lambda \in K$ also, hence w_λ belongs to the closure of $\{w_1\chi_A + w_2\chi_{[0,1]\setminus A} \mid A \subset [0\text{-}1] \text{ measurable}\}$ and the latter set is bounded. So the closure of it is also bounded and convex and thus also weakly sequentially closed. Therefore the set:
$D = \left\{(u, a) \mid (u, v) \in K,\ a = \int_0^1 vdt\right\}$ is closed and convex and it follows from (i) that it is bounded from below thus there is $\varphi \in L_\infty$ such that

$$-a + \int_0^1 \langle \varphi(t), u(t)\rangle dt \leq \text{const} < +\infty$$

for each (u, a) D. The latter inequality and Lemma 1 of Section 2 implies (5.2), where

$$\psi(t) = \operatorname*{ess\,sup}_{(u,v)\in K} (-v(t) + \langle \varphi(t), u(t)\rangle$$

(ii) ⇒ (iii) There is a denumerable subset $\{w_\alpha\}$ dense in L_1-norm topology in K. Put $\tilde{Q}(t) = \text{clco}\{w_\alpha(t)\}$. Obviously, if $w \in K$ then $w(t) \in \tilde{Q}(t)$ a.e. in $[0, 1]$, and vice versa, if $w(t) \in \tilde{Q}(t)$ a.e. in $[0, 1]$ then $w \in \text{cl}\{w_\alpha\} = K$, where the closure is in L_1-norm topology. If the latter were not true then we should be able to show that w(t) on a set of positive measure is outside $\tilde{Q}(t)$. Now, two measurable set-valued functions have to be equal a.e. if they have the same set of integrable selectors. Representation (5.3) follows from (5.2) (that Φ is non-empty) and from the following facts: let $A_N(t)$ be defined: $A_N(t) = \{p \mid \sup_{(q, q_0) \in \tilde{Q}(t)} (-q_0 + \langle p, q\rangle) \le N + \psi(t), \ |p - \varphi(t)| \le N\}$ where N is an integer and ψ are from (5.2). $A_N(t)$ is closed, convex and measurable in t, therefore there is a sequence of measurable functions $p_{N,i}$ such that $A_N(t) = \text{cl}\{p_{N,i}(t)\}$ for each t and N fixed. Put $\Phi = \bigcup_N \{p_{N,i}\}$. Then Φ is the desired set for which (5.3) holds.

(iii) ⇒ (iv) There is f(t, u) such that Q(t) is the epigraph of f(t, ·) for each t fixed. It is obvious that f has the desired properties. From (5.3), $I(u) > -\infty$, thus I(u) is well defined and either finite or $+\infty$. Let $u_n \to u_0$ weakly, then for each $\varphi \in \Phi \subset L_\infty$

$$-f(t, u_n(t)) + \langle \varphi(t), u_n(t)\rangle \le \psi_\varphi(t)$$

Similarly, as in the proof of Theorem 4.1, we deduce that $v_n = f(t, u_n(t))$ contains a subsequence converging to a measure ν and $v_0 = d\nu_a/dt$ satisfies the inequality

$$-v_0(t) + \langle \varphi(t), u_0(t)\rangle \le \psi_\varphi(t) \tag{5.5}$$

while the singular part ν_s is non-negative. In particular if $I(u_n) \to a$ then

$$a = \int_0^1 v_0(t)dt + \nu_s([0, 1]) \ge \int_0^1 v_0(t)dt \tag{5.6}$$

Now, (5.5) together with (5.3), implies that $(u_0(t), v_0(t)) \in Q(t)$ a.e. in $[0, 1]$, therefore $v_0(t) \ge f(t, u_0(t))$. Hence, by (5.6), $I(u_0) \le \liminf I(u_n)$, which was to be proved.

(iv) ⇒ (i) The infimum in (i) is equal to $I(u_0)$ if Q is the epigraph of f as in the case if (iv) holds. From l.s.c. of I(u) it follows that the set

$$D = \{(u, a) \mid u \in L_1, \ a \ge I(u)\}$$

is closed. But $D = \left\{(u, a) \mid a = \int_0^1 v(t)dt, \ v(t) \ge f(t, u(t)) \text{ a.e. in } [0, 1]\right\}$. Therefore, let $(u_n(t), v_n(t)) \in Q(t)$ a.e. in $[0, 1]$ for each n, and let $u_n(t) \to u_0$ weakly and $\int_0^1 v_n(t)dt \le a_0$, then $I(u_0) \le \liminf I(u_n) \le a$. Hence $v_0 = f(t, u_0(t))$ is such that $\int_0^1 v_0(t)dt \le a_0$ and $(u_0(t), v_0(t)) \in Q(t)$ a.e. in $[0, 1]$. Hence the lower closure property holds and the proof of Theorem 5.1 is completed.

It is clear that convexity of $Q(t, x)$ is also a necessary condition for lower closure property of orientor field

$$(u(t), v(t)) \quad Q(t, x(t)) \tag{5.7}$$

when the convergence in u-space is L_1-weak.

In the proof of Theorem 5.1 (part (ii) ⇒ (iii)), we used the following results concerning measurable set-valued maps without proving them: The closed set-valued map Q is measurable if and only if there is a denumerable sequence of measurable selectors $\{p_i\}$ such that $Q(t) = \mathrm{cl}\{p_i(t)\}$. In Theorem 5.1, we assumed that Q has $\mathscr{L}\times\mathscr{B}$-measurable graph. This implies measurability of Q in the sense of Definition 3.1 and is equivalent to if values of Q(t) are closed. We have concluded in the proof above that $\widetilde{Q}(t) = Q(t)$ a.e. in [0, 1] from the fact that the sets of measurable selections are the same. There we used the fact that $\widetilde{Q}(t)\setminus Q(t)$ or $Q(t)\setminus\widetilde{Q}(t)$ have graphs $\mathscr{L}\times\mathscr{B}$-measurable. Thus $A = \{t\,|\,\widetilde{Q}(t)\setminus Q(t) \neq \emptyset\}$ is Lebesgue measurable and if the measure of this set were not zero then there would be a selection w of $\widetilde{Q}$ such that $w(t) \notin Q(t)$ on A. There we made a use of the following selection theorem: If $Q : [0, 1] \to$ subsets of R^n has $\mathscr{L}\times\mathscr{B}$-measurable graph then there is measurable selection of Q.

For Q measurable in the sense of Definition 3.1, such a selection theorem holds provided the values of Q are closed.

Theorem 5.2. Assume that the map $Q(t, x)$ of $[0, 1] \times R^n$ into closed and convex subsets of R^{n+1} has the properties: (i) if $(q, q_0) \in Q(t, x)$ and $r > q_0$ then $(q, r) \in Q(t, x)$, (ii) the graph of Q is $\mathscr{L}\times\mathscr{B}$-measurable subset of $[0, 1] \times R^n \times R^{n+1}$, (iii) for each fixed t the graph of $Q(t, \cdot)$ is closed. Assume further that there is ψ: $[0, 1] \to R$ integrable and constant M such that for each integrable x: $[0, 1] \to R^n$ there is measurable p: $[0, 1] \to R^n$ such that $|p(t)| \leq M$ and

$$-r + \langle q, p(t)\rangle \leq \psi(t) + M_1 |x(t)| \quad \text{for each } (q.\ r) \in Q(t, x(t))$$

Then (5.7) has lower closure property with respect to strong convergence in L_1 for x, weak sequential convergence in L_1 for u. If $M_1 = 0$ then the conclusion holds also for pointwise convergence for x.

In this paper we shall not include the detailed proof of the above result but only mention the main steps.

Suppose $x_n(t) \to x_0(t)$ pointwise and in L_1-norm, $u_n \to u_0$ weakly in L_1, $\int_0^1 v_n(t)dt \to a$ and $(u_n(t), v_n(t)) \in Q(t, x_n(t))$ for each n. To prove the Theorem 5.2 we need to show that there is $v_0 \in L_1$ such that $\int_0^1 v_0(t)dt \leq a$ and $(u_0(t), v_0(t)) \in Q(t, x_0(t))$. To the latter set-valued function we apply Theorem 5.1, parts (ii) and (iii). In particular, for $Q(t, x_0(t))$, (5.3) holds. Let $\varphi \in \Phi$ be fixed. Using the assumption of lower semicontinuity of $Q(t, x)$, one can show that there is $\varphi_n \in L_\infty$, ψ_n, $\psi \in L_1$ such that

$$-v_n(t) + \langle \varphi_n(t), u_n(t)\rangle \leq \psi_n(t) \leq \psi(t)$$

$$\limsup \psi_n(t) = \psi_\varphi(t)$$

$$\varphi_n \to \varphi \text{ in } L_1\text{-norm.}$$

The above conditions imply that there is v_0 independent on φ such that $\int_0^1 v_0(t)dt \leq a$ and $-v_0(t) + \langle \varphi(t), u_0(t) \rangle \leq \psi_\varphi(t)$ a.e. in $[0,1]$, which together with (5.3) proves the Theorem. Thus, the main difficulties in the proof lie in the construction of φ_n and ψ_n.

An equivalent formulation of Theorem 5.2 one gets if one notices that $Q(t,x)$ = epigraph $f(t,x,\cdot)$ and that l.c.p. of (5.7) is equivalent to lower semicontinuity of the functional

$$I(x,u) = \int_0^1 f(t, x(t), u(t))dt$$

6. COMMENTS

The existence theorem 1.1 can be found in the by now classical paper [19] of LaSalle for U being a cube and also in book [35]. Earlier results are due to R.V. Gamcrelidze for U being one-dimensional and to Bushaw for two-dimensional systems.

The exposition of Section 2 follows the author's papers on the subject [22-27]. Lemma 1 is an extension of a result obtained by Blackwell [7]. It can also be found in a different form in the paper by Borges [8]. The main difference between our method of obtaining Theorems 2.1 and 2.2 and that of other authors is contained in Lemma 3. The simple idea of the proof of this lemma is also responsible for the form of Theorem 2.1 more general than that usually to be seen in the literature; by this we mean the second part of it, which states that we may restrict ourselves to piecewise extremal selectors without changing the integral S(t). To prove this lemma, we use the fact that the Lebesgue integral on the interval $[0, t]$ is continuous with respect to t. As has been noted in one of our first publications on this subject [23], working with a Lebesgue measure on an interval is not much of a restriction. What is really needed in the case of an abstract measure space (T, μ) is a one-parameter family of sets T_t such that $T_t \subset T_s$ if $t < s$ and that, for each integrable function v, the integral $\int_{T_t} v(\tau)d\mu(\tau)$ is a continuous function of t. In fact, if (T, μ) does not admit atoms and the measure is finite, then such a family does exist.

The proofs of the lemmas in Section 2 are different from those in the author's previous papers. They are more "co-ordinate-free". This seems to make them simpler. In particular, the proof of Lemma 1 is simpler. Notice that, there, we essentially prove the well-known result that exposed points of a convex closed set are dense in the profile of this set. For propositions concerning convex sets and their extremal structure, we refer to book [37] and also to a paper by Klee [17]. In particular, proposition (ϵ) is due to Klee.

There are reasons why the set of integrals of integrable selectors to a set-valued function is called the integral of the latter: One can define an integral of a set-valued function by imitating the definitions of the integral of a single-valued function. Many of such definitions have been given and it appears that, in general, they are equivalent to the one we used here and which sometimes is referred to in the literature as the Aumann integral [2] For further references and discussion of the integral of set-valued mapping, we refer the reader to paper [27].

Concerning measurable set-valued maps, selection theorem and related question, the literature is quite extensive. The interested reader may look up the lecture notes by Parthasarathy [33]. The paper of Rockafellar [38] on the subject is well written and is a good source of information. In particular, the characterization of a closed set-valued map used in one proof of Theorem 5.1 (part (ii) ⇒ (iii)) (see comments following the proof) has been taken therefrom. In the same place, we have used a recent selection theorem found by Aumann [3]. Proposition 3.1 called often the Filippov lemma [14] has been also proved and used by Ważewski [44]. Proposition 3.2 for compact set-valued maps is proved in Ref. [24].

One of the first generalizations of the existence theorem for linear systems of form (1.1) to non-linear systems in the control parameter of form (3.6) is due to Neustadt [21]. Theorem 3.1, in that generality, is essensially due to the author and can be found in Ref. [25]. In the same paper, the extension of the "bang-bang" principle stated in Theorem 3.2 can be found. This theorem can be used to obtain a generalization of Theorem 1 to systems of form (3.6). We mean to formulate and to prove the existence and/or the "bang-bang" principle in the class of piecewise continuous control functions. Such a result was given by Levinson [20] and Halkin and Hendricks [15]. For Theorem 3.3 and similar results and applications, we refer to Refs [26, 27]. The latter paper contains some more references and discussions and could be consulted in connection with the material contained in Sections 2 and 3. Among more recent publications on this subject we would like to mention Berliocchi and Lasry [6], Wagner and Stone [41], Arstein [1], and Cesari [13]. A general reference for time-optimal control problems is a very nicely written monograph [16] by Hermes and LaSalle.

The first existence theorem for non-linear control problems is due to Filippov [14]. It is concerned with the time-optimal problem and deals with the case where the reachable set is compact. Similar cases were also treated by Roxin [40]. These results are connected with (or follow from) the theory of contingent equations developed in the thirties by Zaremba and Marchaud. It was Ważewski who noticed and explored this connection giving a foundation for, as he called them himself, orientor fields and which also is known under other names as differential equations with multi-valued right-hand side, differential inclusion and some others. In Ref. [43], the interested reader can find the basic theory of orientor fields of Ważewski and references to his papers.

Cesari was the first to investigate the unbounded case. He connected the Tonelli-McShane-Naguno existence theory with optimal control problems and obtained a series of quite general results. His contribution to existence theory cannot be overestimated. A selection of the multitube of papers which he published on the subject in the last ten years is included in the list of references. In particular, existence results of the type of Theorem 4.1 can be found in Ref. [9] with uniform growth condition. This theorem also appears in the author's paper [28] and the proof is taken from that paper.

For Theorem 4.2 we refer to Ref. [29]. In the papers quoted the reader will also find some other existence theorems. For various formulations of optimal control and its relation with classical variational problems, see Rockafellar's work, in particular Ref. [39]. Among others, integral functionals with the integrand assuming also the value $+\infty$ are treated systematically in an elegant and convincing way in this paper.

Theorem 5.1 was stated in Ref. [31]; it generalizes some results contained in Poljak's paper [34]. In the latter paper, also the necessary and sufficient condition for L_1-weak sequential lower semicontinuity is obtained (equivalence between (ii) and (iv)) only for f(t, q) continuous in q, while here we are able to prove it only by assuming that it is Borel-measurable. The proof of Theorem 5.1 appears here for the first time. For related results which are, however, concerned with different weak topology, see the author's papers [30, 31]. Theorem 5.2 was also stated in Ref. [31], and the detailed proof will appear in Ref. [32]. This theorem is essentially due to Berkovitz [5] and Cesari [10]. Our formulation is slightly more general, and the proof is different.

In this contribution, we were not able to cover all results connected with existence theory. The two existence theorems in Section 4 and two results concerning lower closure in Section 5 are only examples. Moreover, also the list of references is far from being complete. The important theory of generalized solutions of Young or sliding regimes of Gamcrelidze or relaxed controls of Warga, which is very closely related with the existence of optimal solutions, has not been mentioned, at all. We refer the reader to monographs by Young [45] and Warga [42]. The idea of starting this paper with the Perron paradox was borrowed from Ref. [45]. Finally, we would also like to mention a recent book by Ioffe and Tichomirov [36], where the existence of optimal solutions is treated in some detail. Many more references to this subject can be found there.

REFERENCES

[1] ARTSTEIN, Z., On a variational problem, J. Math. Anal. Appl. 45 (1974) 404-415.

[2] AUMANN, R.J., Integrals of set-valued functions, J. Math. Anal. Appl., 12 (1965) 1-12.

[3] AUMANN, R.J., Measurable utility and the measurable choice theorem, Proc. Int. Coll. C.N.R.S. "La Decision", Paris (1969) 15-26.

[4] BERKOVITZ, L.D., Existence and lower closure theorems for abstract control problems, SIAM J. on Control, 12 (1974), 27-42.

[5] BERKOVITZ, L.D., Lower semicontinuity of integral functionals, to appear in Trans. A.M.S.

[6] BERLIOCCHI, H., LASRY, J.M., Intégrandes normales et mesures paramétrées en calcul des variations, Bull. Soc. Math. France 101 (1973).

[7] BLACKWELL, D., The range of certain vector integrals, Proc. Am. Math. Soc. 2 (1951) 390-395.

[8] BORGES, R., Ecken des Wertebereiches von Vektorintegralen, Math. Annalen 175 (1967) 53-58.

[9] CESARI, L., Existence theorems for weak and usual optimal solutions in Lagrange problems with unilateral constraints, I and II, Trans. A.M.S., 124 (1966) 396-412 and 413-470.

[10] CESARI, L., Lower semicontinuity and lower closure theorems without seminormality conditions, Ann. Mat. Pura Appl. 98 (1974).

[11] CESARI, L., Closure theorems for orientor fields and weak convergence, Arch. Ration. Mech. Anal., to appear.

[12] CESARI, L., LA PALM, J.R., SANCHEZ, D.A., An existence theorem for Lagrange problems with unbounded controls and a slender set of exceptional points, SIAM J. Control, 9 (1971), 590-605.

[13] CESARI, L., An existence theorem without convexity conditions, SIAM J. on Control, 12 (1974) 319-331.

[14] FILIPPOV, A.F., On certain questions in the theory of optimal control, Vestnik Moskov. Univ., Ser. Math. Astron. 2 (1959) 25-32.

[15] HALKIN, H., HENDRICKS, E.C., Sub-integrals of set-valued functions with semianalytic graphs, Proc. Nation. Acad. Sc., 59 (1968) 365-367.

[16] HERMES, H., LASALLE, J.P., Functional Analysis and Time Optimal Control, Academic Press (1969).

[17] KLEE, V., Extremal structure of convex sets, Arch. Math., 8 (1957) 234-240.

[18] KURATOWSKI, K., RYLL-NARDZEWSKI, C., A general theorem on selectors, Bull. Acad. Pol. Sci., Ser. Sci. Math. Astron. Phys. 6 (1965) 397-403.

[19] LASALLE, J.P., The time optimal control problem, Contr. to the theory of nonlinear oscillations, 5, Princeton Univ. Press, Princeton (1960) 1-24.

[20] LEVINSON, N., Minimax, Liapunov and "bang-bang", J. Diff. Eq. 2 (1966) 218-241.

[21] NEUSTADT, L.W., The existence of optimal controls in the absence of convexity conditions. J. Math. Anal. Appl. 7 (1963) 110-117.

[22] OLECH, C., A contribution to the time optimal control problem, Abhandlungen der Deutschen Akademie der Wissenschaften zu Berlin, Kl. Physik und Technik 2 (1965) 438-446.

[23] OLECH, C., Extremal solution of a control system. J. Diff. Eqs. 2 (1966) 74-101.

[24] OLECH, C., A note concerning set-valued measurable functions, Bull. Acad. Pol. Sci., Ser. Sci. Math. Astron. Phys. 13 (1965) 317-321.

[25] OLECH, C., "Lexicographical order, range of integrals and bang-bang principle", Math. Theory of Control (BALAKRISHNAN, A.V., NEUSTADT, L.W., Eds), Academic Press, New York (1967) 35-37.

[26] OLECH, C., "Integrals of set-valued functions and linear control problems", IFAC Congress Warsaw, 1969 on Optimal Control, Technical Session 7, 22-35.

[27] OLECH, C., Integrals of set-valued functions and linear optimal control problems, Colloque sur la Théorie Mathématique du Contrôle Optimal, C.B.R.M., Vander, Louvain (1970) 109-125.

[28] OLECH, C., Existence theorems for optimal problems with vector valued cost function, Trans. Am. Math. Soc. 136 (1969) 159-179.

[29] OLECH, C., Existence theorems for optimal control problems involving multiple integrals, J. Diff. Eqs 6 (1966) 512-526.

[30] OLECH, C., The characterization of the weak closure of certain sets of integrable functions, SIAM J. Control 12 (1974) 311-318.

[31] OLECH, C., Existence theory in optimal control problems – the underlying ideas, to appear in the proceedings of a conference held at the University of Southern California, Los Angeles, September 1974.

[32] OLECH, C., Weak lower semicontinuity of integral functionals, to appear in J. Optim. Theor. Appl.

[33] PARTHASARATHY, T., Selection theorems and their applications, Lecture Notes in Mathematics 263, Springer-Verlag (1972).

[34] POLJAK, B.T., Semicontinuity of integral functionals and existence theorems for extremal problems, Mat. Sbor. 78 (1969) 65-84, (in Russian).

[35] PONTRYAGIN, L.S., BOLTYANSKI, V.G., GAMCRELIDZE, R.V., MISHCHENKO, E.F., The Mathematical Theory of Optimal Control, Moscow (1961) in Russian. English translation: Interscience, New York (1962).

[36] IOFFE, A.D., TICHOMIROV, W.M., Theory of extremal problems, (in Russian) Moscow (1974).

[37] ROCKAFELLAR, R.T., Convex Analysis, Princeton Univ. Press (1969).

[38] ROCKAFELLAR, R.T., Measurable dependence of convex sets and functions on parameters, J. Math. Anal. Appl., 28 (1967) 4-25.

[39] ROCKAFELLAR, R.T., Existence theorems for general control problems of Bolza and Lagrange, to appear in Advances in Math.

[40] ROXIN, E., The existence of optimal controls, Michigan Math. J. 9 (1962) 109-119.

[41] WAGNER, D.H., STONE, L.D., Necessity and existence results on constrained optimization of seperable functionals by a multiplier rule, SIAM J. Control 12 (1974) 356-372.

[42] WARGA, J., Optimal control of differential and functional equations, Academic Press, New York (1972).

[43] WAŻEWSKI, T., On an optimal control problem, Proc. Conf. Diff. Equations and their Applications, Prague (1964) 229-242.

[44] WAŻEWSKI, T., Sur une condition d'existence des fonctions implicites mesurables, Bull. Acad. Pol. Sci., Ser. Sci. Math. Astron. Phys. 9 (1961) 861-863.

[45] YOUNG, L.C., Lectures on the Calculus of Variations and Optimal Control Theory, W.B. Saunders Company, Philadelphia-London-Toronto (1969).

IAEA-SMR-17/48

ASYMPTOTIC CONTROL

R. CONTI
Istituto Matematico "Ulisse Dini",
Università degli Studi,
Florence, Italy

Abstract

ASYMPTOTIC CONTROL.

Asymptotic control is discussed in the framework of general control theory, special emphasis being placed on stability (including bounded-input bounded-state stability), affine control systems and stabilization problems.

1. PRELIMINARIES

We shall first recall a few more or less well-known facts about "linear" ordinary differential equations, so as to render this paper as self-contained as possible.

Let us denote by:

$J =]\alpha, \omega[$ an open interval of the real line $\mathbb{R}$, with $-\infty \leqq \alpha < \omega \leqq +\infty$;

$A : t \to A(t)$ an $n \times n$ matrix function of $t \in J$, Lebesgue measurable and locally integrable on J;

$\chi : t \to \chi(t)$ an n-vector function of $t \in J$, continuous on J.

Given any $\theta \in J$, the Volterra integral equation

$$x(t) = \chi(t) + \int_\theta^t A(s)\, x(s)\, ds$$

has a single solution $x : t \to x(t)$ defined on J by

$$x(t) = \lim_k \Big[\chi(t) + \int_\theta^t A(t_1)\, \chi(t_1)\, dt_1 + \ldots$$

$$\ldots + \int_\theta^t \ldots \int_\theta^{t_{k-1}} A(t_1) \ldots A(t_k)\, \chi(t_k)\, dt_k \ldots dt_1 \Big]$$

When $\chi(t) = \chi$, a constant n-vector, this can be written

$$x(t) = \lim_k \Big[I + \int_\theta^t A(t_1)\, dt_1 + \ldots$$

$$\ldots + \int_\theta^t \ldots \int_\theta^{t_{k-1}} A(t_1) \ldots A(t_k)\, dt_k \ldots dt_1 \Big] \chi \qquad (1.1)$$

where I denotes the $n \times n$ unit matrix.

The function $t \to x(t)$ defined by (1.1) satisfies the condition

$$x(\theta) = \chi \tag{C}$$

It is locally absolutely continuous on J and such that

$$\frac{d\,x(t)}{dt} - A(t)\,x(t) = 0, \text{ a.e. } t \in J$$

Therefore, we shall call it the (Carathéodory) solution of the linear ordinary differential equation

$$\dot{x} - A(t)\,x = 0 \tag{E_0}$$

satisfying (C).

Since the limit appearing in Eq. (1.1) exists for an arbitrary θ we can define the $n \times n$ matrix

$$G(t, \theta) = \lim_k \left[I + \int_\theta^t A(t_1)\,dt_1 + \ldots \right.$$

$$\left. \ldots + \int_\theta^t \ldots \int_\theta^{t_{k-1}} A(t_1) \ldots A(t_k)\,dt_k \ldots dt_1 \right] \tag{G}$$

which is called the transition matrix of (E_0).

Then we can replace (1.1) by the more compact formula

$$x(t) = G(t, \theta)\,x(\theta), \qquad t, \theta \in J \tag{1.2}$$

From (1.2) it is easy to derive the algebraic properties of G:

$$G(t, \theta)\,G(\theta, \tau) = G(t, \tau), \quad \theta, \tau, t \in J \tag{1.3}$$

$$G(t, t) = I, \quad t \in J \tag{1.4}$$

$$G^{-1}(t, \theta) = G(\theta, t), \qquad \theta, t \in J \tag{1.5}$$

and the differential properties

$$\frac{\partial G(t, \theta)}{\partial t} - A(t)\,G(t, \theta) = 0, \quad \theta \in J, \quad \text{a.e. } t \in J \tag{1.6}$$

$$\frac{\partial G(t, \theta)}{\partial \theta} + G(t, \theta)\,A(\theta) = 0, \quad t \in J, \quad \text{a.e. } \theta \in J \tag{1.7}$$

The norm of the matrix $G(t, \theta)$ is the number

$$|G(t, \theta)| = \sup \left\{ |G(t, \theta)\,\chi|_2 : |\chi|_2 \leq 1 \right\} \tag{1.8}$$

where $|\ |_2$ is the Euclidean norm in $\mathbb{R}^n$.

It is easy to show that $(t,\theta) \to G(t,\theta)$ is a continuous function on $J \times J$ with respect to the norm.

By virtue of (1.2), the number (1.8) is equivalent to

$$|G(t,\theta)| = \sup\left\{|x(t)|_2 \ : \ x \in \Sigma_0, \ |x(\theta)|_2 \leq 1\right\} \tag{1.9}$$

where Σ_0 denotes the set of solutions of (E_0).

Let us now recall some inequalities to be used later on.

Let us denote by $\lambda_H(t)$ and $\mu_H(t)$ the least and the greatest eigenvalue, respectively, of the Hermitian matrix

$$H(t) = \frac{1}{2} A(t) + \frac{1}{2} A^*(t) \tag{1.10}$$

where A^* is the transpose of A.

Then it can be shown that $t \to \lambda_H(t)$ and $t \to \mu_H(t)$ are measurable and locally integrable on J and we have

$$\begin{cases} e^{\int_\theta^t \lambda_H(s)\,ds} \leq |G(t,\theta)| \leq e^{\int_\theta^t \mu_H(s)\,ds}, & \theta \leq t \\ e^{-\int_\theta^t \mu_H(s)\,ds} \leq |G(\theta,t)| \leq e^{-\int_\theta^t \lambda_H(s)\,ds}, & \theta \leq t \end{cases} \tag{1.11}$$

From these inequalities it follows that

$$e^{-\int_\theta^t |H(s)|\,ds} \leq \left\{ \begin{matrix} |G(t,\theta)| \\ |G(\theta,t)| \end{matrix} \right\} \leq e^{\int_\theta^t |H(s)|\,ds}, \quad \theta \leq t \tag{1.12}$$

and

$$e^{-\int_\theta^t |A(s)|\,ds} \leq \left\{ \begin{matrix} |G(t,\theta)| \\ |G(\theta,t)| \end{matrix} \right\} \leq e^{\int_\theta^t |A(s)|\,ds}, \quad \theta \leq t \tag{1.13}$$

For a fixed θ, (1.2) represents an isomorphism between $\mathbb{R}^n$ and the set Σ_0 of solutions of (E_0), so that Σ_0 is a (real) vector space of dimension n.

Let us denote by $X : t \to X(t)$ any $n \times n$ matrix function whose columns $x^1, \ldots, x^n$ are solutions of (E_0). From (1.2) we have

$$X(t) = G(t,\theta)\, X(\theta), \quad t, \theta \in J$$

hence

$$\det X(t) = \det G(t,\theta) \det X(\theta)$$

Since $\det G(t,\theta) \neq 0$ we have that either $X(t)$ is non-singular for every $t \in J$ or it is singular for every $t \in J$. In the first case, we can write

$$G(t,\theta) = X(t)\, X^{-1}(\theta), \qquad t,\ \theta \in J \tag{1.14}$$

This yields easily

$$\det G(t,\theta) = e^{\int_\theta^t \operatorname{tr} A(s)\, ds} \qquad t,\ \theta \in J \tag{1.15}$$

The equation

$$\dot{z} + A^*(t)\, z = 0 \tag{E*}$$

is the adjoint to (E_0). Using (1.14) it is easy to show that the transition matrix Γ of (E_0) is given by

$$\Gamma(t,\theta) = G^*(\theta, t) = G^{*-1}(t,\theta)\ , \qquad t,\ \theta \in J \tag{1.16}$$

Equation (E_0) is said to be autonomous when A is independent of t. In this case $J = \mathbb{R}$ and

$$G(t,\theta) = e^{(t-\theta)A}, \quad t,\ \theta \in \mathbb{R} \tag{1.17}$$

It follows that the solutions of (E_0) are analytic functions of $t \in \mathbb{R}$. More precisely, let $\lambda_1, \ldots, \lambda_k$ denote the distinct eigenvalues of A, $n_1, \ldots, n_k$ their respective multiplicities. Then for every $\chi \in \mathbb{C}^n$ there are $\chi^1, \ldots, \chi^k \in \mathbb{C}^N$ such that

$$e^{tA}\chi = \sum_{j=1}^{k} e^{\lambda_j t} \left[\sum_{r=0}^{n_j - 1} \frac{t^r}{r!} (A - \lambda_j I)^r \right] \chi^j \tag{1.18}$$

2. STABILITY I

From now on, we shall assume $\omega = +\infty$, i.e. $J =]\alpha, +\infty[$, $-\infty \leq \alpha$, and we shall study the behaviour of the solutions of

$$\dot{x} - A(t)\, x = 0 \tag{E_0}$$

as $t \to +\infty$.

Because of

$$x(t) = G(t,\theta)\, x(\theta) \tag{1.2}$$

this amounts to studying the behaviour of $G : (t,\theta) \to G(t,\theta)$ as $t \to +\infty$.

As a matter of notation, we shall write $x(t,\theta,\chi)$ instead of $x(t)$ to denote the solution of (E_0) such that $x(\theta) = \chi$, so that (1.2) will be replaced by

$$x(t,\theta,\chi) = G(t,\theta)\chi \tag{2.1}$$

It is well known that the zero solution of (E_0) is stable (according to Liapunov, as $t \to +\infty$) iff for every $\epsilon > 0$ and $\tau > \alpha$ there exist $\delta = \delta(\tau, \epsilon)$ such that

$$|\chi|_2 < \delta, \quad \alpha < \tau \leqq t \Rightarrow |x(t, \tau, \chi)|_2 < \epsilon \tag{2.2}$$

It is easy to see that this holds iff G has the property S: for every $\tau > \alpha$ there exist $\gamma(\tau) > 0$ such that

$$\alpha < \tau \leqq t \Rightarrow |G(t, \tau)| < \gamma(\tau) \tag{2.3}$$

In fact, if (2.3) holds, for any $\epsilon > 0$, $\tau > \alpha$ we have

$$|x(t, \tau, \chi)|_2 = |G(t, \tau)\chi|_2 < \gamma(\tau)|\chi|_2, \qquad \alpha < \tau \leqq t$$

if $|\chi|_2 < \delta(\tau, \epsilon) = \gamma^{-1}(\tau)\epsilon$. Conversely, if (2.2) holds, for $\epsilon = 1$, $\tau > \alpha$, there exist $\delta = \delta(\tau) > 0$ such that

$$|\chi|_2 < \delta, \quad \alpha < \tau \leqq t \quad \Rightarrow \quad |G(t, \tau)\chi|_2 < 1$$

whence

$$\alpha < \tau \leqq t \quad \Rightarrow \quad |G(t, \tau)| < \delta^{-1}(\tau)$$

Property S means that for every fixed $\tau > \alpha$ the function $t \to |G(t, \tau)|$ is bounded for $\tau \leqq t$. A more restrictive condition is satisfied when for every $\theta > \alpha$ the function $(t, \tau) \to |G(t, \tau)|$ is bounded for $\theta \leqq \tau \leqq t$. This can be expressed by saying that G has the property US: for every $\theta > \alpha$ there exist $\gamma(\theta) > 0$ such that

$$\theta \leqq \tau \leqq t \Rightarrow |G(t, \tau)| < \gamma(\theta) \tag{2.4}$$

It is readily seen (by the same arguments used to prove the equivalence between stability and property S) that property US is equivalent to the uniform stability of the zero solution of (E_0) as $t \to +\infty$. This means that for every $\epsilon > 0$, $\theta > \alpha$, there exist $\delta = \delta(\theta, \epsilon) > 0$ such that

$$|\chi|_2 < \delta, \quad \alpha < \theta \leqq \tau \leqq t \Rightarrow |x(t, \tau, \chi)|_2 < \epsilon \tag{2.5}$$

Another property of G more restrictive than property S is property AS: for every $\tau > \alpha$ we have

$$\lim_{t \to +\infty} |G(t, \tau)| = 0 \tag{2.6}$$

Clearly, this is equivalent to

$$\tau > \alpha, \quad \chi \in \mathbb{R}^n \Rightarrow \lim_{t \to +\infty} |x(t, \tau, \chi)|_2 = 0 \tag{2.7}$$

i.e. to the asymptotic stability of the zero solution of (E_0) as $t \to +\infty$.

The zero solution of (E_0) is said to be exponentially asymptotically stable as $t \to +\infty$, when (2.7) is reinforced by

$$\alpha < \theta \leqq \tau \leqq t, \ \ \chi \in \mathbb{R}^n \Rightarrow |x(t,\tau,\chi)|_2 \leqq \gamma(\theta) \ e^{-\mu(\theta)(t-\tau)} |\chi|_2 \tag{2.8}$$

for some $\gamma(\theta) > 0$ and $\mu(\theta) > 0$.

This is clearly equivalent to property EAS of G: for every $\theta > \alpha$ there exist $\gamma(\theta) > 0$, $\mu(\theta) > 0$ such that

$$\alpha < \theta \leqq \tau \leqq t \quad \Rightarrow \quad |G(t,\tau)| < \gamma(\theta) \ e^{-\mu(\theta)(t-\tau)} \tag{2.9}$$

Looking at the definitions we see immediately that the following implications among the four properties S, US, AS, EAS of G are valid:

$$\begin{array}{ccccc} & \nearrow & US & \searrow & \\ EAS & & & & S \\ & \searrow & AS & \nearrow & \end{array}$$

None of these are reversible and the two properties US, AS are independent of each other, as examples show. To this effect, let us consider the scalar ($n = 1$) equation

$$\dot{x} - \frac{\dot{f}(t)}{f(t)} \ x = 0 \tag{2.10}$$

where $f : t \to f(t)$ is a function defined for $t > \alpha$, positive and locally absolutely continuous. We immediately see that G is given by

$$G(t,\tau) = f(t)/f(\tau)$$

Example 2.1

Let $\alpha = 0$ and let $f(t) = 1/t$ for $0 < t \leqq 2$, while for $t \geqq 2$ the graph of f is the polygonal with subsequent vertices at $(2,1/2)$, $(3,1)$, $(4,1/4)$, $(5,1)$, ... Such an f is bounded but it does not tend to zero so that G has the property S but not AS. Nor has it property US since $G(2k+1,2k) = 2k$, $k = 1, 2, \ldots$

Example 2.2

Let $\alpha = -\infty$, $f(t) = 1 + e^{-t}$. Then G has property US, but not AS.

Example 2.3

Let $\alpha = 0$, $f(t) = t^{-2+\cos t}$. G has the property AS but not US, since $G(2k\pi, (2k-1)\pi) = \pi^2(2k-1)^3/(2k)$, $k = 1, 2, \ldots$

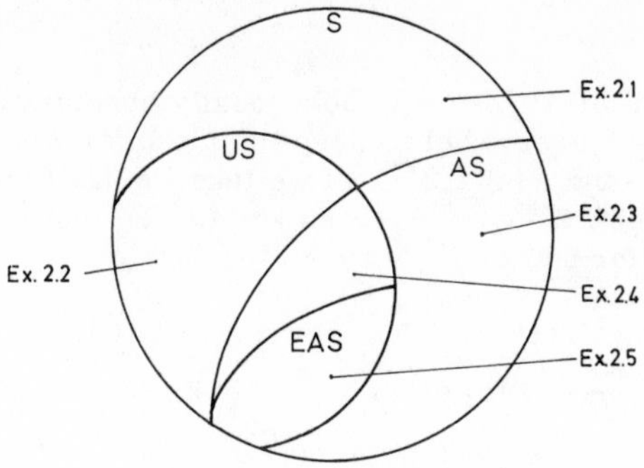

FIG.1. Relationship of properties S, US, AS and EAS.

Example 2.4

Let $\alpha = 0$, $f(t) = 1/t$. Then G has both properties US and AS but not EAS.

Example 2.5

Let $\alpha = -\infty$, $f(t) = e^{-t}$. G has the property EAS.

What we have said can be visualized by a scheme showing the implications among the properties S, US, AS, EAS as is shown in Fig. 1.

When (E_0) is autonomous, then it can be shown that properties S and US coincide and are characterized by: a) all the eigenvalues of A have real parts ≤ 0, and b) those eigenvalues λ_j which have zero real part (if any) have a multiplicity n_j equal to n-rank $(A - \lambda_j I)$.

Furthermore, properties AS and EAS coincide and are equivalent to: c) all the eigenvalues of A have real parts < 0 (strictly).

3. STABILITY, II

Let us now consider a property of G of integral type, namely property IS: for every $\theta > \alpha$ let there exist $k(\theta) > 0$ such that

$$\alpha < \theta \leq t \Rightarrow \int_\theta^t |G(t, s)|\, ds < k(\theta) \tag{3.1}$$

This property is independent of property US.

In fact, the G of Example 2.4 is defined by $G(t, s) = s/t$, so that it has property US but not IS since

$$\int_\theta^t |G(t, s)|\, ds = (t^2 - \theta^2)/(2t) \to +\infty \quad \text{as} \quad t \to +\infty$$

On the other hand, the G of the next example has the property IS but does not have property US.

Example 3.1

Let $\alpha = -\infty$ and let $\lambda : t \to \lambda(t)$ be a locally absolutely continuous function of $t \in \mathbb{R}$, $= 1$ everywhere except on the intervals $J_k = [k - 2^{-4k}, k + 2^{-4k}]$ where $1 \leq \lambda(t) \leq 2^{2k}$ and $\lambda(k) = 2^{2k}$. If we take Eq. (2.10) with $f(t) = e^{-t}/\lambda(t)$ we have $G(k + 2^{-4k}, k) = 2^{2k} e^{-2^{-4k}} \to +\infty$ as $k \to +\infty$, so that US does not hold. On the other hand, for $\theta < 1 - 2^{-4}$ we have

$$\int_\theta^t |G(t, s)| \, ds = \frac{e^{-t}}{\lambda(t)} \int_\theta^t \lambda(s) e^s \, ds \leq e^{-t} \int_\theta^t e^s \, ds + \sum_{k=1}^{[t+1]} \int_{J_k} \lambda(s) \, ds$$

$$\leq e^{-t}(e^t - e^\theta) + 2 \sum_{k=1}^{\infty} 2^{2k} 2^{-4k} \leq 1 + 2/3$$

so that G has the property IS.

Example 2.3 shows that

$$AS \not\Rightarrow IS \tag{3.2}$$

In fact, since $\cos s \leq 1$ we have $s^{2 - \cos s} > s$, hence for $t \geq \theta > 0$

$$\int_\theta^t |G(t, s)| \, ds = t^{-2 + \cos t} \int_\theta^t s^{2 - \cos s} \, ds$$

$$> t^{-2 + \cos t} \int_\theta^t s \, ds = t^{\cos t} (1 - (\theta/t)^2)/2$$

so that

$$\int_\theta^{2k\pi} |G(2k\pi, s)| \, ds = k\pi(1 - (\theta/2k\pi)^2) \to +\infty \quad \text{as} \quad k \to +\infty.$$

On the other hand, we have

$$IS \Rightarrow AS \tag{3.3}$$

To prove this, let $\theta > \alpha$ and let $\theta' = \begin{cases} \theta - 1 & \text{if } \alpha = -\infty \\ (\alpha + \theta)/2 & \text{if } \alpha > -\infty \end{cases}$

Let, further,

$$\varphi(t) = |G(t, \tau)|^{-1}, \qquad \tau \leq t \tag{3.4}$$

$$\psi(t) = \int_{\theta'}^t \varphi(s) \, ds, \qquad \theta' \leq t \tag{3.5}$$

so that $\varphi(t) > 0$, $\psi(t) > 0$, and, in particular, $\psi(\theta) > 0$. Then we have

$$\psi(t)\,\varphi^{-1}(t) = \psi(t)\left|G(t,\tau)\right| = \left|\psi(t)\,G(t,\tau)\right|$$

$$= \left|\left(\int_{\theta'}^{t}\varphi(s)\,ds\right)G(t,\tau)\right| = \left|\int_{\theta'}^{t}\varphi(s)\,G(t,\tau)\,ds\right|$$

$$= \left|\int_{\theta'}^{t}\varphi(s)\,G(t,s)\,G(s,\tau)\,ds\right| \leqq \int_{\theta'}^{t}\varphi(s)\left|G(t,s)\right|\left|G(s,\tau)\right|\,ds$$

$$= \int_{\theta'}^{t}\left|G(t,s)\right|\,ds < k(\theta')$$

by virtue of property IS. Owing to the definition of θ', we can replace $k(\theta')$ by $k(\theta)$ and write

$$\psi(t)\,\varphi^{-1}(t) \leqq k(\theta), \qquad \theta \leqq t \tag{3.6}$$

From this follows $(d/dt)\psi(t)\,e^{-k^{-1}(\theta)(t-\tau)} \geqq 0$, $\theta \leqq \tau \leqq t$, hence, integrating between θ and t

$$\psi(t)\,e^{-k^{-1}(\theta)\,(t-\tau)} \geqq \psi(\theta)\,e^{-k^{-1}(\theta)\,(\theta-\tau)}$$

that is

$$\psi^{-1}(t) \leqq \psi^{-1}(\theta)\,e^{-k^{-1}(\theta)(t-\theta)}$$

$$\leqq \psi^{-1}(\theta)\,e^{-k^{-1}(\theta)(t-\tau)}, \quad \theta \leqq \tau \leqq t$$

Since (3.4), (3.6) give

$$\left|G(t,\tau)\right| = \varphi^{-1}(t) \leqq k(\theta)\,\psi^{-1}(t)$$

we have

$$\left|G(t,\tau)\right| \leqq k(\theta)\,\psi^{-1}(\theta)\,e^{-k^{-1}(\theta)(t-\tau)} \tag{3.7}$$

from which property AS follows.

It should be noted that (3.7) does not imply property EAS since $\psi^{-1}(\theta)$, by definition, depends on τ. However, if G has also the property US then for every $\theta > \alpha$ there exists $\gamma_0(\theta) > 0$ such that

$$\theta \leqq \tau \leqq t \Rightarrow \left|G(t,\tau)\right| < \gamma_0(\theta)$$

whence

$$\theta' \leqq \tau \leqq t \Rightarrow \varphi(t) = |G(t,\tau)|^{-1} > \gamma_0^{-1}(\theta)$$

so that

$$\psi(t) = \int_\theta^t \varphi(s)\, ds > \gamma_0^{-1}(\theta')\,(t - \theta')$$

In particular, $\psi(\theta) > \gamma_0^{-1}(\theta')\,(\theta - \theta')$, that is $\psi^{-1}(\theta) < \gamma_0(\theta')/(\theta - \theta')$, and by definition of θ',

$$\psi^{-1}(\theta) < \gamma_1(\theta) \tag{3.8}$$

where $\gamma_1(\theta) > 0$ is independent of τ.
From (3.8) and (3.7) it follows

$$\theta \leqq \tau \leqq t \Rightarrow |G(t,\tau)| \leqq k(\theta)\,\gamma_1(\theta)\, e^{-k^{-1}(\theta)\,(t-\tau)}$$

that is, property EAS holds with $\gamma(\theta) = k(\theta)\,\gamma_1(\theta)$, $\mu(\theta) = k^{-1}(\theta)$. We have thus proved

$$\text{IS plus US} \Rightarrow \text{EAS} \tag{3.9}$$

The converse is also true. In fac, EAS ⇒ US is obvious. Also, if EAS holds we have

$$\int_\theta^t |G(t, s)|\, ds < \gamma(\theta) \int_\theta^t e^{-\mu(\theta)\,(t-s)}\, ds = \gamma(\theta)\,\mu^{-1}(\theta)\,[1 - e^{-\mu(\theta)(t-\theta)}]$$

that is, IS holds with $k(\theta) = \gamma(\theta)\,\mu^{-1}(\theta)$.

Therefore

$$\text{IS plus US} \Longleftrightarrow \text{EAS} \tag{3.10}$$

and the scheme of Section 2 is now completed as is shown in Fig. 2.

4. STABILITY, III

We shall now try to characterize property EAS, i.e. the exponential asymptotic stability of the zero solution of

$$\dot{x} - A(t)\, x = 0 \tag{E_0}$$

by means of a "Liapunov function".

We start with the case of a constant A.

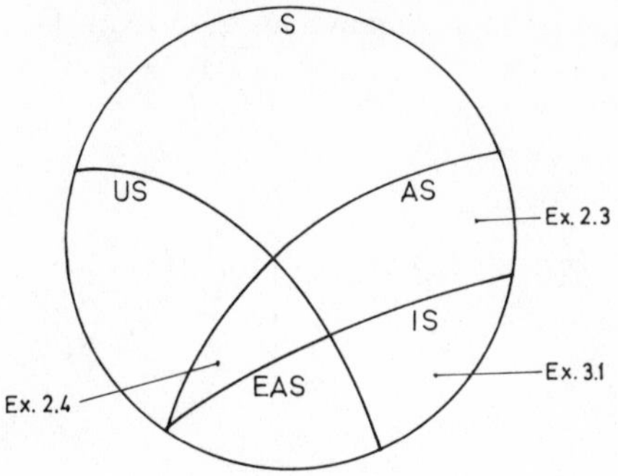

FIG. 2. Relationship of properties S, US, AS, IS and EAS.

Definition 4.1

We say that the constant matrix A admits a Liapunov function iff there are two $n \times n$ matrices Λ, Q which are positive definite:

$$\Lambda = \Lambda^* > 0 \tag{4.1}$$

$$Q = Q^* > 0 \tag{4.2}$$

and such that

$$A^* \Lambda + \Lambda A = -Q \tag{Λ}$$

When this happens the quadratic form

$$\chi \to \chi^* \Lambda \chi$$

is a Liapunov function of A and (Λ) is a Liapunov matrix equation associated with A.

We want to prove

Theorem 4.1

If A has a Liapunov function, then G has the property EAS.

Proof. Let

$$y(t) = e^{t(A+\mu I)} \chi \tag{4.3}$$

with $\mu > 0$, so that $dy(t)/dt - (A + \mu I)\, y(t) = 0$. From (Λ) it follows

$$d\, y^*(t)\, \Lambda y(t)/dt = y^*(t)\, [-Q + 2\, \mu \Lambda]\, y(t)$$

If we denote by μ_Λ the greatest eigenvalue of Λ, from (4.1) we have

$$\chi^* \Lambda \chi \leq \mu_\Lambda |\chi|_2^2$$

Also, denoting by λ_Q the least eigenvalue of Q from (4.2) we have

$$-\chi^* Q\chi \leqq -\lambda_Q |\chi|_2^2$$

Therefore,

$$y^*(t)\,[\,-Q + 2\,\mu\Lambda]\; y(t) \leqq [\,-\lambda_Q + 2\,\mu\mu_\Lambda]\; |y(t)|_2^2$$

and if

$$\mu < \frac{\lambda_Q}{2\mu_\Lambda} \tag{4.4}$$

we have $d\,y^*(t)\Lambda y(t)/dt < 0$, whence integrating between 0 and t

$$y^*(t)\;\Lambda\; y(t) < \chi^*\Lambda\chi$$

But again from (4.1) we have

$$\lambda_\Lambda\,|y(t)|_2^2 \leqq y^*(t)\;\Lambda\; y(t)$$

where λ_Λ is the least eigenvalue of Λ and

$$\chi^*\Lambda\chi \leqq \mu_\Lambda\,|\chi|_2^2$$

Therefore

$$\lambda_\Lambda |y(t)|_2^2 < \mu_\Lambda\,|\chi|_2^2$$

and (4.3) gives

$$|e^{tA}| \leqq \sqrt{\frac{\mu_\Lambda}{\lambda_\Lambda}}\; e^{-\mu t}$$

for μ satisfying (4.4).

Remark 4.1

Inequality (4.4) says that the exponential decay of the solutions of (E_0) is not faster than

$$t \to e^{-\frac{\lambda_Q}{2\gamma_\Lambda} t}$$

The converse of Theorem 4.1 is also valid.

Theorem 4.2

If G has the property EAS then A has a Liapunov function.
Proof. By assumption

$$|e^{tA}| < \gamma\, e^{-\mu t}, \qquad t \in \mathbb{R} \tag{4.5}$$

for some $\gamma > 0$ and $\mu > 0$.

On the other hand, for $\chi \in \mathbb{R}^n$ and any $Q = Q^* > 0$

$$\chi^*\left(\int_0^{+\infty} e^{sA^*} Q\, e^{sA}\, ds\right)\chi = \int_0^{+\infty} (e^{sA}\chi)^*\, Q\, (e^{sA}\chi)\, ds$$

$$\leq \mu_Q \int_0^{+\infty} (e^{sA}\chi)^* (e^{sA}\chi)\, ds = \mu_Q \int_0^{+\infty} |e^{sA}\chi|_2^2\, ds \leq \mu_Q\left(\int_0^{+\infty} |e^{sA}|^2\, ds\right)|\chi|_2^2$$

where μ_Q is the greatest eigenvalue of Q.

Therefore, by virtue of (4.5),

$$\chi^*\left(\int_0^{+\infty} e^{sA^*} Q\, e^{sA}\, ds\right)\chi \leq \frac{\gamma^2 \mu_Q}{2\mu}$$

This means that we can define

$$\Lambda = \int_0^{+\infty} e^{sA^*} Q\, e^{sA}\, ds \tag{4.6}$$

and we have $\Lambda = \Lambda^* > 0$ by virtue of (4.2) and

$$A^*\Lambda + \Lambda A = \int_0^{+\infty} \left[\frac{d}{ds}\, e^{sA^*} Q\, e^{sA}\right] ds = -Q$$

i.e. Λ is a solution of (Λ).

When A does depend on t, it is no longer reasonable to assume that the Λ and the Q in the definition of a Liapunov function are constant. Therefore we have

Definition 4.2

We say that $t \to A(t)$ has a Liapunov function iff for every $\theta > \alpha$ there are two $n \times n$ matrix-valued functions

$$t \to \Lambda_\theta(t), \qquad t \to Q_\theta(t)$$

such that

$$\Lambda_\theta(t) = \Lambda_\theta^*(t) \tag{4.7}$$

$$\alpha(\theta)\, I \leq \Lambda_\theta(t) \leq \beta(\theta)\, I, \qquad \theta \leq t \tag{4.8}$$

for some $\alpha(\theta) > 0$, $\beta(\theta) \geq \alpha(\theta)$;

$$Q_\theta(t) = Q^*_\theta(t) \tag{4.9}$$

$$\gamma(\theta)\, I \leq Q_\theta(t) \leq \delta(\theta)\, I, \quad \theta \leq t \tag{4.10}$$

for some $\gamma(\theta) > 0$, $\delta(\theta) \geq \gamma(\theta)$, and

$$\dot{\Lambda}_\theta(t) + A^*(t)\Lambda_\theta(t) + \Lambda_\theta(t)\, A(t) = -Q_\theta(t), \; \theta \leq t \tag{Λ_θ}$$

Along the same lines followed to prove Theorem 4.1 this can be extended into

Theorem 4.3

If $A : t \rightarrow A(t)$ has a Liapunov function $\chi \rightarrow \chi^* \Lambda_\theta(t)\chi$ then G has the property EAS.

The proof shows that the rate of exponential decay of $|G|$ is

$$\mu = \mu(\theta) < \frac{\gamma(\theta)}{2\beta(\theta)} \tag{4.11}$$

A partial converse of Theorem 4.3 can also be proved to extend Theorem 4.2, namely

Theorem 4.4

If A is bounded and G has the property EAS then A admits a Liapunov function with

$$\Lambda_\theta(t) = \int_t^{+\infty} G^*(s,t)\, Q_\theta(s)\, G(s,t)\, ds \tag{4.12}$$

5. AFFINE CONTROL SYSTEMS. BIBS STABILITY

Let us denote by $t \rightarrow f(t)$ an n-vector function of $t \in J$ measurable and locally integrable on J.

For each $\theta \in J$, $\chi \in \mathbb{R}^n$, there is a unique (Caratheodory) solution of the affine ordinary differential equation

$$\dot{x} - A(t)\, x = f(t) \tag{E_f}$$

such that

$$x(\theta) = \chi \tag{C}$$

i.e. there is a unique n-vector function $t \rightarrow x(t)$, satisfying (C), locally absolutely continuous on J and such that

$$dx(t)/dt - A(t)\, x(t) = f(t), \quad \text{a.e.} \quad t \in J$$

This solution is represented by the Lagrange formula

$$x(t) = G(t,\theta)\,\chi + \int_{\theta}^{t} G(t,s)\,f(s)\,ds \tag{L}$$

Let us now consider a family of affine ordinary differential equations, namely

$$\dot{x} - A(t)\,x = B(t)\,u(t) \tag{U}$$

where $t \to u(t)$ is an m-vector function of $t \in J$ belonging to a given set $U \subset L_{loc}(J)$. This means that each $u \in U$ is measurable and (essentially) bounded on every finite interval $\subset J$. Consequently, $B(t)$ is an $n \times m$ matrix and we shall assume that $B \in L^1_{loc}(J)$.

The family (U) of affine ordinary differential equations, depending on the index $u \in U$ is an affine control system.

A solution of (U) depends on u, as well as on t; it also depends on θ and χ and using the Lagrange formula we shall write

$$x(t,\theta,\chi,u) = G(t,\theta)\,\chi + \int_{\theta}^{t} G(t,s)\,B(s)\,u(s)\,ds \tag{X}$$

Using the terminology of Control Theory the n-vector $x(t,\theta,\chi,u)$ is the state of the system, the n-vector function $t \to u(t)$ is the control (or steering) function or input, the n-vector function $t \to x(t,\theta,\chi,u)$ is the response to u. The variable t is usually interpreted as time.

If u is bounded as $t \to +\infty$ the corresponding response need not be bounded. We want to characterize such control systems for which bounded inputs yield bounded states.

Definition 5.1

The control system (U) is bounded-input bounded-state (BIBS) stable as $t \to +\infty$ iff : a) G has the property US, i.e. for every $\theta > \alpha$ there is $\gamma(\theta) > 0$ such that

$$\theta \leq \tau \leq t \Rightarrow |G(t,\tau)| < \gamma(\theta) \tag{5.1}$$

b) for each u measurable and bounded on $[\theta, +\infty[$ there is some $k(\theta,u) > 0$ such that

$$\theta \leq t \Rightarrow \left| \int_{\theta}^{t} G(t,s)\,B(s)\,u(s)\,ds \right|_2 < k(\theta,u) \tag{5.2}$$

From (5.1), (5.2) and the Lagrange formula (X) it follows

$$\theta \leq t \Rightarrow |x(t,\theta,\chi,u)|_2 < \gamma(\theta)\,|\chi|_2 + k(\theta,u)$$

i.e. to a bounded input u there corresponds a bounded state x.

It is easy to prove:

Theorem 5.1

The control system (U) is BIBS stable if: a) G has the property US; b') for every $\theta > \alpha$ there exist $k_B(\theta) > 0$ such that

$$\theta \leq t \Rightarrow \int_{\theta}^{t} |G(t, s)\ B(s)|\, ds < k_B(\theta) \tag{5.3}$$

Proof. Since b') clearly implies b), a) + b') implies BIBS stability.

Since

$$\int_{\theta}^{t} |G(t, s)\ B(s)|\ ds \leq \int_{\theta}^{t} |G(t, s)|\ |B(s)|\ ds$$

it follows that if B is bounded on $[\theta, +\infty[$ for every $\theta > \alpha$ then the property US plus the property IS are sufficient for BIBS stability. On the other hand, the simultaneous validity of US and IS is equivalent to property EAS. Therefore

Theorem 5.2

Let B be bounded on $[\theta, +\infty[$ for each $\theta > \alpha$. Then (U) is BIBS stable if G has the property EAS.

The converse is not true as is shown by taking $A = 0$, $B = 0$.

6. STABILIZABILITY,I

We start with

Definition 6.1

We say that the control system

$$\dot{x} - A(t)\ x = B(t)\ u(t) \tag{U}$$

is stabilizable (to zero) iff there is an $m \times n$-matrix valued function $F : t \rightarrow F(t)$ of $t \in J$, such that BF is measurable and locally integrable on J and such that

$$\dot{y} - (A(t) + B(t)\ F(t))\ y = 0 \tag{F}$$

is asymptotically stable.

If (F) is exponentially asymptotically stable we say that (U) is exponentially stabilizable.

To explain the meaning of this Definition let us denote by y_χ the solution of (F) such that

$$y_\chi(\theta) = \chi$$

for a given $\theta \in J$ and a given $\chi \in \mathbb{R}^n$. Then we have $y_\chi(t) \to 0$ as $t \to +\infty$, but since

$$y_\chi(t) = G(t,\theta)\chi + \int_\theta^t G(t,s)\, B(s)\, F(s)\, y_\chi(s)\, ds$$

this means that also the solution of (U) corresponding to $u = Fy_\chi$, i.e. $t \to x(t,\theta,\chi,Fy_\chi)$ tends to zero as $t \to +\infty$. In other words, if (U) is stabilizable it is possible to bring the state x of the system from any initial position χ to zero in an infinite time.

In what follows we shall give conditions for stabilizability.

We shall presently start with the case of constant A and B. Then it makes sense to look for constant F such that (F) is (exponentially) stable.

A condition for exponential stabilizability arises from the solution of the so-called "regulator problem". For each initial state χ one looks for a control u such that the "cost" function

$$C(u) = \int_0^{+\infty} [x^* L x + x^* M u + u^* M^* x + u^* N u]\, dt$$

is minimum. Here $x = x(t, 0, \chi, u)$ while L, M, N are three matrices, respectively of types $n \times n$, $m \times n$ and $m \times m$ such that

$$L = L^*, \qquad N = N^* \tag{6.1}$$

The current assumption is

$$\begin{pmatrix} L & M \\ M & N \end{pmatrix} > 0 \tag{6.2}$$

which means that for $x \in \mathbb{R}^n$, $u \in \mathbb{R}^m$

$$x^* L x + x^* M u + u^* M^* x + u^* N u > \alpha\, (x^* x + u^* u)$$

for some $\alpha > 0$. From this follows $(u = 0)$

$$L > 0$$

and $(x = 0)$

$$N > 0$$

so that N^{-1} exists, and also that $(u = -N^{-1}M^* x)$

$$L - MN^{-1}M^* > 0$$

For a given triplet L, M, N satisfying (6.1) and (6.2) we associate with the control system (U) the Kalman matrix equation

$$(L - MN^{-1}M^*) + (A - BN^{-1}M^*)^* \Lambda \tag{6.3}$$

$$+ \Lambda(A - BN^{-1}M^*) - \Lambda(BN^{-1}B^*)\Lambda = 0$$

When $B = 0$, (U) reduces to (E_0) and (6.3) reduces to the Liapunov equation (Λ) with

$$Q = L - MN^{-1}M^* > 0 \tag{6.4}$$

Assume that (6.3) has a solution

$$\Lambda_\infty = \Lambda_\infty^* > 0$$

If we take

$$F = -N^{-1}\left(M^* + \frac{1}{2} B^* \Lambda_\infty\right) \tag{6.5}$$

(6.3) with $\Lambda = \Lambda_\infty$ can be written

$$Q + (A + BF)^* \Lambda_\infty + \Lambda_\infty (A + BF) = 0 \tag{6.6}$$

so that Λ_∞ is also a solution of the Liapunov equation associated to $A + BF$, which means (Theorem 4.1) that (F) is exponentially stable. Therefore we have

Theorem 6.1

Let A and B be constant. Then (U) is exponentially stabilizable if for some triplet L, M, N of matrices satisfying (6.1), (6.2) the Kalman equation (6.3) has a solution $\Lambda_\infty = \Lambda_\infty^* > 0$. In this case a stabilizing F is defined by (6.5).

Remark 6.1

According to Remark 4.1 the exponential decay of the solutions of (F) with F defined by (6.5) is not faster than

$$t \to e^{\frac{-\lambda_Q}{2\mu_{\Lambda_\infty}} t}$$

where λ_Q is the least eigenvalue of $Q = L - MN^{-1}M^*$ and μ_{Λ_∞} is the greatest eigenvalue of Λ_∞.

Remark 6.2

The last assertion of Theorem 6.1 needs a comment. In fact, if $\Lambda_\infty = \Lambda_\infty^* > 0$ is a solution of (6.3) we have

$$Q_\alpha = L - MN^{-1}M^* + \alpha\Lambda_\infty(BN^{-1}B^*)\Lambda_\infty = Q_\alpha^* > 0 \tag{6.7}$$

for all $\alpha \geqq 0$. Therefore, replacing (6.5) by

$$F_\alpha = -N^{-1}\left(M^* + \frac{1+\alpha}{2} B^*\Lambda_\infty\right) \tag{6.8}$$

and (6.6) by

$$Q_\alpha + (A + BF_\alpha)^*\Lambda_\infty + \Lambda_\infty(A + BF_\alpha) = 0$$

we still have a Liapunov equation, and we conclude that not only $F = F_0$, but also F_α, $\alpha \geqq 0$ are stabilizing matrices.

Remark 6.3

Theorem 6.1 has an inverse (D.L. Lukes) in the sense that if (U) is stabilizable then for every triplet L, M, N satisfying (6.1) and (6.2), the Kalman equation (6.3) must have a unique solution $\Lambda_\infty = \Lambda_\infty^* > 0$.

The stabilizability criterion represented by Theorem 6.1 can be extended to the case of time-dependent A and B by using Theorem 4.3 instead of Theorem 4.1.

In this case, L, M, N are also time-dependent, and the Kalman matrix equation is replaced by a Riccati-matrix differential equation

$$\dot{\Lambda} + (L(t) - M(t)N^{-1}(t)M^*(t)) + (A(t) - B(t)N^{-1}(t)M^*(t))^*\Lambda$$
$$+ \Lambda(A(t) - B(t)N^{-1}(t)M^*(t)) - \Lambda(B(t)N^{-1}(t)B^*(t))\Lambda = 0 \tag{6.9}$$

and F is replaced by

$$F(t) = -N^{-1}(t)\left(M^*(t) + \frac{1}{2} B^*(t)\,\Lambda_\infty(t)\right) \tag{6.10}$$

where Λ_∞ is an appropriate solution of (6.9).

7. STABILIZABILITY, II

Let us now consider another criterion of stabilizability for the case of constant A and B (D.L. Lukes, W.A. Coppel).

Theorem 7.1

Let A and B be constant. Let there exist some $T > 0$ such that

$$\Omega = \int_0^T e^{-tA} \, B B^* \, e^{-tA^*} \, dt > 0 \qquad\qquad (7.1)$$

Then if

$$F = -B^* \Omega^{-1} \qquad\qquad (7.2)$$

the equation

$$\dot{x} - (A + B F) x = 0 \qquad\qquad (F)$$

is exponentially stable.

Proof. From (7.1) we have

$$A\Omega + \Omega A^* = -\int_0^T \left[\frac{d}{dt} e^{-tA} \, B B^* \, e^{-tA^*} \right] dt = -e^{-TA} B B^* e^{-TA^*} + B B^*$$

hence, by (7.2),

$$(A + B F)\Omega + \Omega (A + B F)^* = -(e^{-TA} B B^* e^{-TA^*} + B B^*) \leq 0$$

Let λ be an eigenvalue of $(A + B F)^*$ and ζ a corresponding eigenvector,

$$(A + B F)^* \zeta = \lambda \zeta$$

We have

$$(\lambda + \overline{\lambda}) \zeta^* \Omega \zeta = \zeta^* [(A + B F)\Omega + \Omega (A + B F)^*] \zeta$$

$$= -\zeta^* (e^{-TA} B B^* e^{-TA^*} + B B^*) \zeta \leq 0$$

Since $\Omega > 0$ it follows $\text{Re}\lambda \leq 0$. Moreover, if the equality holds then $\zeta^* B B^* \zeta = 0$, i.e. $\left| B^* \zeta \right|_2^2 = 0$, i.e. $B^* \zeta = 0$, hence $(A + B F)^* \zeta = A^* \zeta = \lambda \zeta$ and therefore

$$e^{-tA^*} \zeta = e^{-\lambda t} \zeta$$

which gives

$$\Omega \zeta = \int_0^T e^{-tA} \, B B^* \, \zeta \, e^{-\lambda t} \, dt = 0$$

and this contradicts (7.1) since $\zeta \neq 0$.

Remark 7.1

If (7.1) holds for some $T > 0$ then it holds for every $T > 0$. In fact, let there exist some $\tilde{T} > 0$, $\tilde{\xi} \neq 0$ such that

$$0 = \tilde{\xi}\left(\int_0^{\tilde{T}} e^{-tA} B B^* e^{-tA} \, dt\right)\tilde{\xi} = \int_0^{\tilde{T}} |B^* e^{-tA^*} \tilde{\xi}|_2^2 \, dt$$

This means

$$B^* e^{-tA^*} \tilde{\xi} = 0 , \qquad t \in [0, \tilde{T}]$$

and, since $t \to B^* e^{-tA^*} \tilde{\xi}$ is an analytic function, we have

$$B^* e^{-tA^*} \tilde{\xi} = 0 , \qquad t \in \mathbb{R}$$

It follows that if (7.1) holds for some $T > 0$ then we can define $\Omega = \Omega(T)$, hence $F = F(T)$ for every $T > 0$, so that we have a family, depending on $T > 0$, of stabilizing matrices.

It can be shown that (7.1) is, in fact, equivalent with the existence of a matrix F such that $A + BF$ has any prescribed set of eigenvalues (C. Langenhop, W.M. Wonham).

In other words, (7.1) is equivalent to the property that the "closed-loop transfer matrix"

$$\rho \to [(A + BF) - \rho I]^{-1} B$$

can be assigned an arbitrary set of poles by a suitable choice of the "feedback gain matrix" F.

On the other hand, (7.1) is equivalent to the complete controllability of the control system (U). This means that (7.1) is a necessary and sufficient condition in order that the state x of the system can be transferred from any initial position at $t = 0$ to any final position at $t = T$.

Remark 7.1 means that if this is possible for some $T > 0$ then it is also possible for any $T > 0$.

This remark leads to the notion of "uniform complete controllability" for time-dependent control systems as a sufficient condition to ensure a very strong kind of stabilizability for such systems.

8. UNIFORM COMPLETE CONTROLLABILITY

To define the notion of uniform complete controllability we start by considering the $n \times n$ matrix

$$H(\tau, T) = \int_{\tau}^{T} G(\tau, t) B(t) B^*(t) G^*(\tau, t) \, dt \qquad \text{(H)}$$

which reduces to Ω of Section 7 when $\tau = 0$ and A and B are constant. Of course, integrability of B is not enough for the existence of the above integral, so we have to assume that $B \in L^2_{loc}(J)$, i.e. the elements of B are locally square integrable functions.

We shall also consider the $n \times n$ matrix

$$K(\tau, T) = \int_\tau^T G(T, t) B(t) B^*(t) G^*(T, t)\, dt = G(T, \tau) H(\tau, T) G^*(T, \tau) \qquad (K)$$

Clearly,

$$H(\tau, T) = H^*(\tau, T) \geqq 0, \quad K(\tau, T) = K^*(\tau, T) \geqq 0$$

These inequalities are (both) strict if and only if the control system (U) is completely controllable on $[\tau, T]$, i.e. iff for every pair v, $w \in \mathbb{R}^n$, there is a control $u_{vw} \in L^\infty([\tau, T])$ such that

$$x(\tau, \tau, v, u_{vw}) = v, \quad x(T, \tau, v, u_{vw}) = w \qquad (8.1)$$

It is readily verified that we can take, for instance

$$u_{vw}(t) = B^*(t) G^*(\tau, t) H^{-1}(\tau, T) [G(\tau, T)w - v]$$

$$= B^*(t) G^*(T, t) K^{-1}(\tau, T) [w - G(T, \tau)v] \qquad (8.2)$$

Among the control functions acting the transfer from v to w the one defined by (8.2) has an important property. In fact, if u is another control function satisfying (8.1) we have

$$\int_\tau^T G(T, t) B(t) [u(t) - u_{vw}(t)]\, dt = 0$$

hence

$$\int_\tau^T [u(t) - u_{vw}(t)]^* B^*(t) G^*(T, t) K^{-1}(\tau, T) [w - G(T, \tau)v]\, dt$$

$$= \int_\tau^T [u(t) - u_{vw}(t)]^* u_{vw}(t)\, dt = 0$$

Therefore

$$\int_\tau^T |u_{vw}(t)|_2^2\, dt = \int_\tau^T |u(t)|_2^2\, dt - \int_\tau^T |u(t) - u_{vw}(t)|_2^2 \leq \int_\tau^T |u(t)|_2^2\, dt$$

This means that u_{vw} transfers the state of the system from v at $t = \tau$ to w at $t = T$ at the expense of a minimum amount of energy, namely

$$\int_{\tau}^{T} |u_{vw}(t)|_2^2 \, dt = [G(\tau, T)w - w]^* H^{-1}(\tau, T) [G(\tau, T)w - v]$$

$$= [w - G(T,\tau)v]^* K^{-1}(\tau, T) [w - G(T,\tau)v] \tag{8.3}$$

Let us now consider the following example. Let the scalar control system be defined by

$$\dot{x} + tx = \sqrt{2(t - 1)}\, e^{-t+1/2} u(t)$$

with $t \in J =]1, +\infty[$. It is easy to see that

$$G(t, s) = e^{s^2/2 - t^2/2}$$

$$H(\tau, \tau + \sigma) = e^{2(\sigma - 1)\tau + (\sigma - 1)^2} - e^{-2\tau + 1}$$

$$K(\tau, \tau + \sigma) = e^{-2\tau - 2\sigma + 1} - e^{-2(\sigma + 1)\tau + 1 - \sigma^2}$$

According to (8.3) we see that the transfer from v at $t = \tau$ to $w = 0$ at $t = \tau + \sigma$ takes a minimum amount of energy which $\to +\infty$ as $\tau \to +\infty$ if $\sigma < 1$, while $\to 0$ as $\tau \to +\infty$ if $\sigma > 1$. The transfer from $v = 0$ at $t = \tau$ to w at $t = \tau + \sigma$ takes a minimum amount of energy which $\to +\infty$ as $\tau \to +\infty$, no matter what the length σ of the time interval is.

To avoid such unpleasant circumstances we need a special class of control systems, namely those which are called uniformly completely controllable, according to R. Kalman.

Definition 8.1

The control system (U) is said to be uniformly completely controllable (u.c.c.) iff there exist

$$\sigma > 0, \; 0 < h_1 \leqq h_2, \quad 0 < k_1 \leqq k_2$$

such that for every $\tau > \alpha$

$$h_1 I \leqq H(\tau, \tau + \sigma) \leqq h_2 I \tag{8.4}$$

$$k_1 I \leqq K(\tau, \tau + \sigma) \leqq k_2 I \tag{8.5}$$

These conditions imply, respectively

$$h_2^{-1}\ I \leqq H^{-1}(\tau,\tau+\sigma) \leqq h_1^{-1}\ I$$

$$k_2^{-1}\ I \leqq K^{-1}(\tau,\tau+\sigma) \leqq k_1^{-1}\ I$$

so that, according to (8.2), the amount of energy spent for the transfer of v to zero (or from zero to w) in a time interval $[\tau,\tau+\sigma]$ does not depend on the initial time τ.

Remark 8.1

It can be shown that if (8.4), (8.5) hold for some $\sigma_0 > 0$, then they are valid for every $\sigma > \sigma_0$.

Remark 8.2

It can also be shown that if (8.4), (8.5) hold, then there exist

$$\rho \to g_1(\rho), \quad \rho \to g_2(\rho), \qquad 0 < g_1(\rho) \leqq g_2(\rho)$$

such that

$$0 < g_1(|t-s|) \leqq |G(t,s)| \leqq g_2(|t-s|) \tag{8.6}$$

for $t, s > \alpha$, $|t-s| > \sigma$.
Since

$$\begin{aligned} K(\tau,\tau+\sigma) &= G(\tau+\sigma,\tau)\, H(\tau,\tau+\sigma)\, G^*(\tau+\sigma,\tau) \\ H(\tau,\tau+\sigma) &= G(\tau,\tau+\sigma)\, K(\tau,\tau+\sigma)\, G^*(\tau,\tau+\sigma) \end{aligned} \tag{8.7}$$

it follows that if any two of the relations (8.4), (8.5), (8.6) hold, the remaining relation is also true.

Remark 8.3

(8.4) can be written $(\chi \in \mathbb{R}^n)$

$$0 < h_1|\chi|_2^2 \leqq \chi^* H(\tau,\tau+\sigma)\chi \leqq h_2|\chi|_2^2$$

so that if we take $\chi = G^*(\tau+\sigma,\tau)\eta$ we have

$$h_1|\chi|_2^2 \leqq \eta^* K(\tau,\tau+\sigma)\eta \leqq h_2|\chi|_2^2$$

On the other hand

$$|\chi|_2^2 \leq |G(\tau + \sigma, \tau)|^2 \, |\eta|_2^2$$

and since $\eta = G^*(\tau, \tau + \sigma)\chi$ we also have

$$|\chi|_2^2 \geq |G(\tau, \tau + \sigma)|^{-2} \, |\eta|_2^2$$

Therefore from (8.4) we have

$$h_1 |G(\tau, \tau + \sigma)|^{-2} \, I \leq K(\tau, \tau + \sigma) \leq h_2 |G(\tau + \sigma, \tau)|^2 \, I \tag{8.8}$$

which, in general, does not imply (8.5) because the bounds, left and right, will depend on τ.

However, if A is bounded, i.e., if

$$|A(t)| < a, \quad \theta \leq t \tag{8.9}$$

for some $a > 0$, then, recalling (1.13), we have

$$e^{-a\sigma} \leq |G(\tau, \tau + \sigma)| \leq e^{a\sigma}$$

and from (8.8) we have (8.5) with h_1, h_2 replaced by $h_1 e^{-a\sigma}$, $h_2 e^{a\sigma}$, respectively.

Therefore when A is bounded we can suppress (8.5) from the definition of u.c.c.

We have further

$$\chi^* H(\tau, \tau + \sigma)\chi \leq \left(\int_\tau^{\tau+\sigma} |G(\tau, s)|^2 \, |B(s)|^2 \, ds \right) |\chi|_2^2$$

and since, from (1.13) again:

$$|G(\tau, s)| \leq e^{\int_\tau^s |A(r)| \, dr}$$

it follows, when A is bounded

$$\chi^* H(\tau, \tau + \sigma)\chi \leq e^{2a\sigma} \left(\int_\tau^{\tau+\sigma} |B(s)|^2 \, ds \right) |\chi|_2^2$$

Therefore, if not only A but also B is bounded, i.e., if

$$|B(s)| < b \qquad \theta < s$$

for some $b > 0$, the second inequality of (8.4) can be suppressed and we conclude that when both A and B are bounded the control system (U) is u.c.c. if and only if there are some $\sigma > 0$, $h_1 > 0$ such that

$$h_1 I \leq H(\tau, \tau + \sigma) \quad \text{for all } \tau \tag{8.10}$$

In particular, when A and B are constant, since $G(t,s) = e^{(t-s)A}$ we have $H(\tau, \tau+\sigma) = H(0,\sigma)$ so that (8.10) will reduce simply to

$$H(0,\sigma) > 0$$

i.e. to complete controllability.

9. UNIFORM COMPLETE STABILIZABILITY

M. Ikeda — H. Maeda — S. Kodama, in prosecution of Kalman's work, have recently introduced the notion of uniformly completely stabilizable control systems.

Definition 9.1

The control system (U) is uniformly completely stabilizable (u.c.s.) iff for every $\mu > 0$ there is a matrix F_μ such that

$$\dot{x} - (A(t) + B(t)\, F_\mu(t))\, x = 0 \tag{F}$$

is exponentially stable with exponent $> \mu$.

It can be proved

Theorem 9.1

If (U) is u.c.c. then it is u.c.s.

u.c.c. serves to prove the existence for every $\mu > 0$ of a solution Λ_μ of the Riccati equation

$$\dot{\Lambda} + (A(t) + \mu I)^* \Lambda + \Lambda (A(t) + \mu I) - \Lambda B(t) B^*(t) \Lambda = -I$$

such that

$$\Lambda_\mu(t) = \Lambda_\mu^*(t)$$

$$\alpha I \leq \Lambda_\mu(t) \leq \beta I, \quad t \in \mathbb{R}$$

for some $\alpha > 0$, $\beta \geq \alpha$, independent of t.

Then, according to Definition 4.2, Λ_μ is a Liapunov function for

$$t \to A(t) + \mu I + B(t)\, F_\mu(t)$$

where

$$F_\mu(t) = -\frac{1}{2} B(t) \Lambda_\mu(t)$$

From this follows the exponential stability of

$$\dot{x} - (A(t) + \mu I + B(t) F_\mu(t)) x = 0$$

hence that of (F_μ) with an exponent $> \mu$.

Theorem 9.1 has a partial converse, i.e.:

Theorem 9.2

Let A and B be bounded. Then (U) is u.c.c. if it is u.c.s. by means of bounded F_μ.

Proof. According to Remark 8.3, since A and B are bounded if (U) is not u.c.c., then for every $\sigma > 0$ and every $\rho > 0$ there will exist some vector $\tilde{\chi} = \tilde{\chi}(\rho, \sigma) \neq 0$ and some $\tilde{\tau} > 0$ such that

$$\tilde{\chi}^* H(\tilde{\tau}, \tilde{\tau} + \sigma)\tilde{\chi} < \rho |\tilde{\chi}|_2^2$$

i.e. such that

$$\int_{\tilde{\tau}}^{\tilde{\tau}+\sigma} |B^*(s)G^*(\tilde{\tau}, s)\tilde{\chi}|_2^2 \, ds \leq \rho |\tilde{\chi}|_2^2$$

Hence, using the Schwartz inequality

$$\int_{\tilde{\tau}}^{\tilde{\tau}+\sigma} |B^*(s)G^*(\tilde{\tau}, s)\tilde{\chi}|_2 \, ds \leq \sqrt{\sigma\rho} \, |\tilde{\chi}|_2 \tag{9.1}$$

On the other hand, for every $\tau \in J$, $\chi \in \mathbb{R}^n$, the solution x_μ of (F_μ) such that $x_\mu(\tau) = \chi$ is given by

$$x_\mu(t) = G(t, \tau)\chi + \int_\tau^t G(t, s)B(s) F_\mu(s) x_\mu(s) \, ds$$

hence, denoting by G_μ the evolution matrix of $A + BF_\mu$, we have

$$G_\mu(t, \tau)\chi = G(t, \tau)\chi + \int_\tau^t G(t, s)B(s)F_\mu(s)G_\mu(s, \tau)\chi \, ds$$

Since χ is an arbitrary vector it follows

$$G_\mu(t,\tau) = G(t,\tau) + \int_\tau^t G(t,s)B(s)F_\mu(s)G_\mu(s,\tau)\,ds$$

whence

$$G_\mu^*(t,\tau)G^*(\tau,t) = I + \int_\tau^t G_\mu^*(s,\tau)F_\mu^*(s)B^*(s)G^*(\tau,s)\,ds$$

If we apply this to $\tilde{\chi}$ for $\tau = \tilde{\tau}$ we obtain

$$G_\mu^*(t,\tilde{\tau})G^*(\tilde{\tau},t)\tilde{\chi} = \tilde{\chi} + \int_{\tilde{\tau}}^t G_\mu^*(s,\tilde{\tau})F_\mu^*(s)B^*(s)G^*(\tilde{\tau},s)\tilde{\chi}\,ds$$

whence

$$|G_\mu^*(t,\tilde{\tau})G^*(\tilde{\tau},t)\tilde{\chi}|_2 = |\tilde{\chi} + \int_{\tilde{\tau}}^t G_\mu^*(s,\tilde{\tau})F_\mu^*(s)B^*(s)G^*(\tilde{\tau},s)\tilde{\chi}\,ds|_2$$

$$|G_\mu^*(t,\tilde{\tau})|\;|G^*(\tilde{\tau},t)|\,|\tilde{\chi}|_2 \geq |\tilde{\chi}|_2 - \int_{\tilde{\tau}}^t |G_\mu(s,\tilde{\tau})|\;|F_\mu(s)|\;|B^*(s)G^*(\tilde{\tau},s)\tilde{\chi}|_2\,ds$$

and using (9.1), for $t = \tilde{\tau} + \sigma$

$$|G_\mu(\tilde{\tau}+\sigma,\tilde{\tau})|\;|G(\tilde{\tau},\tilde{\tau}+\sigma)| \geq 1 - \sqrt{\sigma\rho}\int_{\tilde{\tau}}^{\tilde{\tau}+\sigma} |G_\mu(s,\tilde{\tau})|\;|F_\mu(s)|\,ds$$

Now,

$$|G_\mu(\tilde{\tau}+\sigma,\tilde{\tau})| \leq \gamma\, e^{-\mu\sigma}$$

$$|G(\tilde{\tau},\tilde{\tau}+\sigma)| \leq e^{a\sigma}$$

$|F_\mu(s)| \leq f_\mu$, for some $f_\mu > 0$

so that

$$\gamma\, e^{(-\mu+a)\sigma} \geq 1 - \sqrt{\sigma\rho}\,\gamma f_\mu$$

and if we take

$$\mu = 2a, \quad \rho = (9\gamma^2 f_\mu^2 \sigma)^{-1}$$

we have

$$\gamma\, e^{-a\sigma} \geqq 2/3$$

which is a contradiction since we can take σ arbitrarily large.

10. B*IBS STABILITY

In Section 5 we considered BIBS stability and we proved (Theorem 5.2) that

$$\text{property EAS} \Rightarrow \text{BIBS stability} \tag{10.1}$$

provided that B is bounded on $[\theta, +\infty[$, i.e.

$$|B(t)| < \beta, \quad \theta \leqq t \tag{10.2}$$

for some $\beta > 0$.

It is easy to see that the validity of (10.1) can be established under a condition different from (10.2), like

$$\int_\theta^t |B(t)|^2\, dt < \beta(\theta), \quad \theta \leqq t \tag{10.3}$$

for some $\beta(\theta) > 0$.
In fact, to prove (10.1) we have to prove

$$\theta \leqq t \Rightarrow \int_\theta^t |G(t, s)\, B(s)|\, ds < k_B(\theta) \tag{5.3}$$

Since EAS is equivalent to US plus IS we have

$$\int_\theta^t |G(t, s)|^2\; ds \leqq \gamma(\theta)\, k(\theta)$$

hence

$$\int_\theta^t |G(t,s)\, B(s)|\; ds \leqq \left(\int_\theta^t |G(t, s)|^2\; ds\right)^{\frac{1}{2}} \left(\int_\theta^t |B(s)|^2\; ds\right)^{\frac{1}{2}}$$

$$\leqq [\,\gamma(\theta)\, k(\theta)\, \beta(\theta)\,]^{\frac{1}{2}}$$

B.D.O. Anderson – J.B. Moore replaced BIBS stability by B*IBS stability. This is obtained by requiring that (5.2) in Definition 5.1 is satisfied not only for all bounded u, but also for all the bounded* u, which means that there are

$$\sigma_u > 0, \quad \omega_u > 0$$

such that

$$\int_{\tau}^{\tau+\sigma} |u(t)|_2^2 \; dt < \omega_u \quad \text{for all } \tau > \alpha \tag{10.4}$$

Therefore

B*IBS stability ⇒ BIBS stability

but the converse is not true.

Anderson and Moore have studied the relationship between B*IBS stability and the property EAS and found that they are equivalent provided that (U) is u.c.c. We shall now prove the first part of this implication, namely

Theorem 10.1

Let (U) be u.c.c. Then if G has the property EAS, (U) is B*IBS stable.

Proof. Let $\theta \leq t$ and take an integer $k = k(t, \theta)$ such that

$$k + 1 \geq \frac{t - \theta}{\sigma} \tag{10.5}$$

where σ is the one appearing in the Definition of u.c.c. Then we have

$$\int_{\theta}^{t} G(t, s) B(s) u(s) ds = G(t, \theta + \sigma) \int_{\theta}^{\theta+\sigma} G(\theta + \sigma, s) B(s) u(s) ds + \dots$$

$$\dots + G(t, \theta + j\sigma) \int_{\theta+(j-1)\sigma}^{\theta+j\sigma} G(\theta + j\sigma, s)\, B(s)\, u(s) ds + \dots$$

$$\dots + G(t, \theta + k\sigma) \int_{\theta+(k-1)\sigma}^{\theta+k\sigma} G(\theta + k\sigma, s)\, B(s)\, u(s) ds + \int_{\theta+k\sigma}^{t} G(t, s)\, B(s)\, u(s) \tag{10.6}$$

If u is bounded* then (10.4 is valid also with σ replacing σ_u and we have for $j = 1, \dots, k$

$$\left| \int_{\theta+(j-1)\sigma}^{\theta+j\sigma} G(\theta + j\sigma, s)\, B(s)\, u(s)\, ds \right|_2 \leq \int_{\theta+(j-1)\sigma}^{\theta+j\sigma} |G(\theta + j\sigma, s)\, B(s)|\; |u(s)|_2\, ds$$

$$\leq \left[\int_{\theta+(j-1)\sigma}^{\theta+j\sigma} |G(\theta + j\sigma, s)\, B(s)|^2\, ds \right]^{\frac{1}{2}} \omega_u^{\frac{1}{2}} \tag{10.7}$$

Also, because of (10.5),

$$\left| \int_{\theta+k\sigma}^{t} G(t, s)\, B(s)\, u(s)\, ds \right|_2 \leq \int_{\theta+k\sigma}^{t} |G(t, s)\, B(s)|\; |u(s)|_2\, ds$$

$$\leq \int_{t-\sigma}^{t} |G(t, s)\, B(s)|\; |u(s)|_2\, ds \leq \left[\int_{t-\sigma}^{t} |G(t, s)\, B(s)|^2\, ds \right]^{\frac{1}{2}} \omega_u^{\frac{1}{2}} \tag{10.8}$$

From the second half of (8.5) we have

$$\int_{\tau}^{\tau+\sigma} |G(\tau + \sigma, s)\, B(s)|^2\, ds \leq n\, k_2 \tag{10.9}$$

From this and (10.7), (10.8) we have, respectively

$$\begin{cases} \left| \int_{\theta+(j-1)\sigma}^{\theta+j\sigma} G(\theta + j\sigma, s)\, B(s)\, u(s)\, ds \right|_2 \leq (n\, k_2\, \omega_u)^{\frac{1}{2}}, \quad j = 1, \ldots, k \\ \left| \int_{\theta+k\sigma}^{t} G(t, s)\, B(s)\, u(s)\, ds \right|_2 \leq (n\, k_2\, \omega_u)^{\frac{1}{2}} \end{cases} \tag{10.10}$$

From (10.6), for a bounded* u it follows

$$\left| \int_{\theta}^{t} G(t, s)\, B(s)\, u(s)\, ds \right|_2$$

$$\leq (n\, k_2\, \omega_u)^{\frac{1}{2}} \Big[1 + |G(t, \theta + k\sigma)| + \ldots + |G(t, \theta + j\sigma)| + \ldots$$

$$\ldots + |G(t, \theta + \sigma)| \Big]$$

From property EAS we have ($\gamma = \gamma(\theta) > 0$, $\mu = \mu(\theta) > 0$)

$$|G(t, \theta + j\sigma)| = |G(t, \theta + k\sigma)\, G(t + k\sigma, \theta + j\sigma)|$$

$$\leq |G(t, \theta + k\sigma)|\, |G(\theta + k\sigma, \theta + j\sigma)| \leq \gamma |G(t, \theta + k\sigma)|\, (e^{-\mu\sigma})^{k-j}$$

$$\leq \gamma^2 (e^{-\mu\sigma})^{k-j}, \quad j = 1, \ldots, k$$

Hence

$$\left| \int_\theta^t G(t, s)\, B(s)\, u(s)\, ds \right|_2 \leq (n k_2 \omega_u)^{\frac{1}{2}} \left[1 + \gamma^2 \sum_{j=1}^{k} (e^{-\mu\sigma})^{k-j} \right]$$

$$< (n k_2 \omega_u)^{\frac{1}{2}} \left[1 + \gamma^2 \frac{1}{1 - e^{-\mu\sigma}} \right]$$

i.e. (5.2).

Note that we used only a part of u.c.c.

BIBLIOGRAPHY

ANDERSON, B. D. O., MOORE, J. B., New results in linear system stability, SIAM J. Control, 7 (1969) 398.

COPPEL, W. A., Matrix quadratic equations, Bull. Austral. Math. Soc. 10 (1974) 377.

IKEDA, M., MAEDA, H., KODAMA, Stabilization of linear systems, SIAM J. Control, 10 (1972) 716.

KALMAN, R. E., Contributions to the theory of optimal control, Bol. Soc. Mat. Mexicana, (2) 5 (1960) 102.

LUKES, D. L., Stabilizability and optimal control, Funkc. Ekv. 11 (1968) 39.

SILVERMAN, L. M., ANDERSON, B. D. O., Controllability, observability and stability of linear systems, SIAM J. Control 6 (1968) 121.

WONHAM, W. M., On pole assignment in multi-input controllable linear systems, IEEE Trans. Autom. Control, AC 12 (1967) 600.

CONTROLLABILITY OF NON-LINEAR CONTROL DYNAMICAL SYSTEMS

C. LOBRY
Université de Bordeaux I,
France

Abstract

CONTROLLABILITY OF NON-LINEAR CONTROL DYNAMICAL SYSTEMS.

In this paper, the results obtained in the problem of controllability of non-linear systems from 1970 to 1974 are reviewed. After introduction of the idea of a control dynamical system, the abstract theory and applications are treated in a concise way, references to the original sources replacing detailed proofs to a great extent.

INTRODUCTION

This is a review paper on the results obtained in the problem of controllability of non-linear systems from 1970 to 1974. Let us just remark that the ideas developed during the last five years were initiated by Hermann [18], Hermes [19], Kučera [36, 37] and Markus [48] and probably some other people. The idea of this survey is not to write the history of the subject but to present a set of references available to the reader.

In Chapter 1, we introduce the idea of control dynamical systems, which is a new formulation of what is currently called a non-linear control system. Chapter 2 is devoted to the abstract theory and Chapter 3 to applications. At the end, references to other topics of control theory where the use of these geometric tools seems to be useful are given.

1. CONTROL DYNAMIC SYSTEMS

Here, we introduce the concept of a control dynamical system. We hope that Examples 1.3 and 1.4 given at the end of the section will convince the reader that control dynamical systems are standard objects in nature.

1.1. Vector fields on a manifold M

A manifold is something like a "surface" but of a dimension not necessarily equal to 2 (see Milnor [50] for a rapid introduction to the subject). A vector field is a mapping

$$x \to X(x)$$

which associates with x a vector in the tangent space $T_x M$, to M at the point x. Under very reasonable assumptions the Cauchy problem

$$\begin{cases} \dfrac{dx}{dt} = X(x) \\ \\ x(0) = x_0 \end{cases}$$

has a unique solution whose value at time t is denoted by $X_t(x_0)$. When this value is defined for every t we say that the vector field is complete. The smooth mapping

$$(t, x) \to X_t(x)$$

defines a group action.

$$X_0(x) = x$$

$$X_{t_1+t_2}(x) = X_{t_1} \circ X_{t_2}(x)$$

1.2. Control dynamical systems

Let us consider a family $\mathscr{D}$ of complete vector fields on the manifold M. If $\mathscr{D}$ reduces to one element we have a group action of $\mathbb{R}$ on M which is defined by the integration of the differential equation

$$\frac{dx}{dt} = X(x)$$

Consider the collection of symbols of the form

$$(t_1, X^1)(t_2, X^2) \dots (t_p, X^p); \quad p \in N, X^i \in \mathscr{D}, t_i \in \mathbb{R}$$

Take concatenation as the law of composition and adopt the two simplification rules:

i) if $X^i = X^{i+1}$ then replace $(t_i, X^i)(t_{i+1}, X^{i+1})$ by $(t_i + t_{i+1}, X^i)$

ii) suppress terms of the form (O, X).

The set of all irreducible sequences is called the control group associated to $\mathscr{D}$ and denoted by $G(\mathscr{D})$. We define the group action by

$$(t_1, X^1)(t_2, X^2) \dots (t_i, X^i) \dots (t_p, X^p)x = X^1_{t_1} \circ X^2_{t_2} \dots \circ X^i_{t_i} \circ \dots \circ X^p_{t_p}(x)$$

where the $X^i_{t_i}$ are defined in section 1.1. The verification that $G(\mathscr{D})$ is a group and that the above action is a group action is trivial (see Refs [42,46]).

1.2.1. Definition: A family $\mathscr{D}$ of vector fields on M is called a "Control Dynamical System", the (non-commutative) group defined above, $G(\mathscr{D})$, is the associated control group. In $G(\mathscr{D})$ we consider the subset $G^+(\mathscr{D})$ of those sequences:

$$(t_1, X^1)\ (t_2, X^2)\ (\dots)\ (t_i, X^i)\ \dots\ (t_p, X^p)$$

for which the reals $t_1, t_2, \dots, t_i, \dots, t_p$ are positive. The orbit $G(\mathscr{D}).x$ of a point is the set:

$$G(\mathscr{D}).x = \{X^1_{t_1} \circ X^2_{t_2} \circ \dots \circ X^i_{t_i} \circ \dots \circ X^p_{t_p}(x);\ X^i \in \mathscr{D};\ t_i \in \mathbb{R},\ p \in \mathbb{N}\}$$

The positive orbit is the set

$$G^+(\mathscr{D}).x = \{X^1_{t_1} \circ X^2_{t_2} \circ \dots \circ X^i_{t_i} \circ \dots \circ X^p_{t_p}(x);\ X^i \in \mathscr{D};\ t_i \in \mathbb{R}^+,\ p \in \mathbb{N}\}$$

The positive orbit at time t is

$$G_t(\mathscr{D}).x = \left\{X^1_{t_1} \circ X^2_{t_2} \circ \dots \circ X^i_{t_i} \circ \dots \circ X^p_{t_p}(x);\ X^i \in \mathscr{D};\ t_i \in \mathbb{R}^p, \sum_{i=1}^{p} t_i = t,\ p \in \mathbb{N}\right\}$$

The questions which we now want to answer are:

What is the structure of $G(\mathscr{D}).x$?
What is the structure of $G^+(\mathscr{D}).x$?

As we shall see, most of the essential questions are solved except two:

When is $G^+(\mathscr{D}).x$ or $G_t(\mathscr{D}).x$ closed?
When is $G^+(\mathscr{D}).x$ the whole manifold M?

By an answer to these questions we mean a "computable algorithm" in terms of the known data which gives the answer. Unfortunately (or furtunately), the two last questions are the most pertinent ones for control purposes.

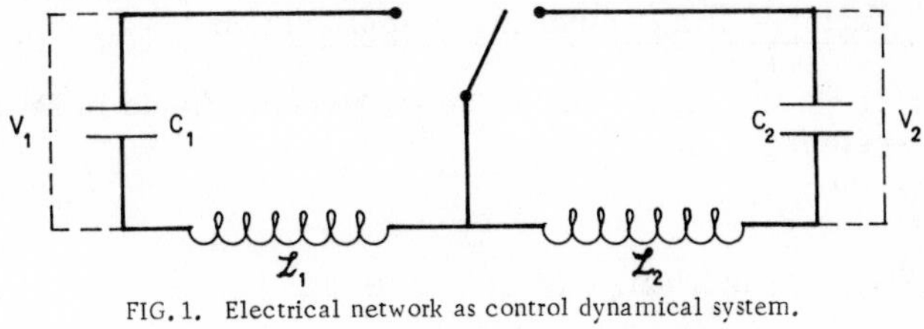

FIG. 1. Electrical network as control dynamical system.

1.3. First example of a control dynamical system (taken from Ref.[3])

Consider the electrical network shown in Fig.1. Here the switch is closed either on the right or on the left. What are the equations of the motion?

a) The switch is on the left:

$$\frac{dV_1}{dt} = -\frac{1}{c_1} i$$

$$\frac{dV_2}{dt} = 0$$

$$\frac{di}{dt} = \frac{1}{\mathscr{L}_1} V_1$$

b) The switch is on the right:

$$\frac{dV_1}{dt} = 0$$

$$\frac{dV_2}{dt} = -\frac{1}{c_2} i$$

$$\frac{di}{dt} = \frac{1}{\mathscr{L}_2} V_2$$

Now assume, for simplicity, that the constants c_1, c_2, $\mathscr{L}_1$, $\mathscr{L}_2$, are equal to 1; denoting V_1 by x_1, V_2 by x_2 and i by x_3 we have:

$$\frac{dx_1}{dt} = -x_3 \qquad \frac{dx_1}{dt} = 0$$

$$\frac{dx_2}{dt} = 0 \qquad \text{or} \qquad \frac{dx_2}{dt} = -x_3$$

$$\frac{dx_3}{dt} = x_1 \qquad \frac{dx_3}{dt} = x_2$$

For each position of the switch the velocity is orthogonal to the position vector, thus all the motions, whatever we are doing with the switch, are restricted to a sphere whose radius is the norm of the initial state. Thus, the network defines a family $\mathscr{D}$ of two vector fields on a sphere S. This is a control dynamical system. If $x^0 = (x_1^0 x_2^0 x_3^0)$ is an initial state then $G^+\mathscr{D} . x^0$ is the set of all possible states from x^0 when we control the switch from right to left and vice versa without any restriction.

It is also noted in Ref.[3] that in the problem of the rotation of a rigid body around its centre of mass we are dealing with a control dynamical system on the tangent bundle to the set of orthogonal matrices.

1.4. Second example of a control dynamical system

Let X^i; (i = 1, 2, ..., p) and X^0 be p+1 vector fields in $\mathbb{R}^n$ and consider the control system

$$\frac{dx}{dt} = X^0(x) + \sum_{i=1}^{p} u_i X^i(x), \; x \in \mathbb{R}^n; \; u_i \in \mathbb{R}$$

and look for B.B. controls, that is piecewise constant controls $t \to u_i(t)$ with value +1 or -1. Denote by $A(t, x_0)$ the accessible set at time t in the usual meaning of control theory and by $A(x_0) = \bigcup_{t>0} A(t, x_0)$ the whole accessible set. Then we have

$$A(t, x_0) = G_t(\mathscr{D}).x_0$$
$$A(x_0) = G^+(\mathscr{D}).x_0$$

where the family $\mathscr{D}$ is the family

$$\mathscr{D} = \left\{X^0 + \sum_{i=1}^{p} \epsilon_i X^i ; \ \epsilon_i = \pm 1\right\}$$

Actually, any system defined by

$$\frac{dx}{dt} = f(x, u), \ x \in \mathbb{R}^n, \ u \in \mathbb{R}^p$$

where the controls are piecewise smooth can be considered to be a control dynamical system [42].

2. GENERAL STRUCTURE OF ORBITS OF CONTROL DYNAMICAL SYSTEMS

In this section, we consider a control dynamical system defined by a collection of vector fields $\mathscr{D}$ on an n-dimensional, connected, para-compact, smooth manifold M. For the sake of simplicity, they are supposed to be complete.

2.1. The rank of a system

If X and Y are two vector fields on a manifold M, we denote by [X, Y] their bracket. We recall that in a local co-ordinate system one has

$$[X, Y](x) = DY(x)X(x) - DX(x)Y(x)$$

where DX(x) and DY(x) denote the Jacobian matrix of X and Y at point x, which reduces to:

$$\left[\frac{\partial}{\partial x}, Y\right](x) = \frac{\partial Y}{\partial x}(x)$$

if the vector field X is $\partial/\partial x$. If $\mathscr{D}$ is a family of vector fields, then $[\mathscr{D}]^\infty$ denotes the smallest family, closed under bracket operation, which contains $\mathscr{D}$.

2.1.1. Definition: The rank of a family $\mathscr{D}$ of vector fields at point x is the dimension of the linear hull of the set of vectors V(x) in TM when V ranges over $[\mathscr{D}]^\infty$. We denote it by $r_{\mathscr{D}}(x)$. We have:

$$r_{\mathscr{D}}(x) = \dim(\mathscr{L}(\{V(x);\ V \in [\mathscr{D}]^\infty\})).$$

The importance of Jacoby brackets and the rank of a system for the controllability of systems seems to have been noticed first by Hermann [17] and Hermes [19]. The systematic use of this tool started in 1970 (see Haynes-Hermes [16], Lobry [40]).

Notice that the rank is unchanged if we replace $[\mathscr{D}]^\infty$ by the algebra generated by $\mathscr{D}$.

2.1.2. A few remarks:

The rank $r_{\mathscr{D}}(x)$, when $\mathscr{D}$ has just one element, is always 0 or 1.

As soon as we have more than one element in $\mathscr{D}$, the rank may be very large. For instance, in $\mathbb{R}^n$, the rank at 0 of the system defined by the two vector fields

$$X = \frac{\partial}{\partial x_i};\quad Y = \sum_{i=0}^{n} x_1^i \frac{\partial}{\partial x_i}$$

is equal to n, because one clearly has

$$[X[X \ldots\ldots X[XY]] \ldots](o) = p!e_p$$

where $\{e_p\}$, p = 1, ... n, is the canonical basis.

2.2. Structure of orbits of C^∞ systems

We are first concerned with the structure of $G(\mathscr{D}).x$. The classical Frobenius theorem tells us [40]:

2.2.1. Proposition: Assume that for every point x in M the dimension of $\mathscr{D}$, (dimension of $\mathscr{L}(\{X(x);\ X \in \mathscr{D}\})$ is equal to the rank and constant equal to p. Such a control system is said to be involutive and of constant dimension. Then there exists a unique manifold structure on $G(\mathscr{D}).x$ such that the mappings

$$(t_1,.t.\ t_i,\ldots,\ t_p) \to X^1_{t_1} \circ \ldots \circ X^i_{t_i} \circ \ldots \circ X^p_{t_p}(x)\quad X^i \in \mathscr{D};\ \ p \in \mathbb{N}\ \ t_i \in \mathbb{R}$$

are differentiable. With this structure $G(\mathscr{D}).x$ is p-dimensional. This theorem is quite inadequate for control purposes. There is no reason for the rank to be equal to the dimension. Chow's theorem [7,30,40] gives us a considerable improvement of this result which is the following proposition:

2.2.2. Proposition: (Chow — Frobenius) Assume that the rank of sytem is constant and equal to p. Then $G(\mathscr{D}).x$ has a unique manifold structure for which the mappings

$$(t_1, \ldots, t_i, \ldots t_p) \to X^1_{t_1} \circ \ldots \circ X^i_{t_i} \circ \ldots \circ X^p_{t_p}(x) X^i \in \mathscr{D};\ p \in \mathbb{N};\ t_i \in \mathbb{R}$$

are differentiable. With this structure $G(\mathscr{D}).x$ is p-dimensional.

Proof. Take the family $[\mathscr{D}]^\infty$, this control dynamical system satisfies the assumption of proposition 2.2.1; thus, $G([\mathscr{D}]^\infty).x$ has a manifold structure of dimension p. It is connected; by standard argument, the conclusion derives from Theorem 2.2.3 below applied in $G([\mathscr{D}]^\infty).x$ which locally looks like $\mathbb{R}_p$.

2.2.3. Theorem (Chow [7]). In $\mathbb{R}^n$, consider a control dynamical system $\mathscr{D}$ of rank n at the origin; then:

i) in every neighbourhood of O the set $G^+(\mathscr{D}).O$ has interior points.
ii) the set $G(\mathscr{D}).O$ is a neighbourhood of O.

Proof. Actually, from Chow's original treatment it is not possible to obtain i), but just ii). Moreover, it does not work directly for families of vector fields but for algebras of vector fields. There are many proofs of Theorem 2.2.3 in control theory literature: Hermann [18], Lobry [40,42,46], Sussmann-Jurdjevic [65], Stefan [55,56]. Below we give Krener's proof, which is very simple (see Ref.[30]).

2.2.4. Krener's proof of 2.2.3:

Let $\mathscr{D}$ be a control dynamical system in $\mathbb{R}^n$, assuming that the rank of $\mathscr{D}$ is n at point O. Then $G^+(\mathscr{D}).O$ has interior points in every neighbourhood of O.

Proof. At least one vector in $\mathscr{D}$ is not zero at the origin, for, if not, by the formulas $[X,Y](0) = DX(0)X(0) - DY(0)X(0)$ every bracket would vanish at O and then the rank is 0. Take X^1 in $\mathscr{D}$ with $X^1 \neq 0$. For ϵ small enough, the set

$$S^1 = \left\{X^1_{t_1}(0);\ 0 < t_1 < \epsilon_1\right\}$$

is a smooth sub-manifold of $\mathbb{R}^n$. There exists a point $p_1 = X^1_{\theta_1}(0)$ in S^1, and a vector X^2 in $\mathscr{D}$ such that $X^2(p_1)$ and $X^1(p_1)$ are independent, for, if not, this would imply that all the vectors of $\mathscr{D}$ are tangent of S^1 and thus the rank would be 1.

For ϵ_2 small enough, the set

$$S^2 = \left\{X^2_{t_2} \circ X^1_{t_1}(0);\ \mathscr{O}_1 - \epsilon_2 < t_1 \ll \mathscr{O}_1 + \epsilon_2,\ 0 \ll t_2 \ll \epsilon_2\right\}$$

is a sub-manifold of $\mathbb{R}^n$. If 2 is smaller than n, there exists a point $p_2 = X^2_{\mu_2} \circ X^1_{\mu_1}$ in S^2 and a vector X^3 (which may be equal to X^1) such that the vectors $S(p_2) X^2(p)_2 X^3(p_2)$ are linearly independent for the same reason than before; one defines S^3 in the same way, etc. up to S^n which is an open subset of $\mathbb{R}^n$ and is contained by construction in G^+0. Since the $\epsilon_1, \epsilon_2 \dots \epsilon_n$ can be chosen arbitrarily small, everything works in any neighbourhood of O. Theorem 2.2.3 is proved. Actually, Theorem 2.2.3 tells us much more than 2.2.2. It says:

2.2.5. Theorem: If the rank at x of the control dynamical system defined by is equal to n then interior points of $G^+(\mathscr{D}).x$ are dense in $G^+(\mathscr{D}).x$:

$$G^+(\mathscr{D}).x \subset \mathrm{Clos}\,(\mathrm{Int}(G^+(\mathscr{D}).x))$$

Proof: Trivial from 2.2.3.

This last result, in view of the stability question (Section 4) seems to be very significant in control theory despite the curious way in which it is stated. Moreover, this property is generic because the rank condition, as the next theorem states, has the following property:

2.2.6. Theorem: Let V(M) denote the set of all vector fields on M; in the C^k-topology of Whitney on $(V(M))^p$, for $p \geq 2$ and k large enough (it depends on p, and the dimension of M), the set of those families $\mathscr{D}$ of p-vector fields which satisfy the rank condition $r_{\mathscr{D}}(x) = n$ everywhere in M, contains an open dense subset.

Proof: See Ref.[41] or Ref.[46]. The proof is carried out by means of Thom's Lemma and the standard techniques in differential topology.

2.2.7. Generic dynamical control systems are "trivial". By "trivial" we mean that for every x in M, $G(\mathscr{D}).x = M$.

Proof: For a generic control dynamical system $r_{\mathscr{D}}(x)$ is n at every point (2.2.6) then $G(\mathscr{D}).x$ is a neighbourhood of x (2.2.3) and thus every orbit is open and hence must be the full space.

Thus, we know the structure of the orbits in the generic case (trivial); in the case of constant rank (Chow + Frobenius), it remains the non-constant case which has been solved recently by Sussmann [59].

2.2.8. Theorem (Sussmann) (Take a family $\mathscr{D}$ of vector fields on M, without any assumption on $\mathscr{D}$ (except that vector fields are smooth) there exists a unique manifold structure on $G(\mathscr{D}).x$ for which the mappings:

$$(t_1, \dots, t_i, \dots, t_p) \to X^1_{t_1} \circ \dots \circ X^i_{t_i} \circ \dots \circ X^p_{t_p}(x) \quad X^i \in \mathscr{D};\ t_i \in \mathbb{R};\ p \in N$$

are differentiable.

Proof: This proof is completely "by hand" in the sense that it does not refer to any other classical result. Thus, it is a true generalization of the Frobenius theorem. It is too technical to be exposed here; see Ref.[59] or Ref.[46].

As a conclusion, we remark that the theory of action of control dynamical systems, as considered from the point of view of orbits, is now complete and actually trivial. Fortunately, many problems on the positive orbits are not yet solved.

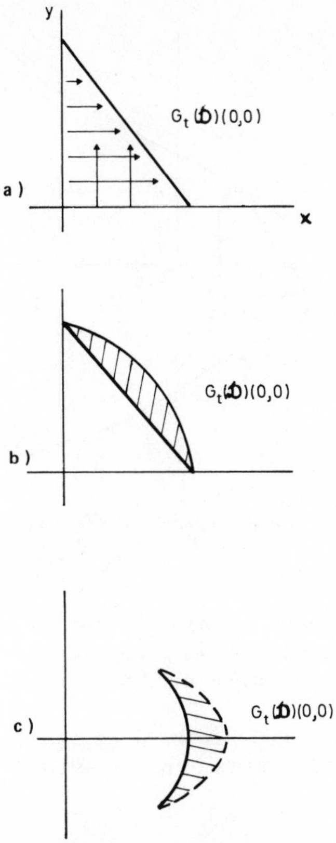

FIG. 2. Control dynamical systems in $\mathbb{R}^2$.

2.2.9. Examples. Consider in $\mathbb{R}^2$ the control dynamical systems defined by (Fig.2):

a)
$$X^1 = \frac{\partial}{\partial x}$$
$$X^2 = \frac{\partial}{\partial y}$$

b)
$$X^1 = \frac{\partial}{\partial x}$$
$$X^2 = (1+x)\frac{\partial}{\partial y}$$

c)
$$X^1 = (1-y^2)\frac{\partial}{\partial x} + \frac{\partial}{\partial y}$$
$$X^2 = (1-y^2)\frac{\partial}{\partial x} - \frac{\partial}{\partial y}$$

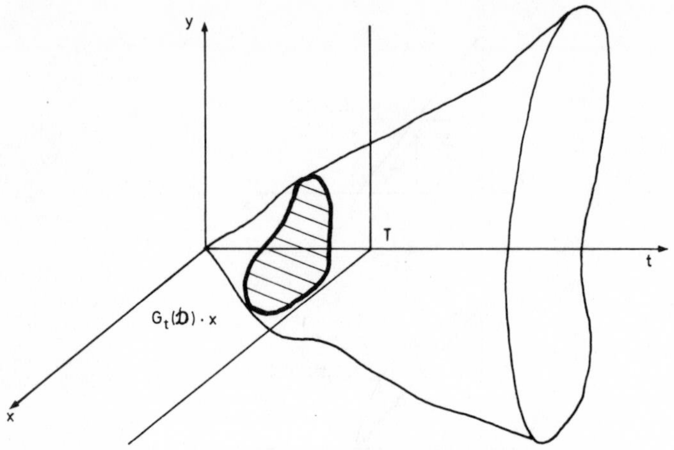

FIG. 3. Proof of Proposition 2.2.10.

In case a) $G_t^+(\mathcal{D}).x$ is a closed set without interior, in case b), it has an interior and is closed, and in case c) it has an interior but is not closed. The question of interior is easily solved by the following:

2.2.10. Proposition: The set $G_t(\mathcal{D}).x$ has a non-empty interior for every t provided that the rank at point (x,0) of the system $\tilde{\mathcal{D}}$ in $M \times \mathbb{R}^n$ defined below is equal to n.

$$\tilde{\mathcal{D}} = \left\{X \oplus \frac{\partial}{\partial t};\ X \in \mathcal{D}\right\}.\left(X(x,t) \oplus \frac{\partial}{\partial t}(x,t) = \begin{pmatrix} X(x) \\ 1 \end{pmatrix}\right)$$

Proof: It is trivial; see Fig.3. Details can be found in Ref.[65].
The question of closedness is not yet solved in this setting. Actually, it is very closely related to questions of Bang-Bang controllability or singular arcs, and we have now a good knowledge of this problem. (See Section 4). Thus, it seems that we are going to obtain a reasonable answer in the near future (if it has not yet been obtained, by now).

We conclude this section by a sort of inverse question: "How many vector fields do we need to obtain $G^+(\mathcal{D}).x' = M$?" The answer is two. See Sussmann [64] and Sussmann-Levitt [39].

2.3. Structure of orbits of analytic systems

In this section, we shall assume that the control dynamical system $\mathcal{D}$ is composed of analytic vector fields on an analytic manifold M. In this case, one can expect "global results" from the knowledge of the "coefficients" of the system at point x. This is actually the case. The following theorem was proved by Nagano [51] in a completely different framework. The proof given in Ref.[40] is false. This was pointed out by P. Stefan [54].

2.3.1. Theorem (Nagano): Let $\mathscr{D}$ be an analytic control system. Let p be the rank of the system at point x. Then the manifold structure defined in 2.2.8 has dimension p.

Proof: See Refs [51,54]. We now turn to the study of $G^+(\mathscr{D}).x$ and $G_t(\mathscr{D}).x$.

Suppose now that the rank of the system at point x is equal to n. This is not a restriction because from 2.3.1. this assumption is satisfied in the restriction to $G(\mathscr{D}).x$. Consider now in $M \times \mathbb{R}$ the system $\tilde{\mathscr{D}}$ defined by: $\tilde{\mathscr{D}} = \{X \oplus \partial/\partial t, X \in \mathscr{D}\}$ and compute its rank at point x. A simple computation shows that

$$\left[X \oplus \frac{\partial}{\partial t}, Y \oplus \frac{\partial}{\partial t}\right] = [X, Y] \oplus 0$$

and, indirectly, all the brackets are of the same form. More precisely, the following holds:

$$[\tilde{\mathscr{D}}]^\infty = \tilde{\mathscr{D}} \oplus [\mathscr{D},[\mathscr{D}]^\infty]$$

and we are interested in the dimension of $\mathscr{L}([\tilde{\mathscr{D}}]^\infty(x,0))$. Clearly, (each vector in $\tilde{\mathscr{D}}(x)$ has a component equal to 1) this linear subspace is not contained in the subspace $(\xi_1, \xi_2, \ldots, \xi_n, 0)$; from this one easily deduces the following

2.3.3. Lemma: The rank of $\tilde{\mathscr{D}}$ is n + 1 at point (x, 0) if and only if the linear space generated by the vectors of the form:

$$\sum_{i=1}^{p} \lambda_i X^i(x) + Y(x); \quad \sum_{i=1}^{p} \lambda_i = 0, \; Y \in [\mathscr{D},[\mathscr{D}]^\infty]$$

is of dimension n.
From this lemma, from 2.2.10, and from 2.3.1 we deduce immediately:

2.3.4. Proposition: Assume that the system $\mathscr{D}$ has rank n at point x. Then $G_t^+(\mathscr{D}).x$ has a non-empty interior (actually it has dense interior points) for every t, if and only if the linear subspace generated by the vectors of the form:

$$\sum_{i=1}^{p} \lambda_i X^i(x) + Y(x); \quad \sum_{i=1}^{p} \lambda_i = 0 \qquad Y \in [\mathscr{D}[\mathscr{D}]^\infty]$$

has dimension n.

2.3.5. We can say a little bit more about $G_t^+(\mathscr{D}).x$ when the dimension is not n. Let us denote by $\mathscr{D}_0$ the collection of vectors of the form

$$\mathscr{D}_0 = \left\{\sum_{i=1}^{p} \lambda_i X^i(x) + Y(x); \quad \sum_{i=1}^{p} \lambda_i = 0 \qquad Y \in [\mathscr{D}[\mathscr{D}]^\infty]\right\}$$

Consider in $M \times \mathbb{R}$ the section at time t, denoted by M_t:

$$M_t = M_x\{t\}$$

From the considerations before Lemma 2.3.3, we know that the sub-manifold $G(\tilde{\mathscr{D}})$ (x, 0) of $M \times \mathbb{R}$ meets each M_t transversally. Hence, the intersection

$$S_t = M_t \cap G(\tilde{\mathscr{D}})\ (x, 0)$$

can be considered to be a sub-manifold of M. From this construction, it is clear that for any X in $\mathscr{D}$ one has

$$X_t(S_0) = S_t$$

and, on the other hand, S_t contains $G_t^+(\mathscr{D}).x$. Moreover, the system $(\tilde{\mathscr{D}})$ restricted to $G^+(\tilde{\mathscr{D}})$ (x, 0) is of maximum rank, thus in the topology of $G(\tilde{\mathscr{D}})$ (x, 0), the set $G^+(\tilde{\mathscr{D}})$ (x, 0) has interior points everywhere dense and because it meets S_t transversally $G_t^+(\mathscr{D}).x$ has an everywhere dense interior in the topology of S_t. To conclude, we know that at every point x in M the family $\mathscr{D}_0$ is tangent to S_0 and of rank n-1. This proves that $S_0 = G(\mathscr{D}_0)x$. We have proved (see Sussmann-Jurdjevic [65] if more details are needed) the following theorem.

<u>2.3.6. Theorem</u> (Sussmann-Jurdjevic) Let $\mathscr{D}$ be an analytic control system. Define by $\mathscr{D}_0$ a new control system, by 2.3.5.

i) if rank of $\mathscr{D}_0$ at point x is n then interior points of $G_t^+(\mathscr{D}).x$ are dense in $G_t^+(\mathscr{D}).x$

ii) if rank of $\mathscr{D}_0$ at point x is less than n then it is n-1. The sub-manifold $G(\mathscr{D}_0).x$ has codimension 1. The set $G_t^+(\mathscr{D}).x$ is contained in $X_t(G(\mathscr{D}_0))$ for any X in $\mathscr{D}$ and in the topology of the sub-manifold $X_t(G(\mathscr{D}_0))$ interior points of $G^+(\mathscr{D}).x$ are dense in $G^+(\mathscr{D}).x$

There is another interesting point in Sussmann-Jurdjevic [65]. We just state it without comments on the proof:

<u>2.3.7. Proposition:</u> Let M be a manifold whose fundamental group has no elements of infinite order; then

$$G^+(\mathscr{D}).x = M \rightarrow G_t^+(\mathscr{D}) \text{ has a non-empty interior for some t.}$$

<u>2.3.8. Proposition:</u> Let M be a manifold whose universal covering is compact, then

$G^+(\mathscr{D}).x$ has a non-empty interior $\rightarrow$ $G_t^+(\mathscr{D}).x$ has a non-empty interior for some t.

3. CONTROLLABILITY OF NON-LINEAR SYSTEMS

In this section we restrict our attention to systems of the following form:

$$\frac{dx}{dt} = X^0(x) + \sum_{i=1}^{p} u_i X^i, \qquad x \in M. \quad (u_1, u_2, \ldots, u_p) \in \mathbb{R}^p \qquad (1)$$

There is some arbitrariness in choosing such systems where the control is a p-dimensional input which enters into the equation linearly. But it is well-known that the essential difficulties are in the space non-linearities and, moreover, this form is a good balance between reasonable generality in order to obtain interesting cases and a simple exposition.

All the controls which we consider are locally bounded measurable controls and, following Jurdjevic-Sussmann [29], we denote by:

U_u the set of controls which take arbitrary values in $\mathbb{R}^p$: the set of "unrestricted controls".
U_r the set of controls $t \to (\mathcal{U}_1(t), \mathcal{U}_2(t), \ldots, \mathcal{U}_i(t), \ldots, \mathcal{U}_p(t))$ such that: $|\mathcal{U}_1(t))| \leq 1$. The sett of "restricted controls".
U_b the set of piecewise constant controls which take just the values +1 or -1: The set of "Bang Bang controls".

The accessible set at time t from some initial condition x is denoted by:

$$A_u(x,t), \qquad A_r(x,t), \qquad A_b(x,t)$$

depending on the fact that the admissible controls are "unrestricted", "restricted" or "Bang Bang". And we introduce:

$$A_u(x,[0,t]) = \bigcup_{s \in [0,t]} A_u(x,t)$$

$$\overline{}$$

$$A_r(x,[0,t]) = \bigcup_{s \in [0,t]} A_r(x,t)$$

$$A_b(x,[0,t]) = \bigcup_{s \in [0,t]} A_b(x,t)$$

We shall study the properties of these sets when M is an arbitrary manifold, smooth or analytic, a compact Riemannian manifold and the control dynamical system conservative, a Lie group and the control dynamical system is right invariant and finally in $\mathbb{R}^n$.

3.1. The case of an arbitrary manifold

We consider a system (1) on a manifold M. When M is an analytic manifold and the vector fields are analytic we say that we are in the analytic case.

3.1.1. Notation: Consider the system (1):

$$\frac{dx}{dt} = X^0(x) + \sum_{i=1}^{p} u_i X^i(x)$$

We denote by L the Lie algebra generated by the family $(X^0, X^1, \ldots, X^p)$, by L_0 the ideal generated by $(X^1, X^2, \ldots, X^p)$ and by ℓ the Lie algebra generated by $(X^1, X^2, \ldots, X^p)$.

3.1.2. Proposition: If (and only if in the analytic case) the dimension of L at point x is equal to the dimension n of the manifold M one has:
For every $(t > 0)$ interior points of $A_\alpha(x,[0,t])$ are dense in $A_\alpha(x,[0,t])$ where the symbol α can take the values u, r, or b.

Proof: This follows from Theorems 2.2.3 and 2.3.1 as soon as we have realized, by a simple computation, that the rank of the family

$$\mathscr{D} = \left\{X^0 + \sum_{i=1}^{p} u_i X^i;\ (u_1, u_2, \ldots, u_p) \in U_b\right\}$$

at point x (see definition 2.1.1) is exactly the dimension of L at point x. Details are left to the reader. Remark that in the analytic case of Theorem 2.3.1 the same conclusion holds in the topology of the manifold $G(\mathscr{D}).x$. Now we turn our attention to the reachable set at time t.

3.1.3. Proposition: If (and only if in the analytic case) the dimension of L_0 at point x is equal to n we have:
For every $t (t > 0)$ interior points of $A_\alpha(x,t)$ are dense in $A_\alpha(x,t)$ where the symbol α can take the values u, r or b. Moreover, if the dimension of L at x is n, the dimension of L_0 at x is either n or n-1, and, in this last case, if the system is analytic, the same conclusion holds in restriction to some sub-manifold of dimension n-1 for the case α = r or α = b.

Proof: It follows from the fact that the dimension of L_0 is the same as the dimension of the family described in Lemma 2.3.3 (with respect to $\mathscr{D}$).

We turn now to controllability results, which are not very strong in this context.

3.1.4. Proposition: Assume that the system is homogeneous: i.e. $X^0 \equiv 0$. Then if (and only if in the analytic case) the dimension of ℓ is equal to n at every point of M (which is assumed to be connected) then
For every x in M one has $A_r(x,[0,\infty]) = A_b(x,[0,\infty]) = M$
For every x in M and $t(t > 0)$ one has $A_u(x,t) = M$.

Proof: Since in this case we can "reverse time" (the trajectory corresponding to some constant control $(u_1, \ldots, u_p)$ can be followed backward using the control $(-u_1, -u_2, \ldots, -u_p)$, it turns out that this result comes directly, see 2.2.4).

3.1.4. Proposition: Assume that the rank of ℓ is n everywhere, then:
For every x in M one has $A_u(x,[0,\infty] = M$.

Proof: Take a point y; a neighbourhood of this point is accessible at time t, for arbitrary small t, from x under the homogeneous system (see above). During a very short time the introduction of the non-homogeneous term X^0 introduces just a "small error". One concludes by a fixed-point theorem.

The correct proof of this last result can be found in Brunovsky-Lobry [6] where this idea is used to prove various results of controllability.

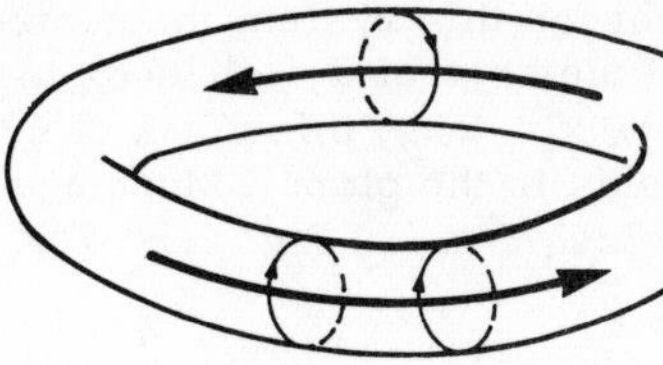

FIG. 4. Torus illustrating Proposition 3.1.5.

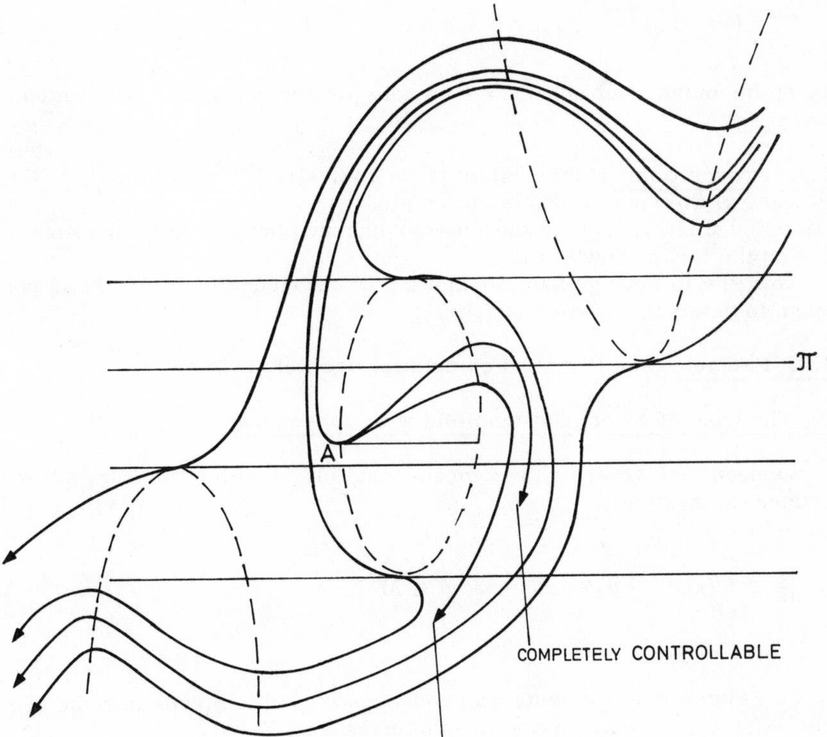

FIG. 5. Knowledge of Σ^+ and Σ^- does not determine controllability. Controllability of A depends on whether the trajectory is above or below line π.

We conclude this paragraph by some comments taken from the work of Gerbier [11, 12]. We look for the controllability of a very simple system, i.e. the system:

$$\frac{dx}{dt} = X^0(x) + u\, X^1(x) \qquad x = (x_1, x_2) \in \mathbb{R}^2 \tag{2}$$

$$X^0 = \frac{1}{2}\,(1 + f(x_1, x_2))\,\frac{\partial}{\partial x_1} + \frac{1}{2}\,g(x_1\, x_2)\,\frac{\partial}{\partial x_2}$$

$$X^1 = \frac{1}{2}\,(1 - f(x_1, x_2))\,\frac{\partial}{\partial x_1} - \frac{1}{2}\,g(x_1, x_2)\,\frac{\partial}{\partial x_2}$$

Notice that for this system the control u = +1 leads to the vector field $\partial/\partial x_1$ and the control u = -1 leads to the vector field $f(x_1 x_2)\partial/\partial x_1 +$ $+ g(x_1 x_2)\partial/\partial x_2$; thus, this class of systems is parametrized by the set of vector fields in the plane. Moreover, we restrict ourselves to non-vanishing vector fields: $f^2(x_1 x_2) + g^2(x_1 x_2) > 0$. It turns out that the two sets

$$\Sigma^+ = \{(x_1, x_2);\ g(x_1, x_2) = 0 \text{ and } f(x_1, x_2) > 0\}$$

$$\Sigma^- = \{(x_1, x_2);\ g(x_1, x_2) = 0 \text{ and } f(x_1, x_2) < 0\}$$

play an important, but non-exclusive role for controllability, i.e. we can prove:

3.1.5. Proposition: If the system (2) is controllable: ($A_r(x,[0,\infty]) = \mathbb{R}^2$ for every x) then the set Σ^- is not empty.
Notice that this depends on the Euler-Poincaré characteristic because it is false on the torus shown in Fig.4.

Now, the following example shows how knowledge of Σ^+ and Σ^- is not enough to determine controllability.

3.1.6. Example (see Fig.5). For details, see Ref.[11].

3.2. The case of a compact manifold with a measure

Suppose that we are on a compact manifold M with a measure μ. We consider the system:

$$\frac{dx}{dt} = X^0(x) + \sum_{i=1}^{p} u_i X^i(x) \qquad x \in M \tag{1}$$

and we assume that the system is conservative; this means that the vector fields $X^0 \pm X^i$, i = 1 ...p, perserve the measure μ.

$$\mu((X^0 \pm X^i)_t(B)) = \mu(B)$$

We denote by L the Lie algebra generated by $(X^0, X^1, \ldots, X^p)$.

3.2.1. Proposition: If (and only if in the analytic case) the rank of the dimension of L is n (the dimension of M) at point x then one has:

$$A_r(x,T) = A_b(x,T) = M \text{ for some sufficiently large } T.$$

Proof: It uses the fact that "Poisson-stable" points are dense for conservative dynamics. See Lobry [44].

Compactness plus maximum rank is not sufficient for controllability as is shown by the following example:

3.2.2. Example: Consider, on the sphere, the system described by Fig.6.

FIG. 6. Compactness plus maximum rank not enough for controllability.

3.3. The case of a Lie group

The case where the state space is a Lie group G has been extensively studied by Brockett [2,4], Jurdjevic [28], Jurdjevic-Sussmann [29], and Sussmann [58]. The technique consists in the use of general results from section 3.1, together with some advantages due to the group structure. The systems considered are right-invariant:

$$\frac{dx}{dt} = X^0(x) + \sum_{i=1}^{p} u_i X^i(x), \qquad x \in G \tag{1}$$

which means that the vector fields X^i; $i = 0, 1, \ldots, p$, are right-invariant vector fields. The Lie algebras L, L_0, and ℓ, with notation of 3.1.1, determine Lie sub-groups S, S_0, and s. In this case, the question of controllability reduces to a knowledge of whether or not $A(e,[0,\infty])$ (where e is the neutral element of G) is a sub-group. An answer is given by the following theorem:

3.3.1. Theorem: If G is compact the system (1) is completely controllable (i.e. $A_\alpha(e,[0,\infty]) = G$; $\alpha = u, r, b$) if and only if the Lie algebra L is the Lie algebra of G.

Proof: Because of the compactness of G, one can prove, in this case, that $A_\alpha(e,[0,\infty])$ is a sub-group [29]. Another way of proving this is to remark that, for flows generated by right-invariant vector fields on a compact Lie group, all points are Poisson-stable and then the result of section 3.2 can be applied. Actually, in this compact case, one can say even more:

3.3.2. Theorem: Assume that G is compact and the Lie algebra L of (1) is the Lie algebra of G. Then for each class of controls U_α($\alpha = u, r, b$) there exists a $T > 0$ such that for every g in G $A_\alpha(g,[0T]) = G$. If, moreover, G is semi-simple the same statement holds with $A_\alpha(g,T)$ in place of $A_\alpha(g,[0T])$.

Proof: see Ref.[29].

3.3.3. The group $GL(n,\mathbb{R})$ of invertible matrices turns out to be a Lie group. In this case, the tangent space at each point is identified with the set $M(n,\mathbb{R})$ of n-by-n matrices. A right-invariant vector field is of the form:

$$X \to AX \qquad X \in GL(n,\mathbb{R}); \quad A \in M(n,\mathbb{R})$$

Thus for every sub-group G of GL(n, ℝ) the "matrix differential system":

$$\frac{dX}{dt} = [A^0 + \sum_{i=1}^{p} u_i A^i] X, \quad X \in G; \; A^i \in L(G) \; i = 0, 1, \ldots, p$$

is a particular case of right-invariant systems on Lie groups. In this case, the bracket of the two vector fields

$$X \to A^i X$$
$$X \to A^j X$$

is the vector field

$$X \to (A^i A^j - A^j A^i)X$$

or

$$X \to [A^i, A^j]X$$

with the usual notation:

$$[A, B] = AB - BA.$$

3.3.4. We conclude this paragraph by the following remark due to Hirshorn [24,25]. By a theorem of Palais [52], if the Lie algebra of a family of vector fields is finitely generated, then it can be realized as the Lie algebra of a Lie group of diffeomorphisms of the manifold. Then our problem reduces to a control problem on a Lie group.

3.4. The manifold is $\mathbb{R}^n$

We show how classical results on controllability in $\mathbb{R}^n$ are consequences of the previous results. But first we state a result by Brunovsky whose proof is quite delicate and does not reduce to 3.1, 3.2, or 3.3.

3.4.1. Proposition: The control system (1)

$$\frac{dx}{dt} = X^0(x) + \sum_{i=1}^{p} u_i X^i(x), \quad i = 1, 2, \ldots, p$$

is said to be "odd" in case we have $X^0(-x) = -X^0(x)$. Then the set $A(0,[0,T])$ is a neighbourhood of the origin for every positive T if (and only if in the analytical case) the dimension of the Lie algebra L generated by $(X^0, X^1, \ldots, X^p)$ is n.

Proof: see Ref.[5].

Let us look now for "bilinear systems":

3.4.2. Definition: A bilinear system on $\mathbb{R}^n$ is a system of the form:

$$\frac{dx}{dt} = \left(A^0 + \sum_{i=1}^{p} u_i A^i\right)x \qquad x \in \mathbb{R}^n \qquad (2)$$

It is clear that the origin is a rest point for systems (2): thus we look for controllability in $\mathbb{R}^n \setminus \{0\}$. Because the solution at time t of system (2), from initial condition x_0 is given by $X(t).X_0$, where $X(t)$ is the solution of the matrix equation:

$$\frac{dX}{dt} = \left(A^0 + \sum_{i=1}^{p} u_i A^i\right)X$$

the problem of controllability reduces to the problem of controllability on Lie groups of matrices (see paragraph 3.3).

We conclude by the celebrated Kalman criterium:

3.4.3. Theorem: The system:

$$\frac{dx}{dt} = Ax + Bu, \; x \in \mathbb{R}^n, \quad u \in \mathbb{R}^p$$

is completely controllable if and only if the rank of the matrix

$$(B, AB, A^2B, \ldots\ldots\ldots\ldots, A^{n-1}B)$$

is equal to n.

Proof: The reachable set from any point is a linear subspace with a non-empty interior, i.e. the full space, iff the rank of the system is n at every point. Compute the brackets:

$$[Ax + Bu_1, Ax - Bu_1] \text{ with } u_1 = (1, 0, 0, 0)$$

$$[Ax + Bu_1, Ax - Bu_1] = 2\,A\,Bu_1$$

Now compute:

$$[Ax + Bu_1, A\,Bu_1]$$

We obtain

$$[Ax + Bu_1, A\,Bu_1] = A^2Bu_1$$

etc. One easily sees that the rank of the system is precisely the rank of the matrix $(B, AB, \ldots, A^{n-1}B)$.

4. OTHER RELATED QUESTIONS

To conclude this survey, we just mention here some related questions which are object of current research.

4.1. Closedness of $G^{+}(\mathscr{D}).x$:

Consider the system:

$$\frac{dx}{dt} = X^0(x) + \sum_{i=1}^{p} u_i X^i(x); \quad x \in \mathbb{R}^n$$

With the notations of Chapter 3 we have:

$$A_r(x,t) = \overline{A_b(x,t)}$$

and clearly if $A_b(x,t)$ is closed this means:

$$A_r(x,t) = A_b(x,t)$$

that is to say, everything you can do you can do with a bang-bang control. This is a bang-bang principle. It is known that bang-bang principle, singular arcs and maximum principle are very closely related questions. On these topics, we mention papers by Hermes [20,21] Krener [30,34,35], Lobry [43,45].

4.2. Realization theory

Assume that you have a control system of the standard form:

$$\frac{dx}{dt} = X^0(x) + \sum_{i=1}^{p} u_i X^i(x) \qquad x \in M; \quad (u_1 u_2 \dots u_p) \in \mathbb{R}^p$$

on some manifold M. Let $\varphi : M \to N$ be a mapping from M into some other manifold N (the observation). Let us say that two states x_0 and x_1 are equivalent if for any control $t \to \mathscr{U}(t)$ we have:

$$\varphi(x(x_0, t, \mathscr{U})) = \varphi(x(x_1, t, \mathscr{U})), \ t \geq 0$$

where $x(x_0, t, \mathscr{U})$, $x(x, t, \mathscr{U})$ denote, respectively, the response issued from x_0 and x_1. To speak of minimal realization, it is necessary to make the quotient of M under the above equivalence and to define a manifold structure on the quotient. It turns out that this is possible by a theorem of Sussmann [60], which generalizes the classical "closed-sub-group theorem". Applications are given in Sussmann [61,62]. Another approach concerning local realizations of non-linear systems by bilinear ones can be found in Krener [33].

4.3. Structural stability and classification

Let us say that a control system is structurally stable if a little change in the data does not very much change the general behaviour of the system. What are structurally stable systems? Are they generic? Try to classify them. There are some attempts to work in this direction in Gerbier-Lobry [12,13].

4.4. Connection with diffusion process

Consider the diffusion process:

$$d\xi_t = X^0(\xi_t) + \sum_{i=1}^{p} X^i(\xi_t)\, dB_i$$

where X^0, X^1 ... X^p are vector fields and the B_i are independent one-dimensional Brownian motions. By a result of Stroock and Varadhan, the support of the diffusion process is connected to the controllability problem for the control system:

$$\frac{dx}{dt} = X^0(x) + \sum_{i=1}^{p} u_i X^i(x).$$

On these topics, see Elliot [8,9], Kunita [38], Strooke-Varadhan [57].

REFERENCES

[1] BOOTHBY, W., A transitivity problem from control theory.

[2] BROCKETT, R., Lie theory and control systems defined on spheres, SIAM J. Appl. Math. 24 5 (1973).

[3] BROCKETT, R., System theory on group manifolds and cosetspaces, SIAM J. Control 10 2 (1972).

[4] BROCKETT, R., "Lie algebras and Lie groups in control theory, Geometric Methods in System Theory", Reidel (1973).

[5] BRUNOVSKY, P., Local controllability of odd systems, to appear.

[6] BRUNOVSKY, P., LOBRY, C., "Contrôlabilité Bang-Bang, contrôlabilité différentielle et perturbation des systèmes non linéaires", (à paraître dans Ann. Matem. pura ed appl.)

[7] CHOW, W.L., "Über Systeme von linearen partiellen Differentialgleichungen erster Ordnung", Math. Ann. 117 (1939) 98-105.

[8] ELLIOTT, D., A consequence of controllability, J. Diff. Equ. 10 (1971) 364-370.

[9] ELLIOTT, D., "Diffusions on manifolds arising from controllable systems", Geometric Methods in System Theory, (loc. cit. see Ref.[4]).

[10] ELLIOTT, D., TARN, J.T., Controllability and observation for bilinear systems (unpublished).

[11] GERBIER, Y.,"Classification des couples de systèmes dynamiques du plan. Application à la théorie de la commande", Thèse, 3è cycle, Bordeaux 1974.

[12] GERBIER, Y., "Classification de certains systèmes dynamiques contrôlés du plan", C.R. Acad. Sc. Paris 280 (20 janvier 1975) 109-112.

[13] GERBIER, Y., LOBRY, C., "On the structural stability of dynamical control systems". Communication to 1975 I.F.A.C. Congress.

[14] GROTE, J., "Problems in Geodesic Control", Geometric Methods in Systems theory (loc. cit. see Ref.[4]).

[15] GROTE, J., "La théorie des connections et contrôle". Publications Mathématiques de l'Université de Bordeaux, Année 73-74, Fascicule 3, 7-13.

[16] HAYNES, G.W., HERMES, H., Nonlinear controllability via Lie theory, SIAM J. Control 8 (1970) 450-460.

[17] HERMANN, R., "On the Accessibility problem in control theory". Internat. Sym. Non-linear Differential Equations and Nonlinear Mechanics, Academic Press New York (1963).
[18] HERMANN, R., Differential Geometry and the Calculus of Variations, Academic Press, New York (1968).
[19] HERMES, H., Controllability and the singular problem, SIAM J. Control 2 (1965) 241-260.
[20] HERMES, H., On local and global controllability, SIAM J. Control, 12 (1974) 252-261.
[21] HERMES, H., "On necessary and sufficient conditions for local controllability among a reference trajectory", Geometric Methods in System Theory (loc. cit. see Ref.[4]).
[22] HERMES, H., "Local controllability and sufficient conditions in singular problems" (to appear).
[23] HERMES, H., HAYNES, G.W., On the nonlinear control problem with control appearing linearly, SIAM J. Control 1 (1963) 85-108.
[24] HIRSHORN, R., "Topological groups and the control of nonlinear Systems" (to appear).
[25] HIRSHORN, R., Topological semi-groups, sets of generators and controllability, Duke Math. J. 40 4 (1973) 937-947.
[26] HIRSHORN, R., "Controllability in Nonlinear Systems", Geometric Methods in System Theory (loc. cit. see Ref.[4]).
[27] HIRSHORN, R., "Topological semi groups and controllability in bilinear systems", Ph. D. Thesis, Div. of Eng. and Appl. Phys., Harvard Univ. (Sept. 1973).
[28] JURDJEVIC, V., "Certain controllability property of analytic control systems", SIAM J. Control 10 (1972) 354-360.
[29] JURDJEVIC, V., SUSSMANN, H., "Control systems on Lie groups", J. Diff. Eqs 12 2 (1972).
[30] KRENER, A., "A generalization of Chow's theorem and the Bang-Bang theorem to nonlinear control systems", SIAM J. Control 12 (1974) 43-52.
[31] KRENER, A., "On the equivalence of control systems and the linearization of nonlinear control systems", SIAM J. Control 11 (1973) 670-676.
[32] KRENER, A., "Bilinear and nonlinear realizations of input output maps" (to appear).
[33] KRENER, A., "Local approximation of control systems" (to appear in J. Diff. Eqs)
[34] KRENER, A., "The high order maximum principle" (to appear).
[35] KRENER, A., "The high order maximum principle", Geometric Methods in System Theory (loc. cit. see Ref.[4]).
[36] KUČERA, Y., "Solution in large of control problem $\dot{x} = (A(1-u) + B)x$". Czech. Math. J. 16 (91) (1966) 600-623.
[37] KUČERA, Y., "Solution in large of control problem $\dot{x} = (Au + Bv)x$". Czech. Math. J. 17 (92) (1967) 91-96.
[38] KUNITA, Sh., Diffusion process and control systems, cours de D.E.A. Univ. Paris VI, Lab. de Calcul des probabilités (1973-74).
[39] LEVITT, M., SUSSMANN, H., On controllability by means of two vector fields (to appear in SIAM).
[40] LOBRY, C., Controlabilité des systèmes non linéaires, SIAM J. Control 8 (1970) 573-605.
[41] LOBRY, C., Une propriété générique des couples de champs de vecteurs, J. Math. Czech. 22 (1972) 230-237.
[42] LOBRY, C., Quelques aspects qualitatifs de la théorie de la commande, Thèse-Grenoble (1972).
[43] LOBRY, C., Geometric structure of Dynamical polysystems. Warwick Control Theory Center report 19 (1972).
[44] LOBRY, C., Controllability of nonlinear systems on compact manifolds, SIAM J. Control 12-1 (1974) 1-4.
[45] LOBRY, C., "Deux remarques sur la commande Bang Bang des systemes semi linéaires". (Proc. Conf. Zakopane, Poland).
[46] LOBRY, C., "Dynamical Polysystems and Control Theory", Geometric Methods in systems theory" (loc. cit. see Ref. [4]).
[47] MARKUS, L., "Control dynamical systems", Math. Systems Theory 3 (1969) 179-185.
[48] MARKUS, L., SELL, G.R., "Capture and control in conservative dynamical systems" Arch. Ration. Mech. Anal. 31 (1968) 271-287.
[49] MARKUS, L., SELL, G.R., "Control in conservative dynamical systems: Recurrence and capture in aperiodic fields". J. Diff. Eqs 16 (1974) 472-505.
[50] MILNOR, J.W., Topology from the Differential View Point, The University Press of Virginia, Charlottesville (1965).
[51] NAGANO, T., Linear differential systems with singularities and application to transitive Lie algebras. J. Math. Soc. Japan 18 (1966) 398-404.
[52] PALAIS, R., A global formulation of the Lie theory of transformation groups, Mem. A.M.S. 22 (1957).
[53] REBHUHN, D., On the set of attainability, PhD-thesis Univ. of Illinois Urbana - Champaign (1974).
[54] STEFAN, P., Accessibility and singular foliations, PhD-thesis, Univ. of Warwick (1973).

[55] STEFAN, P., Integrability of systems of vector fields (to appear).
[56] STEFAN, P., "Two proofs of Chow's theorem" Geometric Methods in Systems Theory (loc. cit., see Ref. [4]).
[57] STROOKE, P., VARADHAN, S., "On the support of diffusion Processes, with applications to the strong maximum principle", 6th Berkeley Symp. Mathematical Statistics and Probability.
[58] SUSSMANN, H., The Bang-Bang problem for certain systems in G.L.(n, ℝ), SIAM J. Control (to appear).
[59] SUSSMANN, H., "Orbits of families of vector fields and integrability of systems with singularities". Trans. Amer. Math. Soc. 180(1973) 171-188.
[60] SUSSMANN, H., "On quotients of manifolds: a generalization of the dosed subgroup theorem" (to appear in J. Diff. Geom.).
[61] SUSSMANN, H., Minimal realizations of nonlinear systems, Geometric Methods in Systems Theory (loc. cit., see Ref. [4]).
[62] SUSSMANN, H., "Observable realization of finite dimensional non-linear systems" (to appear in SIAM J. control.)
[63] SUSSMANN, H., "Existence and uniqueness of minimal realizations of nonlinear systems I: Initialized systems" (to appear).
[64] SUSSMANN, H., "On the number of vector fields needed to achieve controllability"(to appear).
[65] SUSSMANN, H., JURDJEVIC, V., "Controllability of nonlinear systems", J. Diff. Eqs 12 (1972) 95-116.
[66] SWAMY-TARN, "Optimal control of discrete bilinear systems", Geometric Methods in System Theory (loc. cit., see Ref. [4]).
[67] SWAMY, "Optimal control of single input bilinear systems", D.S.C. Dissertation (Dec. 1973) Washington, University St. Louis, Missouri.

INTRODUCTION TO CONVEX ANALYSIS

J.P. CECCONI
Centro di Studio per la Matematica e la Fisica Teorica,
C.N.R.,
Genova, Italy

Abstract

INTRODUCTION TO CONVEX ANALYSIS.

This paper considers the main elements of convex analysis in infinite-dimensional spaces, that is the most essential properties of convex functions valued in $\overline{R}$ (the extended real numbers) such as: continuity property, duality, sub-differentiability, properties concerning the minimization of such functions, and, finally, the connections of these with the monotonic-operators theory and the variational-inequalities theory. Convexity has always played a fundamental role in the study of variational problems, but systematic studies on convexity have been carried out only in recent times.

1. CONVEX SETS, SEPARATION THEOREMS

Let X be a vector space over R (abbreviated v. s.) if $x, y \in X$, the set

$$[x, y] = \{\lambda x + (1-\lambda) y : 0 \leq \lambda \leq 1\}$$

is called the closed line segment joining x and y.

A subset K of X is convex if $x, y \in K$ imply that $[x, y] \in K$. It is immediate that X and $\emptyset$, the empty set, are convex sets. The intersection of convex sets is a convex set. The union of convex sets is not, in general, a convex set. If A is a subset of X the convex hull of A, denoted by co A, is the smallest convex subset containing A. Evidently

$$\text{co } A = \left\{ \sum_i^n \alpha_i x_i : n \in N; \ \sum_i^n \alpha_i = 1, \ \alpha_i \geq 0, \ x_i \in X; \ i = 1 \ldots n \right\}$$

Let $H = \{x : f(x) = \alpha\}$, where f is a non-zero linear form on X and $\alpha \in R$, be a real affine hyperplane in X; the convex sets

$$F_\alpha = \{x : f(x) < \alpha\}, \quad F^\alpha = \{x : f(x) > \alpha\}$$

are called the algebraically open semispaces determined by H, and the convex sets

$$G_\alpha = \{x : f(x) \leq \alpha\}, \quad G^\alpha = \{x : f(x) \geq \alpha\}$$

are the algebraically closed semispaces determined by H. Two non-empty subsets A, B of X are said to be algebraically separated (respectively strictly separated) by the real hyperplane H if either $A \subset G_\alpha$, $B \subset G^\alpha$ or $A \subset G^\alpha$, $B \subset G_\alpha$ (respectively, if either $A \subset F_\alpha$, $B \subset F^\alpha$ or $A \subset F^\alpha$, $B \subset F_\alpha$). It is obvious that the hyperplane $H = \{x \in X : f(x) = \alpha\}$ separates A and B if and only if

$$f(x) \leqq \alpha \text{ (or } f(x) \geqq \alpha) \ \forall x \in A$$

and

$$f(y) \geqq \alpha \text{ (or } f(y) \leqq \alpha) \ \forall y \in B$$

We recall now the Hahn-Banach theorem (in its analytical and geometrical form) and the separation theorems for convex sets.

Theorem 1 (Hahn-Banach, analytic form). In a v.s. X let $p : X \to R$ be a sublinear real function, i.e. a real function such that

$$p(x+y) \leqq p(x) + p(y); \qquad x, y \in X$$

$$p(\alpha x) = \alpha p(x); \qquad x \in X, \ \alpha \in R_+ \cup \{0\}$$

Let A be a linear subspace of X and let $f : A \to R$ be a linear real function on A with

$$f(x) \leqq p(x), \qquad x \in A$$

Then there exists a linear real function $F : X \to R$ for which

$$F(x) = f(x) \qquad x \in A$$

$$F(x) \leqq p(x) \qquad x \in X$$

Proof. Consider the family $\mathscr{E}$ of all real linear extensions g of f for which the inequality $g(x) \leqq p(x)$ holds for x in the domain of g. Let the relation $h > g$ be defined in $\mathscr{E}$ to mean that h is an extension of g; this relation partially orders $\mathscr{E}$. Let us show that every totally ordered subset of $\mathscr{C} = \{g\alpha : \alpha \in \mathscr{A}\}$ has an upper bound on $\mathscr{E}$. Let us define

$$B = \bigcup_{\alpha \in \mathscr{A}} A_\alpha, \text{ where } A_\alpha \text{ is the domain of } g_\alpha$$

$$g(x) = g_\alpha(x) \text{ if } g_\alpha \in \mathscr{C}, \ x \in A_\alpha$$

This definition is not ambiguous, for if g_{α_1}, $g_{\alpha_2} \in \mathscr{C}$, then either $g_{\alpha_1} < g_{\alpha_2}$ or $g_{\alpha_2} < g_{\alpha_1}$. At any rate, if $x \in A_{\alpha_1} \cap A_{\alpha_2}$, then $g_{\alpha_1}(\alpha) = g_{\alpha_2}(x)$. Clearly $g \in \mathscr{E}$, and it is an upper bound for $\mathscr{C}$ in $\mathscr{E}$. Then from Zorn's lemma follows the existence of a maximal extension F of f for which the inequality $F(x) \leq p(x)$ holds for every x in the domain of F. It remains to be shown that the domain M of F is X. Otherwise we can prove that there would be a $g \in \mathscr{E}$ such that $F < g$ and $F \neq g$ and this would violate the maximality of F.

Let us assume the existence of an $x_0 \in X - M$. Every vector x in the space $M_1 = M \oplus \{x_0\}$ spanned by M and x_0 has a unique representation in the form $x = m + \alpha x_0$ with $m \in M$, $\alpha \in R$. For any constant c, the function g defined on M by

$$g(m + \alpha x_0) = F(m) + \alpha c$$

is a proper extension of F. The desired contradiction will be made when it is shown that c may be chosen so that $g \in \mathscr{E}$. Let m, n be vectors in M. Then from

$$F(m) - F(n) = F(m - n) \leq p(m - n) \leq p(m + x_0) + p(-n - x_0)$$

it follows that

$$-p(-n-x_0) - F(n) \leq p(m + x_0) - F(m)$$

Since the left side of this inequality is independent of m and the right side is independent of n, there is a constant c with

$$-p(-m-x_0) - F(m) \leq c \qquad m \in M$$

$$c \leq p(m + x_0) - F(m) \qquad m \in M$$

We show that, with this c, it results for every $x = m + \alpha x_0 \in M_1$

$$g(m + \alpha x_0) = F(m) + \alpha c \leq p(m + \alpha x_0)$$

so that $g \in \mathscr{E}$. In fact, if $\alpha > 0$, then

$$g(m + \alpha x_0) = \alpha c + F(m) = \alpha \left\{c + F\left(\frac{m}{\alpha}\right)\right\} \leq \alpha \left\{p\left(\frac{m}{\alpha} + x_0\right) - F\left(\frac{m}{\alpha}\right)\right.$$

$$\left. + F\left(\frac{m}{\alpha}\right)\right\} \leq \alpha p\left(\frac{m}{\alpha} + x_0\right) = p(m + \alpha x_0)$$

if $\alpha = -\beta < 0$, then

$$g(m+\alpha x_0) = F(m) + \alpha c = F(m) - \beta c = \beta\left\{F\left(\frac{m}{\beta}\right) - c\right\}$$

$$\leq \beta\left\{F\left(\frac{m}{\beta}\right) + F\left(-\frac{m}{\beta}\right) + p\left(+\frac{m}{\beta} - x_0\right)\right\} = p(+m - \beta x_0)$$

$$\leq p(m+\alpha x_0)$$

The proof of the theorem is so completed.

Corollary 1. In a v. s. X let $p: X \to R$ be a seminorm in X, i.e. a real function such that

$$p(x+y) \leq p(x) + p(y) \qquad x, y \in X$$

$$p(\alpha x) = |\alpha| \cdot p(x) \qquad x \in X, \quad \alpha \in R$$

Let A be a linear subspace of X and let $f: A \to R$ be a linear real function on A with

$$|f(x)| \leq p(x), \quad x \in A$$

Then there exists a linear real function $F: X \to R$ for which

$$F(x) = f(x), \qquad x \in A$$

$$|F(x)| \leq p(x), \qquad x \in X$$

Proof. From the theorem follows the existence of $F: X \to R$ for which

$$F(x) = f(x); \qquad x \in A$$

$$F(x) \leq p(x); \qquad x \subset X$$

Therefore

$$-F(x) = F(-x) \leq p(-x) = p(x)$$

Definition 1. Let K be a convex set in a v. s. X. A point $x \in X$ is called an internal point of K if for every $y \in X$ there exists $\epsilon \in R_+$ such that for every $\delta \in R$, $|\delta| < \epsilon$ it results $x + \delta y \in K$.

Definition 2. Let K be a convex set on a v. s. X and let the origin 0 of X be an internal point of K. For each $x \in X$ let $I(x) = \{a \in R_+ : (x/a) \in K\}$ and $k(x) = \inf I(x)$. The function k(x) is called the gauge function of K.

Theorem 2. Let K be a convex set containing the origin of X as an internal point and be k its gauge function. Then

(a) k is a sublinear non-negative function in X;
(b) if $x \in K$ then $k(x) \leq 1$;
(c) the set of the internal points of K is characterized by the condition $k(x) < 1$.

Proof. Statement (a) follows from the fact that the origin is an internal point of K, from the fact that $I(\alpha x) = \alpha I(x)$ for every $x \in X$ and $\alpha \in R_+$, and from the fact that, for every $x \in X$, $\epsilon \in R_+$

$$k(x) + \epsilon \in I(x)$$

It follows that if $x, y \in K$ then

$$\frac{x}{k(x) + \epsilon}, \quad \frac{y}{k(y) + \epsilon} \in K$$

and from the convexity of K

$$\frac{x+y}{k(x) + k(y) + 2\epsilon} = \frac{x}{k(x)+\epsilon} \frac{k(x) + \epsilon}{k(x) + k(y) + 2\epsilon} + \frac{y}{k(y) + \epsilon} \frac{k(y) + \epsilon}{k(x) + k(y) + 2\epsilon} \in K$$

Therefore

$$k(x) + k(y) + 2\epsilon \in I(x + y)$$

and consequently

$$k(x+y) = \inf I(x+y) \leq k(x) + k(y) + 2\epsilon$$

Hence

$$k(x+y) \leq k(x) + k(y)$$

Statement (b) is self-evident.

If x is an internal point of K then $x + \epsilon x \in K = x(1+\epsilon) \in K$ for some sufficiently small $\epsilon \in R_+$ so that $\frac{1}{1+\epsilon} \in I(x)$. Hence $k(x) = \inf I(x) \leq \frac{1}{1+\epsilon} < 1$.

Conversely, if $k(x) < 1$ let $\epsilon = 1 - k(x)$ and for every $y \in X$ let $|\delta| \{k(y) + k(-y)\} < \epsilon$. Then

$$k(x + \delta y) \leq k(x) + k(\delta y) \leq (1 - \epsilon) + \delta k(y) < (1 - \epsilon) + \epsilon = 1$$

$$k(x + \delta y) \leq k(x) + k(\delta y) \leq (1 - \epsilon) - \delta k(-y) < (1 - \epsilon) + \epsilon = 1$$

according to δ is positive or negative, so that $x + \delta y \in K$. Therefore x is an internal point of K.

We can prove now the following separation theorem for convex sets:

Theorem 3. Let A and B be disjoint non-empty convex subsets of a v. s. X and let A have an internal point. Then there exists a hyperplane H which separates K and H.

Proof. If a is an internal point of A, then the origin 0 of X is an internal point of the convex set $A - a = \{x \in X: x = z - a,\ z \in A\}$. The affine hyperplane $H = \{x \in X : f(x) = \alpha\}$ separates A and B if and only if the affine hyperplane $H' = \{x \in X : f(x) = \alpha - f(a)\}$ separates the sets A-a and B-a. Thus it suffices to prove the theorem under the additional assumption that 0 is an internal point of A.

Let b be any point of B so that -b is an internal point of

$$A - B = \{z \in X : z = x - y,\ \ x \in A,\ \ y \in B\}$$

and 0 is an internal point of the convex set

$$K = A - B + b = \{z \in X : z = x - y + b,\ \ x \in A,\ \ y \in B\}$$

Since A and B are disjoint, the set A-B does not contain 0; hence K does not contain b. Let k be the gauge function of the convex set K, so $k(b) \geq 1$.

If for every $\alpha \in R$ we put $f_0(\alpha b) = \alpha k(b)$ then f_0 is a linear function defined on the one-dimensional subspace of X which consists of real multiples of b. Moreover, $f_0(\alpha b) \leq k(\alpha b)$ for all $\alpha \in R$, since for $\alpha \geq 0$ we have $f_0(\alpha b) = k(\alpha b)$ while for $\alpha < 0$ we have $f_0(\alpha b) = \alpha f_0(b) < 0 \leq k(\alpha b)$. By the Hahn-Banach theorem f_0 can be extended to a real function f such that $f(x) \leq k(x)$ for every $x \in X$.

It follows $f(x) \leq 1$ for every $x \in K$, while $f(b) \geq 1$. Thus $K = A - B + b \subset \{x : f(x) \leq 1\}$, $f(b) \geq 1$. For every $\alpha \in A$, $\beta \in B$ we have

$$f(\alpha) - f(\beta) + f(b) \leq 1$$

and also

$$f(\alpha) \leq f(\beta) + 1 - f(b) \leq f(\beta)$$

Let us call $\gamma = \sup\{f(\alpha) : \alpha \in A\}$. Hence results

$$f(\alpha) \leqq \gamma, \quad \alpha \in A$$

$$f(\beta) \geqq \gamma, \quad \beta \in B$$

Thence

$$A \subset \{x : f(x) \leqq \gamma\}, \quad B \subset \{x : f(x) \geqq \gamma\}$$

and the proof is completed.

In the particular case where B is a linear affine subspace of X we have

Corollary 2 (Hahn-Banach theorem in geometrical form). If A is a convex subset of a v.s. X, which has an internal point, and if B is a linear affine subspace of X such that $A \cap B = \emptyset$, then there exists an affine hyperplane of X which contains B and does not intersect A.

Proof. Let $H = \{x : h(x) = \alpha\}$ the affine hyperplane of X that separates A and B in consequence of Theorem 3; and let be $A \subset \{h(x) \leqq \alpha\}$, $B \supset \{h(x) \geqq \alpha\}$. If b is a point in the linear affine space B we have $h(y) = h(b) = \gamma$ for every $y \in B$. In fact, if $h(b') \neq h(b)$, $b' \in B$ and $B_0 = B - b$, $b' = b + u$, $u \in B_0$, it results $b + \lambda u \in B$ for every $\lambda \in R$ and $h(u) \neq 0$ so that

$$h(b + \lambda u) = h(b) + \lambda b(u) < \alpha$$

for convenient λ; a contradiction.

Therefore $h(x) \leqq \alpha \leqq h(b)$ for every $x \in A$ and $h(y) = h(b)$ for every $y \in B$ so that the hyperplane $H' = \{x : h(x) = h(b)\}$, which contains B, separates A and B.

Definition 3. A real v.s. endowed with a topology $\mathscr{T}$ for which the two axioms

(1) $(x, y) \to x + y$ is a continuous map on $X \times X$ into X

(2) $(\alpha, x) \to \alpha x$ is a continuous map on $R \times X$ into X

are satisfied, is called a real topological vector space (abbreviated t.v.s.). If $\mathscr{N} = \{U\}$ is a basis of neighbourhood of the origin 0 of X in $\mathscr{T}$ and $x_0 \in X$, then $x_0 + U$; $U \in \mathscr{N}$ is a base of neighbourhood of x_0 in the topology $\mathscr{T}$. If K is a convex set in a t.v.s. X, the closure $\overline{K}$ and the interior K° of K are convex sets and $\overline{K} = \overline{(K^\circ)}$ if $K \neq \emptyset$; moreover, if $x \in K^\circ$, then x is an internal point of K; if $x \in K^\circ$ and $y \in \overline{K}$, then every point in the open segment $[x, y[$ is interior to K. If $K^\circ \neq \emptyset$ and x is an internal point of the convex set K, from $\overline{K} = \overline{(K^\circ)}$ it follows that $x \in K^\circ$.

If A is a subset of a t. v. s. X, the intersection of all closed convex sets containing A is called the closed convex hull of A and is denoted by

$\overline{\text{co}}$ K

If K is a t. v. s. and H = $\{x : f(x) = \alpha\}$ is an affine hyperplane of X, then H is a closed subset of X if the linear real function f is continuous on X. In this case the semispaces F_α, F^α [or G_α, G^α] are open [or closed] subsets of X.

Conversely, if for an $\alpha \in R$ and a linear function $f: X \to R$ F_α, F^α are open subsets of X, then f is a continuous function. In fact we have

Theorem 4. If H = $\{x : f(x) = \alpha\}$ is an affine hyperplane of X which separates two non-empty subsets A, B of X, one of which has an interior point, then the linear real function $f: X \to R$ is continuous.

Proof. Let x_0 be an interior point of A. Then there exists a neighbourhood U of the origin 0 in X such that $x_0 + U \in A$. In the hypothesis that $A \subset G_\alpha$, $B \subset G^\alpha$, then we have $f(x_0 + U) \leqq \alpha$ and therefore, for every $x \in U$, $f(x) \leqq \alpha - f(x_0)$. Let $V = U \cap (-U)$; then V and -V are neighbourhoods of the origin of X in $\mathscr{T}$ and for every $y \in V$ it results

$$f(y) \leqq \alpha - f(x_0)$$

$$f(-y) \leqq \alpha - f(x_0)$$

so that

$$|f(y)| \leqq \alpha - f(x_0)$$

Then, for every $\epsilon \in R_+$, we have

$$|f(y)| \leqq \epsilon$$

if $y \in \frac{\epsilon}{\alpha - f(x_0)} V = V'$ and V' is a neighbourhood of the origin 0 of X in $\mathscr{T}$. Therefore f is continuous at the origin 0 of X. Since f is linear and continuous at 0 it is continuous everywhere.

Theorems 3 and 4 yield the following separation theorem in t. v. s.

Theorem 5. Let A and B be non-empty disjoint convex subsets of a t. v. s. X, one of them being an interior point. Then there exists an affine closed hyperplane which separates A and B.

Definition 4. A t. v. s. X is called locally convex (abbreviated l. c. s.) if it is a Hausdorff space such that every neighbourhood of the origin X contains a convex neighbourhood of 0.

Every normed space is obviously a l. c. s. In general, every v. s. endowed with a family $\{P_\gamma : \gamma \in \Gamma\}$ of seminorms in X such that for every $x \in X$ a seminorm p_γ exists such that $P_\gamma(x) \neq 0$, is a l. c. s.; its base of neighbourhoods of the origin 0 are all finite intersections of sets $U_{\gamma,\epsilon} = \{x : P_\gamma(x) \leq \epsilon\}$, $\gamma \in \Gamma$, $\epsilon \in R_+$; this topology on X is called the topology generated by the family of seminorms P_γ. In this topology each of the given seminorms P_γ is continuous on X. In a l. c. s. we can derive from the preceding theorems of separation the following theorem which has become a standard tool of the theory.

Theorem 6. Let A and B be non-empty disjoint convex closed subsets of a l. c. s. X, one of which is compact. Then there exists a closed affine hyperplane which strictly separates A and B.

Proof. A-B is a convex closed set and $(A-B) \cap \{0\} = \emptyset$. Therefore there exists a convex neighbourhood U of the origin in X such that $(A-B) \cap U = \emptyset$. From Theorem 5 it follows that there exists a closed hyperplane H which separates A-B and U. Let $f : X \to R$ be the non-zero continuous linear function such that

$$H = \{x : f(x) = \alpha,\ \alpha \in R\} \text{ and } f(A-B) \geq \alpha,\ f(U) \leq \alpha$$

Since f is a non-zero function, there exists an $x \in X$ such that $f(x) \neq 0$; since U is a neighbourhood of 0, there exists a $\delta \in R_+$ such that $kx \in U$, $|k| \leq \delta$ and consequently $f(kx) \leq \alpha$ if $|k| \leq \delta$. Therefore there exists an $\epsilon \in R_+$ such that $[-\epsilon, \epsilon] \subset \{kf(x) : |k| \leq \delta\} \subset f(U)$ so that $\alpha \geq \epsilon$. Hence $f(a-b) = f(a) - f(b) \geq \alpha > \epsilon$ for every $a \in A$, $b \in B$.

Let $c = \inf\{f(a),\ a \in A\}$, then $c \in R$ because A is compact and f is continuous on X. We have also

$$f(a) \geq c \geq f(b) + \alpha > f(b) + \epsilon,\quad a \in A,\quad b \in B$$

and therefore

$$f(a) \geq c > c - \epsilon > f(b),\quad a \in A,\quad b \in B$$

Two important consequences of these theorems are

Corollary 3. If K is a convex subset of a t. v. s. for which $K^\circ \neq \emptyset$, then through every boundary point of K passes a closed supporting hyperplane H; i. e. a closed hyperplane H such that $K \cap H \neq \emptyset$ and K is contained in one of the closed semispaces determined by H. Moreover, K is the intersection of the closed semispaces which contain K and are determined by the supporting hyperplanes of K.

Corollary 4. If K is a non-empty convex closed subset of a l. c. s., then K is the intersection of all closed semispaces containing K.

Proof of Corollary 3. To see that through every boundary point x of K passes at least one supporting hyperplane it suffices to apply theorem 5 to the convex sets K° and x_0. Indeed there exists an affine closed hyperplane $H = \{x : f(x) = \alpha\}$ such that $f(x) \leqq \alpha$ for every $x \in K°$ and $f(x_0) \geqq \alpha$. Then for every $y \in K$ it results $f(y) \leqq \alpha$ and consequently $f(x_0) = \alpha$. To prove the second assertion we prove preliminarly that there exists no supporting hyperplane of K containing an interior point of K. In fact: assume that $x \in K° \cap H$ where $H = \{x : f(x) = \alpha\}$ is a supporting hyperplane of K such that $K \subset G_\alpha$. There exists $y \in K°$ with $f(y) < \alpha$, since H cannot contain K°. Now

$$f\{x+\epsilon(x-y)\} = f(x)\,(1+\epsilon) - \epsilon f(y) = \alpha(1+\epsilon) - \epsilon f(y) > \alpha\,(1+\epsilon) - \epsilon\alpha = \alpha$$

for every $\epsilon \in R_+$. But from $x \in K°$ it follows that $x + \epsilon(x-y) \in K$ for convenient $\epsilon \in R_+$ and therefore $f(x+\epsilon(x-y)) \leqq \alpha$ for convenient $\epsilon \in R_+$. This is a contradiction. From the first assertion we know that $K \subset \bigcap_{G_\alpha \in \mathscr{S}} G_\alpha$ where $\mathscr{S}$ is the family of all supporting closed hyperplanes H of K and $K \subset G_\alpha$ with $H = \{x : f(x) = \alpha\}$. It remains to prove that if $y \notin K$, then there exists a closed supporting hyperplane $H = \{x : f(x) = \alpha\}$ such that $K \subset G_\alpha$, $f(y) > \alpha$. Let $y \notin K$, $x \in K°$; the open segment $]x, y[$ contains exactly one boundary point x_0 of K. There exists a closed supporting hyperplane $H = \{x : f(x) = \alpha\}$ passing through x_0 such that $K \subset G_\alpha$; H does not contain y; otherwise it would contain x and this is a contradiction.

Proof of corollary 4. It follows obviously from theorem 6 and the fact that sets containing exactly one point are compact.

All the properties now proved are of fundamental importance in functional analysis. For example, if X is a l.c.s., from the Hahn-Banach theorem follows the existence of linear continuous non-zero real functions on X: it suffices to consider two points $x, y \in X$; $x \neq y$ and to separate them by a closed hyperplane H; if $H = \{x : f(x) = \alpha\}$ then $f: X \to R$ is continuous and $f(x) \neq f(y)$. If X is a t.v.s., we call topological dual of X the v.s. X^* of all the continuous, real linear functions on X endowed with the usual structure.

It is convenient to denote elements of X^* with notations x^*, y^* and so on, and to denote the value of the linear continuous real function $x^* \in X^*$ on x with the notation $\langle x, x^* \rangle$.

Then the map $(x, x^*) \to \langle x, x^* \rangle$ is a real bilinear function on $X \times X^*$ and X^* [resp. X] is a v.s. of linear real functions on X [resp. X^*]; moreover, every $x^* \in X^*$ [resp. $x \in X$] is the linear real function $x \to \langle x, x^* \rangle$ on X [resp. $x^* \to \langle x, x^* \rangle$ on X^*]. We can introduce on X^* [resp. on X] the $\sigma(X^*, X)$-topology [resp. the $\sigma(X, X^*)$-topology] by taking as a base at 0 the family of all sets of the form $U(A, \epsilon) = \{x^* \in X^* : |\langle x, x^* \rangle| < \epsilon, x \in A\}$ where A is a finite subset of X, $\epsilon \in R_+$ [resp. $U(A, \epsilon) = \{x \in X : |\langle x, x^* \rangle| < \epsilon, x^* \in A\}$ where A is a finite subset of X^*, $\epsilon \in R_+$].

These topologies are called weak topologies on X^* [resp. on X] associated with the duality between X and X^*. Since $x^* \to |\langle x, x^* \rangle|$ is a seminorm for $\forall x \in X$ and for every pair x^*, $y^* \in X^*$, $x^* \neq y^*$ there exists an x such that $\langle x, x^* \rangle \neq \langle x, y^* \rangle$, it results that X^* endowed with $\sigma(X^*, X)$ is a

l. c. s. If X (with the initial topology) is a l. c. s., then also X endowed with the $\sigma(X, X^*)$ is a l. c. s.; in fact $x \to |\langle x, x^* \rangle|$ is a seminorm for every $x^* \in X^*$ and for every pair $x, y \in X$, $x \neq y$ there exists, in the hypothesis that X is a l. c. s., an $x^* \in X^*$ such that $\langle x, x^* \rangle \neq \langle y, x^* \rangle$.

The (X, X^*) topology on X is coarser than the initial topology on X; therefore the subsets of X closed in the $\sigma(X, X^*)$ topology are closed also in the initial topology.

It is of fundamental importance that from corollary 4 follows

Corollary 5. In a l. c. s. X every closed convex set is closed in the $\sigma(X, X^*)$ topology; therefore a convex subset in a l. c. s. X is closed if and only if it is weakly closed.

Proof. From corollary 5 it follows that $K = \bigcap_{G_\alpha \in S}$ where S is the family of all closed semispaces $G_\alpha = \{x : f(x) \leq \alpha\}$ such that $K \subset G_\alpha$. But every f considered in the definition of a $G_\alpha \in S$ is also continuous in the $\sigma(X, X^*)$ topology of X; therefore every $G_\alpha \in S$ is also closed in that topology. Then from $K = \bigcap_{G_\alpha \in S} G_\alpha$ follows that K is closed in the $\sigma(X, X^*)$ topology.

In the following we shall need also two theorems concerning normed spaces.

Definition 5. Let X be a normed v. s. and $f : X \to R$ a linear continuous function on X. We call $\|f\| = \inf F$ when $F = \{\beta \in R_+ : |f(x)| \leq \beta \|x\|,\ x \in X\}$. It is obvious that F is not empty because f is continuous. $\|f\|$ is called the norm of f.

Theorem 7. Let X be a normed v. s., M a linear subspace of x and $f : M \to R$ a continuous linear function on M. Then there exists a continuous linear function $F : X \to R$ such that $F(x) = f(x)$ on X and $\|F\| = \|f\|$.

Proof. Let $\|f\|$ be the norm of $f : M \to R$ and $p(x) = \|f\|\ \|x\|$. Then $p : X \to R$ is a seminorm on X such that $|f(x)| \leq p(x)$ on M. Then by corollary 1 there exists a linear extension $F : X \to R$ of f such that $|F(x)| \leq p(x) = \|f\|\ \|x\|$. Thus $\|F\| \leq \|f\|$. On the other hand, since F is an extension of f we must have $\|f\| \leq \|F\|$. This completes the proof of the theorem.

Theorem 8. (Mazur) Let X be a normal v. s. and $\{x_n\}_{n \in N}$ a sequence in X which converges to x in the weak topology of X. Then there exists a sequence

$$y_n = \sum_{i=n}^{p_n} \alpha_i x_i; \quad p_n \in N, \quad \alpha_i \geq 0, \quad \sum_{n}^{p_n} \alpha_i = 1$$

whose elements are convex combinations of given x_n, which converge in norm to x.

Proof. For each $n \in N$ let us define $A_n = \bigcup_{i=n}^{\infty} \{x_i\}$. Then x belongs to the weak closure of A_n, for every $n \in N$. Therefore x belongs to the weak closure of $co(A_n)$ $\forall n \in N$. But, for corollary 5, $x \in \overline{co(A_n)}$ $\forall n \in N$. Then for each $n \in N$ we can take a $y_n \in co(A_n)$ such that $\|x - y_n\| < 1/n$. The theorem is proved.

2. CONVEX FUNCTIONS

Let X be a (real) v. s. and $\overline{R} = R \cup \{-\infty\} \cup \{+\infty\}$.

Definition 1. If K is a convex set in X and $f : K \to \overline{R}$ is a function on K whose values are real numbers, either $-\infty$ or $+\infty$, we say that f is convex on K if for any $x \in X$, $y \in X$ we have

$$f(\lambda x + (1 - \lambda)y) \leqq \lambda f(x) + (1 - \lambda) f(y), \quad \lambda \in [0, 1] \tag{2.1}$$

provided the second member is defined; i.e. the inequality subsists in all x, $y \in K$ with the exception of the points where $f(x) = -f(y) = \pm\infty$ with the usual computation rules in $\overline{R}$

$$\alpha + \infty = +\infty + \alpha = +\infty \text{ if } \alpha \neq -\infty$$

$$\alpha - \infty = -\infty + \alpha = -\infty \text{ if } \alpha \neq +\infty$$

$$\alpha \cdot \infty = \infty, \quad \alpha \cdot (-\infty) = -\infty \text{ if } \alpha \in R_+, \; 0 \cdot \infty = 0 \cdot (-\infty) = 0$$

We say that f is strictly convex if (2.1) is a strict inequality for $x \neq y$, $\lambda \in]0, 1[$.

From definition 1 it follows obviously that, if $x_1, x_2 \ldots, x_n \in K$,

$$\alpha_i \geqq 0; \quad i = 1 \ldots n; \quad \sum_1^n \alpha_i = 1$$

then

$$f\left(\sum_1^n \alpha_i x_i\right) \leqq \sum_1^n \alpha_i f(x_i)$$

provided the second member is defined.

If $f : K \to \overline{R}$ is a convex function on the convex set K, then for every $\alpha \in \overline{R}$ the sets

$$G_\alpha = \{x \in X : f(x) \leqq \alpha\}, \quad F_\alpha = \{x \in X : f(x) < \alpha\}$$

are convex; but the fact that G_α [resp. F_α] is a convex set for every $\alpha \in \overline{R}$, does not imply that f is a convex function.

For example, if $K = R$, any increasing $f : R \to \overline{R}$ is such that G_α [resp. F_α] is convex for every $\alpha \in \overline{R}$.

For any function $f: X \to \overline{R}$ we call effective domain of f the convex set

$$\text{dom } f = \{x \in X : f(x) < +\infty\}$$

By a proper convex function on X we shall mean a convex function with values in $R \cup \{+\infty\}$ which is not merely the constant function $f(x) = +\infty$. Therefore, if $f: X \to \overline{R}$ is a proper convex function, the dom f is a non-empty convex set on which f is real-valued. Conversely, if K is a non-empty convex subset of X and f is a real-valued function on K which is convex, i.e. satisfies (2.1) when $x \in K$, $y \in K$, then one can obtain a proper convex function on X by setting $f(x) = +\infty$ for every $x \notin K$.

A very useful example of a proper convex function is the indicator function X of a non-empty subset K of X, which is defined by

$$\mathscr{X}_K(x) = \begin{cases} 0 & \text{if } x \in K \\ +\infty & \text{if } x \notin K \end{cases}$$

It is obvious that a subset K of X is convex if and only if $\mathscr{X}_A$ is a convex function on X. So the theory of convex subsets of X is a part of the theory of the proper convex functions on X.

Definition 2. If $f: X \to \overline{R}$ is an extended real-valued function on X, the set

$$\text{epi } f = \{(x, a) \in X \times R : f(x) \leq a\}$$

is called the epigraph of f.

It is obvious that a subset $G \subset X \times R$ is the epigraph of some function if and only if for each $x \in X$ the set $\{a \in R;\ (x, a) \in G\}$ is either R, $\emptyset$ or a closed internal of the type $[d, +\infty[$. Then the function f is obtained from G = epi f by

$$f(x) = \inf \{a \in R : (x, a) \in G\}$$

with the usual rule $\inf \emptyset = +\infty$.

We have the following characterization of convex functions:

Theorem 1. In a v. s. X a real extended function $f: X \to \overline{R}$ is convex if and only if epi f is a convex subset of $X \times R$.

Proof. Let $f: X \to \overline{R}$ be a convex function on X and let (x, a), (y, b) be in epi f. We prove that $\lambda(x, a) + (1 - \lambda)(y, b) \in$ epi f for every $\lambda \in [0, 1]$. In fact, $f(x) \leq a < +\infty$, $f(y) \leq b < +\infty$ and for every $\lambda \in [0, 1]$ it results

$$f[\lambda x + (1 - \lambda) y] \leq \lambda f(x) + (1 - \lambda) f(y) \leq \lambda a + (1 - \lambda) b$$

so that

$$\lambda (x, a) + (1 - \lambda)(y, b) = (\lambda x + (1 - \lambda) y, \quad \lambda a + (1 - \lambda) b) \in \text{epi } f$$

Conversely, let us suppose that epi f is a convex subset of $X \times R$. Then

$$\text{dom } f = \{x \in X : f(x) \neq +\infty\}$$

is convex.

In fact, if $x, y \in$ dom f, then there exists $a, b \in R$ such that $f(x) \leq a$, $f(y) \leq b$ and therefore $(x, a) \in$ epi f, $(y, b) \in$ epi f. From the convexity of epi f it follows that for every $\lambda \in [0, 1]$

$$\lambda (x, a) + (1 - \lambda) (y, b) = (\lambda x + (1 - \lambda) y, \ \lambda a + (1 - \lambda) b) \in \text{epi } f$$

and so

$$f(x\lambda + (1 - \lambda) y) \leq \lambda a + (1 - \lambda) b \in R \tag{2.1a}$$

and dom f is convex.

Let us prove that $f : \text{dom } f \to R \cup \{-\infty\}$ is convex. This follows from (2.1a) by making in it $a = f(x)$, $b = f(y)$ if $f(x), f(y) \in R$ and by passing by the limit for $a \to -\infty$ (or $b \to -\infty$) if $f(x) = -\infty$ (or $f(y) = -\infty$). Now we can easily say that $f : X \to \overline{R}$ is convex: in fact, (2.1) subsists if $x, y \in$ dom f; and it follows from the computation rules in $\overline{R}$ if $x, y \notin$ dom f. If, finally, $x \in$ dom f, $y \notin$ dom f (2.1) is not requested if $f(x) = -\infty$ and it follows from the computation rules in $\overline{R}$ if $f(x) \in R$.

The class of all convex functions on a v. s. X has the following closure properties which imply moreover that the class of all convex functions is a convex class.

Theorem 2. The supremum of any family $\{f_\alpha : \alpha \in A\}$ of convex functions on a v. s. X is a convex function on X. The sum of two convex functions f on a v. s. X is a convex function on X if we let $(f + g)(x) = +\infty$ when $f(x) = -g(x) = -\infty$. If f is a convex function in a v. s. X and $\alpha \in R_+$ then αf is a convex function on X.

Proof. The first assertion follows from the fact that

$$\text{epi } (\sup f_\alpha : \alpha \in A) = \bigcap_{\alpha \in A} \text{epi } f_\alpha$$

and from theorem 1. To prove the second assertion, let f, g be convex functions on X, $x, y \in X$. For every $\lambda \in [0,1]$ we have

$$f[\lambda x + (1 - \lambda) y] \leq \lambda f(x) + (1 - \lambda) f(y)$$

$$g[\lambda x + (1 - \lambda) y] \leq \lambda g(x) + (1 - \lambda) g(y)$$

and, by addition, the corresponding statement for f + g; in fact,this is obvious if additions are done in R, and it follows from calculation rules incluse $(+\infty) + (-\infty) = +\infty$ if additions are done in $\overline{R}$. The third assertion is obvious.

The most interesting part of the theory of convex functions is done in a l. c. s. X as we suppose from now on.

We recall that a function $f : X \to \overline{R}$ is said to be lower-semicontinuous on X (abbreviated l. s. c.) if, for every $x_0 \in X$, we have

$$f(x_0) \leqq \underline{\lim}_{x \to x_0} f(x)$$

A function f is said to be upper-semicontinuous on X (abbreviated u. s. c.) if -f is l. s. c. on X. If f is l. s. c. and u. s. c. on X it is continuous on X. Obviously the indicator function $\mathscr{X}_A$ of a set A in X is l. s. c. [resp. u. s. c.] if and only if A is a closed [resp. open] set. The following theorem gives a characterization of l. s. c. functions.

Theorem 3. Let $f : X \to \overline{R}$ be a real extended function on a l. c. s. Then the following conditions are equivalent:

(1) f is l. s. c. on X
(2) the sets $F_\alpha = \{x \in X : f(x) > \alpha\}$ are open for every $\alpha \in R$
(3) epi f is a closed subset of $X \times R$.

Proof. (1) implies (2). In fact, if $y \in F_\alpha$, it follows from (1) that $\underline{\lim}_{x \to y} f(x) \geqq f(y) > \alpha$. Therefore there exists a neighbourhood U(y) of y where $f(x) > \alpha$. Hence $U(y) \subset F_\alpha$ and F_α is open. (2) implies (1). Let $x_0 \in X, f(x_0) \neq -\infty$ and $\alpha < f(x_0)$, $\alpha \in R$, then $x_0 \in F_\alpha$ and F_α is open. Therefore, $\inf \{f(x) : x \in F_\alpha\} \neq -\infty$ and also $\underline{\lim}_{x \to x_0} f(x) \geqq \alpha$, and for the arbitrariness of α, $\underline{\lim}_{x \to x_0} f(x) \geqq f(x_0)$. Nothing must be proved if $f(x_0) = -\infty$.

To prove the equivalence between (2) and (3) we define the map $\emptyset : X \times R \to R$ which value $\emptyset(x, a) = f(x) - a$ and observe that f is l. s. c. on X if and only if $\emptyset$ is l. s. c. on $X \times R$. But as we have proved, $\emptyset$ is l. s. c. on $X \times R$ if and only if the set $\{(x, a) \in X \times R : \emptyset(x, a) = f(x) - a \leqq \alpha\}$ is closed for every $\alpha \in R$. But this set coincides with the set $\{(x, a) \in X \times R : f(x) - \alpha \leqq a\}$ and this set is a translation of epi f and therefore it is closed if and only if epi f is closed. The proof of the theorem is so completed.

Corollary 1. If $f : X \to \overline{R}$ is a convex l. s. c. function on a l. c. s. X, then f is l. s. c. in the weak topology $\sigma(X, X^*)$.

For every $\alpha \in R$ the set $G_\alpha = \{x \in X : f(x) \leqq \alpha\}$ is convex and closed, therefore it is convex and closed in the weak $\sigma(X, X^*)$ topology. So f is l. s. c. in this topology.

The following theorems give us conditions under which a convex function is continuous.

Theorem 4. If a proper convex function $f : X \to \overline{R}$ on a t. v. s. is bounded above by a real constant in a neighbourhood of a point $x \in X$, then f is continuous on X.

Proof. We can suppose that $x = 0$ and that $f(0) = 0$. Let U be a neighbourhood of the origin 0 where $f(x) \leq a \in R$. Let $V = U \cap (-U)$ and observe that V is a symmetric neighbourhood of the origin. Let $\epsilon \in]0,1[$. If $x \in V$ we have, by the convexity of f, $(x/\epsilon) \in V$ and therefore

$$f(x) = f((1-\epsilon)0 + \epsilon\left(\frac{x}{\epsilon}\right) \leq (1-\epsilon)\, f(0) + \epsilon\, f\left(\frac{x}{\epsilon}\right) \leq \epsilon a - \frac{x}{\epsilon} \in V$$

and therefore

$$f(0) = f\left(\frac{\epsilon}{1+\epsilon}\,\frac{-x}{\epsilon} + x\,\frac{1}{1+\epsilon}\right) \leq \frac{\epsilon}{1+\epsilon}\, f\left(-\frac{x}{\epsilon}\right) + \frac{1}{1+\epsilon}\, f(x)$$

$$f(x) \geq -\epsilon f\left(-\frac{x}{\epsilon}\right) \geq -\epsilon a$$

We have also $|f(x)| \leq \epsilon a$ if $x \in \epsilon V$ and f is so continuous in x.

Theorem 5. Let $f: X \to \overline{R}$ be a proper convex function on a t. v. s. X. If there exists an open non-empty set U where f is bounded above by a real constant, then $(\mathrm{dom}\, f)^\circ \supset U$ and f is continuous on $(\mathrm{dom}\, f)^\circ$.

Proof. From the preceding theorem follows that f is continuous at every $x \in U$, moreover it is obvious that $U \subset (\mathrm{dom}\, f)^\circ$.

For every $y \in (\mathrm{dom}\, f)^\circ$ and a fixed $x_0 \in U$ let $p > 1$ be such that $z = x_0 + p(y - x_0) \in (\mathrm{dom}\, f)^\circ$. Let $t = 1 - (1/p) \in]0,1[$ and consider the map $\varphi: x \in X \to tx + (1-t)z \in X$.

This is a continuous bijective map of X onto X whose inverse φ^{-1} is also continuous. Therefore $\varphi(U)$ is an open subset of X containing

$$\varphi(x_0) = \left(1 - \frac{1}{p}\right)x_0 + \frac{1}{p}\,z = \frac{1}{p}\,(z + (p-1)x_0) = y.$$

For every $w = \varphi(x)$, $x \in U$ we have

$$f(w) = f(tx + (1-t)z) \leq t\,f(x) + (1-t)\,f(z) \leq tM + (1-t)\,f(z) < +\infty$$

since $z \in (\mathrm{dom}\, f)^\circ$.

We have so proved that f is bounded above in the neighbourhood $\varphi(U)$ of y. Therefore, by theorem 4, it is continuous on y. The theorem is completely proved. In the particular case, when X is a normed space, we have:

Theorem 6. If $f: X \to \overline{R}$ is a proper convex function on a normed space X and if there exists a subset U of X where f is bounded above by a real M, then $(\mathrm{dom}\, f)^\circ \neq \emptyset$ and f is locally Lipschitzian on $(\mathrm{dom}\, f)^\circ$.

Proof. By theorem 5, f is continuous on $(\mathrm{dom}\, f)^\circ$. Therefore, if $x \in (\mathrm{dom}\, f)^\circ$ there exists $r_0 \in R_+$ such that if $\|y-x\| \leq r_0$ it is $|f(y)-f(x)| \leq 1$ and consequently there are real constants m, M such that $m \leq f(y) \leq M$ where $\|y-x\| < r_0$. Let us consider an $r \in]0, r_0[$ and a point y_1 such that $\|y_1 - x\| \leq r$. If we let $G(z) = f(y_1 + z) - f(y_1)$ for $\|z\| \leq r_0 - r$, then $G(0) = 0$ and $G(z) \leq M-m$ if $\|z\| \leq r_0 - r$.

Another consequence of theorem 4 is that, for every $\epsilon \in]0,1[$, $|G(z)| \leq \epsilon (M - m)$ if $\|z\| \leq \epsilon (r_0 - r)$. If we take $y \in X$ such that $\|y - y_1\| \leq r_0 r$ then $\|y - y_1\| = \frac{\|y - y_1\|}{r_0 - r} (r_0 - r)$ with $\frac{y - y_1}{r_0 - r} \in]0,1[$

so that

$$|f(y) - f(y_1)| = |G(y - y_1)| \leq \frac{\|y - y_1\|}{r_0 - r} (M - m)$$

If y_2 is another point in X for which $\|y_2 - x\| \leq r$ we choose on the segment $[y_1, y_2]$ a convenient number of points $z_1 = y_1, z_2, z, \ldots z_p = y_2$ such that

$$\|z_{i+1} - z_i\| = \frac{1}{p-1} \|y_2 - y_1\| < r_0 - r, \quad i = 1, 2 \ldots p-1$$

Then, for every consecutive pair z_i, z_{i+1}, we have

$$|f(z_{i+1}) - f(z_i)| \leq \frac{M-m}{r_0 - r} \|z_{i+1} - z_i\|$$

and finally, summing up, we see that

$$|f(y_2) - f(y_0)| \leq \frac{M-m}{r_0 - r} \|y_2 - y_1\|$$

The theorem is completely proved.

In the special case where $X = R_n$ we have

Theorem 7. If $f : R_n \to \overline{R}$ is a convex proper function, then it is continuous in $(\mathrm{dom}\, f)^\circ$.

Proof. If $(\mathrm{dom}\, f)^\circ \neq \emptyset$ it contains $(n+1)$ points $x_1, x_2 \ldots x_{n+1}$ which are linearly independent. If x is in the open convex set

$$(\mathrm{co}\, \{x_1 \ldots x_{n+1}\})^\circ \text{ we have } x = \sum_1^{n+1} \alpha_i x_i, \quad \alpha_i > 0, \quad \sum_1^{n+1} \alpha_i = 1$$

and from the convexity of f

$$f(x) = f\left(\sum_1^{n+1} \alpha_i x_i \right) \leq \sum_i^n \alpha_i f(x_i)$$

our assertion follows now from theorem 4.

In the first section we proved that every convex closed set in a l. c. s. can be obtained as intersection of all semispaces containing it. Now we investigate similar properties for convex l. s. c. functions in a l. c. s.

Definition 3. In a t. v. s. X we call affine continuous function an $f: X \to R$ such that $f(x) = \langle x, x^* \rangle + \alpha$ where $x^* \in X^*$ and $\alpha \in R$.

Theorem 7. A proper convex l. s. c. function $f: X \to \bar{R}$ in a l. c. s. X is the pointwise supremum of all affine continuous functions $h: X \to R$ such that $h(x) \leq f(x)$ for every $x \in X$.

Proof. It suffices to prove that for every $\bar{x} \in X$ and every real $\bar{a} < f(\bar{x})$ there exists an affine continuous function h in X such that $\bar{a} < h(\bar{x})$ and $h(x) \leq f(x)\ \forall x \in X$. In order to prove it, we observe that epi f is a convex closed subset of $X \times R$ and that $(\bar{x}, \bar{a}) \notin$ epi f.

By means of the separation theorem we can strictly separate epi f and $(\bar{x}, \bar{a})$ using an affine closed hyperplane

$H = \{(x, a) : \langle x, x^* \rangle + \alpha \cdot a = \gamma\}$ of $X \times R$ where $x^* \in X^*$, $\alpha, \gamma \in R$ and $\langle x, x^* \rangle + \alpha a$ is a non-zero linear function on $X \times R$. It is also

$$\langle \bar{x}, x^* \rangle + \bar{a}\alpha < \gamma \tag{2.2}$$

$$\langle x, x^* \rangle + a\alpha > \gamma \quad \text{if} \quad (x, a) \in \text{epi } f \tag{2.3}$$

Now if $f(\bar{x}) \in R$ $(\bar{x}, f(\bar{x})) \in$ epi f and the second inequality gives us

$$\langle \bar{x}, x^* \rangle + f(\bar{x})\alpha > \gamma \tag{2.4}$$

so that

$$(\bar{a} - f(\bar{x}))\,\alpha < 0$$

It must be $\alpha > 0$ and by dividing (2.2), (2.4) by α we obtain

$$\bar{a} < \frac{\gamma}{\alpha} - \left\langle \bar{x}, \frac{x^*}{\alpha} \right\rangle, \quad \frac{\gamma}{\alpha} - \left\langle x, \frac{x^*}{\alpha} \right\rangle < f(x) \quad \forall\, x \text{ dom } f$$

and this proves that the affine continuous function $h(x) = \frac{\gamma}{\alpha} - \left\langle x, \frac{x^*}{\alpha} \right\rangle$ has the required property. If $f(\bar{x}) = +\infty$ and $\alpha \neq 0$ we can conclude as in the preceding case.

Let now $f(\bar{x}) = +\infty$ and $\alpha = 0$. In this case (2.2), (2.3) say that the affine continuous function $K(x) = \gamma - \langle x, x^* \rangle$ on X is such that $K(\bar{x}) > 0$, $K(x) < 0$ if $x \in$ dom f. Let us take now $(\bar{y}, \bar{b})$ such that $\bar{y} \in$ dom f, $\bar{b} > f(\bar{y})$ and construct, as in the first case, an affine continuous function

$K'(x) = \beta - \langle x, y^* \rangle$ such that $\bar{b} < K'(\bar{y})$ and $K'(x) \leq f(x) \quad \forall x \in X$. Then, for every $c \in R_+$, the affine continuous function

$$\ell(x) = \beta - \langle x, y^* \rangle + cK(x) \text{ is such that } \ell(x) \leq f(x) \text{ if } x \in \text{dom } f$$

At this point let us remember that $K(\bar{x}) > 0$ and choose c so great that $\ell(\bar{x}) > \bar{a}$.

Thus the theorem is completely proved.

If we recall that the supremum of a family of convex l. s. c. functions in a t. v. s. X is a convex l. s. c. function, we have also proved

Theorem 9. A proper function $f : X \to \overline{R}$ in a l. c. s. X is convex l. s. c., if and only if it is the supremum of a family of affine continuous functions. Let us introduce the following definition:

Definition 4. In a l. c. s. X let $f : X \to \overline{R}$ be an extended real-valued function. We call regularized of f (abbreviated $\Gamma(f)$) the function

$$\Gamma(f) = \sup \{\varphi : \varphi \text{ affine continuous function on X such that } \varphi(x) \leq f(x) \text{ in } X\}$$

Obviously $\Gamma(f)$ is a convex l. s. c. function on X and it results

$$(\Gamma f)(x) \leq f(x) \quad \forall x \in X$$

In theorem 8 we have also proved this.

Theorem 10. If $f : X \to R$ is a proper convex function on a l. c. s. X, then $f = \Gamma(f)$ if and only if f is a convex l. s. c. function.

A useful approach to the regularization is provided by the notion of the conjugate of a given function.

Definition 5. In a l. c. s. X let $f : X \to \overline{R}$ be an extended real function. For every $x^* \in X^*$ let

$$f^*(x^*) = \sup \{\langle x, x^* \rangle - f(x); \ x \in X\}$$

The function $f^* : X^* \to \overline{R}$ so defined is called the conjugate function of f. It is immediately seen that an affine continuous function on X, $h(x) = \langle x, x^* \rangle = \alpha$, satisfies the condition $h(x) \leq f(x)$ on X if and only if $\langle x, x^* \rangle - f(x) \leq \alpha$, i. e. if and only if $f^*(x^*) \leq \alpha$. It is also immediate that for every $x^* \in X^*$ it results $f^*(x^*) = \sup \{\langle x, x^* \rangle - f(x) : x \in \text{dom } f\}$. The proof of the following theorem is obvious.

Theorem 11. If $f, g : X \to \overline{R}$ are real extended functions in a l. c. s. X, then

(i) $f^*(0) = -\inf \{f(x) : x \in X\}$

(ii) $f(x) \leq g(x)$ for every $x \in X$ implies $f^*(x^*) \geq g^*(x^*)$ for every $x^* \in X^*$

(iii) for every $\lambda \in R_+$ we have $(\lambda f^*)(x^*) = \lambda f^*\left(\frac{x^*}{\lambda}\right)$

(iv) for every $\alpha \in R$ we have $(f+\alpha)^*(x^*) = f^*(x^*) - \alpha$

(v) $\langle x, x^* \rangle \leq f(x) + f^*(x^*)$ for every $x \in X$, $x^* \in X^*$

Once f^* is defined we can consider, for every $x \in X$, $f^{**}(x) = \sup_{x^* \in X^*} \{\langle x, x^* \rangle f^*(x^*)\}$. The function $f^{**}: X \to \overline{R}$ so defined is called the second conjugate of f. This function f^{**} is, as we know, a convex l. s. c. function on X. If we compare f^{**} with f we have

Theorem 12. If $f: X \to \overline{R}$ is an extended-valued function on the l. c. s. X, then for its second conjugate we have: $f^{**} = \Gamma(f)$. Therefore, if f is a proper convex l. s. c. function, then $f^{**} = f$.

Proof. From the definitions of $\Gamma(f)$ and f^{**} we can derive

$$(\Gamma(f))(x) = \sup \{\langle x, x^* \rangle - \alpha : \alpha \in R, \quad x^* \in X^* \text{ such that } \langle x, x^* \rangle - \alpha \leq f(x) \text{ on } X\}$$

$$f^{**}(x) = \sup \{\langle x, x^* \rangle - f^*(x^*) : x^* \in X^*\}$$

and

$$f^*(x^*) = \sup_{\alpha \in X} \{\langle x, x^* \rangle - f(x)\}$$ so that by a fixed $x^* \in X^*$ it is $\langle x, x^* \rangle - \alpha \leq f(x)$ on X if and only if $f^*(x^*) \leq \alpha$. Therefore

$$(\Gamma(f))(x) = \sup \{\langle x, x^* \rangle - f^*(x^*) - \epsilon, \quad x^* \in X^*, \quad \epsilon \in R_+ \cup \{0\}\}$$

It follows that

$$(\Gamma f)(x) = f^{**}(x)$$

If we define $f^{***}(x^*) = \sup \{\langle x, x^* \rangle - f^{**}(x) \; x \in X\} = (f^{**})^*(x^*)$ then

Theorem 13. For every $f: X \to \overline{R}$ it results $f^*(x^*) = f^{***}(x^*)$ on X^*.

Proof. In fact, from $f(x) \geq (\Gamma f)(x) = f^{**}(x)$ we obtain, by theorem 11, $f^*(x^*) \leq f^{***}(x^*)$ on X^*. But from the definition of f^{**}, we have also for every $x \in X$, $x^* \in X^*$:

$$\langle x, x^* \rangle \leq f^{**}(x) + f^*(x^*)$$

so that

$$\langle x, x^* \rangle - f^{**}(x) \leq f^*(x^*) \;\; \forall x \in X, \quad x^* \in X^*$$

From the definition of $f^{***}(x^*)$ we obtain therefore $f^{***}(x^*) \leq f^*(x^*)$ on X^* and the proof is complete.

The following theorem fixes a significant geometric property of $\Gamma(f)$.

Theorem 14. If $f: X \to \overline{R}$ is a proper real extended function on a l. c. s. X, then $\text{epi}\, \Gamma(f) = \overline{\text{co}}\,(\text{epi}\, f)$.

Proof. We prove preliminarly that, for the given f, there exists a $g: X \to \overline{R}$ such that $\text{epi}\, g = \text{co}\,(\text{epi}\, f)$. In fact, if $(x, a) \in \text{co}\,(\text{epi}\, f)$ and $b > a$, then $(x, b) \in \text{co}\,(\text{epi}\, f)$ so that, for every $x \in X$, if we let $g(x) = \inf \{\lambda \in \overline{R} : (x, \lambda) \in \text{co}(\text{epi}\, f)\}$ it results obviously $\text{epi}\, g = \text{co}(\text{epi}\, f)$.

Let $(x, a) \in \text{co}(\text{epi}\, f)$ and $b > a$. Then there exists $(x_i, a_i) \in \text{epi} f : i = 1 \dots n$ $\lambda_i \in R_+$, such that

$$(x, a) = \sum_i^n \lambda_i\, (x_i, a_i), \quad \sum_i^n \lambda_i = 1$$

Since $f(x_i) \leq a_i$ we have $f(x_i) < a_i + \dfrac{b-a}{\lambda_i}$ and therefore $\left(x_i, a_i + \dfrac{{}^{i}b-a}{\lambda_i}\right) \in \text{epi}\, f$.

$$\text{Then } \sum_1^n \lambda_i\left(x_i, a_i + \frac{b-a}{\lambda_i}\right) = \left[\sum_1^n (\lambda_i x_i), \sum_1^n \lambda_i a_i + (b-a)\right] = (x, b) \in \text{co}(\text{epi}\, f).$$

Successively we prove that, for the obtained $g: X \to \overline{R}$, there exists a $G: X \to \overline{R}$ such that $\text{epi}\, G = \overline{\text{epi}\, g}$.

In fact, if $(x, a) \in \overline{\text{epi}\, g}$ and $b > a$, then $(x, b) \in \overline{\text{epi}\, g}$, and then, if we let for every $x \in X$, $G(x) = \inf \{\lambda \in \overline{R} : (x, \lambda) \in \overline{\text{epi}\, g}\}$, we have obviously $\text{epi}\, G = \overline{\text{epi}\, g}$.

To prove that if $(x, a) \in \overline{\text{epi}\, g}$, and $b > a$, then $(x, b) \in \overline{\text{epi}\, g}$, let us consider a generalized sequence (x_α, a_α); $\alpha \in a$, such that $(x_\alpha, a_\alpha) \in \text{epi}\, g$, for every $\alpha \in a$ and (x_α, a_α) tends to (x, a) in $X \times R$. Then, for a convenient $\bar{\alpha} \in a$, it results $a_\alpha < b$ when $\alpha > \bar{\alpha}$ and, since $g(x_\alpha) \leq a_\alpha$, it results also $g(x_\alpha) \leq b$ when $\alpha > \bar{\alpha}$. Then $(x_\alpha, b) \in \text{epi}\, g$ when $\alpha > \bar{\alpha}$, and therefore $(x, b) \in \overline{\text{epi}\, g}$.

It is so proved that, for the given $f: X \to \overline{R}$, there exists a $G: X \to \overline{R}$ such that $\text{epi}\, G = \overline{\text{co}}\,(\text{epi}\, f)$. We prove that this G coincides with $\Gamma(f)$. Since it follows from the definition of $\Gamma(f)$ that $\Gamma(f)(x) \leq f(x)$ for every $x \in X$ we have $\text{epi}\,\Gamma(f) \supset \text{epi}\, f$ and therefore

$$\text{epi}\,(\Gamma(f)) = \overline{\text{co}}\,(\text{epi}\,\Gamma(f)) \supset \overline{\text{co}}\,(\text{epi}\, f) \supset \text{epi}\, f$$

We obtain so $\text{epi}\,\Gamma(f) \supset \text{epi}\, G \supset \text{epi}\, f$ and therefore

$$f^{**}(x) = \Gamma(f)(x) \leq G(x) \leq f(x) \quad \forall x \in X$$

Then $(f^{**})^*(x^*) \geq G^*(x^*) \geq f^*(x^*)$ and, using theorem 13, we see that

$$(f^{**})^*(x^*) = G^*(x^*) = f^*(x^*) \quad \forall x^* \in X^*$$

Then $(f^{***})^*(x)\ (f^{**})^{**}(x) = G^{**}(x) = f^{**}(x)$ for every $x \in X$. But as G and f^{**} are convex l. s. c. functions on X,

$$f^{**}(x) = (f^{**})^{**}(x) = G^{**}(x) = G(x) \quad \forall x \in X$$

and also

$$\text{epi}\,\Gamma(f) = \text{epi}\, f^{**} = \text{epi}\, G = \overline{\text{co}}\,(\text{epi}\, f)$$

The theorem is so completely proved.

Let us consider a few examples.

Example 1. If $f(x) = +\infty$ [resp. $-\infty$] $\forall x \in X^*$, then $f^*(x^*) = -\infty$ [resp. $+\infty$] for every $x^* \in X^*$.

Example 2. If A is a subset of X and $\mathscr{X}_A$ is the corresponding indicator function (i. e. $\mathscr{X}_A(x) = 0$ if $x \in A$, $\mathscr{X}_A(x) = +\infty$ if $x \notin A$), then

$$(\mathscr{X}_A)^*(x^*) = \sup\{\langle x, x^*\rangle - \mathscr{X}_A(x) : x \in X\} = \sup\{\langle x, x^*\rangle - \mathscr{X}_A(x);\ x \in A\}$$

is called the support function of A. We have $\text{epi}\,\mathscr{X}_A^{**} = \overline{\text{co}}\,(\text{epi}\,\mathscr{X}_A)$.

Example 3. If $f(x) = \langle x, x_0^*\rangle$, then

$$f^*(x^*) = \sup\{\langle x, x^*\rangle - \langle x, x_0^*\rangle;\ x \in X\} = \sup\{\langle x, x^* - x_0^*\rangle;\ x \in X\}$$

$$= \begin{cases} +\infty & \text{if } x^* \neq x_0^* \\ 0 & \text{if } x^* = x_0^* \end{cases}$$

Example 4. Let $X = R$ and $p \in R$ such that $p > 1$. Let $1/p + 1/q = 1$ and consider $f: R \to R$ such that $f(x) = |x|^p/p$. Then f is a convex continuous function on R. Obviously $X^* = R^* = R$ and hence results

$$f^*(y) = \sup\left\{\langle x, y\rangle - \frac{|x|^p}{p} : x \in R\right\}$$

$$= \sup\left\{yx - \frac{|x|^p}{p} : x \in R\right\} \quad \forall y \in R$$

But for every $y \in R$, $\varphi(x) = yx - |x|^p/p$ is a continuous function such that

$$\lim_{x \to +\infty} \varphi(x) = -\infty \text{ and } \varphi(x) = y - |x|^{p-1}\,\text{sig}(x), \quad \text{if } x \neq 0$$

We have therefore

$$f^*(y) = \sup\left\{xy - \frac{|x|^p}{p} : x \in R\right\} = \left(x|x|^{p-1}\,\mathrm{sig}(x) - \frac{|x|^p}{p}\right)_{y=|x|^{p-1}\,\mathrm{sig}(x)}$$

$$= \left(|x|^p\left(1-\frac{1}{p}\right)\right)_{|y|=|x|^{p-1}} = \frac{1}{q}|y|^{\frac{p}{p-1}} = \frac{1}{q}|y|^q$$

From theorem (v) we have also

$$xy \leq \frac{|x|^p}{p} + \frac{|y|^q}{q} \quad \forall\, x, y \in R$$

the well-known Hölder's inequality.

3. SUB-DIFFERENTIAL OF A CONVEX FUNCTION

Definition 1. Let $f : X \to \overline{R}$ be a proper real extended function on a l. c. s. X and $x^* \in X^*$. We say that x^* is a sub-gradient of f at a point $x_0 \in X$ where $f(x_0) \in R$ if

$$f(x) \geq f(x_0) + \langle x - x_0, x^* \rangle \quad \forall\, x \in X \tag{3.1}$$

If a sub-gradient exists we say that f is sub-differentiable at the point x_0. The set of all sub-gradients x^* of f at x_0 is denoted by $\partial f(x)$. The multi-valued mapping $f : x \in X \to \partial f(x) \subset X^*$ is called the sub-differential of f. If f is not differentiable at x we have also $\partial f(x) = \emptyset$.

Geometrically, the condition that x^* is a sub-gradient of f at x_0 means that the affine closed hyperplane in $X \times R$ $\{(x, a) \in X \times R : \langle x - x_0, x^* \rangle + f(x_0) = a\}$ is a supporting hyperplane of epi f.

The following properties of $\partial f(x)$ are immediate consequences of definition 1.

Theorem 1. If $f : X \to \overline{R}$ is a real extended function on a l. c. s. X, then

(i) $f(x_0) = \min\{f(x) : x \in X\}$ if and only if $0 \in \partial f(x_0)$

(ii) $x^* \in \partial f(x_0)$ if and only if $f(x_0) + f^*(x^*) = \langle x_0, x^* \rangle$

(iii) $\partial f(x_0) \neq \emptyset$ implies $f(x_0) = f^{**}(x_0)$

(iv) $\partial f(x_0)$ is a convex subset of X^* which is closed in the $\sigma(X^*, X)$ topology of X^*.

Proof. To prove assertion (i) we observe that, if $f(x) \geq f(x_0)$ $\forall\, x \in X$, then $f(x) \geq f(x_0) + \langle x - x_0, 0 \rangle$ so that $0 \in \partial f(x_0)$.

Conversely, if $0 \in \partial f(x_0)$, then (3.1) holds with $x^* = 0$ so that $f(x) \geq f(x_0)$ $\forall\, x \in X$.

To prove (ii) we observe that, if $x^* \in \partial f(x_0)$, then we have $f(x) \geq f(x_0) + \langle x - x_0, x_0^* \rangle$ $\forall\, x \in X$, and also $-f(x) + \langle x, x^* \rangle \leq -f(x_0) + \langle x_0, x^* \rangle$ $\forall\, x \in X$, so that

$$f^*(x^*) = \sup \{\langle x, x^* \rangle - f(x) : x \in X\} = \langle x_0, x^* \rangle - f(x_0)$$

If, conversely, $f^*(x^*) = \langle x_0, x^* \rangle - f(x_0)$, then $\langle x_0, x^* \rangle - f(x_0) \geq f(x) - \langle x, x^* \rangle$ for $\forall\, x \in X$ and (3.1) holds.

To prove (iii) we observe that, if $x^* \in \partial f(x_0)$, then the affine continuous function $f(x_0) + \langle x - x_0, x^* \rangle$ is a minorant of $f(x)$ and therefore $f(x_0) + \langle x - x_0, x^* \rangle \leq \Gamma(f)(x) = f^{**}(x)$ $\forall\, x \in X$. Then $f(x_0) \leq f^{**}(x_0) \leq f(x_0)$ and the assertion is proved.

To prove (iv) we recall that $\langle x, x^* \rangle \leq f(x) + f^*(x^*)$ $\forall\, x \in X$, $x^* \in X^*$. Therefore, by (ii) $\partial f(x_0) = \{x^* \in X^* : f^*(x^*) - \langle x_0, x^* \rangle \leq -f(x_0)$ and this set is convex and $\sigma(X^*, X)$ is closed because $f^*(x^*) - \langle x_0, x^* \rangle$ is a convex l. s. c. function on X^*.

For convex functions we have the following theorem of differentiability:

Theorem 2. Let $f : X \to \overline{R}$ be a proper convex real extended function on a l. c. s. X which is continuous at a point $x_0 \in X$. Then $\partial f(x) \neq \emptyset$ $\forall\, x \in (\text{dom } f)^\circ$. Moreover we know that $x_0 \in (\text{dom } f)^\circ$ so that $\partial f(x_0) = \emptyset$.

Proof. Since f is continuous at x_0, there exists an open set containing x_0 where f is bounded above. By theorem 5 of section 2 we obtain that $(\text{dom } f)^\circ \neq \emptyset$, and that f is continuous on $(\text{dom } f)^\circ$. Therefore it suffices to prove that $\partial f(x_0) \neq \emptyset$.

Since f is convex, epi f is a convex subset of $X \times R$; since f is continuous at x_0, $(\text{epi } f)^\circ \neq \emptyset$. In fact, there exists an open subset U of X such that $x_0 \in U$ and $f(x) \leq K$ on U, and therefore $V = U \times \,]K, +\infty[$ is an open subset of $X \times R$ such that $V \subset \text{epi } f$. Since $(x_0, f(x_0))$ belongs to the boundary of epi f, then by corollary 3 of section 1 we can separate $(x_0, f(x_0))$ and $(\text{epi } f)^\circ$ with an affine closed hyperplane $H = \{(x, a) \in X \times R : \langle x, x^* \rangle + \alpha a = \gamma\}$ of $X \times R$, where $x^* \in X^*$, α, $\gamma \in R$ and $(x, a) \to \langle x, x^* \rangle + \alpha a$ is a non-zero real function on $X \times R$. We have therefore $\langle x, x^* \rangle + \alpha a \geq \gamma$ $\forall$ $(x, a) \in \text{epi } f$ and $\langle x_0, x^* \rangle + \alpha f(x_0) = \gamma$ and hence

$$\langle x_0, x^* \rangle + \alpha\,(f(x_0) + \epsilon) \geq \langle x_0, x^* \rangle + \alpha f(x_0); \quad \epsilon \in R_+$$

so that $\alpha \geq 0$.

If $\alpha = 0$ we would have $\langle x - x_0, x^* \rangle \geq 0$ $\forall\, x \in \text{dom } f$ and, since $(\text{dom } f)^\circ \neq \emptyset$, $x^* = 0$, so that $(x, a) \to \langle x, x^* \rangle + \alpha a = 0$ on $X \times R$; this is a contradiction. Therefore it is $\alpha > 0$. Then we have

$$\langle x, x^* \rangle + \alpha f(x) \geq \gamma = \langle x_0, x^* \rangle + \alpha f(x_0) \;\; \forall\, x \in \text{dom } f$$

so that

$$f(x) \geq f(x_0) + \left\langle x_0 - x, -\frac{x^*}{\alpha} \right\rangle \;\; \forall\, x \in X$$

This proves that $-\dfrac{x^*}{\alpha} \in \partial f(x_0)$. The theorem is so proved.

Let us recall the definition of the Gateaux differential of a real extended function in a l. c. s. X and investigate the connections between the Gateaux differential and sub-gradients.

Definition 2. Let $f: X \to \bar{R}$ be a real extended function on a l. c. s. X, and let $y \in X$. We call a derivative of f at $x_0 \in X$ in the direction y and denote it by $f'(x_0, y)$:

$$\lim_{t \to 0^+} \frac{f(x_0 + ty) - f(x_0)}{t}$$

provided the limit exists in R. If there exists an $x^* \in X^*$ such that $f'(x_0, y) = \langle y, x^* \rangle$ for every $y \in X$ we say that f is Gateaux-differentiable at the point x_0 and we call x^* the Gateaux derivative of f at the point x_0 and denote it by $f'(x_0)$. Obviously the Gateaux derivative is unique, provided it exists.

Theorem 3. Let $f: X \to \bar{R}$ be a proper convex function on a l. c. s. X. If f is Gateaux-differentiable at $x_0 \in X$, then $\partial f(x_0) = \{f'(x_0)\}$. Conversely, if at $x_0 \in X$, $f(x_0) \in R$, f is continuous and $\partial f(x_0) = \{x^*\}$, then f is Gateaux-differentiable at x_0 and $f'(x_0) = x^*$.

Proof. If f is proper convex and Gateaux-differentiable at x_0 and, for every $y \in X$, we consider $\lambda \to \varphi(\lambda) = f(x_0 + \lambda y)$, then φ is a proper convex function on R, differentiable at x_0, and $\varphi'(0) = \langle x_0, f'(x_0) \rangle$. We have therefore

$$f(x_0 + y) - f(x_0) = \varphi(1) - \varphi(0) \geq \varphi'(0) = \langle x_0, f'(x_0) \rangle \; y \in X$$

so that

$$f'(x_0) \in \partial f(x_0)$$

On the other hand, if $x^* \in \partial f(x_0)$ we have $\forall\, y \in X$ and $\lambda \in R_+$

$$f(x_0 + \lambda y) \geq f(x_0) + \langle \lambda y, x^* \rangle$$

so that

$$\frac{f(x_0 + \lambda y) - f(x_0)}{\lambda} \geq \langle y, x^* \rangle$$

therefore, going to the limit for $\lambda \to 0^+$,

$$\langle y, f'(x_0) \rangle = \langle y, x^* \rangle \; \forall\, y \in X$$

It follows that $\langle y, f'(x_0) - x^* \rangle = 0 \;\forall\; y \in X$ and therefore $x^* = f'(x_0)$. Thus it is proved that $\partial f(x_0) = f'(x_0)$.

Conversely, let f be convex and let $\partial f(x_0)$ contain a unique element. Since f is proper convex, $f(x_0) \in R$ and $f(x)$ is continuous at x_0, we have $\forall\, y \in X,\; f(x_0 + \lambda y) - f(x_0) \geq f'(x_0, \lambda y) \;\forall\; \lambda$ such that $x_0 + \lambda y \in (\text{dom } f)^\circ$ and therefore this holds $\forall\,\lambda$. It follows that the one-dimensional subspace $A = \{x_0 + \lambda y,\; f(x_0) + \lambda f'(x_0, y) : \lambda \in R\}$ of $X \times R$ does not intersect $(\text{epi } f)^\circ$ which is a convex, non-empty set of $X \times R$ since f is continuous at x_0. By corollary 2 and theorem 4 (of section 1) there exists a closed affine hyperplane H containing A such that $H \cap (\text{epi } f)^\circ = \emptyset$. If $H = \{(x, a) \in X \times R : \langle x, x^* \rangle + \alpha a = \gamma\}$, where $x^* \in X^*$, $\alpha, \gamma \in R$, and $(x, a) \to \langle x, x^* \rangle + \alpha a$ is a non-zero function on $X \times R$, we can suppose that $\langle x, x^* \rangle + \alpha a \geq \gamma \;\forall\; (x, a) \in (\text{epi } f)^\circ$ and

$$\langle x_0 + \lambda y,\; x^* \rangle + \alpha(f(x_0) + \lambda f'(x_0)y) = \gamma$$

Let us observe that $\alpha \neq 0$; in fact, if $\alpha = 0$, then

$$\langle x_0, x^* \rangle + \lambda \langle y, x^* \rangle = \gamma \;\forall\; y \in X,\; \lambda \in R$$

so that $x^* = 0$ and, consequently, $\langle x, x^* \rangle + a\alpha = 0 \;\forall\; (x, a)$; a contradiction.

Then we have $\langle x, x^* \rangle + \alpha\,(f(x) + \epsilon) \geq \gamma = \langle x_0, x^* \rangle + \alpha f(x_0) \;\forall\; x \in (\text{dom } f)$ and therefore, by taking $x = x_0$, $\alpha > 0$. So we have

$$f(x) \geq \left\langle x_0 - x,\; \frac{x^*}{\alpha} \right\rangle + f(x_0) \;\;\forall x \in (\text{dom } f)$$

and therefore $\forall\, x \in X$. We have so proved that $-\dfrac{x^*}{\alpha} \in \partial f(x_0)$ and therefore $\partial f(x_0) = \left\{-\dfrac{x^*}{\alpha}\right\}$. Let us recall that

$$\langle x_0 + \lambda y,\; x^* \rangle + \alpha\,(f(x_0) + \lambda f'(x_0, y)) = \gamma \;\forall\; \lambda \in R,\;\; y \in X$$

then we obtain $\langle y, x^* \rangle = -\alpha f'(x_0, y)$ for every $y \in X$ so that f is Gateaux-differentiable with the Gateaux derivative $-x^*/\alpha$.

For a Gateaux-differentiable function the convexity can be characterized in the following form:

Theorem 4. Let $f : K \to R$ be a real valued function on a convex non-empty subset of a l. c. s. X which is Gateaux-differentiable at every $x \in K$ in the following sense: there exists an $f'(x) \in X^*$ (not necessarily unique if $x \notin K^\circ$) such that for every $y \in X$ for which there exists an $\epsilon \in R_+$ such that $x + \delta y \in K$ if $\delta \in \,]0, \epsilon[$, then $\lim\limits_{t \to 0^+} \dfrac{f(x + ty) - f(x)}{t} = \langle x, f'(x) \rangle$. Then f is convex on K if and only if $f(x) \geq f(x_0) + \langle x - x_0,\; f'(x) \rangle \;\forall\; x_0,\, x \in K$; f is strictly convex if and only if $f(x) > f(x_0) + \langle x - x_0,\; f'(x_0) \rangle \;\forall\; x \in K,\; x \neq x_0$.

Proof. If f is convex on K and Gateaux-differentiable, then the arguments used in the proof of theorem 3 also prove that

$$f(x) \geq f(x_0) + \langle x - x_0, f'(x_0)\rangle \text{ for } x, x_0 \in K$$

Conversely, if f is Gateaux-differentiable at every $x \in K$ and

$$f(x) \geq f(x_0) + \langle x - x_0, f'(x_0)\rangle \text{ for } x, x_0 \in K$$

then we have for every $x, y \in K$, $\lambda \in [0,1]$

$$f(x) \geq f[x(1-\alpha) + y] + \alpha\langle x - y, f'[x(1-\alpha) + \alpha y]\rangle$$

$$f(y) \geq f[x(1-\alpha) + \alpha y] + (1-\alpha)\langle y - x, f'[(1-\alpha)x + \alpha y]\rangle$$

We multiply the first inequality by $(1-\alpha)$ and the second by α and sum up to obtain

$$(1-\alpha) f(x) + \alpha f(y) \geq f[(1-\alpha)x + \alpha y]$$

so that f is convex in K.

If it is strictly convex and Gateaux-differentiable we have for every $x_0, x \in K$, $\alpha \in]0,1[$, $x_0 \neq x$,

$$\langle x - x_0, f'(x_0)\rangle \leq \frac{f[x_0 + \alpha(x - x_0)] - f(x_0)}{\alpha}$$

$$< \frac{\alpha f(x) + (1-\alpha) f(x_0) - f(x_0)}{\alpha} = f(x) - f(x_0)$$

Conversely, if f is Gateaux-differentiable and $f(x) > f(x_0) + \langle x - x_0, f'(x_0)\rangle$ $\forall x \neq x_0$, then the argument used above proves that

$$f[\alpha x + (1-\alpha)y)] > \alpha f(x) + (1-\alpha) f(y) \text{ if } x \neq y, \ \alpha \in]0,1[$$

The theorem is completely proved.

DEFINITION 3. Let $\varphi: X \to X^*$, a map of a l.c.s. X in the space X^*. We say that φ is monotone if, $\forall\ x, y \in X$,

$$\langle x - y, \varphi(x) - \varphi(y)\rangle \geq 0$$

With this definition we can prove that

Theorem 5. Let $f: K \to R$ be a real-valued function on a convex subset K of a l. c. s. X, which is Gateaux-differentiable at every $x \in K$. Then f is convex if and only if $f'(x)$ is monotone on K.

Proof. If f is Gateaux-differentiable at every $x \in K$ and convex, then for every $x, y \in K$ there exist $f'(x), f'(y) \in X^*$ such that

$$f(x) \geq f(y) + \langle x - y, f'(y) \rangle$$

$$f(y) \geq f(x) + \langle y - x, f'(x) \rangle$$

Adding these inequalities yields

$$\langle x - y, f'(x') - f'(y) \rangle \geq 0 \quad x, y \in K$$

and f' is monotone on K.

Conversely, if f is Gateaux-differentiable at every $x \in K$ and f' is monotone on K let, $\forall \lambda \in [0, 1]$, $x, y \in K$, $\varphi(\lambda) = f[x + \lambda(y - x)]$ and consider $\varphi: [0, 1] \to R$. According to the hypothesis of Gateaux-differentiability of f, φ is differentiable at every $\lambda \in [0, 1]$ and

$$\varphi'(\lambda) = \lim_{h \to 0} \frac{f[x + (\lambda + h)(y - x)] - f[x + \lambda(y - x)]}{h}$$

$$= \langle y - x, f'[x + \lambda(y - x)] \rangle$$

Therefore

$$(\lambda - \mu)\{\varphi'(\lambda) - \varphi'(\mu)\} = (\lambda - \mu)\langle y - x, f'[x + \lambda(y - x)] - f'[x + \mu(y - x)] \rangle$$

$$= \langle x + \lambda(y - x) - \{x + \mu(y - x)\}, f'[x + \lambda(y - x)] - f'[x + \mu(y - x)] \rangle \geq 0$$

so that $\varphi'(\lambda)$ is an increasing function on $[0, 1]$ and φ is a convex function on $[0, 1]$. Then

$$\varphi(\lambda) \leq \lambda\varphi(1) + (1 - \lambda)\varphi(0) \quad \forall \lambda \in [0, 1]$$

and therefore

$$f[x(1 - \lambda) + \lambda y] = \varphi(\lambda) \leq \lambda\varphi(1) + (1 - \lambda)\varphi(0) = \lambda f(y) + (1 - \lambda) f(x)$$

The theorem is so proved.

We conclude these preliminary topics on convex-functions theory with the following theorem which, for convex functions, connects the property of being l. s. c. with the property of being Gateaux-differentiable:

Theorem 6. If $f: X \to \overline{R}$ is a proper convex function on a l. c. s. X which is Gateaux-differentiable at x, then f is l. s. c. in the $\sigma(X, X^*)$ topology at x.

Proof. Let x converge to x_0 in the $\sigma(x, x^*)$ topology; we must prove that $\underline{\lim}_{x \to x_0} f(x) \geq f(x_0)$. From theorem 3 follows that $\forall x \in X$

$$f(x) \geq f(x_0) + \langle x - x_0, f'(x_0) \rangle$$

therefore, if we go to the limit in the topology $\sigma(X, X^*)$, we obtain $\lim_{x \to x_0} \langle x - x_0, f'(x_0) \rangle = 0$ and consequently

$$\underline{\lim}_{x \to x_0} f(x) \geq f(x_0)$$

The theorem is so proved.

4. MINIMIZATION OF CONVEX FUNCTIONS

In this last section we give the most essential results on the minimization of real-valued convex functions on Banach spaces.

Theorem 1. Let K be a closed convex set in a Banach space and let $f: K \to R$, a real-valued convex l. s. c. function on K.

We suppose furthermore that either

(i) K is bounded, or

(ii) $\lim_{\substack{\|x\| \to \infty \\ x \in K}} f(x) = +\infty$

Then there exists an $x_0 \in K$ where f(x) has a minimum in K, i. e.

$$f(x_0) \leq \inf \{f(x) : x \in K\}$$

If f is strictly convex then x_0 is unique in K.

Proof. Let us consider $\lambda = \inf \{f(x) : x \in K\}$; then $\lambda \in [-\infty, +\infty[$. Let x_n be a minimizing sequence in K, i. e. a sequence such that

$$\lim_{n \to \infty} f(x_n) = \lambda$$

The set $A = \{x_n : n \in N\}$ is bounded in X: in fact, in the case (i), $A = \{x_n : n \in N\} \subset K$ and K is bounded; in the case (ii) A is bounded since the sequence $\{f(x_n)\}$ is bounded above.

Then there exists a sub-sequence $\{x_{n_i}\}$ of x_n which converges in the $\sigma(X, X^*)$ topology to an $x_0 \in X$. But K, being closed and convex, is closed also in the $\sigma(X, X^*)$ topology so that $x_0 \in K$. Moreover, f, being convex and l. s. c. in the initial topology of X, remains l. s. c. in the $\sigma(X, X^*)$ topology. We have also

$$f(x_0) \leq \underline{\lim}_{i \to \infty} f(x_{n_i}) = \lambda$$

this proves that $f(x_0) = \lambda = \inf\{f(x) : x \in K\}$ and also that $\lambda \neq -\infty$. If $x_1, x_2 \in X$ are such that $f(x_1) = f(x_2) = \lambda$, then from the convexity of f we have $f[\frac{1}{2}(x_1 + x_2)] \leq \frac{1}{2}(f(x_1) + f(x_2)) = \lambda$ so that also $\frac{x_1 + x_2}{2} \in K$ is a point of minimum for f in K. Therefore, if f is strictly convex we cannot have $f(x_1) = f(x_2) = \lambda$ with $x_1, x_2 \in K$ and therefore there exists a unique $x_0 \in K$ where f has its minimum.

Theorem 2. Let $f : K \to R$ be a convex real-valued function on a closed convex subset K of a Banach space which is Gateaux-differentiable at every $x \in K$. Moreover, $\lim_{\substack{\|x\| \to \infty \\ x \in K}} f(x) = +\infty$ if K is not bounded. Then there exists an $x_0 \in K$ such that $f(x_0) = \inf\{f(x) : x \in K\}$; moreover

$$\langle x - x_0, f'(x_0) \rangle \geq 0 \qquad x \in K \tag{4.1}$$

$$\langle x - x_0, f'(x) \rangle \geq 0 \quad \forall x \in K \tag{4.2}$$

Conversely, if $x_0 \in K$ satisfies (4.1) [resp. (4.2) and f' is continuous on K], then $f(x_0) = \inf\{f(x) : x \in K\}$.

Proof. From theorem 6 of the preceding section follows that f is l. s. c. on K in the $\sigma(X, X^*)$ topology. Therefore from theorem 1 it follows that there exists an $x_0 \in K$ such that $f(x_0) = \inf\{f(x) : x \in K\}$. Then, for every $x \in K$ and $\lambda \in [0,1]$

$$f(x_0) \leq f[(1-\lambda)x_0 + \lambda x]$$

and consequently

$$\frac{f[x_0 + \lambda(x - x_0)] - f(x_0)}{\lambda} \geq 0 \qquad \lambda \in [0,1]$$

Going to the limit for $\lambda \to 0^+$, we have

$$\langle x - x_0, f'(x_0) \rangle \geq 0, \quad x \in K$$

(4.1) is so proved. But from theorem 5 of the preceding section it follows that

$$\langle x - x_0,\ f'(x) - f'(x_0)\rangle \geq 0 \quad \forall\ x, x_0 \in K$$

Addition yields

$$\langle x - x_0,\ f'(x)\rangle \geq 0, \quad x \in K$$

and (4.2) is proved.

Conversely, let $x_0 \in K$ satisfy (4.1). Then from the convexity of f in K follows $\forall\, x \in K,\ \lambda \in\ ]0,1[$

$$f(x) - f(x_0) \geq \frac{1}{\lambda}\ \{f\,[(1-\lambda)x_0 + \lambda x\,] - f(x_0)\}$$

and going to the limit $\lambda \to 0^+$

$$f(x) - f(x_0) \geq \langle x - x_0,\ f'(x_0)\rangle \geq 0$$

so that $f(x_0) = \inf\{f(x) : x \in K\}$ and consequently (4.2) is satisfied. Finally let $x_0 \in K$ satisfy (4.2) and let $f'(x)$ be unique $\forall\ x \in K$ and continuous on K. By taking in (4.2)

$$x = (1-\lambda)x_0 + \lambda y, \quad y \in K, \quad \lambda \in\ ]0,1[$$

we obtain

$$\lambda\,\langle y - x_0,\ f'[x_0(1-\lambda) + \lambda y\,]\rangle \geq 0$$

and therefore

$$\langle y - x_0,\ f'[x_0(1-\lambda)x_0 + \lambda y\,]\rangle \geq 0, \quad y \in K; \quad \lambda \in\ ]0,1[$$

In the limit $\lambda \to 0^+$, from the continuity of f' we have (4.1).

<u>Remark 1.</u> If the point x_0 in the hypothesis of the theorem is such that $x_0 \in K^\circ$ then $f'(x_0) = 0$ according to theorem 1 of section 3.

<u>Remark 2.</u> If x_0 satisfies $\langle x - x_0,\ f'(x_0)\rangle \geq 0\ \ \forall\ x \in K$, we say that it satisfies a variational inequality. Therefore theorem 2 is an existence theorem for solutions of some variational inequalities.

In order to indicate a general existence theorem for variational inequalities we state the following (without proof):

Theorem. (Minty-Browder) Let $f: K \to X^*$ be a monotone map of a closed convex subset K of X on X^*, we suppose that

(i) f is monotone on X
(ii) $\forall x, y \in X$ the real function $t \to \langle x - y, f[x + t(x - y)] \rangle$ is continuous on R.
(iii) there exists a $\zeta \in K$ such that

$$\lim_{\|x\| \to \infty} \frac{\langle x - \zeta, f(x) \rangle}{\|x\|} = +\infty$$

Then $\forall x^* \in X^*$ there exists an $x_0 \in X$ such that

$$\langle x - x_0, f(x) - x^* \rangle \geq 0, \quad x \in K$$

Example 1. Let a(x, y) be a bilinear symmetric continuous form in a Banach space X, i.e. a map $(x, y) \in X \times X \to a(x, y) \in R$ which is linear in both variables x, y, and let it be such that $|a(x, y)| \leq M \|x\| \|y\| \ \forall x, y \in X$ and a convenient M in R.

Let us suppose that a(x, y) is coercive, i.e. there exists an $\alpha \in R_+$ such that $|a(x, x)| \geq \alpha \|x\|^2 \ \forall x \in X$.

Let $x^* \in X^*$ and consider the real function $f(x) = a(x, x) = 2\langle x, x^* \rangle$ on X.

Then f is a strictly convex function on X. In fact, from (4.3) $a(x - y, x - y) \geq 0$, if $x, y \in X$, we obtain

$$a(x, x) - 2a(x, y) + a(y, y) \geq 0 \tag{4.3}$$

and therefore

$$2a(x, y) \leq a(x, x) + a(y, y) \tag{4.4}$$

moreover, the equality in (4.3), (4.4) holds if and only if x = y.

From (4.4) follows that, $\forall x, y \in X, \ \lambda \in [0,1]$

$$a[\lambda x + (1 - \lambda)y; \ \lambda x + (1 - \lambda)y]$$

$$= \lambda^2 a(x, x) + 2\lambda(1 - \lambda)\, a(x, y) + (1 - \lambda)^2\, a(y, y)$$

$$\leq \lambda^2 a(x, x) + \lambda(1 - \lambda)\{a(x, x) + a(y, y)\} + (1 - \lambda)^2 a(y, y)$$

$$\leq \lambda a(x, x) + (1 - \lambda)\, a(y, y)$$

so that, from the linearity of $\langle x, x^* \rangle$ on X, it follows that

$$f[\lambda x + (1-\lambda)y] \leqq \lambda f(x) + (1-\lambda) f(y)$$

and the equality holds if and only if $x \neq y$.

Obviously the function f is continuous (and consequently l.s.c.) on X and such that $\lim_{\|x\|\to\infty} f(x) = +\infty$. In fact, for every $x \in X$, from the coercivity of $a(x, y)$ it follows that

$$f(x) = a(x,x) - 2\langle x, x^*\rangle \geqq \alpha \|x\|^2 - 2\|x\| \cdot \|x^*\|$$

and since

$$\left(\frac{\sqrt{\alpha}}{2}\|x\| - \frac{2}{\sqrt{\alpha}}\|x^*\|\right)^2 = \frac{\alpha}{2}\|x\|^2 + \frac{4}{\alpha}\|x^*\|^2 - 2\|x\|\,\|x^*\|$$

we obtain also

$$f(x) \geqq \frac{\alpha}{2}\|x\|^2 - \frac{4}{\alpha}\|x^*\|^2$$

Hence

$$\lim_{\|x\|\to+\infty} f(x) = +\infty$$

Therefore, if K is a closed convex subset of X, there exists, by theorem 1, a unique $x_0 \in K$ such that $f(x_0) = \inf\{f(x) : x \in K\}$.

Since f is obviously Gateaux-differentiable at every $x \in X$ with the Gateaux-derivative $f'(x)$ such that $\langle y, f'(x)\rangle = 2\,a(x,y) - \langle y, x^*\rangle$, theorem 2 entails that at x_0

$$a(x_0, x - x_0) - \langle x - x_0, x^*\rangle \geqq 0 \;\; \forall \; x \in K \qquad (4.5)$$

We have so proved that the variational inequality (4.5) has one and only one solution in K.

BIBLIOGRAPHY

EKELAND J., TEMAM, R., Analyse convexe et problèmes variationnels, Dunod, Paris (1974).

LIONS, J.L., Quelques methodes de risolution de problèmes aux limites non linéaires, Dunod, Paris (1969).

ROCKAFELLAR, E.T., Convex analysis, Princeton Univ. Press (1970).

ASPLUND, E., "Topics in the theory of convex functions", Theory and Applications of Monotone Operators, A. Ghizzetti (1965).

IOFFE, A.D., TIKHOMIROV, V.M., "Duality in problems of the calculus of variations", Trudy Moskov. Obs. 18 (1968) 187.

IOFFE, A.D., TIKHOMIROV, V.M., "Duality of convex functions and extremum problems", Uspekhi Mat. Nauk (1970).

LIONS, J.L., "Partial differential inequalities", Uspekhi Math. Nauk (1972).

ROCKAFELLAR, E.T., "Convex functions, monotone operators and variational inequalities", Theory and Applications of Monotone Operators, A. Ghizzetti (1969).

AN INTRODUCTION TO PROBABILITY THEORY

J. ZABCZYK
Institute of Mathematics,
Polish Academy of Sciences,
Warsaw, Poland

Abstract

AN INTRODUCTION TO PROBABILITY THEORY.

The purpose of this paper is to give the probabilistic foundations of stochastic control theory and to show some applications of probability theory to functional analysis. Such topics as martingales, conditional expectations, Wiener process, linear stochastic equations and Ito's integral are treated in a rigorous way. Only the elements of integration theory and normed spaces are taken for granted.

INTRODUCTION

The purpose of this paper is to give the probabilistic foundations to stochastic control theory and at the same time to present some applications of probability theory to functional analysis.

The formal prerequisite of the paper is familiarity with the concept of (probability) measure and Lebesgue's integral and a knowledge of such theorems (from integration theory) as Lebesgue's theorem, Fatou's lemma, Lebesgue's monotone convergence theorem, although all these theorems are formulated in the text. To make use of elementary parts of the measure theory only, we omit all kinds of "extension theorems" such as the Carathéodory or Kolmogorov theorems and instead we take for granted the Lebesgue's measure on the Euclidean space $\mathbb{R}^n$. We also assume as known the elementary properties of Hilbert and normed spaces. With these exceptions, the paper is self-contained and complete although some proofs are given through problems.

The text is divided into five chapters: Preliminaries, Martingales, Conditioning, Wiener Process, Ito's Stochastic Integral. Preliminaries contain basic information on σ-fields, measurability and independence. Examples of normal distributions, independent random variables are also introduced. As far as applications to functional analysis are concerned the next chapter on Martingales is most important. "The basic definitions (of martingale theory) are inspired by crude notions of gambling but the theory has become a sophisticated tool of modern mathematics drawing from and contributing to other fields" [1]. We apply martingales to prove fundamental properties of Haar and Rademacher systems and the Radon-Nikodym theorem. The main results of this chapter are Doob's inequalities and the Martingale convergence theorem. Some relations between martingales and Markov chains (important in stochastic stability) are also indicated. The material of the chapter Conditioning is extensively used in stochastic control theory (see the lectures "Stochastic control of discrete-time systems"). An effort is made to link general definitions with concrete examples. Chapters 4 and 5 treat of a Wiener process and Ito's integral. They can be considered a starting point for stochastic differential equations and for stochastic control of continuous-time systems. Linear stochastic equations and their physical interpretation ("white noise") are also considered.

Preparing the paper, we have used many existing books and articles on probability and martingales, among others an article on martingales by J.L. Doob [1] and the books "Probability" by J. Lamperti, "Probability and Potentials" by P.A. Meyer, "Stochastic Integrals" by H.P. McKean Jr. [2, 3, 4].

1. PRELIMINARIES

1.1. Random variables and generated σ-fields

Let Ω be a set. A collection $\mathscr{F}$ of subsets of Ω is said to be a σ-field (σ-algebra) if

1) $\Omega \in \mathscr{F}$
2) If $A \in \mathscr{F}$ then A^c, the complement of A, belongs to $\mathscr{F}$ too,
3) If $A_n \in \mathscr{F}$, $n = 1, 2, \ldots$ then $\bigcup_{n=1}^{+\infty} A_n \in \mathscr{F}$

The pair $(\Omega, \mathscr{F})$ is called a measurable space.

Let $(\Omega, \mathscr{F})$, $(E, \mathscr{E})$ be two measurable spaces. A mapping X from Ω into E such that for all $A \in \mathscr{E}$ the set $\{\omega : X(\omega) \in A\}$ belongs to $\mathscr{F}$ is called a measurable mapping or a random variable.

Problem 1. Show that the composition of two random variables is also a random variable.

Let $\mathscr{M}$ be a collection of subsets of Ω. The smallest σ-field on Ω, which contains $\mathscr{M}$, is denoted by $\sigma(\mathscr{M})$ and is called a σ-field generated by $\mathscr{M}$. It is obtained as the intersection of all σ-fields on Ω containing $\mathscr{M}$. Analogously, let $(X_i)_{i \in I}$ be a family of mappings from Ω into $(E, \mathscr{E})$ then the smallest σ-field on Ω with respect to which all functions X_i are measurable is called a σ-field generated by $(X_i)_{i \in I}$ and is denoted by $\sigma(X_i : i \in I)$.

Example 1. Let X be a mapping from Ω into $(E, \mathscr{E})$ then $\sigma(X) = \{A_a : a \in \mathscr{E}\}$ where $A_a = \{\omega : X(\omega) \in a\}$.

Let E be a metric space. The σ-field on E generated by all open (or all closed) subset of E is called a Borel σ-field and denoted $\mathscr{B}(E)$.

Let $\mathbb{R}^1$ be the real line. A real-valued random variable is a random variable with values in $(\mathbb{R}^1, \mathscr{B}(\mathbb{R}^1))$.

A random variable which takes only a finite number of values is called a simple (or elementary) random variable. The proofs of the following two propositions are easy and left to the reader (see also Ref. [5]).

Proposition 1. Let $X, Y, X_1, X_2, \ldots$ be real-valued random variables and $\alpha, \beta \in \mathbb{R}^1$, then $X \vee Y = \max(X, Y)$, $X \wedge Y = \min(X, Y)$, $\alpha X + \beta Y$ and $\limsup_n X_n$, $\liminf_n X_n$ are also random variables (in general, with values in $(\overline{\mathbb{R}}, \mathscr{B}(\overline{\mathbb{R}}))$, where $\overline{\mathbb{R}} = \mathbb{R} \cup \{-\infty, +\infty\}$).

Proposition 2. If X is a non-negative random variable (r.v.), then there exists an increasing sequence of elementary non-negative r.v.'s that converges to X.

Let X be a measurable mapping from $(\Omega, \mathscr{F})$ into $(E, \mathscr{E})$. Very often it is necessary to consider real-valued r.v.'s which are measurable with respect to $\sigma(X)$. The following lemma shows that such r.v.'s are measurable functions of X.

Lemma 1. A real-valued function Y defined on Ω is measurable with respect to $\sigma(X)$ if and only if it is of the form $Y = f(X)$ for some real-valued r.v. f defined on $(E, \mathscr{E})$.

Proof. If f is a real-valued $\mathscr{E}$-measurable function then f(X) is $\sigma(X)$-measurable as a composition of two random variables (see Problem 1). Let now Y be any $\sigma(X)$-measurable function. We can assume that X is non-negative. If Y is the indicator I_A of a set $A \in \sigma(X)$ then (see Example 1) $A = \{X \in a\}$ for some $a \in \mathscr{E}$ and, therefore, $Y = I_a(X)$. Consequently, the lemma is true in this case. Since every simple r.v. is a linear combination of indicators, the lemma is true for all simple r.v.'s. If $Y = \lim_n f_n(X)$ and f_n are $\mathscr{E}$-measurable then $Y = f(Y)$ where $f = \lim_n \inf f_n$. Application of Propositions 1 and 2 finishes the proof.

Let $(E_1, \mathscr{E}_1), \ldots, (E_n, \mathscr{E}_n)$ be measurable spaces then by $\mathscr{E}_1 \times \ldots \times \mathscr{E}_n$ we denote the smallest σ-field of subsets of $E_1 \times E_2 \times \ldots \times E_n$ which contains all sets of the form $A_1 \times \ldots \times A_n$, $A_i \in \mathscr{E}_i$, $i = 1, 2, \ldots, n$.

Problem 2. Let X_1 and X_2 be mappings from Ω into E_1 and E_2, respectively. Show that the mapping (X_1, X_2) from Ω into $E_1 \times E_2$ is a random variable if the mappings X_1 and X_2 are random variables. Hint: The family of all sets $A \in \mathscr{E}_1 \times \mathscr{E}_2$ such that $\{\omega : (X_1(\omega), X_2(\omega)) \in A\} \in \mathscr{F}$ is a σ-field.

Lemma 2. Let $\mathscr{M}_1, \ldots, \mathscr{M}_n$ be some families of subsets of E_1, $E_2, \ldots, E_n$ respectively and let $\mathscr{M}$ be the collection of all sets $A_1 \times \ldots \times A_n$, $A_1 \in \mathscr{E}_1, \ldots, A_n \in \mathscr{E}_n$. Then

$$\sigma(\mathscr{M}) = \sigma(\mathscr{M}_1) \times \ldots \times \sigma(\mathscr{M}_n)$$

Proof. Since $\mathscr{M} \subset \sigma(\mathscr{M}_1) \times \ldots \times \sigma(\mathscr{M}_n)$ therefore $\sigma(\mathscr{M}) \subset \sigma(\mathscr{M}_1) \times \ldots \times \sigma(\mathscr{M}_n)$. On the other hand let $\mathscr{F}_i$ be the family of all sets $A_i \in \sigma(\mathscr{M}_i)$ for which $E_1 \times \ldots \times E_{i-1} \times A_i \times E_{i+1} \times \ldots \times E_n \in \sigma(\mathscr{M})$. Then $\mathscr{F}_i$ is a σ-field and contains all sets from $\mathscr{M}_i$, thus $\mathscr{F}_i = \sigma(\mathscr{M}_i)$. From this if $A_i \in \sigma(\mathscr{M}_i)$, $i = 1, 2, \ldots, n$ then $A_1 \times \ldots \times A_n = \bigcap_{i=1}^{n} E_1 \times \ldots \times E_{i-1} \times A_i \times E_{i+1} \times \ldots \times E_n \in \sigma(\mathscr{M})$ and $\sigma(\mathscr{M}_1) \times \ldots \times \sigma(\mathscr{M}_n) \subset \sigma(\mathscr{M})$.

Corollary 1. For any natural numbers $p_1, p_2, \ldots, p_n$

$$\mathscr{B}(\mathbb{R}^{p_1} \times \ldots \times \mathbb{R}^{p_n}) = \mathscr{B}(\mathbb{R}^{p_1}) \times \ldots \times \mathscr{B}(\mathbb{R}^{p_n})$$

Corollary 2. Let $X = (X_1, \ldots, X_n)$, where X_i are random variables with values in $(\mathbb{R}^p, \mathscr{B}(\mathbb{R}^p))$, then any real-valued function Y, $\sigma(X)$-measurable, is of the form $Y = f(X_1, X_2, \ldots, X_n)$ where f is a Borel function on $\mathbb{R}^{np}$.

A collection $\mathscr{M}$ of subsets of Ω is said to be a π-system if, $\phi \in \mathscr{M}$ and if $A, B \in \mathscr{M}$ then $A \cap B \in \mathscr{M}$.

Example 2. If E is a matric space then the family of all open (closed) subsets of E is a π-system. If $E = \mathbb{R}$ or $E = [0, 1)$ then respectively $\{(-\infty, b] : b \in \mathbb{R}\}$ and $\{[a, b) : 0 \leq a \leq b < 1\}$ are π-systems.

The lemma below will be used frequently in the sequel.

Lemma 3. If $\mathcal{M}$ is a π-system and $\mathcal{G}$ the smallest family of subsets of Ω such that

1) $\mathcal{M} \subset \mathcal{G}$
2) If $A \in \mathcal{G}$ then $A^c \in \mathcal{G}$
3) If $A_1, A_2, \ldots, \in \mathcal{G}$ and $A_n \cap A_m = \emptyset$ for $n \neq m$ then $\bigcup_{n=1}^{+\infty} A_n \in \mathcal{G}$,

then $\mathcal{G} = \sigma(\mathcal{M})$.

Proof. Since $\sigma(\mathcal{M})$ satisfies 1), 2) and 3), $\mathcal{G} \subset \sigma(\mathcal{M})$. To prove the opposite inclusion, we show first that $\mathcal{G}$ is a π-system. Let $A \in \mathcal{G}$ and define $\mathcal{G}_A = \{B : B \in \mathcal{G}$ and $A \cap B \in \mathcal{G}\}$. It is easy to check that $\mathcal{G}_A$ satisfies 2) and 3) and if $A \in \mathcal{M}$ then the condition 1) is also satisfied. Thus for $A \in \mathcal{M}$, $\mathcal{G}_A = \mathcal{G}$ and we have proved that if $A \in \mathcal{M}$ and $B \in \mathcal{G}$ then $A \cap B \in \mathcal{G}$. But this implies $\mathcal{G}_B \supset \mathcal{M}$ and, consequently, $\mathcal{G}_B = \mathcal{G}$ for any $B \in \mathcal{G}$. Now the application of the following problem finishes the proof.

Problem 3. If a π-system $\mathcal{G}$ satisfies 2) and 3) then $\mathcal{G}$ is a σ-field.

Lemma 4. Let f be a measurable mapping from $(E_1 \times E_2, \mathcal{E}_1 \times \mathcal{E}_2)$ into $(E, \mathcal{E})$. For every $x_1 \in E_1$, $f(x_1, \cdot)$ is a measurable mapping from $(E_2, \mathcal{E}_2)$ into $(E, \mathcal{E})$.

Proof. Assume first that $E = \mathbb{R}$ and $\mathcal{E} \doteq \mathcal{B}(\mathbb{R})$ and let $\mathcal{G}$ be a family of all sets $A \in \mathcal{E}_1 \times \mathcal{E}_2$ such that for all $x_1 \in E_1$ the function $I_A(x_1, \cdot)$ is a real-valued r.v. on $(E_2, \mathcal{E}_2)$. If $A = A_1 \times A_2$ where $A_1 \in \mathcal{E}_1$, $A_2 \in \mathcal{E}_2$ then clearly $A \in \mathcal{G}$. Moreover $\mathcal{G}$ satisfies conditions 2) and 3) of lemma 3, therefore $\mathcal{G} = \mathcal{E}_1 \times \mathcal{E}_2$. Let f be a simple function then $f(x_1, \cdot) = \sum_{i=1}^{n} \alpha_1 I_{A_i}(x_1, \cdot)$ for some disjoint sets $A_i \in \mathcal{E}_1 \times \mathcal{E}_2$ real numbers $\alpha_i \in \mathbb{R}$ $i = 1, 2, \ldots, n$ and all $x_1 \in E_1$. Taking into account that a linear combination of measurable functions is also a measurable function, we see that $f(x_1, \cdot)$ is measurable in this case, too. But every non-negative measurable function f is a limit of simple functions, therefore the lemma is valid for all non-negative, thus for all measurable function. If $(E, \mathcal{E})$ is an arbitrary measurable space and $A \in \mathcal{E}$, then the composition $I_A(f)$ (I_A is the indicator function of A) is a random variable on $E_1 \times E_2$. Applying the first part of the proof, we see that for every $x_1 \in E$ the function $I_A(f(x_1, \cdot))$ is $\mathcal{E}_2$-measurable. But then $\{x_2 : f(x_1, x_2) \in A\} = \{x_2 : I_A f(x_1, x_2) = 1\} \in \mathcal{F}_2$ and the proof is complete.

1.2. Integration

The definition of the integral $\int_\Omega X dP$ of a real-valued r.v. is taken for granted (see Refs [3, 5]). We only recall that the integral is well defined if either $\int_\Omega X^+ dP < +\infty$ or $\int_\Omega X^- dP < +\infty$, where $X^+ = X \vee 0$, $X^- = (-X) \vee 0$ and that $\int_\Omega X dP = \int_\Omega X^+ dP - \int_\Omega X^- dP$. If $\int_\Omega |X| dP < +\infty$ we say that X is an integrable random variable. In probability theory, the integral of a random variable X is called the expectation of X and is denoted E(X).

For the proof of the following two theorems we refer to Ref. [5].

Lebesgue's theorem. Let X_1, X_2, be real-valued r.v.'s. If for some integrable r.v. Y, $|X_n| \leq Y$ almost surely, n = 1,2,..., and the sequence (X_n) converges (a.s.) to X, then $E(X_n) \to E(X)$.

Fatou's lemma. Let X_1, X_2, be non-negative r.v.'s, then $\liminf_n E(X_n) \geq E(\liminf_n X_n)$. If, in addition, the sequence (X_n) is increasing then $E(X_n) \to E(X)$. This last property is called Lebesgue's monotone convergence theorem.

1.3. Independence

Let $(\Omega, \mathscr{F}, p)$ be a probability space and let $(\mathscr{F}_i)_{i \in I}$ be a family of sub-σ-fields of $\mathscr{F}$. These σ-fields are said to be independent if, for every finite subset $J \subset I$ and every family $(A_i)_{i \in J}$ such that $A_i \in \mathscr{F}_i$, $i \in J$, $P(\bigcap_{i \in J} A_i) = \prod_{i \in J} P(A_i)$. Random variable $(X_i)_{i \in I}$ are independent if the σ-fields $(\sigma(X_i))_{i \in I}$ are independent.

Problem 1. Let X_1 and X_2 be two real-valued r.v.'s, X_1 measurable in the respect to $\mathscr{F}_1$, X_2 with respect to $\mathscr{F}_2$. If $\mathscr{F}_1$, $\mathscr{F}_2$ are independent σ-fields and the expectation $E(X_1X_2)$ is well defined, then

$$E(X_1X_2) = E(X_1)E(X_2) \tag{1}$$

Hint: Assume that X_1, X_2 are non-negative; show first that (1) holds for simple real-valued r.v.'s. Use Proposition 1, Section 1.1. and Lebesgue's monotone convergence theorem (Section 1.2).

Example 1. Let us consider the probability space $([0,1), \mathscr{B}[0,1), P)$ where P is the Lebesgue's measure on the interval [0,1) and define for every n = 1,2,... and k = 1,2,...,2^n intervals $I_k^n = [\frac{k-1}{2^n}, \frac{k}{2^n})$. Random variables

$$X_n(\omega) = \begin{cases} 0 \text{ if } \omega \in I_k^n, & k \text{ odd} \\ 1 \text{ if } \omega \in I_k^n, & k \text{ even} \end{cases}$$

are independent.

Proof. It follows, by induction on n, that for every $\epsilon_i = 0$ or 1, i = 1,2,...,n,

$$\{\omega : X_1(\omega) = \epsilon_1, \ldots, X_n(\omega) = \epsilon_n\} = \left[\frac{\epsilon_1}{2^1} + \ldots + \frac{\epsilon_n}{2^n}, \frac{\epsilon_1}{2^1} + \ldots + \frac{\epsilon_n}{2^n} + \frac{1}{2^n}\right).$$

This implies that $P(X_1 = \epsilon_1, \ldots, X_n = \epsilon_n) = \frac{1}{2^n} = \prod_{i=1}^{n} P(X_i = \epsilon_i)$. An application of the definition of independence and the fact that $A_i \in \sigma(X_i)$ if $A_i = \{\omega : X_i \in a_i\}$ for some Borel set a_i (see Example 1 (Section 1.1)) finishes the proof.

Problem 2. Show that for every $\omega \in [0,1)$, $\omega = \sum_{k=1} \frac{X_k(\omega)}{2^k}$, (dyadic expension of ω). Hint: $\sum_{k=1}^{n} \frac{X_k(\omega)}{2^k} = \frac{\ell-1}{2^n}$ for $\omega \in \left[\frac{\ell-1}{2^n}, \frac{\ell}{2^n}\right)$, $\ell = 1, \ldots, 2^n$.

Lemma 1. Let $\mathscr{M}_i$ be a π-system on Ω and let $\mathscr{F}_i = \sigma(\mathscr{M}_i)$, $i \in I$. The σ-fields $(\mathscr{F}_i)_{i \in I}$ are independent if for every finite set $J \subset I$ and sets $A_i \in \mathscr{M}_i$, $i \in J$, $P(\bigcap_{i \in J} A) = \prod_{i \in J} P(A_i)$.

Proof. Assume, without loss of generality, that $I = J = \{1, 2, \ldots, n\}$. Let us fix the sets $A_2, A_3, \ldots, A_n$ and denote by $\mathscr{G}_1$ the family of all sets $A_1 \in \mathscr{F}_1$ for which

$$P(\bigcap_{k=1}^{n} A_k) = \prod_{k=1}^{n} P(A_k) \tag{2}$$

The family $\mathscr{G}_1$ and the π-system $\mathscr{M}_1$ satisfy the conditions 1), 2) and 3) of Lemma 3 (Section 1.1) therefore $\mathscr{G}_1 = \sigma(\mathscr{M}_1) = \mathscr{F}_1$. Analogously let us fix sets $A_1 \in \mathscr{F}_1$ and $A_i \in \mathscr{M}_i$, $i = 3, 4, \ldots, n$ and denote by $\mathscr{G}_2$ the family of all $A_2 \in \mathscr{F}_2$ which satisfy (2). Then $\mathscr{G}_2 = \sigma(\mathscr{M}_2) = \mathscr{F}_2$. Easy induction shows that (2) holds for all $A_i \in \mathscr{F}_i$, $i = 1, \ldots, n$.

Corollary 1. Random variables X_i with values in $(E_i, \sigma(\mathscr{M}_i))$ where $\mathscr{M}_i$ are π-systems on E_i, $i \in I$, are independent if $P(\bigcap_{i \in J} \{X_i \in A_i\}) = \prod_{i \in J} P(X_i \in A_i)$ for any finite subset $J \subset I$ and sets $A_i \in \mathscr{M}_i$, $i \in J$.

Proof. It is sufficient to remark that

$$\sigma(X_i) = \sigma(\{X_i \in A_i\}: \ A_i \in \sigma(\mathscr{M}_i))$$

$$= \sigma(\{X_i \in A_i\}: \ A_i \in \mathscr{M}_i)$$

Corollary 2. Since all intervals $(-\infty, x]$, $x \in \mathbb{R}$ form a π-system which generates $\mathscr{B}(\mathbb{R})$ therefore real-valued r.v.'s $X_1, X_2, \ldots, X_n$ are independent if and only if

$$P(\bigcap_{i=1}^{n} \{X_i \leq x_i\}) = \prod_{i=1}^{n} P(X_i \leq x_i)$$

for any real numbers $x_1, x_2, \ldots, x_n$.

Problem 3. Let $X_1, X_2, \ldots$ be independent random variables with values in $(E, \mathscr{E})$ and let $J_1, J_2, \ldots$ be a finite or infinite sequence of disjoint subsets of $\{1, 2, \ldots\}$. Then the σ-fields $\mathscr{F}_i = \sigma(X_k;\ k \in J_i)$, $i = 1, 2, \ldots$ are independent Hint: σ-fields $\mathscr{F}_i$ are generated by π-systems $\mathscr{M}_i$ of all sets $\bigcap_{k \in J_i} \{X_k \in A_k\}$ where $A_k \in \mathscr{E}$ and only for a finite number of k, $A_k \neq E$.

The lemma below is an important generalization of the Problem 1. The proof is analogous to that of Lemma 4 (Sec.1.1).

Lemma 2. Let f be a real-valued random variable defined on $(E_1 \times E_2, \mathscr{E}_1 \times \mathscr{E}_2)$ and let X_1, X_2 be independent random variables from

$(\Omega, \mathcal{F}, P)$ into, respectively, $(E_1, \mathcal{E}_1)$, $(E_2, \mathcal{E}_2)$. If the expectation $E(f(X_1, X_2))$ is well defined then the function $f_1(\cdot) = E(f(\cdot, X_2))$ is $\mathcal{E}_1$-measurable and $E(f(X_1, X_2)) = E(f_1(X_1))$.

Proof. It is sufficient to prove Lemma 2 for functions $f = I_A$ where $A \in \mathcal{E}_1 \times \mathcal{E}_2$. Denote by $\mathcal{G}$ the family of all sets A such that for $f = I_A$ the lemma holds. Then $\mathcal{G} \supset \mathcal{M}$, where $\mathcal{M}$ is the π-system of all sets $B_1 \times B_2$, $B_1 \in \mathcal{E}_1$, $B_2 \in \mathcal{E}_2$. Since $\mathcal{E}_1 \times \mathcal{E}_2 = \sigma(\mathcal{M})$ and the family $\mathcal{G}$ satisfies the assumptions 2) and 3) of Lemma 3, Section 1.1; therefore, this yields $\mathcal{G} = \mathcal{E}_1 \times \mathcal{E}_2$.

1.4. Distributions of random variables

Let $(\Omega, \mathcal{F}, P)$ be a probability space, $(E, \mathcal{E})$ a measurable space and X a random variable from Ω into E. The distribution of the r.v. X is a probability law μ_X on $(E, \mathcal{E})$ defined as $\mu_X(A) = P(\omega : X(\omega) \in A)$, $A \in \mathcal{E}$.

Lemma 1. Let μ and ν be two probability measures on $(E, \sigma(\mathcal{M}))$ where $\mathcal{M}$ is a π-system. If

$$\mu(A) = \nu(A) \tag{1}$$

for $A \in \mathcal{M}$ then (1) holds for all $A \in \sigma(\mathcal{M})$.

Proof. Denote by $\mathcal{G}$ the family of all sets $A \in \sigma(\mathcal{M})$ for which (1) holds. Then $\mathcal{G}$ satisfies assumptions 1), 2) and 3) of Lemma 3 (Section 1.1) and, consequently, $\mathcal{G} = \sigma(\mathcal{M})$.

Corollary 1. Let μ be a probability measure on $(\mathbb{R}^p, \mathcal{B}(\mathbb{R}^p))$. Then the distribution function F_μ of μ defined as

$$F_\mu(x) = \mu(y : y \leq x), \quad x \in \mathbb{R}^p$$

determines μ uniquely, $((y_1, \ldots, y_p) \leq (x_1, \ldots, x_p)$ means $x_1 \leq y_1, \ldots, x_p \leq y_p$.

Proof. Since the family $\mathcal{M}$ of all sets $\{y : y \leq x\}$, $x \in \mathbb{R}^p$ is a π-system and $\mathcal{B}(\mathbb{R}^p) = \sigma(\mathcal{M})$ therefore Corollary 1 follows from Lemma 3, Section 1.1.

Thus if X is a r.v. with values in $\mathbb{R}^p$ then the distribution function of X defined as $F_X = F_{\mu_X}$ determines the distribution of X.

Example 1. If X is a real-valued r.v. such that

$$F_X(t) = \begin{cases} 0 & \text{for } t < 0 \\ t & \text{for } 0 \leq t < 1 \\ 1 & \text{for } 1 \leq t \end{cases}$$

then the distribution of X is the Lebesgue's measure restricted to [0,1]. We say, in this case, that X has a uniform distribution on [0,1].

The concept of distribution is often applied through the following lemma:

Lemma 2. If f is a non-negative random variable defined on $(E, \mathcal{E})$ then

$$E(f(X)) = \int_E f(x)\, \mu_X(dx) \tag{2}$$

Proof. Formula (2) holds if $f = I_A$, $A \in \mathscr{E}$ and thus for any simple random variable. If f is a non-negative r.v. then $f = \lim_n f_n$ for some increasing sequence of simple non-negative r.v. $f_1, f_2, \ldots$. Since $\int_\Omega f_n(X)dP = \int_E f_n(x)\mu_X(dx)$; therefore, (2) holds because of Lebesgue's monotone convergence theorem (Section 1.2).

Corollary 2. The random variable f(X) is integrable if and only if the function f is μ_X-integrable.

Let μ be a probability measure on $(\mathbb{R}^n, \mathscr{B}(\mathbb{R}^n))$ such that $\int_{\mathbb{R}^n} |x|^2 \mu(dx) < +\infty$, $(|x|^2 = x_1^2 + \ldots + x_n^2)$. The column vector m with components $m_i = \int_{\mathbb{R}^n} x_i \mu(dx)$, $i = 1, \ldots, n$ is called the mean vector of μ and the matrix $Q = (\sigma_{i,j})_{i,j=1,\ldots,n}$ where $\sigma_{i,j} = \int_{\mathbb{R}^n} (x_i - m_i)(x_j - m_j)\mu(dx)$ is called the covariance matrix of μ.

Let us remark that if $\mu = \mu_X$ is the distribution of a random variable X with components $X_1, X_2, \ldots, X_n$ and $E(X_i^2) < +\infty$, $i = 1, \ldots, n$ then, by virtue of Lemma 2, $m_i = E(X_i)$, $\sigma_{i,j} = E(X_i - m_i)(X_j - m_j)$ $i, j = 1, \ldots, n$. Vector m and matrix Q are called in this case the mean vector and the covariance matrix of X.

If T is a linear mapping from R^n into R^k (or, equivalently, T is a $k \times n$ matrix) and m is a vector from R^n, then T' and m' denote, respectively, the conjugate mapping (the transpose matrix) and the transpose vector.

Problem 1. Show that the mean vector and the covariance matrix of the random variable TX are equal Tm and TQT'. Hint: $E(TX) = TE(X)$ and $E(TX-Tm)(TX-Tm)' = T(E(X-m)(X-m)')T'$.

1.5. Normal distributions

Let μ be a measure on $(\mathbb{R}^n, \mathscr{B}(\mathbb{R}^n))$ and g a Borel function such that

$$\mu(A) = \int_A g(x)dx \quad \text{for all } A \in \mathscr{B}(\mathbb{R}^n)$$

then the function g is called a density of the measure μ (with respect to the n-dimensional Lebesgue's measure which is taken for granted).

Here, we introduce the normal distribution through the following problems (see Ref. [6, vol.2]):

Problem 1. Let R be an $n \times n$ symmetric matrix, then the function $\exp\{-(1/2)\langle Rx, x\rangle\}$, $x \in \mathbb{R}^n$ is integrable on $\mathbb{R}^n$ if and only if $\langle Rx, x\rangle > 0$ for all $x \neq 0$, i.e. if the matrix R is positive-definite. Hint:

$$\int_{-\infty}^{+\infty} e^{-\gamma x^2} dx < +\infty$$

if and only if $\gamma > 0$. Use an appropriate system of co-ordinates on $\mathbb{R}^n$.

Problem 2. Let Q be an $n \times n$ positive-definite matrix and m a vector from $\mathbb{R}^n$, then the function $g_{m,Q}$:

$$g_{m,Q}(x) = ((2\pi)^n \det Q)^{-\frac{1}{2}} \exp\{-(1/2)\langle Q^{-1}(x-m), x-m\rangle\}$$

is a density and a probability law on $\mathbb{R}^n$. Hint: Use the formula

$$\frac{1}{\sqrt{2\pi\lambda}} \int_{-\infty}^{+\infty} e^{-\frac{x^2}{2\lambda}} dx = 1, \ \lambda > 0.$$

This probability law is called a (non-degenerate) normal distribution.

Problem 3. The normal distribution with the density $g_{m,Q}$ has the mean vector m and the covariance matrix Q. Hint: See Section 1.4.

Problem 4. Let a random variable X with components $X_1, \dots, X_n$ be normally distributed, then, for every $k \leq n$, the random vector with components $X_1, X_2, \dots, X_k$ is also normally distributed on $\mathbb{R}^k$. Hint: use induction on $k = n, n-1, \dots, 1$.

If μ is a measure on $\mathbb{R}^n$ and T is a linear map from $\mathbb{R}^n$ into $\mathbb{R}^k$ then the measure $T\mu$ is defined by the formula

$$T\mu(A) = \mu(T^{-1}A), \quad A \in \mathscr{B}(\mathbb{R}^k).$$

If a random variable X has the distribution μ then the random variable TX has the distribution $T\mu$. Any measure of the form $T\mu$ where μ is a normal distribution is called a general normal distribution.

Problem 5. The measure $T\mu$ has the mean vector Tm and the covariance matrix TQT'. Hint: Define on $\Omega = \mathbb{R}^n$ the random variable $X(x) = x$, $x \in \mathbb{R}^n$ and use Problem 1 (Section 1.4).

Problem 6. The general normal distribution $T\mu$ is a non-degenerate normal distribution if and only if the linear mapping T is onto or, equivalently, if the matrix TQT' is positive-definite. Hint: See Problem 4.

Corollary 1. A general normal distribution is a non-degenerate normal distribution if and only if its covariance matrix is positive definite.

Problem 7. The components $X_1, \dots, X_n$ of a normally distributed r.v. X are independent if and only if the covariance matrix Q is diagonal (if $\sigma_{i,j} = 0$ for $i \neq j$).

Problem 8. Let $X_1, X_2, \dots$ be random variables normally distributed (in the general sense) with parameters (m_1, Q_1), $(m_2, Q_2), \dots$. Assume that the sequence (X_k) converges almost surely to a random variable X, $(m_k) \to m$, $(Q_k) \to Q$ and Q is positive definite, then X is normally distributed with parameters (m, Q).

1.6. Sequences of independent random variables

Classical probability theory deals with sequences of independent real-valued random variables. In this section, we show that such sequences can be easily constructed if the Lebesgue's measure on $[0,1)$ is taken for granted. The construction goes back to H. Steinhaus.

The theorem below will be used as a source of concrete examples in the next chapters.

Theorem. Let $\mu_1, \mu_2, \ldots$ be a sequence of probability measures on $\mathbb{R}^1$. There exists a sequence (Y_n) of independent real-valued random variables defined on $([0,1), \mathscr{B}[0,1), P)$ where P is the Lebesgue's measure on $[0,1)$ such that the distributions of Y_n are exactly the measures μ_n, $n = 1, 2, \ldots$.

Proof. Let $X_1, X_2, \ldots$ be independent random variables constructed in Example 1, Section 1.3 and let $J_i = \{n_{i,j} : j = 1, 2 \ldots\}$, $i = 1, 2, \ldots$ be disjoint subsets of the set of natural numbers. Then, because of Problem 3, Section 1.3, the random variables

$$Z_i = \sum_{j=1}^{+\infty} X_{n_{i,j}} / 2^j \quad i = 1, 2, \ldots \tag{1}$$

are independent. We show that they have uniform distribution on $[0,1)$. Let, for instance, $i = 1$ and define

$$S_n = \sum_{j=1}^{n} X_{n_{1,j}} / 2^j$$

Then $P(S_n = k/2^n) = 1/2^n$, $k = 0, 1, 2, \ldots, 2^n - 1$, (compare Example 1, Section 1.3) and, therefore, for $t \in [0,1)$, $P(S_n \leq t) \to t$. On the other hand, $P(S_n \leq t) \to P(Z_1 \leq t)$. Thus $P(Z_1 \leq t) = t$. An application of Example 1 (Section 1.4) proves that Z_i has the uniform distribution on $[0,1)$.

Now let μ be a probability measure on $\mathbb{R}^1$ and let $F = F_\mu$ be its distribution function. Define $F^+(s) = \inf\{t : s \leq F(t)\}$, $s \in [0,1)$. If Z has uniform distribution on $[0,1)$, then the distribution of $F^+(Z)$ is exactly μ. Really, from the definition of F^+, for $s \in [0,1)$ and $t \in (-\infty, +\infty)$, $s \leq F(F^+(s))$ and $F^+(F(t)) \leq t$. Therefore, $\{s : F^+(s) \leq t\} = [0, F(t)]$ and, consequently, $P(\omega : F^+(Z(\omega)) \leq t) = P(\omega : Z(\omega) \leq F(t)) = F(t)$.

To finish the proof of the theorem, it is sufficient to remark that if $F_1^+, F_2^+, \ldots$ are functions (defined as above) corresponding to the measures $\mu_1, \mu_2, \ldots$ and $Z_1, Z_2, \ldots$ are real-valued random variables defined in Chapter 1, then the sequence $F_1^+(Z_1), F_2^+(Z_2), \ldots$ has all properties required.

2. MARTINGALES

2.1. Definition of martingales and supermartingales

Let T be an arbitrary subset of $\overline{\mathbb{R}}$ ordered by the relation $\leq$. Let $(\Omega, \mathscr{F}, P)$ be a probability space and $(\mathscr{F}_t)_{t \in T}$ an increasing family of sub-σ-fields of $\mathscr{F}$.

A family $(X_t)_{t\in T}$ of finite real-valued random variables adapted to the family $(\mathscr{F}_t)_{t\in T}$, (i.e. X_t are $\mathscr{F}_t$-measurable, $t\in T$), is said to be a martingale (or, respectively, a supermartingale, a submartingale) with respect to the family $(\mathscr{F}_t)_{t\in T}$ if

1) X_t are integrable random variables, $t\in T$;
2) If $s \leqq t$, then for every event $A \in \mathscr{F}_s$

$$\int_A X_t\, dP = \int_A X_s\, dP$$

or, respectively

$$\int_A X_t\, dP \leqq \int_A X_s\, dP$$

$$\int_A X_t\, dP \geqq \int_A X_s\, dP$$

Every real constant (or, respectively, decreasing, increasing) function defined on T is a martingale (or, respectively, a supermartingale, a submartingale).

Let $(X_t)_{t\in T}$ be a supermartingale. The process $(-X_t)_{t\in T}$ is then a submartingale and conversely.

Example 1. Let $(\Omega, \mathscr{F}, P)$ be a probability space and let $Y_1, Y_2, \ldots$ be a sequence of independent random variables (see Section 1.6) defined on Ω such that $E|Y_n| < +\infty$ and $E(Y_n) = 0$, (or, respectively, $\leqq 0$, $\geqq 0$) for all $n = 1, 2, \ldots$; then the sequence (X_n), $X_n = Y_1 + \ldots + Y_n$, is a martingale (or, respectively, a super-, a submartingale) with respect to $(\mathscr{F}_n)$, where $\mathscr{F}_n = \sigma(Y_1, \ldots, Y_n)$. Really, by virtue of Problem 3, (Section 1.3), the σ-fields $\mathscr{F}_n$ and $\sigma(Y_{n+1})$ are independent. Therefore, if $A \in \mathscr{F}_n$, then I_A and Y_{n+1} are independent random variables. Applying Problem 1 (Section 1.3) we see that

$$\int_A (X_{n+1} - X_n)dP = \int_A Y_{n+1}dP = E(I_A Y_{n+1}) = P(A)E(Y_{n+1}) = 0,$$

(or respectively $\leqq 0$, $\geqq 0$).

Problem 1. An urn contains b black and w white balls. Balls are drawn at random without replacement. Let b_n and w_n denote the numbers of black and white balls in the urn before the n-th drawing. Construct an appropriate probability space and show that the sequence

$$X_n = \frac{b_n}{b_n + w_n}, \quad n = 1, \ldots, b+w$$

is a martingale with respect to $\mathscr{F}_n = \sigma(X_1, \ldots, X_n)$.

Lemma 1. If (X_t) and (Y_t) are supermartingales (relative to $(\mathscr{F}_t)$) and α, β are positive numbers, then the processes $(\alpha X_t + \beta Y_t)$ and $(X_t \wedge Y_t)$ are also supermartingales. If (X_t) is a martingale, then $(|X_t|)$ is a submartingale.

Proof. Let $s \leqq t$ and $A \in \mathscr{F}_s$. Since $\alpha \int_A X_s dP \geqq \alpha \int_A X_t dP$ and $\beta \int_A Y_s dP \geqq \beta \int_A Y_t dP$, therefore $\int_A (\alpha X_s + \beta Y_s) dP \geqq \int_A (\alpha X_t + \beta Y_t) dP$. Obviously,

$$\int_A (X_s \wedge Y_s) dP = \int_{A \cap \{X_s < Y_s\}} X_s dP + \int_{A \cap \{X_s \geq Y_s\}} Y_s dP$$

and

$$A \cap \{X_s < Y_s\} \in \mathscr{F}_s, \quad A \cap \{X_s \geqq Y_s\} \in \mathscr{F}_s.$$

By virtue of the definition of a supermartingale, we obtain

$$\int_{A \cap \{X_s < Y_s\}} (X_s - X_t) dP \geqq 0, \quad \int_{A \cap \{X_s \geq Y_s\}} (Y_s - Y_t) dP \geqq 0$$

and, consequently,

$$\int_A X_s \wedge Y_s \, dP \geqq \int_{A \cap \{X_s < Y_s\}} X_t \wedge Y_t \, dP + \int_{A \cap \{X_s \geq Y_s\}} X_t \wedge Y_t \, dP = \int_A X_t \wedge Y_t \, dP.$$

If (X_t) is a martingale, then $(X_t \vee 0)$ and $((-X_t) \vee 0)$ are submartingales, therefore $|X_t| = X_t \vee 0 + (-X_t) \vee 0$ is a submartingale, too.

2.2. Rademacher and Haar systems and martingales

In this section, we introduce two martingales which are important from the point of view of functional analysis. The first martingale is connected with the so-called Rademacher functions. To define these functions, we use the same notations as in Example 1 (Section 1.3).

Thus, $\Omega = [0,1)$, $\mathscr{F} = \mathscr{B}[0,1)$ and P is the Lebesgue's measure on $[0,1)$. Define $\mathscr{F}_n$ as the σ-field generated by the intervals $I_k^n = \left[\frac{k-1}{2^n}, \frac{k}{2^n}\right)$, $k = 1, \ldots, 2^n$, and $n = 1, 2, \ldots$. Clearly, $\mathscr{F}_n \subset \mathscr{F}_{n+1}$. Rademacher functions r_n are defined by the formula $r_n = 1 - 2X_n$ or, explicitly,

$$r_n(\omega) = \begin{cases} 1 & \text{if } \omega \in I_k^n, \quad k \text{ odd} \\ -1 & \text{if } \omega \in I_k^n, \quad k \text{ even} \end{cases}$$

Proposition 1. For an arbitrary sequence of real numbers $\alpha_1, \alpha_2, \ldots$ the sequence $(\alpha_1 r_1 + \ldots + \alpha_n r_n)$ is a martingale with respect to $(\mathscr{F}_n)$.

Proof. It is sufficient to note that $\int_{I_k^n} r_{n+1} dP = 0$, $k = 1, 2, \ldots, 2^n$ or to apply Example 1 (Section 2.1).

Let now $(\Omega, \mathscr{F}, P)$ be an arbitrary probability space and let $(\mathscr{F}_n)_{n=0,1,\ldots}$ be an increasing sequence of σ-fields contained in $\mathscr{F}$ such that every $\mathscr{F}_n$ is generated by exactly $n+1$ disjoint sets $A_0^n, \ldots, A_n^n$, $\bigcup_{k=0}^{n} A_k^n = \Omega$, $P(A_k^n) > 0$

$k = 0, 1, \ldots, n$. (To construct the $n+1$-st partition, we choose a set, say $A^n_{k_n}$, from the n-th partition, divide it into two parts of positive probabilities, and the remaining sets A^n_k, $k \neq k_n$, are left unchanged).

A generalized Haar system is a sequence of random variables $(H_n)_{n=0,1,\ldots}$ adapted to the $(\mathcal{F}_n)_{n=0,1,\ldots}$ such that

1) $H_0 \equiv 1$;

2) $E|H_{n+1}| > 0$, and $H_{n+1} = 0$ in the complement of the set $A^n_{k_n}$;

3) $E(H_n) = 0$, $E(H_n^2) = 1$.

Proposition 2. For an arbitrary sequence of real numbers $\alpha_0, \alpha_1, \alpha_2, \ldots$, the sequence $(\alpha_0 H_0 + \ldots + \alpha_n H_n)$ is an $\mathcal{F}_n$-martingale.

Proof. Let $A \in \mathcal{F}_n$. If $A \cap A^n_{k_n} = \emptyset$, then 2) implies $E(I_A H_{n+1}) = 0$. If $A \cap A^n_{k_n} \neq \emptyset$ then $A = B \cup A^n_{k_n}$, $B \in \mathcal{F}_n$ and $B \cap A^n_{k_n} = \emptyset$ and, therefore, $E(I_A H_{n+1}) = E(I_{A^n_{k_n}} H_{n+1}) = E(H_{n+1}) = 0$ because of 3). This proves the proposition.

2.3. Martingales and densities

The results of this section will be needed in Section 2.10 which is devoted to the Radon-Nikodym theorem.

Let μ and P be two probability (or finite) measures on $(\Omega, \mathcal{G})$. We recall that the measure μ is said to be absolutely continuous on $\mathcal{G}$ with respect to P if for every $A \in \mathcal{G}$ such that $P(A) = 0$, $\mu(A) = 0$ holds. Note that the definition not only depends on P and μ, but also on the σ-field $\mathcal{G}$.

A finite partition of Ω is a sequence $A_1, A_2, \ldots, A_n$ of disjoint subsets of Ω such that

$$\bigcup_{i=1}^{n} A_i = \Omega.$$

Example 1. Let $\mathcal{G} = \sigma(A_1, \ldots, A_n)$, where $(A_1, \ldots, A_n)$ is a finite partition of Ω, and let μ be absolutely continuous with respect to P (on $\mathcal{G}$), then

$$\mu(A) = \int_A g(\omega)\, dP(\omega), \quad \text{for } A \in \mathcal{G} \tag{1}$$

where

$$g(\omega) = \begin{cases} \dfrac{\mu(A_i)}{P(A_i)}, & \text{if } \omega \in A_i,\ P(A_i) > 0 \\ 0, & \text{if } \omega \in A_i,\ P(A_i) = 0 \end{cases}$$

The proof is obvious. Let us remark that the function g is $\mathcal{G}$-measurable, and that (1) implies absolute continuity of μ with respect to P.

More generally, if μ and P are arbitrary probability measures on $(\Omega, \mathcal{G})$ then a $\mathcal{G}$-measurable function g such that (1) holds is called a density of μ with respect to P (on $\mathcal{G}$). If g is a density then a $\mathcal{G}$-measurable function h is a density also if and only if $P(\omega : g(\omega) \neq h(\omega)) = 0$.

Let $\mathscr{F}$ be the collection of all finite partitions of Ω. If $t = (A_1^t, \ldots, A_{n_t}^t) \in \mathscr{F}$ we define $\mathscr{F}_t = \sigma(A_1^t, \ldots, A_{n_t}^t)$. We write $t \leqq s$ if and only if $\mathscr{F}_t \subseteq \mathscr{F}_s$ that means if the s-th partition is a refinement of the s-th partition.

Proposition. Let $t_1 \leqq t_2 \leqq \ldots$ be an increasing sequence of partitions of Ω. Let $(\Omega, \mathscr{F}, P)$ be a probability space and $\mathscr{F}_{t_n} \subset \mathscr{F}$, $n = 1, 2, \ldots$. If a measure μ is absolutely continuous with respect to P on $\mathscr{F}_{t_n}$ and g_n is a corresponding $\mathscr{F}_{t_n}$-measurable density, $n = 1, 2, \ldots$, then the sequence (g_n) is $(\mathscr{F}_{t_n})$-martingale.

Proof. Let $A \in \mathscr{F}_{t_n}$ then $A \in \mathscr{F}_{t_{n+1}}$. Therefore, $\mu(A) = \int_A g_n dP$ and $\mu(A) = \int_A g_{n+1} dP$. Thus $\int_A g_n dP = \int_A g_{n+1} dP$. This proves the result.

Example 2. Let us consider the probability space $([0,1), \mathscr{B}[0,1), P)$ where P is the Lebesgue's measure and a Borel, integrable function g. If μ is a measure with density g and $\mathscr{F}_n = \sigma(I_1^n, \ldots, I_{2^n}^n)$, $I_k^n = \left[\frac{k-1}{2^n}, \frac{k}{2^n}\right)$, $k = 1, \ldots, 2^n$ then the density g_n is a picewise constant (on I_k^n) function given by the formula

$$g_n(\omega) = 2^n \int_{I_k^n} g(s)ds \quad \text{for} \quad \omega \in I_k^n, \quad k = 1, \ldots, 2^n$$

2.4. Martingales and Markov chains

Let $(\Omega, \mathscr{F}, P)$ be a probability space and $(E_1, \mathscr{E}_1)$, $(E_2, \mathscr{E}_2)$ some measurable spaces. A sequence $X_0, X_1, \ldots$ of random variables from Ω into E_1 is said to be a Markov chain if there exists a sequence of independent random variables $\xi_0, \xi_1, \xi_2, \ldots$ from Ω into E_2 and measurable mappings F_n from $E_1 \times E_2 \to E_1$ such that

$$X_{n+1} = F_n(X_n, \xi_n), \qquad n = 0, 1, \ldots \tag{1}$$

and $\xi_0, \xi_1, \xi_2, \ldots$ are independent of X_0.

If the mappings F_n and the distributions of ξ_n do not depend on n, then the Markov chain (X_n) is homogeneous (in time).

Example 1. If $\xi_0, \xi_1, \ldots$ are real-valued r.v.'s then the sequence (X_n), where $X_{n+1} = X_0 + \xi_0 + \ldots + \xi_n = X_n + \xi_n$, is a Markov chain, so called random walk.

Problem 1. Let us define the transition function $(\mathbb{P}_n)$ of the Markov chain (X_n) by the formula

$$\mathbb{P}_n(x, A) = P(F_n(x, \xi_n) \in A) = E(I_A(F_n(x, \xi_n)))$$

then (for $n = 1, 2, \ldots$):

$$\text{If } A \in \mathscr{E}_1,\ \mathbb{P}_n(\cdot, A) \text{ is } \mathscr{E}_1\text{-measurable} \tag{2}$$

$$\text{If } x \in E_1,\ \mathbb{P}_n(x, \cdot) \text{ is a probability measure on } \mathscr{E}_1 \tag{3}$$

Hint: Apply Lemma 4 (Section 1.1) and Lemma 2 (Section 1.3).

Usually, Markov chains are defined starting from a transition function ($\mathbb{P}_n$), (see, e.g. Ref. [6, Vol.1]) but models of stochastic control theory are described often by Eqs (1) (see Refs [7, 8]). From the mathematical point of view, these two approaches are equivalent. In this direction, we propose to solve the following problem:

Problem 2. Let E_1 be 1) a countable (or finite) set and $\mathscr{E}_1$ the family of all subsets of E_1 or 2) $E_1 = \mathbb{R}^1$ and $\mathscr{E}_1 = \mathscr{B}(\mathbb{R}^1)$ and let a function $\mathbb{P}$ satisfy (2) and (3). Then there exists a measurable function F from $E_1 \times [0,1)$ into E_1 such that for any random variable ξ uniformly distributed on $[0,1)$, $P(F(x,\xi)\in A) = \mathbb{P}(x,A)$. Hint: 1) Let $E_1 = \{1,2,\ldots\}$ and $P_{n,m} = \mathbb{P}(n,\{m\})$. Define $F(n,t) = m$ on $[P_{n,1}+\ldots+P_{n,m-1}, P_{n,1}+\ldots+P_{n,m})$. 2) Define $F(x,t) = F_x^+(t)$, as in Section 1.6, where $F_x(t) = \mathbb{P}(x,(-\infty,t])$.

Let us now consider a homogeneous Markov chain (X_n) with the transition function $\mathbb{P}$. A real-valued $\mathscr{E}_1$-measurable function h is said to be harmonic (sub-, superharmonic) for (X_n) if

1) $\mathbb{P}h = h$, (respectively, $\mathbb{P}h \geq h$, $\mathbb{P}h \leq h$);

2) For every initial point $X_0 = x\in E_1$, $E|h(X_n)| < +\infty$, $n = 0,1,2,\ldots$.

Here the operator $\mathbb{P}$ is defined as

$$\mathbb{P}h(x) = \int_{E_1} h(y)P(x,dy), \quad x\in E_1$$

The theorem below has many applications to stability theory.

Theorem. If h is a harmonic (sub-, superharmonic) function for (X_n), then the sequence ($h(X_n)$) is a martingale (sub-, supermartingale) with respect to $\mathscr{F}_n = \sigma(X_0,\ldots,X_n)$ for any initial point $X_0 = x$.

Proof. Let $A\in\sigma(X_0,\ldots,X_n)$ then (see Example 1, Section 1.1) $A = \{\omega : (X_0(\omega),\ldots,X_n(\omega))\in a\}$ for some $a\in\mathscr{E}_1 \times \mathscr{E}_1 \times \ldots \times \mathscr{E}_1$. Therefore, $E(I_A h(X_{n+1})) = E(I_a(X_0,\ldots,X_n)\, h(F(X_n,\xi_n)))$. It is easy to see that ξ_n is independent of $\mathscr{F}_n$ and, consequently, by virtue of Lemma 2, (Section 1.3), $E(I_A h(X_{n+1})) = E(I_a(X_0,\ldots,X_n)\mathbb{P}h(X_n)) = E(I_A\mathbb{P}h(X_n))$. Since $\mathbb{P}h = h$ ($\mathbb{P}h \geq h$ or $\mathbb{P}h \leq h$) we obtain $E(I_A h(X_{n+1})) = E(I_A h(X_n))$, ($E(I_A h(X_{n+1})) \geq E(I_A h(X_n))$, or, respectively $\leq$).

Example 2 (coin tossing). Let $E_1 = \{\ldots,-1,0,1,\ldots\}$ $E_2 = \{-1,1\}$ and let $\xi_0,\xi_1,\ldots$ are independent r.v.'s such that $P(\xi_n = 1) = p$, $P(\xi_n = -1) = 1-p$, $0 < p < 1$. The function h:

$$h(x) = x, \text{ if } p = \frac{1}{2}$$

$$h(x) = \left(\frac{1-p}{p}\right)^x, \text{ if } p \neq \frac{1}{2}, \quad x\in E_1$$

is harmonic for random walk $X_n = X_0 + \xi_0 + \ldots + \xi_n$. Thus, the sequence ($X_n$) in the former and the sequence $((1-p)/p)^{X_n}$ in the latter case are martingales.

For the sake of completeness we propose solution of the following problem (see Chapter 3):

Problem 3 (Markov property). Show that for any bounded, $\mathscr{E}_1$-measurable real function h:

$$E(h(X_{n+1})|X_0,\ldots,X_n) = E(h(X_{n+1})|X_n) = \mathbb{P}_n h(X_n), \text{ a.s.}$$

Hint: $E(h(X_{n+1})|X_0,\ldots,X_n) = E(h(F_n(X_n,\xi_n)|X_0,\ldots,X_n)$.

But ξ_n is independent of $\mathscr{F}_n = \sigma(X_0,\ldots,X_n)$; therefore, using Lemma 1, Section 3.2, $E(h(X_{n+1})|\mathscr{F}_n) = \mathbb{P}_n h(X_n)$ a.s.

2.5. Stopping times

Let $(\Omega,\mathscr{F})$ be a measurable space and $(\mathscr{F}_t)_{t\in T}$ an increasing family of σ-fields, $\mathscr{F}_t \subset \mathscr{F}$, $t\in T$. A function $S:\Omega\to T$ is said to be a stopping time (relative to the $(\mathscr{F}_t)$) if for every $t\in T$ the set $\{\omega : S(\omega)\le t\}$ belongs to $\mathscr{F}_t$, i.e. the condition $S\le t$ is a condition involving only what has happened up to and including time t. Let S be a stopping time. By $\mathscr{F}_S$ we denote the collection of events $A\in\mathscr{F}$ such that $A\cap\{S\le t\}\in\mathscr{F}_t$ for all $t\in T$. It is easy to verify that $\mathscr{F}_S$ is a σ-field, so called σ-field of events prior to S.

Proposition 1 a). If T is a finite or a countable subset of $\mathbb{R}$ then $S:\Omega\to T$ is a stopping time if and only if $\{\omega : S(\omega) = t\}\in\mathscr{F}_t$ for $t\in T$. b) If S_1 and S_2 are stopping times then $S_1\wedge S_2$ and $S_1\vee S_2$ are again stopping times. c) Any stopping time S is $\mathscr{F}_S$-measurable. d) If $S_1\le S_2$, then $\mathscr{F}_{S_1}\subset\mathscr{F}_{S_2}$.

Proof. The properties a), b), c) follow directly from the definitions. To prove d) assume that $A\in\mathscr{F}_{S_1}$, then $\{S_2\le t\}\cap A = \{S_2\le t\}\cap\{S_1\le t\}\cap A$ because $\{S_2\le t\}\cap\{S_1\le t\} = \{S_2\le t\}$. Since $\{S_1\le t\}\cap A\in\mathscr{F}_t$ and $\{S_2\le t\}\in\mathscr{F}_t$ therefore $\{S_2\le t\}\cap\{S_1\le t\}\cap A\in\mathscr{F}_t$.

Example 1. Let (X_n) be a sequence of real-valued r.v.'s adapted to $(\mathscr{F}_n)$, (that means X_n are $\mathscr{F}_n$-measurable) and let $\mathscr{F}_\infty = \mathscr{F}$. Then

$$S = \begin{cases} \text{the least } n \text{ such that } X_n \ge a \\ \quad\cdot\cdot \\ +\infty \text{ if } X_n < a \text{ for all } n = 1,2,\ldots \end{cases}$$

is a stopping time. To see this fix a natural number k then $\{S = k\} = \{X_1 < a,\ldots,X_{k-1} < a,\ X_k \ge a\}\in\sigma(X_1,\ldots,X_k)\subset\mathscr{F}_k$.

The following special case of a theorem proved by Courrege and Priouret [9] is helpful in considering concrete examples.

Proposition 2. Let $X_1, X_2,\ldots$ be random variables with values in a countable set $E\subset\mathbb{R}^1$. Define $\mathscr{F}_n = \sigma(X_1,\ldots,X_n)$, $\mathscr{F}_\infty = \sigma(X_1,X_2,\ldots)$. A mapping $S:\Omega\to\{1,2,\ldots,+\infty\}$ is a stopping time relative to $(\mathscr{F}_n)$ if and only if

1) S is $\mathscr{F}_\infty$-measurable
2) If $S(\omega) = n$ and $X_k(\omega) = X_k(\omega')$ for $k = 1,2,\ldots,n$ then $S(\omega') = n$.

Proof. We prove only that 1) and 2) imply: S is a stopping time. (The opposite case is obvious). Let us define $A = \{\omega;\ S(\omega) = n\}$ and $a = \{(x_1, \ldots, x_n) \in E^n;\ x_i = X_i(\omega)$ for $i = 1, \ldots, n$ and some $\omega \in A\}$. The set a is at most a countable set, therefore it is a Borel one. The property 2) implies $A = \{\omega : (X_1(\omega), \ldots, X_n(\omega)) \in a\}$ and, therefore, $A \in \sigma(X_1, \ldots, X_n) = \mathscr{F}_n$.

2.6. More on stopping times. Marriage problem.

In this section, we try to consider stopping times in a more intuitive way.

Let Ω be the set of all sequences $\omega = (\epsilon_1, \epsilon_2, \ldots, \epsilon_N)$ where $\epsilon_n = -1$ or 1, $n = 1, \ldots, N$. Any sequence $\omega \in \Omega$ can be interpreted as the record of N successive tosses of a coin: -1 stands for heads and 1 for tails. Let us introduce, for any $n = 1, \ldots, N$, a random variable Y_n and a σ-field $\mathscr{F}_n$ by the formulas $Y_n(\epsilon_1, \ldots, \epsilon_N) = \epsilon_n$, $\mathscr{F}_n = (Y_1, \ldots, Y_n)$. Thus Y_n represents the outcome of coin tossing at the moment n and $\mathscr{F}_n$ the class of all past events up and including time n ($\mathscr{F}_n$ is generated by the sets $\{Y_1 = \epsilon_1, \ldots, Y_n = \epsilon_n\}$). Any stopping time S relative to $\mathscr{F}_1, \ldots, \mathscr{F}_N$ can be interpreted as a (non-anticipating) rule for stopping gambling: stop at the moment n if the event $S = n$ has just occurred. Since the information available to the gambler at the moment n is contained in the sequence of outcomes $Y_1, Y_2, \ldots, Y_n$, we require $\{S = n\} \in \mathscr{F}_n$. Conversely any rule for stopping gambling based on the past history only (non-anticipating rule) defines a stopping time. For instance, if we have the following rule: stop at moment 1 if $Y_1 = -1$, continue if $Y_1 = 1$; stop at moment 2 if $Y_1 = 1$ and $Y_2 = -1$, continue if $Y_2 = 1$; stop at moment 3 if $Y_1 = 1$ and $Y_2 = 1$ and $Y_3 = -1$ or $Y_3 = 1$. Then the corresponding stopping time has to be defined as follows: $S(\omega) = 1$, for all ω such that $Y_1(\omega) = -1$; $S(\omega) = 2$ for all ω such that $Y_1(\omega) = 1$ and $Y_2(\omega) = -1$; $S(\omega) = 3$ otherwise.

To define a rule that anticipates the future (or, equivalently, a "stopping time" S which does not satisfy $\{S = n\} \in \mathscr{F}_n$) we introduce a sequence $X_1, X_2, \ldots, X_N$, which describes the fortune of the gambler at moments $1, 2, \ldots, N$ i.e. let $X_n = 1 + Y_1 + \ldots + Y_n$. The following rule: stop at the first moment n such that $X_n = \max(X_1, \ldots, X_N)$ anticipates the future. If for instance $N = 3$ then $\{S = 1\} = \{X_1 \geq X_2$ and $X_1 \geq X_3\} = \{Y_2 = -1\} \notin \mathscr{F}_1$. To stop the game at moment 1, the player should know the outcome at moment 2.

Problem 1. (The space Ω and the sequence $(\mathscr{F}_n)$ as above). Let r_N and d_N denote respectively the number of all stopping times $S \leq N$ and the number of all random variables T with values in $\{1, 2, \ldots, N\}$ and $\mathscr{F}_N$-measurables. Show that $r_1 = 1$, $r_{N+1} = (1 + r_N)^2$, $d_N = (N)^{2^N}$, $N = 1, \ldots$ and that $(r_N/d_N) \leq (1/2)(2/N)^{2^N}$. Hint: Use Proposition 2, Section 2.5 and the estimate $r_N \leq 2^{2^N - 1}$.

We propose also to consider the following "Marriage Problem".

It can be formulated as follows: In an urn, there are N balls. Inside, each ball contains a piece of gold which we do not see and for different balls the amounts of gold are different. We may draw (at random) one and keep it and, of course, we want to draw the one with more gold inside. But we are not allowed to check all balls for the amount of gold they contain. Instead, we are to follow the following rule. We are allowed to draw and check one ball a time and decide either to keep it or to have another drawing. However, all balls which have been drawn, checked and rejected, are not available any longer. Keeping the k-th ball we choose, the probability of maximal gain is $1/N$ for $k = 1, 2, \ldots, N$. Here, we do not use the information available from previous choices. However, it may be shown that, using the information available, one can find a strategy which increases the probability of the best choice to p_N where $p_N \to 1/e$ as $N \to \infty$. This problem one faces in life if one wishes to get married and believes that one will be given N possibilities of choice.

The above problem can be formalized as an optimal stopping-time problem. All details are given in the book [10].

2.7. Doob's optional sampling theorem

Theorem. Let $(X_n)_{n=1,\ldots,k}$ be a supermartingale (a martingale) relative to $(\mathscr{F}_n)_{n=1,\ldots,k}$. Let $S_1, S_2, \ldots, S_m$ be an increasing sequence of $(\mathscr{F}_n)$ stopping times. The sequence $(X_{S_i})_{i=1,\ldots,m}$ is then also a supermartingale (a martingale) with respect to σ-fields $\mathscr{F}_{S_1}, \mathscr{F}_{S_2}, \ldots, \mathscr{F}_{S_m}$.

Proof. Let $A \in \mathscr{F}_{S_1}$. We shall prove that $\int_A (X_{S_1} - X_{S_2})\,dP \geq 0$ in the supermartingale case and $\int_A (X_{S_1} - X_{S_2})\,dP = 0$ in the martingale case. If for every ω, $S_2(\omega) - S_1(\omega) \leq 1$ then

$$\int_A (X_{S_2} - X_{S_1})\,dP = \sum_{r=1}^{k} \int_{A \cap \{S_1 = r\} \cap \{S_2 > r\}} (X_{r+1} - X_r)\,dP$$

but $A \cap \{S_1 = r\}$ and $\{S_2 > r\}$, belong to $\mathscr{F}_r$, therefore $A \cap \{S_1 = r\} \cap \{S_2 > r\} \in \mathscr{F}_r$. The definition of a supermartingale (a martingale) implies that the desired inequality (equality) follows.

Let $r = 0, 1, \ldots, k$ and define stopping times $R_r = S_2 \wedge (S_1 + r)$. Then $S_1 = R_0 \leq R_1 \leq \ldots \leq R_k = S_2$ and $R_{i+1} - R_i \leq 1$, $i = 0, \ldots, k-1$. By virtue of the first part of the proof

$$\int_A X_{S_1}\,dP \geq \int_A X_{R_1}\,dP \geq \ldots \geq \int_A X_{R_k}\,dP = \int_A X_{S_2}\,dP$$

In the martingale case, the above inequalities should be replaced by the equalities and, thus, the theorem is proved.

Let us consider some simple applications of the above Doob's optional sampling theorem (more serious applications will be given in Sections 2.8 and 2.9).

Example 1. (Ruin problem). Let (X_n) be the random walk (defined in Section 2.4) which starts at the origine ($X_0 = 0$). Thus $X_{n+1} = Y_0 + \ldots + Y_n$ where Y_0, Y_1, are independent random variables $P(Y_n = -1) = 1 - p$, $P(Y_n = 1) = p$, $n = 0, 1, 2, \ldots$. Let a and b be natural numbers and let $S = \inf\{n : X_n = -a \text{ or } X_n = b\}$. In gambling language: a gambler with initial capital a plays against an adversary with initial capital b; in each game, he can win or lose a dollar with probability p and $q = 1-p$, respectively, and the games are independent; X_n is the gambler's total winnings after the n-th game and the game ends exactly at S, when he or his adversary lost all his money. The ruin problem consists in finding the probability that the gambler will be ruined, i.e. the probability that $X_S = -a$.

Problem 1. Show that $P(S < +\infty) = 1$. Hint: See Remark 3, Section 2.9.

Assume that $p = 1/2$ (fair game) and define stopping times $S_n = S \wedge n$, $n = 1, 2, \ldots$. Since (X_n) is a martingale, by Doob's theorem (applied to stopping times: 0 and S_n) $E(X_{S_n}) = E(X_0) = 0$. But $S < +\infty$ (a.s.) and, therefore, the sequence X_{S_n} is convergent:

$$X_{S_n} \to -a \quad \text{on} \quad \{X_S = -a\}$$

$$X_{S_n} \to b \quad \text{on} \quad \{X_S = b\}$$

Lebesgue's theorem (Section 1.2) implies

$$0 = E(X_{S_n}) \to E(\lim_n X_{S_n}) = -aP(X_S = -a) + b(1 - P(X_S = -a))$$

Consequently,

$$P(X_S = -a) = b/(a+b), \quad P(X_S = b) = a/(a+b)$$

Problem 2. Consider the case $p \neq 1/2$. Hint: Sequence $(q/p)^{X_n}$ is a martingale (Section 2.4). Answer

$$P(X_S = -a) = (1 - (q/p)^b)/(p/q)^a - (q/p)^b$$

Example 2 (Strategy for a schoolboy). Let us consider the martingale introduced in Example 1, Section 2.1. Again by Doob's theorem, we obtain that, for any stopping time S (relative to $(\mathscr{F}_n)$),

$$E(X_S) = E(X_0) = b/(b+w)$$

The above identity can be interpreted as follows: During an examination, every schoolboy draws at random (without replacement) one question he has to answer. At the beginning of the examination, there are b + w questions, and some pupil is able to answer in a satisfactory way exactly w of them. At any moment, he knows which questions have been drawn and he wants to find an optimal moment for drawing question to minimize the probability of choosing a "bad" question. But for any rule S, $E(X_S) = b/(b+w)$; therefore, all possible "strategies" are of the same value for him.

2.8. Fundamental inequalities

Theorem 1. Let $(X_n)_{n=1,\ldots,k}$ be a supermartingale and c a non-negative constant. Then we have

$$1) \quad cP(\sup_n X_n \geq c) \leq E(X_1) - \int_{\{\sup X_n < c\}} X_k dP$$

$$2) \quad \leq E(X_1) + E(X_k^-)$$

3) $cP(\inf_n X_n \leq -c) \leq -\int_{\{\inf X_n \leq -c\}} X_k dP$

4) $\leq E(X_k^-)$

Proof. To prove the inequalities 1), 2), define $S(\omega) = \inf\{n : X_n(\omega) \geq c\}$, or $S(\omega) = k$ if $\sup_n X_n(\omega) < c$. S is a stopping time and $S \geq 1$. Thus, by Doob's optional sampling theorem, we obtain

$$E(X_1) \geq E(X_S) = \int_{\{\sup X_n \geq c\}} X_S dP + \int_{\{\sup X_n < c\}} X_S dP \geq cP(\sup X_n \geq c) + \int_{\{\sup X_n < c\}} X_k dP$$

or equivalently

$$cP(\sup_n X_n \geq c) \leq E(X_1) - \int_{\{\sup X_n < c\}} X_k dP$$

This is exactly inequality 1).
Since $-X_k \leq X_k^-$, inequality 2) follows, too. To establish the relations 3) and 4), we introduce an analogous stopping time S, $S = \inf\{n : X_n \leq -c\}$ or $S = k$ if $\inf X_n > -c$. Since $S \leq k$, therefore

$$E(X_k) \leq E(X_S) = \int_{\{\inf X_n \leq -c\}} X_S dP + \int_{\{\inf X_n > -c\}} X_S dP$$

$$\leq -cP(\inf_n X_n \leq -c) + \int_{\{\inf X_n > -c\}} X_k dP$$

These inequalities imply 3) and 4).

Problem 1. (Doob-Kolmogorov's inequality). Let $(X_n)_{n=1,\ldots,k}$ be a martingale then for every $c > 0$

$$P(\sup_n |X_n| \geq c) \leq \frac{E(|X_k|)}{c}$$

Hint: If (X_n) is a martingale then $(|X_n|)$ is a submartingale (Lemma 1, Section 2.1). Use 4), Theorem 1.

To formulate and prove the next theorem, we have to introduce some new definitions and notations.

Let $x = (x_1, \ldots, x_n)$ be a sequence of real numbers and let $a < b$. Let S_1 be the first of the numbers $1, 2, \ldots, n$ such that $X_{S_1} \geq b$ or n if there exists no such number. Let S_k be, for every even (respectively odd) integer $k > 1$ the first of the numbers $1, 2, \ldots, n$ such that $S_k > S_{k-1}$ and $x_{S_k} \leq a$ (respectively $x_{S_k} \geq b$). If no such number exists, we set $s_K = n$. In this way, the sequence $S_1, S_2, S_3, \ldots$ is defined. The number $\mathscr{D}_n(x; (a,b))$ of down-crossings by the sequence x of the interval (a,b) is defined as the greatest integer k such that one actually has $x_{S_{2k-1}} \geq b$ and $x_{S_{2k}} \leq a$. If not such integer exists, we put $\mathscr{D}_n(x; (a,b)) = 0$. Let us remark that the "intervals":

$[S_1, S_2], \ldots, [S_{2k-1}, S_{2k}]$ represent the periods of time when the sequence x is descending from b to a. Analogously the number $\mathscr{U}_n(x; (a,b))$ of upcrossings by the sequence x of the interval (a, b) can be defined. In fact $\mathscr{U}_n(x; (a,b)) = \mathscr{D}_n(-x, (-b, -a))$. Numbers $S_1, S_2, \ldots$ which correspond to the sequence -x and the interval (-b, -a) will be denoted as $R_1, R_2, \ldots$.

Theorem 2 (Doob's inequalities). Let $(X_m)_{m=1,\ldots,n}$ be a supermartingale relative to $(\mathscr{F}_m)_{m=1,\ldots,n}$ and let $a < b$ be two real numbers. Then the following inequalities hold:

1) $E(\mathscr{D}_n(a,b)) \leq \frac{1}{b-a}(E(X_1 \wedge b) - E(X_n \wedge b))$

2) $E(\mathscr{U}_n(a,b)) \leq \frac{1}{b-a} E((a - X_n)^+)$

Here $\mathscr{D}_n(a,b) = \mathscr{D}_n((X_1, \ldots, X_n); (a,b))$, $\mathscr{U}_n(a,b) = \mathscr{U}_n((X_1, \ldots, X_n); (a,b))$, ($\mathscr{D}_n(a,b)$ and $\mathscr{U}_n(a,b)$ are clearly random variables).

Proof. 1) Since $(X_m \wedge b)_{m=1,\ldots,n}$ is a supermartingale and the numbers of downcrossings corresponding to $(X_m)_{m=1,\ldots,n}$ and $(X_m \wedge b)_{m=1,\ldots,n}$ are the same we can assume that $X_m = X_m \wedge b$, $m = 1, \ldots, n$. Let $S_1, \ldots, S_{2\ell}$, $2\ell > n$ be moments defined above and corresponding to the sequence $(X_1, \ldots, X_n)$. Then $\Sigma \overset{df}{=} (X_{S_1} - X_{S_2}) + \ldots + (X_{S_{2\ell-1}} - X_{S_{2\ell}}) \geq \mathscr{D}_n(a,b)(b-a)$. Indeed, if $\mathscr{D}_n(a,b) = k$ then $\Sigma = (X_{S_1} - X_{S_2}) + \ldots + (X_{S_{2k-1}} - X_{S_{2k}}) + (X_{S_{2k+1}} - X_{S_{2k+2}})$. But $S_{2k+1} = n$ or $S_{2k+1} < n$ and $S_{2k+2} = n$, therefore $X_{S_{2k+1}} - X_{S_{2k+2}} = 0$ or $b - X_{S_{2k+2}} \geq 0$. Thus $X_{S_{2k+1}} - X_{S_{2k+2}} \geq 0$ and $\Sigma \geq k(b-a)$. Consequently,

$$X_{S_1} - X_{S_{2\ell}} = \Sigma + (X_{S_2} - X_{S_3}) + (X_{S_4} - X_{S_5}) + \ldots + (X_{S_{2\ell-2}} - X_{S_{2\ell-1}})$$
$$\geq (b-a)\mathscr{D}_n(a,b) + (X_{S_2} - X_{S_3}) + \ldots + (X_{S_{2\ell-2}} - X_{S_{2\ell-1}})$$

By Doob's optional sampling theorem, the sequence $X_{S_1}, \ldots, X_{S_{2\ell}}$ is a supermartingale and, in particular, $E(X_{S_i}) \geq E(X_{S_j})$ for $i \leq j$. Therefore, $E(X_{S_1}) - E(X_{S_{2\ell}}) \geq (b-a)E(\mathscr{D}_n(a,b))$. But $E(X_{S_{2\ell}}) = E(X_n)$ and $E(X_{S_1}) \leq E(X_1)$ and we conclude that

$$E(\mathscr{D}_n(a,b)) \leq \frac{1}{(b-a)}(E(X_1) - E(X_n))$$
$$\leq \frac{1}{(b-a)}(E(X_1 \wedge b) - E(X_n \wedge b))$$

Proof 2). Let $R_1, \ldots, R_{2\ell}$, $(2\ell > n)$ be moments used in the definition of upcrossings and let $\Sigma_1 = (X_{R_2} - X_{R_1}) + \ldots + (X_{R_{2\ell}} - X_{R_{2\ell-1}})$. Then $\Sigma_1 \geq (b-a)\mathscr{U}_n(a,b) + (X_n - a) \wedge 0$. Assume $\mathscr{U}_n(a,b) = k$, then $\Sigma_1 = (X_{R_2} - X_{R_1}) + \ldots + (X_{R_{2k}} - X_{R_{2k-1}}) + (X_{R_{2k+2}} - X_{R_{2k+1}})$. If $R_{2k+1} = n$ then $X_{R_{2k+2}} - X_{R_{2k+1}} = 0$ and if $R_{2k+1} < n$ and $R_{2k+2} = n$ then $X_{R_{2k+2}} - X_{R_{2k+1}} = X_n - X_{R_{2k+1}} \geq X_n - a$. Thus, in both cases, $X_{R_{2k+2}} - X_{R_{2k+1}} \geq (X_n - a) \wedge 0$.

Again Doob's theorem implies that

$$0 \geq E(\Sigma_1) \geq (b-a)E(\mathcal{U}_n(a,b)) + E((X_n - a) \wedge 0)$$

Finally,

$$(b-a)E(\mathcal{U}_n(a,b)) \leq -E((X_n - a) \wedge 0) = E((a - X_n)^+)$$

2.9. Martingale convergence theorem

Many important applications of martingales are connected with the following convergence theorem:

Theorem. Let $(X_n)_{n=1,2,\ldots}$ be a supermartingale such that $\sup_n E(X_n^-) < +\infty$. Then the sequence (X_n) converges a.s. to an integrable random variable X.

Proof. For every $a < b$ define $\mathcal{U}(a,b) = \sup_n \mathcal{U}_n(a,b) = \lim_n \mathcal{U}_n(a,b)$. Doob's inequality implies that

$$E(\mathcal{U}(a,b)) = \lim_n E(\mathcal{U}_n(a,b)) \leq \frac{1}{(b-a)} \sup_n E((a - X_n)^+)$$

But $(a - X_n)^+ \leq |a| + X_n^-$ and therefore

$$E(\mathcal{U}(a,b)) \leq \frac{|a|}{b-a} + \frac{1}{b-a} \sup_n E(X_n^-) < +\infty$$

On the other hand,

$$\{\liminf X_n < \limsup X_n\} = \bigcup_{a,b \text{ rationals}} \{\liminf_n X_n < a < b < \limsup_n X_n\}$$

$$= \bigcup_{a,b \text{ rationals}} \{\mathcal{U}(a,b) = +\infty\}$$

Since $E(\mathcal{U}(a,b)) < +\infty$ therefore $P(\mathcal{U}(a,b) = +\infty) = 0$. Thus

$$P(\liminf_n X_n < \limsup_n X_n) \leq \sum_{a,b \text{ rationals}} P(\mathcal{U}(a,b) = +\infty) = 0$$

This way we have obtained that almost surely $\liminf_n X_n = \limsup_n X_n = \lim_n X_n = X$. Let us remark that $|X_n| = X_n + 2X_n^-$. But for all n, $E(X_n) \leq E(X_1)$ and, by virtue of Fatou's lemma,

$$E|X| \leq \sup_n E(X_n + 2X_n^-) \leq E(X_1) + 2\sup_n E(X_n^-) < +\infty$$

This finishes the proof.

Remark 1. Since $X_n^- \leq |X_n| \leq X_n + 2X_n^-$, therefore $E(X_n^-) \leq E|X_n| \leq E(X_1) + 2E(X_n^-)$ and the conditions $\sup_n E(X_n^-) < +\infty$, $\sup_n E|X_n| < +\infty$ are equivalent (provided $E(X_1) < +\infty$).

As an example of application of martingale convergence theorem to functional analysis we prove the following proposition:

Proposition. If $\sum_{k=0}^{+\infty} \alpha_k^2 < +\infty$ then the Haar Series $\sum_{k=0}^{+\infty} \alpha_k H_k$ and the Rademacher series $\sum_{k=1}^{+\infty} \alpha_k r_k$ converge almost surely.

Proof. It is easy to check that Haar as well as Rademacher systems are orthonormal systems. Therefore, using Schwarz' inequality

$$E\left|\sum_{k=0}^{n} \alpha_k H_k\right| \leq \left(E\left(\sum_{k=0}^{n} \alpha_k H_k\right)^2\right)^{\frac{1}{2}} = \left(\sum_{k=0}^{n} \alpha_k^2\right)^{\frac{1}{2}}$$

$$E\left|\sum_{k=1}^{n} \alpha_k r_k\right| \leq \left(E\left(\sum_{k=1}^{n} \alpha_k r_k\right)^2\right)^{\frac{1}{2}} = \left(\sum_{k=1}^{n} \alpha_k^2\right)^{\frac{1}{2}}$$

Since $\sum_{k=0}^{+\infty} \alpha_k^2 < +\infty$ we obtain:

$$\sup_n E\left|\sum_{k=0}^{n} \alpha_k H_k\right| < +\infty \quad \text{and} \quad \sup_n E\left|\sum_{k=1}^{n} \alpha_k r_k\right| < +\infty.$$

But sequences $\left(\sum_{k=0}^{n} \alpha_k H_k\right)$ and $\left(\sum_{k=1}^{n} \alpha_k r_k\right)$ are martingales (see Section 2.2) and the application of martingale convergence theorem completes the proof.

Remark 2. Exactly in the same way as the above proposition, it is possible to prove that if $\xi_1, \xi_2, \ldots$ are independent r.v.'s such that

$E(\xi_k) = 0$, $E(\xi_k^2) = 1$, $k = 1, \ldots$ and $\sum_{k=1}^{+\infty} \alpha_k^2 < +\infty$, then the series $\sum_{k=1}^{+\infty} \alpha_k \xi_k$ converges almost surely.

Remark 3. Let us consider Problem 1, Section 2.7. Assume, e.g. $p = \frac{1}{2}$. If $S = +\infty$, then the sequence (X_{S_n}) does not converge. But (X_{S_n}) is a bounded martingale, therefore the convergence theorem implies $P(S = +\infty) = 0$.

2.10. Radon-Nikodym theorem

As an application of martingale theory we prove the Radon-Nikodym theorem:

Theorem. Let μ and ν be finite measures defined on a measurable space $(\Omega, \mathscr{F})$. There exists a function $g \geq 0$, $\mathscr{F}$-measurable and ν-integrable and a set $B \in \mathscr{F}$, $\nu(B) = 0$, such that for every set $A \in \mathscr{F}$

$$\mu(A) = \int_A g\, d\nu + \mu(A \cap B)$$

The proof will be a rather simple application of the martingale convergence theorem (Section 2.9). Relations between the martingale convergence theorem and the Radon-Nikodym theorem in a more general setting (martingales and measures with values in Banach spaces) were intensively investigated by many authors. It turned out, for instance, that these two theorems are equivalent (Chatterji [11]). In the proof of the theorem, we shall need a simple fact concerning "generalized" sequences. A partially ordered set $\mathscr{T}$ is said to be filtering to the right if for any $t_1, t_2 \in \mathscr{T}$ there exists $t \in \mathscr{T}$ such that $t_1 \leq t$, $t_2 \leq t$.

Every mapping $x(\cdot)$ from $\mathscr{T}$ into a matric space E is called a generalized sequence. A generalized sequence $(x(t))_{t \in \mathscr{T}}$ converges to x_0 if for every $\epsilon > 0$ there exist $t(\epsilon) \in \mathscr{T}$ such that for $t \geq t(\epsilon)$, $\rho(x(t), x_0) < \epsilon$. In this case we write $\lim_{t \in \mathscr{T}} x(t) = x_0$. Here ρ is a metric on E.

Lemma 1. Let E be a complete metric space and $(x(t))_{t \in \mathscr{T}}$ a generalized sequence such that for every increasing sequence $t_1 \leq t_2 \leq \ldots$ the sequence $(x(t_n))$ converges in E. Then there exists an element $x_0 \in E$ such that $\lim_{t \in \mathscr{T}} x(t) = x_0$ and moreover $\lim_n x(t_n) = x_0$ for some increasing sequence $t_1 \leq t_2 \leq \ldots$.

Proof. It is easy to see (by contradiction) that for every $\epsilon > 0$ there exists $t(\epsilon) \in \mathscr{T}$ such that for every $t \geq t(\epsilon)$ $\rho(x(t), x(t(\epsilon))) < \epsilon$. If we define an increasing sequence (t_n) such that for $t \geq t_n$, $\rho(x(t), x(t_n)) \leq 1/n$ we obtain that $(x(t_n))$ converges to an element $x_0 \in E$ and moreover that $\lim_{t \in \mathscr{T}} x(t) = x_0$.

Proof of the theorem. Let us define a probability measure $P = \frac{1}{c}(\mu + \nu)$ where $c = \mu(\Omega) + \nu(\Omega)$ and let $\mathscr{T}$ be the collection of all finite partitions $t = (A_1^t, \ldots, A_{n_t}^t)$, $(A_i^t \in \mathscr{F})$, of Ω and let $\mathscr{F}_t = \sigma(A_1^t, \ldots, A_{n_t}^t)$. Denote by $\overline{g}_t$ the density of the measure μ with respect to P (considered on $\mathscr{F}_t$), (see Section 2.3). For every increasing sequence of partitions (t_n), the corresponding sequence of densities $(\overline{g}_{t_n})$ is $(\mathscr{F}_{t_n})$-martingale (see Section 2.3). Since $0 \leq \overline{g}_{t_n} \leq \frac{1}{c}$, the convergence theorem (Section 2.9) implies that the sequence $(\overline{g}_{t_n})$ converges P-almost surely to an integrable random variable and from Lebesgue's theorem we obtain that $(\overline{g}_{t_n})$ converges also in $E = L^1(\Omega, \mathscr{F}, P)$. Applying the above lemma we see that there exists a P-integrable, $\mathscr{F}$-measurable function $\overline{g}$ such that $\lim_{t \in \mathscr{T}} \overline{g}_t = \overline{g}$. If $A \in \mathscr{F}$ then $A \in \mathscr{F}_t$ for some $t \in \mathscr{T}$. From this if $s \geq t$,

$$\int_A \overline{g}_t dP = \int_A \overline{g}_s dP = \int_A \overline{g}\, dP$$

Therefore

$$\mu(A) = \int_A \overline{g}_t dP = \frac{1}{c}\int_A \overline{g} d\mu + \frac{1}{c}\int_A \overline{g} d\nu$$

or equivalently $\int_A (c-\overline{g})d\mu = \int_A \overline{g}d\nu$ for all $A \in \mathscr{F}$. Using the fact that every $\mathscr{F}$-measurable function $h \geq 0$ is a limit of increasing sequence of simple functions as well as Lebesgue's monotone convergence theorem we obtain

$$\int_A h(c-\overline{g})d\mu = \int_A h\overline{g}d\nu$$

To finish the proof of the theorem define the set B and the function g as follows:

$$B = \{\overline{g} = c\} \text{ and } g = 0 \text{ on } B, \; g = \frac{\overline{g}}{c-\overline{g}} \text{ on } B^c$$

Since $\int_B h\overline{g}d\nu = 0$ for any $\mathscr{F}$-measurable function $h \geq 0$ therefore $\nu(B) = 0$. Let $h = 0$ on B and $\frac{1}{c-\overline{g}}$ on B^c then

$$\mu(A\setminus B) = \int_{A\setminus B} \frac{1}{c-\overline{g}}(c-\overline{g})d\mu = \int_{A\setminus B}\frac{1}{c-\overline{g}}\overline{g}d\nu = \int_{A\setminus B} gd\nu = \int_A gd\nu$$

and finally

$$\mu(A) = \mu(A\setminus B) + \mu(A\cap B) = \int_A gd\nu + \mu(A\cap B)$$

Problem 1. Prove that the function g and the set B given in the theorem are unique in the following sense: if g_1 and B_1 satisfy the statement of the theorem, then

$$\nu(g \neq g_1) = 0 \quad \text{and} \quad \mu(B\setminus B_1) + \mu(B_1\setminus B) = 0$$

Corollary 1. Let μ be a measure which is absolutely continuous with respect to ν, then there exists a function $g \geq 0$, $\mathscr{F}$-measurable such that $\mu(A) = \int_A gd\nu$, $A \in \mathscr{F}$.

Moreover, there exists an increasing sequence of partitions $t_1 \leq t_2 \leq \ldots$ such that the corresponding sequence (g_{t_n}) of densities of the measure μ with respect to ν (considered on $\mathscr{F}_{t_n}$), — see Section 2.3 — converges ν-almost surely to the density g.

Proof. Only the second part of the corollary requires a proof. From Lemma 1, we obtain the existence of an increasing sequence of partitions $t_1 \leq t_2 \leq \ldots$ such that $\overline{g}_{t_n} \to \overline{g}$ in E and P-almost surely. But $g = \overline{g}/(c-\overline{g})$ and thus $\overline{g}_{t_n}/(1-\overline{g}_{t_n}) \to g$. It is sufficient to note that $\overline{g}_{t_n}/(c-\overline{g}_{t_n}) = g_{t_n}$.

Problem 2. Let $\mathscr{F}_n$ be a σ-field generated by a partition $(A_1^n, \ldots, A_{k_n}^n)$, $n = 1, 2, \ldots$, $\mathscr{F}_\infty = \sigma(\mathscr{F}_1, \mathscr{F}_2, \ldots)$ and let $\mathscr{F}_n \subset \mathscr{F}_{n+1}$. If measures μ and ν are defined on a measurable space $(\Omega, \mathcal{F}_\infty)$ and μ is absolutely continuous with respect to ν, then the corresponding sequence of densities (g_n) converges ν-almost surely to the density g of μ with respect to ν.

Hint: See the proof of the above corollary.

3. CONDITIONING

3.1. Definition of the conditional expectation

As usual, let $(\Omega, \mathcal{F}, P)$ denote a fixed probability space. Let X be an integrable random variable and $A \in \mathcal{F}$. The conditional expectation of X given that A has occurred is defined, in elementary probability theory, as

$$E(X|A) = \begin{cases} \dfrac{\int_A XdP}{P(A)} & \text{if } P(A) > 0 \\ \text{any number} & \text{if } P(A) = 0 \end{cases}$$

In particular, the conditional probability of an event B given that A has occurred is equal to

$$P(B|A) = E(I_B|A) = \frac{P(A \cap B)}{P(A)}$$

It has been found useful to generalize these definitions and define conditional expectation as random variables, as follows. Let $(A_1, \ldots, A_k)$ be a partition of Ω and let $\mathcal{G} = \sigma(A_1, \ldots, A_k) \subset \mathcal{F}$. Then the conditional expectation of X relative to $\mathcal{G}$ is the random variable with constant value $E(X|A_i)$ on each set A_i, $i = 1, \ldots, k$. More generally, let $\mathcal{G} \subset \mathcal{F}$ be any σ-field. The conditional expectation of X relative to $\mathcal{G}$ is a random variable Y such that

1) Y is $\mathcal{G}$-measurable
2) If $A \in \mathcal{G}$ then $\int_A XdP = \int_A Ydp$

Let us remark that if $\mathcal{G} = \sigma(A_1, \ldots, A_k)$ then the random variable Y with values $E(X|A_i)$ on A_i, $i = 1, \ldots, k$ satisfies 1) and 2). Any random variable Y that satisfies 1) and 2) will be denoted by $E(X|\mathcal{G})$. It is clear that if a random variable Y_1 satisfies 1) and 2) then $Y = Y_1$, except possibly on an ω-set of probability zero. Moreover, as a consequence of the Radon-Nikodym theorem, we obtain the following proposition:

Proposition 1. A random variable Y which satisfies the conditions 1) and 2) always exists. There exists also an increasing sequence of partitions $(A_1^n, \ldots, A_{k_n}^n)$ of Ω such that $E(X|\mathcal{G}_n) \to Y$, almost surely P, as $n \to +\infty$, where $\mathcal{G}_n = \sigma(A_1^n, \ldots, A_{k_n}^n)$.

Proof. Without loss of generality, we may assume $X \geq 0$. Let us define a measure $\mu : \mu(A) = \int_A XdP$, $A \in \mathcal{G}$, then the measure μ is absolutely continuous with respect to P and the Radon-Nikodym theorem implies that there exists a $\mathcal{G}$-measurable function Y such that $\int_A X dP = \mu(A) = \int_A Y dP$. To obtain the latter part of the proposition, it is sufficient to apply Corollary 1 (Section 2.10).

Proposition 2. Let $\mathcal{G}_n$ be σ-fields generated by finite partitions $(A_1^n, \ldots, A_{k_n}^n)$, n = 1, ... and let $\mathcal{G} = \sigma(\mathcal{G}_1, \mathcal{G}_2, \ldots)$, then

$$E(X|\mathcal{G}_n) \to E(X|\mathcal{G}) \text{ almost surely.}$$

Proof. It is sufficient to apply Problem 1, Section 2.10.

Remark 1. By the very definition, if X is a $\mathcal{G}$-measurable r.v. then $E(X|\mathcal{F}) = X$. It is worth to remark that if $X = I_B$, $\mathcal{G} = \sigma(A_1, \ldots, A_k)$ and $A = \Omega$ then the condition 2) implies classical Bayes formula:

$$P(B) = \sum_{i=1}^{k} P(B|A_i)P(A_i)$$

Problem 1. Let Z be a $\mathcal{G}$-measurable r.v. such that $E|XZ| < +\infty$ then $E(XZ|\mathcal{G}) = ZE(X|\mathcal{G})$, P almost surely. Hint: show first that the above formula is true for simple Z, and then pass to the limit.

Problem 2. Let $(H_n)_{n=0,1,\ldots}$ be a generalized Haar system (see Section 2.2) relative to $(\mathcal{F}_n)_{n=0,1,\ldots}$ and X an integrable random variable then

$$E(X|\mathcal{F}_n) = \sum_{k=0}^{n} E(XH_k)H_k$$

Hint: Show first that any $\mathcal{F}_n$-measurable function is of the form

$$\sum_{k=0}^{n} \alpha_k H_k$$

Then use the orthonormality of the Haar system and Problem 1.

Corollary 1. A generalized Haar system is a complete and orthonormal basis in $L^2(\Omega, \mathcal{F}, P)$ where $\mathcal{F} = \sigma(\mathcal{F}_0, \mathcal{F}_1, \ldots)$.

Proof. By Proposition 2 we know that $E(X|\mathcal{F}_n) \to E(X|\mathcal{F})$, (a.s.). If $X \in L^2(\Omega, \mathcal{F}, P)$ then X is $\mathcal{F}$-measurable and therefore $E(X|\mathcal{F}) = X$. Thus the Fourier's series $\sum_{k=0}^{+\infty} E(XH_k)H_k$ converges to X (a.s.) and since it converges also in the sense of $L^2(\Omega, \mathcal{F}, P)$ we conclude that

$$X = \sum_{k=0}^{+\infty} E(XH_k)H_k$$

in $L^2(\Omega, \mathcal{F}, P)$. Corollary 1 will play an important role in the construction of Brownian motion (Section 4.2).

Remark 2. If σ-field $\mathcal{G}$ is generated by a family of random variables, say $\mathcal{G} = \sigma(X_t; t \in T)$ then $E(X|\mathcal{G})$ will be denoted as $E(X|X_t, t \in T)$.

Problem 3. Let $r_1, \ldots, r_n$ be Rademacher functions (Section 2.2) and let s_k be the indicator of the interval $[(k-1)/2n, k/2)$, $k = 1, \ldots, 2^n$. Show that

$$E(X|r_1, \ldots, r_n) = E(X|s_1, \ldots, s_{2^n}) = \sum_{k=1}^{2^n} \alpha_k s_k. \text{ Find } \alpha_k.$$

3.2. Basic properties of the conditional expectation

In this section, we summarize the basic properties of the conditional expectation. Connections with martingales and an estimation problem will be discussed separately in Sections 3.3 and 3.4.

In the proposition below, we list the properties of the conditional expectation which are similar to those of the usual expectation.

Proposition 1. If X, Y are integrable random variables and a, b, c are real numbers then

1) $E(aX + bY + c|\mathcal{G}) = aE(X|\mathcal{G}) + bE(Y|\mathcal{G}) + c$;
2) If in addition, $X \leq Y$ a.s. then $E(X|\mathcal{G}) \leq E(Y|\mathcal{G})$ a.s.;
3) If X_n, $n = 1, \ldots,$ are integrable random variables which increase to X, then

$$\lim_n E(X_n|\mathcal{G}) = E(X|\mathcal{G}), \text{ (a.s.)}$$

4) (Jensen's inequality). Let h be a convex mapping from $\mathbb{R}^1$ into $\mathbb{R}^1$ and let X and h(X) be integrable r.v.'s, then

$$h(E(X|\mathcal{G})) \leq E(h(X)|\mathcal{G}), \text{ (a.s.)}.$$

Proof. We only prove property 4), the remaining properties are left to the reader. Function h is upper envelope of a countable family of affine functions h_n, $h_n(x) = a_n x + b_n$, $x \in \mathbb{R}^1$. The random variables $h_n(X)$ are clearly integrable and the property 1) implies $h_n(E(X|\mathcal{G})) = E(h_n(X)|\mathcal{G})$. Since $h_n \leqslant h$, using 2), we obtain $h_n(E(X|\mathcal{G})) \leq E(h(X)|\mathcal{G})$, (a.s.). Therefore, (the family $h_n(E(X|\mathcal{G}))$ is countable) $h(E(X|\mathcal{G})) = \sup_n h_n(E(X|\mathcal{G})) \leq E(h(X)|\mathcal{G})$, (a.s.).

Corollary 1. Jensen's inequality implies that if $E|X|^p < +\infty$, $1 \leq p \leq +\infty$ then

$$\|E(X|\mathcal{G})\|_p \leq \|X\|_p, \text{ (a.s.)}$$

Proof. Let $1 \leq p < +\infty$, then $|E(X|\mathcal{G})|^p \leq E(|X|^p|\mathcal{G})$ and therefore $E|E(X|\mathcal{G})|^p \leq E|X|^p$. If $p = +\infty$, then $X \leq \|X\|_\infty$, (a.s.) then $E(X|\mathcal{G}) \leq \|X\|_\infty$ and $\|E(X|\mathcal{G})\|_\infty \leq \|X\|_\infty$.

More characteristic properties of the conditional expectation are given in the following proposition:

Proposition 2. Let X be an integrable random variable,

1) If $\mathscr{G} = \{\emptyset, \Omega\}$, then $E(X|\mathscr{G}) = E(X)$;

2) If $\mathscr{G}$ is an arbitrary σ-field, and X is a $\mathscr{G}$-measurable r.v., then $E(X|\mathscr{G}) = X$, (a.s.);

3) If X is independent of $\mathscr{G}$, then $E(X|\mathscr{G}) = E(X)$ a.s.;

4) If $\mathscr{G}$, $\mathscr{G}_1$ are two σ-fields, $\mathscr{G}_1 \subset \mathscr{G} \subset \mathscr{F}$, then $E(E(X|\mathscr{G})|\mathscr{G}_1) = E(X|\mathscr{G}_1)$ and, in particular,

5) $E(E(X|\mathscr{G})) = E(X)$ (Bayes formula).

Proof. We show, e.g. the property 4). Let $A \in \mathscr{G}_1$ then $\int_A E(X|\mathscr{G}_1)dP = \int_A XdP$. On the other hand, since $A \in \mathscr{G}$, we obtain $\int_A E(E(X|\mathscr{G})|\mathscr{G}_1)dP = \int_A E(X|\mathscr{G})dP = \int_A XdP$. Thus 4) follows.

We formulate separately a generalization of the property 3), Proposition 2, because this generalization is very important in applications:

Lemma 1. Let X_1, X_2 be two random variables with values in $(E_1, \mathscr{E}_1)$, $(E_2, \mathscr{E}_2)$, respectively. Assume that X_1 is independent of a σ-field $\mathscr{G}$ and X_2 is $\mathscr{G}$-measurable. If f is a real function defined on $E_1 \times E_2$, $\mathscr{E}_1 \times \mathscr{E}_2$-measurable, then

$$E(f(X_1, X_2)|\mathscr{G}) = f_2(X_2), \text{ (a.s.)}$$

where $f_2(x_2) = E(f(X_1, x_2))$, $x_2 \in E_2$, provided $E|f(X_1, X_2)| < +\infty$.

Proof. The proof is analogous to that of Lemma 4, Section 1.1 and therefore will be omitted, (the usual technique of π-systems does work).

3.3. Conditional expectation and martingales

It is very important that, by using the notion of conditional expectation, it is possible to give a new equivalent definition of a martingale, a supermartingale and a submartingale.

A family $(X_t)_{t \in T}$ of integrable random variables is said to be a martingale (or, respectively, a supermartingale, a submartingale) with respect to an increasing family of σ-fields $(\mathscr{F}_t)_{t \in T}$ if

1) X_t is $\mathscr{F}_t$-measurable for $t \in T$, (adapted to $\mathscr{F}_t$),

2) $E(X_t) = X_s$, a.s. (or, respectively $\leq$, $\geq$), for $t \geq s$.

The proof that the definition of a martingale (or, respectively, a supermartingale, a submartingale) given here is equivalent to that from Section 2.1 is almost immediate and is left to the reader.

We use the new definition to prove the following important Lemma 1

Lemma 1. Let $(X_t)_{t \in T}$ be a martingale such that for some $p \in [1, +\infty)$ and all $t \in T$, $E|X_t|^p < +\infty$. Then

1) The family $(|X_t|^p)_{t\in T}$ is a submartingale

2) If $T = \{1,2,\ldots,n\}$ and c is a positive number, then

$$P(\sup_k |X_k| \geq c) \quad \frac{E|X_n|^p}{c^p}$$

Proof. 1) Let $s < t$ then by Jensen's inequality (Proposition 1, Section 3.2), $E(|X_t|^p|\mathscr{F}_s) \geq |E(X_t|\mathscr{F}_s)|^p \geq |X_s|^p$ (a.s.). 2) Since $(-|X_k|^p)_{k=1,\ldots,n}$ is a supermartingale therefore using inequality 4) from Theorem 1, Section 2.8 we obtain $c^pP(\inf_k(-|X_k|^p) \leq -c^p) \leq E|X_n|^p$. After elementary transformation, 2) follows.

3.4. Conditional expectation and an estimation problem

As usual, $(\Omega, \mathscr{F}, P)$ is a fixed probability space. Let X be a real-valued random variable and Y a random variable with values in a measurable space $(E, \mathscr{E})$. X and Y will be interpreted as unobservable parameter and observable data, respectively. The estimation problem can be stated as follows: knowing Y, estimate X in the best possible way. To be more specific, we assume $E(X^2) < +\infty$ and formulate precisely the so-called:

Least-square estimation problem: Find a real function $\hat{f}$ defined on E and $\mathscr{E}$-measurable such that for any real function f defined E and $\mathscr{E}$-measurable:

$$E(X - \hat{f}(Y))^2 \leq E(X - f(Y))^2 \tag{1}$$

Any function $\hat{f}$ satisfying (1) is called an optimal estimator.

The existence of an optimal estimator is a consequence of the following proposition:

As usual $L^2(\Omega, \mathscr{G}, P)$ denotes the Hilbert space of all $\mathscr{G}$-measurable, square integrable r.v.'s. It can be considered as a closed subspace of $L^2(\Omega, \mathscr{F}, P)$.

Proposition. The conditional expectation $E(X|\mathscr{G})$ is exactly the orthogonal projection of X onto $L^2(\Omega, \mathscr{G}, P)$. Moreover for any $Z \in L^2(\Omega, \mathscr{G}, P)$:

$$E(X - E(X|\mathscr{G}))^2 \leq E(X - Z)^2 \tag{2}$$

Proof. Let $X^\perp$ be the orthogonal projection of X onto $L^2(\Omega, \mathscr{G}, P)$ and $A \in \mathscr{G}$ then $X - X^\perp$ is orthogonal to I_A. Thus $E(I_A(X - X^\perp)) = 0$ or equivalently $\int_A X dP = \int_A X^\perp dP$. This implies $X^\perp = E(X|\mathscr{G})$. The Pythagorean identity completes the proof.

The solution to the Least-square estimation problem is given by the following corollary:

Corollary: Let $\mathscr{G} = \sigma(Y)$. Since every $\mathscr{G}$-measurable real function is of the form f(Y) for some $\mathscr{E}$-measurable f, therefore, for some $\mathscr{E}$-measurable function $\hat{f}$, $E(X|\mathscr{G}) = \hat{f}(Y)$. Inequality (2) says that $\hat{f}$ is the required optimal estimator. Let us note that $\hat{f}$ is, in general, not uniquely determined.

Remark. The above proposition suggests a new proof (independent of the Radon-Nikodym theorem) of the existence of the conditional expectation $E(X|\mathscr{G})$. Without loss of generality we can assume $X \geq 0$. If X is a bounded r.v. then $E(X|\mathscr{G})$ can be defined as the orthogonal projection $X^{\perp}$. If X is unbounded then there exists an increasing sequence of bounded nonnegative random variables $X_n \uparrow X$ and it is sufficient to define $E(X|\mathscr{G}) = \lim_n E(X_n|\mathscr{G})$.

Example 1. Let $H_0, \ldots, H_n$ be Haar functions and define $Y = (H_0, \ldots, H_n)$. If X is a real-valued r.v., $E(X^2) < +\infty$, then the optimal estimator $\hat{f}$ is a function defined on R^{n+1}, $\hat{f}(y_0, \ldots, y_n) = \sum_{k=0}^{n} E(XH_k)y_k$, (see Problem 2, Section 3.1).

Example 2. If $Y = (r_1, \ldots, r_n)$ where r_k are Rademacher functions then the optimal estimator $\hat{f}$ can be defined as $\hat{f}(y_1, \ldots, y_n) = 2^n \int_{I(y_1,\ldots,y_n)} X(\omega)P(d\omega)$ where $I(y_1, \ldots, y_n)$ is the interval $[y_1/2^1 + \ldots + y_n/2^n,$ $y_1/2 + \ldots + y_n/2^n + 1/2^n) \cap [0,1)$, (see Section 2.2).

More important, from the "practical" point of view, examples of optimal estimators will be given in the next section.

3.5. Conditional densities

Let X and Y be two random vectors with components $X_1, \ldots, X_n$ and $Y_1, \ldots, Y_k$, respectively, and let $g(\cdot,\cdot)$ be a (Borel-measurable) density of the joint distribution of (X, Y) and assume (to simplify notations) that g is a positive function on $\mathbb{R}^n \times \mathbb{R}^k$. The function $g(\cdot|\cdot)$ defined as

$$g(x|y) = \frac{g(x,y)}{\int_{\mathbb{R}^n} g(z,y)dz}, \quad x \in \mathbb{R}^n, \; y \in \mathbb{R}^k$$

is called the conditional density of X with respect to Y.

Importance of the conditional densities follows from the following proposition.

Proposition 1. If f is a Borel real-valued function on $\mathbb{R}^n \times \mathbb{R}^k$ then

$$E(f(X,Y)|Y) = \int_{\mathbb{R}^n} f(x,Y)g(x|Y)dx, \quad (a.s) \tag{1}$$

provided f(X, Y) is an integrable r.v..

Proof. First we show that for $A \in \mathscr{B}(\mathbb{R}^n)$

$$P(X \in A|Y) = \int_A g(x|Y)dx \tag{2}$$

Since every set belonging to $\sigma(Y)$ is of the form $\{Y \in B\}$ where $B \in \mathscr{B}(\mathbb{R}^k)$ therefore (2) is equivalent to

$$E(I_A(X)I_B(Y)) = E(I_B(Y) \int_A g(x|Y)dx)$$

But g is the density of the distribution of (X, Y); therefore, see Lemma 2, Section 1.4,

$$E(I_A(X)I_B(Y)) = \int_{A\times B} g(x,y)dxdy$$

On the other hand, the function $\int_{\mathbb{R}^n} g(z,\cdot)dz$ is the density of Y and, consequently,

$$E(I_B(Y)\int_A g(x|Y)dx) = \int_B (\int_A g(x|y)dx)\ (\int_{\mathbb{R}^n} g(z,y)dz)dy$$

$$= \int_{A\times B} g(x,y)dxdy$$

because of the definition of the conditional density $g(\cdot\,|\,\cdot)$. Combining all this together, we obtain (2). From (2), we easily deduce that (1) holds for functions f of the form $f(x,y) = I_A(x)I_B(y)$, $A\in\mathscr{B}(\mathbb{R}^n)$, $B\in\mathscr{B}(\mathbb{R}^k)$. Let us denote by $\mathscr{G}$ the family of all sets $C\in\mathscr{B}(\mathbb{R}^n\times\mathbb{R}^k)$ such that (1) holds for I_C. Then $\mathscr{G}$ satisfies the conditions 1), 2) and 3) of Lemma 3 (Section 1.1) and $\mathscr{G}$ contains the π-system $\mathscr{M}$ of all sets $A\times B$, $A\in\mathscr{B}(\mathbb{R}^n)$, $B\in\mathscr{B}(\mathbb{R}^k)$ therefore the mentioned Lemma 3 implies $\mathscr{G} = \sigma(\mathscr{M}) = \mathscr{B}(\mathbb{R}^n\times\mathbb{R}^k)$. The usual technique of "increasing sequences of simple functions" completes the proof.

Corollary 1. If f depends only on "x" then

$$E(f(X)|Y) = \int_{\mathbb{R}^n} f(x)g(x|Y)dx$$

Corollary 2. If Z is a random variable measurable with respect to $\sigma(Y)$ and takes values in $(\mathbb{R}^m, \mathscr{B}(\mathbb{R}^m))$, then

$$E(f(X,Z)|Y) = \int_{\mathbb{R}^n} f(x,Z)g(x|Y)dx,\quad (a.s)$$

provided f is a real-valued $\mathscr{B}(\mathbb{R}^n)\times\mathscr{B}(\mathbb{R}^m)$-measurable function and f(X, Z) is an integrable r.v. Since $Z = h(Y)$ for some $\mathscr{B}(\mathbb{R}^m)$-measurable h, therefore the above formula follows from (1).

If X is a random vector with components X_i (or a random matrix with components $X_{i,j}$) then $E(X|\mathscr{G})$ is a random vector with components $E(X_i|\mathscr{G})$, (or a random matrix with components $E(X_{i,j}|\mathscr{G})$).

Example 1. Define

$$\hat{m}_i(y) = \int_{\mathbb{R}^n} x_i g(x|y)dx$$

$$\hat{\sigma}_{i,j}(y) = \int_{\mathbb{R}^n} (x_i - \hat{m}_i(y))(x_j - \hat{m}_j(y))g(x|y)dx,\quad i,j = 1,\dots,n,\ y\in\mathbb{R}^k$$

and let $\hat{m}$, $\hat{Q}$ be vector and matrix with components respectively $\hat{m}_i$, $\hat{\sigma}_{i,j}$, then

$$E(X|Y) = \hat{m}(Y)$$

and

$$E((X - \hat{m}(Y))(X - \hat{m}(Y))' \,|\, Y) = \hat{Q}(Y), \text{ (a.s)}$$

The random matrix $\hat{Q}(Y)$ is called the conditional covariance matrix. The above formulas follow directly from Proposition 1.

An important role in the stochastic control theory plays the following proposition:

Proposition 2. Let the random vector (X, Y) be normally distributed with the density $g_{m,Q}$, where $m = (m_1, m_2)$ and

$$Q = \begin{pmatrix} Q_{11}, & Q_{12} \\ Q_{21}, & Q_{22} \end{pmatrix}$$

are, respectively, the mean vector and the covariance matrix of the random variable (X, Y). For every fixed $y \in \mathbb{R}^k$ the conditional density $g_{m,Q}(\cdot \,|y)$ is normal with the mean vector $\hat{m}(y) = m_1 + Q_{12}Q_{22}^{-1}(y - m_2)$ and the conditional covariance matrix $\hat{Q}(y) = Q_{11} - Q_{12}Q_{22}^{-1}Q_{21}$. Let us note that $\hat{Q}$ does not depend on y.

Proof. Let

$$R = \begin{pmatrix} R_{11}, & R_{12} \\ R_{21}, & R_{22} \end{pmatrix}$$

denote the inverse matrix of Q. R is positive-definite and, therefore, R_{11}, R_{22} are also positive-definite matrices. Since QR = I, we obtain $Q_{11}R_{11} + Q_{12}R_{21} = I$, $Q_{21}R_{11} + Q_{22}R_{21} = 0$ and, consequently,

$$R_{22}R_{11}^{-1} = -Q_{22}^{-1}Q_{21}, \; R_{11}^{-1} = Q_{11} - Q_{12}Q_{22}^{-1}Q_{22} \tag{3}$$

From the definition of the conditional densities, we have

$$\begin{aligned} g_{m,Q}(x|y) &= C(y)g_{m,Q}(x, y) \\ &= C_1(y)\exp -\tfrac{1}{2} \langle R_{11}(x - m_1 + R_{11}^{-1}R_{12}(y - m_2)), \\ &\quad (x - m_1 + R_{11}^{-1}R_{12}(y - m_2))\rangle \end{aligned}$$

where C(y), $C_1(y)$ are constants depending on y only. Using (3), we obtain, finally

$$g_{m,Q}(x|y) = C_1(y)\exp\{-\tfrac{1}{2} \langle \hat{Q}^{-1}(x - \hat{m}(y)), \; x - \hat{m}(y)\rangle\}$$

the desired result.

4. WIENER PROCESS

4.1. Definition of a Wiener process

A family of real random variables $(W(t))_{t \geq 0}$ defined on a probability space $(\Omega, \mathscr{F}, P)$ is called a Wiener process or a Brownian motion process if it has the following properties:

1) $W(0,\omega) = 0$, (a.s);
2) If $0 = t_0 < t_1 < \ldots < t_n$, then the random variables $W(t_1) - W(t_0), \ldots, W(t_n) - W(t_{n-1})$ are independent;
3) For every $t, s \geq 0$ the increment $W(t+s) - W(t)$ has a normal distribution with covariance s (and mean 0);
4) For almost all ω, the function $W(\cdot,\omega)$ is continuous on $[0, +\infty)$

Problem 1. Show that a family $(W(t))_{t \geq 0}$ of real r.v.'s is a Wiener process if it satisfies 1) and 4) and the following two conditions

2') If $0 < t_1 < \ldots < t_n$ then the random vector $(W(t_1), \ldots, W(t_{n-1}), W(t_n))$ is normally distributed

3') For every t, s ≥ 0;
$E(W(t)) = 0$, $E(W(t)W(s)) = t \wedge s$,

which means that $(W(t))_{t \geq 0}$ is a Gaussian process with the mean value function $\equiv 0$ and the covariance function $= t \wedge s$. Hint: Use Problem 5, Section 1.5.

It is not obvious that a family $(W(t))_{t \geq 0}$ with properties 1) - 4) actually exists. The first construction was given by Wiener in 1923. We introduce here a different technically simpler construction, the so called Levy-Ciesielski construction (see [2], [4]).

4.2. Levy-Ciesielski construction of Wiener process

In the Levy-Ciesielski construction, the essential role is played by a Haar system (a special case of a generalized Haar system considered in Section 2.2) connected with a dyadic partition of the interval [0,1]. Namely let $h_0 \equiv 1$, and if $2^n \leq k < 2^{n+1}$ then

$$h_k(t) = \begin{cases} 2^{\frac{n}{2}} & \text{if } \frac{k-2^n}{2^n} \leq t < \frac{k-2^n}{2^n} + \frac{1}{2^{n+1}} \\ -2^{\frac{n}{2}} & \text{if } \frac{k-2^n}{2^n} + \frac{1}{2^{n+1}} \leq t < \frac{k-2^n}{2^n} + \frac{1}{2^n} \end{cases}$$

$$h_k(1) = 0$$

From the Corollary 1, Section 3.1, we know that the system $\{h_k;\ k = 0, 1, \ldots\}$ forms an orthonormal and complete basis in the space $L^2([0,1])$.

Theorem 1. Let $(X_k)_{k=0,1,\dots}$ be a sequence of independent random variables normally distributed with mean 0 and covariance 1, defined on a probability space $(\Omega, \mathscr{F}, P)$. Then, for almost all ω, the series

$$\sum_{k=0}^{+\infty} X_k(\omega) \int_0^t h_k(s)ds = W(t,\omega), \quad t \in [0,1]$$

is uniformly convergent on $[0,1]$ and defines a Wiener process on $[0,1]$.

Lemma 1. Let $\epsilon \in (0, \frac{1}{2})$ and $M > 0$. If $|a_k| \leq Mk^{\epsilon}$ for $k = 1, 2, \dots$ then

the series $\sum_{k=0}^{+\infty} a_k \int_0^t h_k(s)ds$ is uniformly convergent in the interval $[0,1]$.

Proof. If $2^n \leq k < 2^{n+1}$, then the Schauder functions $S_k(t) = \int_0^t h_k(s)ds$, $t \in [0,1]$ are non-negative, have disjoint supports and are bounded from above by $2^{-(n+1)} 2^{\frac{n}{2}} = 2^{-\frac{n}{2}-1}$. Let us denote by $b_n = \max(|a_k|;\ 2^n \leq k < 2^{n+1}$ then

$$\sum_{2^n \leq k < 2^{n+1}} |a_k| \, S_k(t) \leq b_n 2^{-\frac{n}{2}-1}$$

for all $t \in [0,1]$ and $n = 0, 1, \dots$. Thus the condition

$$\sum_{n=0}^{+\infty} b_n 2^{-\frac{n}{2}} < +\infty$$

is sufficient for the uniform convergence of the series

$$\sum_{n=0}^{+\infty} \sum_{2^n \leq k < 2^{n+1}} |a_k| \, S_k(t)$$

and therefore for the uniform convergence of

$$\sum_{k=0}^{+\infty} a_k S_k(t)$$

too. From the inequalities $|a_k| \leq Mk^{\epsilon}$ it follows that $b_n \leq 2^{\epsilon} M 2^{n\epsilon}$ for all $n = 0, 1, \dots$ and, consequently,

$$\sum_{n=0}^{+\infty} b_n 2^{-\frac{n}{2}} \leq 2^{\epsilon} M \sum_{n=0}^{+\infty} 2^{n(-\frac{1}{2}+\epsilon)} < +\infty$$

Lemma 2. Let $(X_k)_{k=0,1,\ldots}$ be a sequence of normally distributed random variables with mean 0 and covariance 1 then with probability one the sequence

$$\left(\frac{|X_k|}{\sqrt{\log k}}\right)_{k=2,3,\ldots}$$

is bounded.

Proof. Let c be a fixed positive number, then

$$P(|X_k| \geq c) = \frac{2}{\sqrt{2\pi}} \int_c^{+\infty} e^{-\frac{x^2}{2}} dx \leq \frac{2}{\sqrt{2\pi}} \int_c^{+\infty} \frac{x}{c} e^{-\frac{x^2}{2}} dx \leq \frac{2}{c\sqrt{2\pi}} e^{-\frac{c^2}{2}}$$

From this we obtain that for $c > \sqrt{2}$

$$\sum_{k=2}^{+\infty} P(|X_k| \geq c\sqrt{\log k}) \leq \frac{2}{\sqrt{2\pi}} \sum_{k=2}^{+\infty} \frac{k^{-\frac{c^2}{2}}}{c\sqrt{\log k}} < +\infty$$

Therefore, if $c > \sqrt{2}$ then, with probability one, only for a finite number k, $|X_k| \geq c\sqrt{\log k}$.

Proof of Theorem. Lemma 1 and Lemma 2 imply that the series

$$\sum_{k=2}^{+\infty} X_k(\omega)S_k(t)$$

is for almost all ω uniformly convergent on the interval $[0,1]$. Since the functions S_k are continuous and $S_k(0) = 0$ for $k = 0, 1, \ldots$, the constructed process satisfies properties 1) and 4).

To prove that conditions 2') and 3') are satisfied, it is sufficient to show that the random vector $(W(t_1), \ldots, W(t_n))$ is normally distributed with zero mean-vector and the positive definite covariance matrix $Q = (t_i \wedge t_j)$, $i, j = 1, \ldots, n$. Let us introduce functions I_t: the indicators of the intervals $[0,t] \subset [0,1]$. Clearly, for $i, j = 1, \ldots, n$, $t_i \wedge t_j$

$$= \int_0^1 I_{t_i}(s) I_{t_j}(s) ds = \langle I_{t_i}, I_{t_j} \rangle$$

Problem 1. Show that the matrix $(t_i \wedge t_j)$ is positive definite. Hint:

$$\sum_{i,j=1}^{n} \lambda_i \lambda_j t_i \wedge t_j = \sum_{i,j=1}^{n} \lambda_i \lambda_j \langle I_{t_i}, I_{t_j} \rangle = \int_0^1 (\lambda_1 I_{t_1}(s) + \ldots + \lambda_n I_{t_n}(s))^2 ds.$$

If $\sum_{i,j=1}^{n} \lambda_i \lambda_j t_i \wedge t_j = 0$ prove, by induction, that $\lambda_n, \lambda_{n-1}, \ldots, \lambda_1$ are equal zero.

Applying Parceval's identity we obtain that

$$t_i \wedge t_j = \sum_{k=0}^{+\infty} \langle I_{t_i}, h_k \rangle \langle I_{t_j}, h_k \rangle = \sum_{k=0}^{+\infty} S_k(t_i) S_k(t_j)$$

Let us define $W_N(t) = \sum_{k=0}^{N} X_k S_k(t)$, then $m_N = E(W_N(t)) = 0$ and the covariance matrices Q_N of $(W_N(t_1), \ldots, W_N(t_n))$ are according to straightforward calculations, equal to:

$$Q_N = \left(\sum_{k=0}^{N} S_k(t_i) S_k(t_j) \right)_{i,j=1,\ldots,n}$$

It is also clear that the random vectors $(W_N(t_1), \ldots, W_N(t_n))$ are normally distributed (see Section 1.5), therefore the limit random vector $(W(t_1), \ldots, W(t_n))$ is also normally distributed with the mean vector zero and the covariance matrix $Q = (t_i \wedge t_j)_{i,j=1,\ldots,n}$, (see Problem 6, Section 1.5). This completes the proof.

Problem 2. Construct a Wiener process on $[0, +\infty)$. Hint: Construct a countable family of independent Wiener processes on $[0,1]$ and piece them together.

4.3. "White noise", stochastic equations and Wiener's integral

In this section, we give an elementary introduction to "white noise", stochastic equations and the Wiener integral theory.

In the applications, the following definition of "white noise" is used: "white noise" is a stationary process whose spectral density is a constant function on the whole real line. Since spectral measures (of stationary processes) are finite, the above definition is inconsistent. Nevertheless, we show that by using a Wiener process a good approximation of "white noise" can be constructed. Moreover, starting from this approximation, we give a physical interpretation of the stochastic equation.

Let W_1 and W_2 be two independent Wiener processes on $[0, +\infty)$. A Wiener process W on the whole line $\mathbb{R}$ is defined as

$$W(t) = \begin{cases} W_1(t) & \text{if } t \geq 0 \\ W_2(-t) & \text{if } t \leq 0 \end{cases}$$

Let W be a fixed Wiener process on $\mathbb{R}$ and define for every number $h > 0$ a new process

$$\Delta_h(t) = \frac{1}{h}(W(t+h) - W(t)), \quad t \in \mathbb{R}$$

It is easy to see that the process Δ_h is a Gaussian process (see the definition of a Wiener process) and that $E(\Delta_h(t)) = 0$, $t \in \mathbb{R}$. The proposition below shows that this process can be treated as an approximation of "white noise".

Proposition 1. For any $h > 0$ and all $t, s \in \mathbb{R}$ $E(\Delta_h(t)\Delta_h(s)) = r_h(t-s)$, where

$$r_h(u) = \begin{cases} \frac{1}{h}\left(1 - \frac{u}{h}\right) & \text{if } |u| \leq h \\ 0 & \text{if } |u| > h \end{cases} \tag{1}$$

Thus Δ_h is a stationary process (covariance function depends on t - s). If ρ_h is the spectral density of r_h:

$$\rho_h(x) = \frac{1}{2\pi} \int_{-\infty}^{+\infty} e^{-iux}\, r_h(u)du, \quad x \in \mathbb{R}$$

then

$$\rho_h(x) = \frac{1}{\pi} \frac{(1 - \cos hx)}{(hx)^2}, \; x \neq 0, \; \rho_h(0) = \frac{1}{2\pi} \tag{2}$$

and, consequently, $\rho_h \to 1/2\pi$ uniformly on every finite interval as $h \to 0$.

Proof. Both formulas are a consequence of straightforward calculations. Formula (1) follows from the easily checked relation $E(W(t) - W(s))^2 = |t - s|$, $t, s \in \mathbb{R}$ and (2) by integrating by parts.

Remark 1. The above proposition is also a justification of the statement: "white noise is a derivative of a Wiener process", because if $h \downarrow 0$ then (formally) $\Delta_h \to (d/dt)W(t)$. In fact, the trajectories of a Wiener process are nowhere differentiable functions (see Refs [2, 4]) and, therefore, the sequence Δ_h, $h \downarrow 0$ is divergent. A related result will be proved in the next chapter (Section 5.2).

Let us now consider a mechanical system described by a linear differential equation

$$x^{(n)} + a_1 x^{(n-1)} + \ldots + a_{n-1}\dot{x} + a_n x = b\xi(t) \tag{3}$$

with initial conditions $x^{(k)}(0) = \overline{x}_{k+1}$, $k = 0, \ldots, n-1$ and an outer stochastic force $b\xi$. If $\xi(t)$ is a "white noise" process, then, instead of considering a "formal" equation (3), it is reasonable to deal with its approximate version:

$$x^{(n)} + a_1 x^{(n-1)} + \ldots + a_n x = b\Delta_h \tag{4}$$

$$x^{(k)}(0) = \overline{x}_{k+1}, \; k = 0, 1, \ldots, n-1$$

where the right-hand side is well defined for every "ω". Equation (4) is equivalent to the system of equations

$$\begin{aligned} \dot{x}_1 &= x_2 \\ &\vdots \\ \dot{x}_{n-1} &= x_n \\ \dot{x}_n &= -a_1 x_{n-1} - \ldots - a_{n-1}x_1 + b\,\Delta_h \end{aligned} \tag{5}$$

with the initial conditions: $x_k(0) = \overline{x}_k$, $k = 1, \ldots, n$. We want to generalize slightly the system (5). To do this let us introduce independent Wiener processes $W_1, W_2, \ldots, W_n$ defined for $t \in \mathbb{R}$ and let $\Delta_h^1, \ldots, \Delta_h^n$ be corresponding almost "white noise" processes: $\Delta_h^i(t) = (1/h)(W_i(t+h) - W_i(t))$. $W(t)$ and $\Delta_h(t)$ will denote from now the (column) processes with components respectively $W_i(t)$ and $\Delta_h^i(t)$, $i = 1, \ldots, n$. Let A and B be two $n \times n$ matrices then the system (5) is a special case of the following system:

$$\dot{x} = Ax + B\Delta_h, \; x(0) = \overline{x} \tag{6}$$

Proposition 2. Let $x_h(t)$, $t \geq 0$ be the solution of Eq. (6). If $h \downarrow 0$ then the stochastic processes $(x_h(t))_{t \geq 0}$ tend, for almost all "ω", uniformly on finite intervals to a stochastic process $(X(t))_{t \geq 0}$, which satisfies the stochastic integral equation

$$X(t) = \bar{x} + \int_0^t AX(s)ds + BW(t), \quad t \geq 0 \tag{7}$$

Equation (7) is sometimes written as a stochastic differential equation

$$dX(t) = AX(t)dt + BdW(t), \quad X(0) = \bar{x} \tag{8}$$

The proof will be an easy application of the following important lemma:

Lemma 1. Let $B(\cdot)$ be a continuously differentiable function from $[0, +\infty)$ into the space of $n \times n$ matrices. Then

$$\int_0^t B(s)\Delta_h(s)ds \to B(t)W(t) - \int_0^t \dot{B}(s)W(s)ds, \text{ as } h \downarrow 0 \tag{9}$$

for almost all "ω" and uniformly with respect to t from compact intervals $\subset (0, +\infty)$.

Proof. The proof of the lemma follows from the identity

$$\frac{1}{h}\int_0^t B(s)\Delta_h(s)ds = \int_h^t \frac{1}{h}(B(s-h) - B(s))W(s)ds - \frac{1}{h}\int_0^h B(s)W(s)ds$$

$$+ \frac{1}{h}\int_t^{t+h} B(s-h)W(s)ds, \quad t, h > 0$$

Remark 2. If $B(\cdot)$ is a continuously differentiable function then the Stieltjes integral

$$\int_0^t B(s)dW(s) = B(t)W(t) - \int_0^t \dot{B}(s)W(s)ds \tag{10}$$

and we can write that:

$$\int_0^t B(s)\Delta_h(s)ds \to \int_0^t B(s)dW(s), \quad t \geq 0$$

Paley, Wiener and Zygmund [12] extended the definition of the stochastic integral $\int_0^t B(s)dW(s)$, to all functions $B(\cdot)$ such that $\int_0^t |B(s)|^2 ds < +\infty$. Their approach is presented in the following problem.

Problem 1. Assume $n = 1$. Then the mapping from $L^2([0,1])$ into $L^2(\Omega, \mathscr{F}, P)$ given by $b(\cdot) \to \int_0^1 b(s)dW(s)$ defined for all b continuously differentiable by (10) is an isometry: $E(\int_0^1 b(s)dW(s))^2 = \int_0^1 b^2(s)ds$ and therefore can be extended to an isometry on the whole $L^2([0,1])$. Show this. Hint: $E(\int_0^1 b(s)dW(s))^2 = \int_0^1 \int_0^1 t \wedge s \, \dot{b}(t)\dot{b}(s)dtds$.

The integral $\int_0^1 b(s)dW(s)$ defined in the above Problem 1 is sometimes called Wiener's integral. A different construction of Wiener's integral will be given in the next chapter.

Proof of Proposition 2. Let us start from the observation that explicit formulas for solutions of (6) and (7) can be easily derived. Namely, for all $t \geq 0$ we have:

$$x_h(t) = e^{At}\,\bar{x} + e^{At}\int_0^t e^{-As}\,B\Delta_h(s)ds \quad \text{and}$$

$$X(t) = (\bar{x} + BW(t)) + A\int_0^t e^{A(t-s)}\,(\bar{x} + BW(s))ds$$

$$= e^{At}\,\bar{x} + Ae^{At}\int_0^t e^{-As}\,BW(s)ds + BW(t)$$

And it is sufficient to apply Lemma 1.

Remark 3. By exactly the same method a more general case can be treated. If, e.g. in Eq. (6) the matrix B is a continuously differentiable function of t, then the Proposition 2 holds but Eq. (7) has to be changed:

$$X(t) = \bar{x} + \int_0^t AX(s)ds + \int_0^t B(s)dW(s) \tag{7'}$$

Problem 2. Generalize Proposition (2) to the "time-dependent" case: A and B depend on t. Consider also a non-linear case: $\dot{x} = A(x) + B\Delta_h$, and $A(\cdot)$ a non-linear mapping from $\mathbb{R}^n$ into $\mathbb{R}^n$.

Remark 4. Relations between ordinary differential equations and stochastic differential equations are studied in Ref. [13].

5. ITO'S STOCHASTIC INTEGRAL

5.1. Introduction

In this chapter, following Ito, we define a stochastic integral $\int_0^t f(s)dW(s)$ for a wide class of stochastic processes $(f(t))_{t\geq 0}$ (not only for functions $f \in L^2([0,1])$ as we did in Section 4.3). Such extension is needed if one wants to deal with stochastic integral equations:

$$X(t) = \bar{x} + \int_0^t a(X(s))ds + \int_0^t b(X(s))dW(s), \quad t \geq 0$$

(in differential form:

$$dX(t) = a(X(t))dt + b(X(t))dW(t))$$

where the function b does depend on x, (compare the equations (7) and (7') in Section 4.3).

The integral $\int_0^t f(s)dW(s)$ cannot be defined (even for continuous processes $(f(t))$) as the Stieltjes or Lebesgue-Stieltjes integral because:

Lemma 1. The trajectories of a Wiener process are (a.s.) not rectifiable in any time interval of positive length, therefore, they are not of bounded variation.

Proof. Let

$$\ell_n = \sum_{k=1}^{2^n} |W((k-1)/2^n) - W(k/2^n)|, \quad n = 1, \ldots$$

The $E(e^{-\ell_n}) = (E(e^{-|W(t)|}))^{\frac{1}{t}}$ for $t = 1/2^n$, because the Wiener process $(W(t))$ has independent increments (Property 2). Therefore

$$E(e^{-\ell_n}) = \left(1 - \frac{1}{\sqrt{2\pi t}} \int_{-t}^{t} e^{-\frac{x^2}{2t}} dx\right)^{\frac{1}{t}}$$

Using the estimates $e^{-u} \leqq 1 - u + u^2/2$, $u \geqq 0$ we obtain $E(e^{-\ell_n}) \to 0$ and since (ℓ_n) is an increasing sequence it tends to $+\infty$ (a.s.).

The other possibility is to use the explicit Levy-Ciesielski formula for a Wiener process (see Theorem 1, Section 4.2) and define

$$\int_0^t f(s)dW(s) = \sum_{k=0}^{+\infty} X_k \int_0^t f(s)h_k(s)ds \tag{1}$$

but it is difficult, in this case, to discover the appropriate class of integrands $f(t)$.

Problem 1. Show that if $f \in L^2([0,1])$, then for fixed t the series in (1) is (a.s.) convergent and is equal to Wiener's integral of Section 4.3. Hint: Use Remark 2, Section 2.9.

It follows from the paper by Ito-Nisio [14] that if $f \in L^2[0,1]$ then the series (1) converges (a.s.) uniformly on $[0,1]$.

5.2. Construction of the Ito stochastic integral

The Ito definition of stochastic integral is valid for the so-called non-anticipating Brownian functionals f. To define them, let $(\mathscr{F}_t)_{t \geq 0}$ be an increasing family of σ-fields such that

1) For every $t \geqq 0$, $W(t)$ is $\mathscr{F}_t$-measurable
2) σ-fields $\mathscr{F}_t$ and $\mathscr{G}_t^+ = \sigma(W(s) - W(t): s \geqq t)$ are independent.

For instance, using a standard "method of π-systems", (see Section 1.1) it is possible to prove that σ-fields $\mathscr{G}_t = \sigma(W(s);\ s \leqq t)$, satisfy 1) and 2).

A family $(f(t))_{t \geq 0}$ of real-valued random variables is said to be a non-anticipating Brownian functional if

1) The function $f(t,\omega)$, $(t,\omega) \in [0, +\infty) \times \Omega$ is $\mathscr{B}([0,+\infty)) \times \mathscr{F}$-measurable
2) $f(t)$ is $\mathscr{F}_t$-measurable, $t \geqq 0$
3) $p(\int_0^t f^2(s)ds < +\infty,\ t \geqq 0) = 1$.

$((\Omega, \mathscr{F}, P)$ is a basic probability space).

First, we define stochastic integral $\int_0^t f dW$ for $t \in [0,1]$. Let us denote by $\mathcal{N}$ and $\mathcal{N}_c$ the linear spaces of all non-anticipating (on $[0,1]$) Brownian functionals and all non-anticipating Brownian functionals continuous on $[0,1]$ (for almost all ω), and define

$$|f|_2 = \left(\int_0^1 f^2(s)ds\right)^{\frac{1}{2}} \text{ for } f \in \mathcal{N}$$

$$|f|_c = \sup(|f(s)|,\ 0 \leq s \leq 1), \text{ for } f \in \mathcal{N}_c$$

The linear spaces $\mathcal{N}$, $\mathcal{N}_c$ can be treated as normed spaces with norms:

$$\|f\|_2 = E\left(\frac{|f|_2}{1 + |f|_2}\right), \quad f \in \mathcal{N}$$

$$\|f\|_c = E\left(\frac{|f|_c}{1 + |f|_c}\right), \quad f \in \mathcal{N}_c$$

Problem 2. Show that the spaces $\mathcal{N}$, $\mathcal{N}_c$ are complete (the so-called F-spaces).

Let us denote by $\mathcal{N}_s$ the subspace of $\mathcal{N}$ of all simple Brownian functionals. A process $f \in \mathcal{N}$ is said to be simple if there exist numbers, $0 \leq t_0 \leq t_1 \leq \ldots \leq t_n \leq 1$, such that

$$f(t) = f(t_k), \text{ for } t_k \leq t < t_{k+1},\ k = 0, \ldots, n-1$$

Problem 3. Show that the subspace $\mathcal{N}_s$ is dense in $\mathcal{N}$.

If $f \in \mathcal{N}_s$ the Ito's stochastic integral is defined by the formula

$$\int_0^t f(s)dW(s) = \sum_{t_{k+1} \leq t} f(t_k)(W(t_{k+1}) - W(t_k)) + f(t_l)(W(t) - W(t_l))$$

where $t_l = \max(t_k : t_k \leq t)$, $t \in [0,1]$

To extend the above definition to the whole $\mathcal{N}$, it is convenient to consider the stochastic integral as operator:

$$\int : \mathcal{N}_s \to \mathcal{N}_c,$$

$$\int f(t) = \int_0^t f(s)dW(s), \quad t \in [0,1]$$

We show that $\int$ is a continuous (linear) operator. This follows from the following two lemmas:

Lemma 1. If $f \in \mathcal{N}_s$ and $\alpha \in \mathbb{R}$ then the process

$$X_t = \exp\left(\alpha \int_0^t f dW - \frac{\alpha^2}{2}\int_0^t f^2 ds\right)$$

is $(\mathcal{F}_t)$-martingale for all $\alpha \in \mathbb{R}$.

Proof. We prove the lemma for $f \equiv 1$, because the general case follows then directly by induction. Let us remark that for $t > s$ and $u = t - s$

$$E(\exp(\alpha(W_t - W_s) - \frac{1}{2}\alpha^2(t-s))) = \frac{1}{\sqrt{2\pi u}}\int_{-\infty}^{+\infty}\exp(\alpha x - \frac{1}{2}\alpha^2 u - \frac{x^2}{2u})dx$$

$$= \frac{1}{\sqrt{2\pi u}}\int_{-\infty}^{+\infty}\exp(-\frac{1}{2}\frac{(x-\alpha u)^2}{u})dx = 1$$

Therefore for $A \in \mathscr{F}_s$

$$E(I_A(X_t - X_s)) = E(I_A\exp(\alpha W_s - \frac{\alpha^2}{2}s)(\exp(\alpha(W_t - W_s) - \frac{\alpha^2}{2}u) - 1))$$

$$= E(I_A\exp(\alpha W_s - \frac{\alpha^2}{2}s))\,E(\exp(\alpha(W_t - W_s) - \frac{\alpha^2 u}{2}) - 1) = 0$$

because the random variables $I_A\exp(\alpha W_s - \frac{\alpha^2}{2}s)$, and $\exp(\alpha(W_t - W_s) - \frac{\alpha^2}{2}(t-s))$ are independent as r.v.'s, $\mathscr{F}_s$ and $\mathscr{G}_s^+$ measurable, respectively. This completes the proof.

Lemma 2. If α, β are positive numbers and $f \in \mathscr{N}_s$ then

$$P(\max_{0 \leq t \leq 1}(\int_0^t f dW - \frac{\alpha}{2}\int_0^t f^2 ds) > \beta) \leq e^{-\alpha\beta} \qquad (1)$$

Proof. Let $t_0 \leq t_1 \leq \ldots \leq t_n \leq 1$ be an arbitrary sequence of positive numbers then, by virtue of Lemma 1, the sequence $(X_{t_k})_{k=1,\ldots,n}$ is $(\mathscr{F}_{t_k})_{k=1,\ldots,n}$-martingale and consequently by Doob-Kolmogorov inequality (Problem 1, Section 2.8)

$$P(\max_{1 \leq k \leq n}(\int_0^{t_k} f\,dW - \frac{\alpha}{2}\int_0^{t_k} f^2 ds) \geq \beta)$$

$$= P(\max_{1 \leq k \leq n} X_{t_k} \geq e^{\alpha\beta}) \leq E(X_{t_n})e^{-\alpha\beta}$$

Since the sequence $(t_k)_{k=1,\ldots,n}$ was arbitrary, the required result follows.

Since $\int_0^t f^2 ds \leq \int_0^1 f^2 ds$, for $t \leq 1$ therefore from the Lemma 2 we obtain:

Fundamental inequality:

$$P(|\int f|_c > \beta + \frac{\alpha}{2}|f|_2^2) \leq 2\,e^{-\alpha\beta}$$

Corollary 1. Assume that $f_n \in \mathscr{N}_s$ and that $\|f_n\|_2 \to 0$ as $n \to +\infty$. This is equivalent that for every $\epsilon > 0$, $P(|f_n|_2 > \epsilon) \to 0$. But then the fundamental inequality implies $P(\int |f_n|_c > \epsilon) \to 0$ for every $\epsilon > 0$ and, therefore, $\|\int f_n\|_c \to 0$. This proves that the operator $\int$ is continuous.

Now we are able to finish the definition of the Ito Integral. Let $\int$ be the unique continuous extension of the operator $\int$ to the whole $\mathscr{N}$. Such an extension exists because the operator $\int$ is continuous on $\mathcal{N}_s$. Since $\mathcal{N}_s$ is dense in $\mathcal{N}$, the extension is uniquely determined. If $f \in \mathcal{N}$ then we define

$$\int_0^t f(s)dW(s) = \boldsymbol{\int} f(t), \quad t \in [0, 1]$$

Since the operator $\boldsymbol{\int}$ is into $\mathcal{N}_c$, the stochastic integral defined above for all $f \in \mathcal{N}$ is a continuous function (a.s.) of the parameter t.

Problem 4. Extend the definition of the stochastic integral to all $t \geq 0$.

Remark 1. For further information on stochastic integrals as well as on stochastic differential equations we refer to Refs [4, 15].

REFERENCES

[1] DOOB, J.L., What is a martingale? Amer. Math. Month., 78 (1971) 451-463.
[2] LAMPERTI, J., Probability, Benjamin, New York and Amsterdam (1966).
[3] MEYER, P.A., Probability and Potentials, Blaisdell Publishing Company (1966).
[4] McKEAN, H.P., Jr., Stochastic Integrals, Academic Press, New York and London (1969).
[5] HALMOS, P.R., Measure Theory, Van Nostrand, New York (1950).
[6] FELLER, W., An Introduction to Probability Theory and its Applications, Wiley, New York, 1 (1968); 2 (1971).
[7] ASTROM, K.J., Introduction to Stochastic Control Theory, Academic Press, New York (1970).
[8] KUSHNER, H., Introduction to Stochastic Control, Holt, New York (1971).
[9] COURREGE, P., PRIOURET, P., Temps d'arrêt d'une fonction aléatoire: Relations d'equivalence associées et propriétés de decomposition, Publ. Inst. Statist. Univ. Paris 14 (1965) 245-274.
[10] DYNKIN, E.B., YUSHKEVICH, A., Markov Processes: Theorems and Problems, Plenum, New York (1969).
[11] CHATTERJI, S.D., Martingale convergence and the Radon-Nikodym theorem, Math. Scand. 22 (1968) 21-41.
[12] PALEY, R.E.A.C., WIENER, N., ZYGMUND, A., Note on random functions, Math. Z. 37 (1933), 647-668.
[13] WONG, E., ZAKAI, M., On the relation between ordinary and stochastic differential equations, Int. J. Eng. Sci. 3 (1965) 213-229.
[14] ITO, K., NISIO, M., On the convergence of sums of independent Banach space valued random variables, Osaka J. Math. 5 (1968) 35-48.
[15] LIPCER, P.S., SIRJAJEV, A.N., Statistic of Random Processes, Science, Moscow (1974) in Russian.

SECRETARIAT OF SEMINAR

ORGANIZING COMMITTEE

R. Conti	Mathematics Institute "U. Dini", University of Florence, Italy
L. Markus	Mathematics Institute, University of Warwick, Coventry, United Kingdom and School of Mathematics, University of Minnesota, Minneapolis, USA
C. Olech	Institute of Mathematics, Polish Academy of Sciences, Warsaw, Poland

EDITOR

J.W. Weil	Division of Publications, IAEA, Vienna, Austria

The following conversion table is provided for the convenience of readers and to encourage the use of SI units.

FACTORS FOR CONVERTING UNITS TO SI SYSTEM EQUIVALENTS*

SI base units are the metre (m), kilogram (kg), second (s), ampere (A), kelvin (K), candela (cd) and mole (mol).
[For further information, see International Standards ISO 1000 (1973), and ISO 31/0 (1974) and its several parts]

Multiply		*by*	*to obtain*
Mass			
pound mass (avoirdupois)	1 lbm	= 4.536×10^{-1}	kg
ounce mass (avoirdupois)	1 ozm	= 2.835×10^{1}	g
ton (long) (= 2240 lbm)	1 ton	= 1.016×10^{3}	kg
ton (short) (= 2000 lbm)	1 short ton	= 9.072×10^{2}	kg
tonne (= metric ton)	1 t	= 1.00×10^{3}	kg
Length			
statute mile	1 mile	= 1.609×10^{0}	km
yard	1 yd	= 9.144×10^{-1}	m
foot	1 ft	= 3.048×10^{-1}	m
inch	1 in	= 2.54×10^{-2}	m
mil (= 10^{-3} in)	1 mil	= 2.54×10^{-2}	mm
Area			
hectare	1 ha	= 1.00×10^{4}	m^2
(statute mile)^2	1 mile^2	= 2.590×10^{0}	km^2
acre	1 acre	= 4.047×10^{3}	m^2
yard^2	1 yd^2	= 8.361×10^{-1}	m^2
foot^2	1 ft^2	= 9.290×10^{-2}	m^2
inch^2	1 in^2	= 6.452×10^{2}	mm^2
Volume			
yard^3	1 yd^3	= 7.646×10^{-1}	m^3
foot^3	1 ft^3	= 2.832×10^{-2}	m^3
inch^3	1 in^3	= 1.639×10^{4}	mm^3
gallon (Brit. or Imp.)	1 gal (Brit)	= 4.546×10^{-3}	m^3
gallon (US liquid)	1 gal (US)	= 3.785×10^{-3}	m^3
litre	1 l	= 1.00×10^{-3}	m^3
Force			
dyne	1 dyn	= 1.00×10^{-5}	N
kilogram force	1 kgf	= 9.807×10^{0}	N
poundal	1 pdl	= 1.383×10^{-1}	N
pound force (avoirdupois)	1 lbf	= 4.448×10^{0}	N
ounce force (avoirdupois)	1 ozf	= 2.780×10^{-1}	N
Power			
British thermal unit/second	1 Btu/s	= 1.054×10^{3}	W
calorie/second	1 cal/s	= 4.184×10^{0}	W
foot-pound force/second	1 ft·lbf/s	= 1.356×10^{0}	W
horsepower (electric)	1 hp	= 7.46×10^{2}	W
horsepower (metric) (= ps)	1 ps	= 7.355×10^{2}	W
horsepower (550 ft·lbf/s)	1 hp	= 7.457×10^{2}	W

* Factors are given exactly or to a maximum of 4 significant figures

Multiply		*by*	*to obtain*
Density			
pound mass/inch^3	1 lbm/in^3	= 2.768×10^4	kg/m^3
pound mass/foot^3	1 lbm/ft^3	= 1.602×10^1	kg/m^3
Energy			
British thermal unit	1 Btu	= 1.054×10^3	J
calorie	1 cal	= 4.184×10^0	J
electron-volt	1 eV	$\simeq 1.602 \times 10^{-19}$	J
erg	1 erg	= 1.00×10^{-7}	J
foot-pound force	1 ft·lbf	= 1.356×10^0	J
kilowatt-hour	1 kW·h	= 3.60×10^6	J
Pressure			
newtons/metre^2	1 N/m^2	= 1.00	Pa
atmosphere[a]	1 atm	= 1.013×10^5	Pa
bar	1 bar	= 1.00×10^5	Pa
centimetres of mercury (0°C)	1 cmHg	= 1.333×10^3	Pa
dyne/centimetre^2	1 dyn/cm^2	= 1.00×10^{-1}	Pa
feet of water (4°C)	1 ftH$_2$O	= 2.989×10^3	Pa
inches of mercury (0°C)	1 inHg	= 3.386×10^3	Pa
inches of water (4°C)	1 inH$_2$O	= 2.491×10^2	Pa
kilogram force/centimetre^2	1 kgf/cm^2	= 9.807×10^4	Pa
pound force/foot^2	1 lbf/ft^2	= 4.788×10^1	Pa
pound force/inch^2 (= psi)[b]	1 lbf/in^2	= 6.895×10^3	Pa
torr (0°C) (= mmHg)	1 torr	= 1.333×10^2	Pa
Velocity, acceleration			
inch/second	1 in/s	= 2.54×10^1	mm/s
foot/second (= fps)	1 ft/s	= 3.048×10^{-1}	m/s
foot/minute	1 ft/min	= 5.08×10^{-3}	m/s
mile/hour (= mph)	1 mile/h	= 4.470×10^{-1}	m/s
		1.609×10^0	km/h
knot	1 knot	= 1.852×10^0	km/h
free fall, standard (= g)		= 9.807×10^0	m/s^2
foot/second^2	1 ft/s^2	= 3.048×10^{-1}	m/s^2
Temperature, thermal conductivity, energy/area·time			
Fahrenheit, degrees − 32	°F − 32	$\frac{5}{9}$	°C
Rankine	°R		K
1 Btu·in/ft^2·s·°F		= 5.189×10^2	W/m·K
1 Btu/ft·s·°F		= 6.226×10^1	W/m·K
1 cal/cm·s·°C		= 4.184×10^2	W/m·K
1 Btu/ft^2·s		= 1.135×10^4	W/m^2
1 cal/cm^2·min		= 6.973×10^2	W/m^2
Miscellaneous			
foot^3/second	1 ft^3/s	= 2.832×10^{-2}	m^3/s
foot^3/minute	1 ft^3/min	= 4.719×10^{-4}	m^3/s
rad	rad	= 1.00×10^{-2}	J/kg
roentgen	R	= 2.580×10^{-4}	C/kg
curie	Ci	= 3.70×10^{10}	disintegration/s

[a] atm abs: atmospheres absolute;
atm (g): atmospheres gauge.

[b] lbf/in^2 (g) (= psig): gauge pressure;
lbf/in^2 abs (= psia): absolute pressure.

HOW TO ORDER IAEA PUBLICATIONS

Exclusive sales agents for IAEA publications, to whom all orders and inquiries should be addressed, have been appointed in the following countries:

UNITED KINGDOM	Her Majesty's Stationery Office, P.O. Box 569, London SE 1 9NH
UNITED STATES OF AMERICA	UNIPUB, P.O. Box 433, Murray Hill Station, New York, N.Y. 10016

In the following countries IAEA publications may be purchased from the sales agents or booksellers listed or through your major local booksellers. Payment can be made in local currency or with UNESCO coupons.

ARGENTINA	Comisión Nacional de Energía Atómica, Avenida del Libertador 8250, Buenos Aires
AUSTRALIA	Hunter Publications, 58 A Gipps Street, Collingwood, Victoria 3066
BELGIUM	Service du Courrier de l'UNESCO, 112, Rue du Trône, B-1050 Brussels
CANADA	Information Canada, 171 Slater Street, Ottawa, Ont. K 1 A OS 9
C.S.S.R.	S.N.T.L., Spálená 51, CS-110 00 Prague Alfa, Publishers, Hurbanovo námestie 6, CS-800 00 Bratislava
FRANCE	Office International de Documentation et Librairie, 48, rue Gay-Lussac, F-75005 Paris
HUNGARY	Kultura, Hungarian Trading Company for Books and Newspapers, P.O. Box 149, H-1011 Budapest 62
INDIA	Oxford Book and Stationery Comp., 17, Park Street, Calcutta 16; Oxford Book and Stationery Comp., Scindia House, New Delhi-110001
ISRAEL	Heiliger and Co., 3, Nathan Strauss Str., Jerusalem
ITALY	Libreria Scientifica, Dott. de Biasio Lucio "aeiou", Via Meravigli 16, I-20123 Milan
JAPAN	Maruzen Company, Ltd., P.O.Box 5050, 100-31 Tokyo International
NETHERLANDS	Marinus Nijhoff N.V., Lange Voorhout 9-11, P.O. Box 269, The Hague
PAKISTAN	Mirza Book Agency, 65, The Mall, P.O.Box 729, Lahore-3
POLAND	Ars Polona, Centrala Handlu Zagranicznego, Krakowskie Przedmiescie 7, Warsaw
ROMANIA	Cartimex, 3-5 13 Decembrie Street, P.O.Box 134-135, Bucarest
SOUTH AFRICA	Van Schaik's Bookstore, P.O.Box 724, Pretoria Universitas Books (Pty) Ltd., P.O.Box 1557, Pretoria
SPAIN	Diaz de Santos, Lagasca 95, Madrid-6 Calle Francisco Navacerrada, 8, Madrid-28
SWEDEN	C.E. Fritzes Kungl. Hovbokhandel, Fredsgatan 2, S-103 07 Stockholm
U.S.S.R.	Mezhdunarodnaya Kniga, Smolenskaya-Sennaya 32-34, Moscow G-200
YUGOSLAVIA	Jugoslovenska Knjiga, Terazije 27, YU-11000 Belgrade

---

Orders from countries where sales agents have not yet been appointed and requests for information should be addressed directly to:

Publishing Section,
International Atomic Energy Agency,
Kärntner Ring 11, P.O.Box 590, A-1011 Vienna, Austria